Essential PTC® Mathcad Prime® 3.0

Companion Website

Please visit the companion website for this book. Most of the figures and examples used in this book are posted on the site. It will be very beneficial to download the examples and work through them as you read the book. It will also be beneficial to change the input in many of the examples to see how the results can change. Additional examples are also provided. From time to time, new examples will be posted to the site. To access this content go to http://store.elsevier.com/9780124104105 and click on the Resources tab, and then click on the link for the Online Companion Materials.

Essential PTC® Mathcad Prime® 3.0

A Guide for New and Current Users

Brent Maxfield, P.E.

AMSTERDAM • BOSTON • HEIDELBERG • LONDON
NEW YORK • OXFORD • PARIS • SAN DIEGO
SAN FRANCISCO • SINGAPORE • SYDNEY • TOKYO

Academic Press is an imprint of Elsevier

Academic Press is an imprint of Elsevier
The Boulevard, Langford Lane, Kidlington, Oxford, OX5 1GB
225 Wyman Street, Waltham, MA 02451, USA

First Published 2014

British Library Cataloguing in Publication Data
A catalogue record for this book is available from the British Library

Library of Congress Cataloging-in-Publication Data
A catalog record for this book is available from the Library of Congress

ISBN: 978-0-12-410410-5

For information on all Academic Press publications
visit our website at **store.elsevier.com**

Printed and bound in the United States

14 15 16 17 10 9 8 7 6 5 4 3 2 1

Contents

Preface

This book is a result of feedback from many readers of the books *Engineering with Mathcad: Using Mathcad to Create and Organize your Engineering Calculations,* and *Essential Mathcad for Engineering, Science and Math.* I have appreciated the positive reviews and feedback that have been provided by readers.

PTC Mathcad Prime is a significant departure from Mathcad 15. There is a definite learning curve associated with making the switch from Mathcad 15 to PTC® Mathcad Prime® 3.0. The new features included in PTC Mathcad Prime 3.0 make switching from Mathcad 15 worthwhile.

There are two primary goals of *Essential PTC® Mathcad Prime® 3.0: A Guide for New and Existing Users*. The first is to help those who are new PTC Mathcad users get up-to-speed as quickly and as efficiently as possible. This includes not only learning how to use the software tools, but more importantly how to apply the software tools in creating technical calculations.

A second goal of the book is to help those who are very comfortable with using Mathcad 15 make the switch to PTC Mathcad Prime 3.0. Based on my experience, this is a difficult transition. The book includes a whole chapter (Chapter 2) dedicated just for you. It also includes numerous highlights throughout the book letting you know the differences between Mathcad 15 and PTC Mathcad Prime 3.0.

A challenge with any book is to hit a balance between too little material and too much material. Based on feedback from *Engineering with Mathcad*, the book *Essential Mathcad* was revised to incorporate the recommendations. *Essential Mathcad* added a new Chapter 1, An Introduction to Mathcad, and rearranged other chapters in an attempt to make learning PTC Mathcad even easier. These changes received very favorable feedback. As a result, *Essential PTC® Mathcad Prime® 3.0: A Guide for New and Current Users,* follows a similar format. This book cannot and does not include a discussion of all the many PTC Mathcad functions and features. It does attempt to focus on the functions and features that will be most useful to a majority of readers.

Book Overview

This book uses an analogy of teaching you how to build a house. If you were to learn how to build a house, the final goal would be the completed house. Learning how to use the tools would be a necessary step, but the tools are just a means to help you complete the project. It is the same with this book. The ultimate goal is to teach you how to apply PTC Mathcad to build comprehensive project calculations.

In order to begin building, you need to learn a little about the tools. You also need to have a toolbox where you can put the tools. When building a house, there are simple hand tools and more robust power tools. It is the same with PTC Mathcad. We will learn to use the simple tools before learning about the power tools. After learning about the tools, we learn to build.

This book is divided into four parts:

Part I—Building Your PTC Mathcad Toolbox. This is where you build your basic understanding of PTC Mathcad. It teaches the basics of the PTC Mathcad program. The chapters in this part create a solid foundation upon which to build.

Part II—Hand Tools for Your PTC Mathcad Toolbox. The chapters in this part focus on simple features to get Your comfortable with PTC Mathcad.

Part III—Power Tools for Your PTC Mathcad Toolbox. This part addresses more complex and powerful PTC Mathcad features.

Part IV—Creating and Organizing Your Project Calculations with PTC Mathcad. This is where you start using the tools in your toolbox to build something—project calculations. This part discusses how to assemble calculations from multiple PTC Mathcad files, and files from Microsoft® Excel spreadsheets.

Additional Resources

This book is written as a supplement to the PTC Mathcad Help and the features on the **Getting Started** tab. It adds insights not contained in these resources. You should become familiar with the use of these resources prior to beginning an earnest study of this book. To access PTC Mathcad Help, click the round circle with a question mark on the right side of the ribbon, or press the **F1** key.

In addition to the PTC Mathcad Help and the **Getting Started** tab, the PTC Mathcad Tutorials provide an excellent resource to help learn PTC Mathcad. The PTC Mathcad Tutorials are accessed from the PTC Mathcad Help Center. Take the opportunity to review some of the topics covered by the tutorials.

Teminology

There are a few terms we need to discuss in order to communicate effectively.

The terms "click," "clicking," or "select" mean to click with the left mouse button.

The terms "expression" and "equation" are sometimes used interchangeably. The term "equation" is a subset of the term "expression." When we use the term "equation," it generally means some type of algebraic math equation that is being defined on the right side of the ***definition*** operator ":=". The term "expression" is broader. It usually means anything located to the right of the ***definition*** operator. It can mean "equation," a PTC Mathcad program, a user-defined function, a matrix or vector, or any number of other PTC Mathcad elements. When working with regions, the term "activate" means to click within the region. This places a dashed line around the region. The term "select" means to hold down CTRL and click within the region. This places a shaded area within the region.

Acknowledgements

This book would not have been possible without the help of Jakov Kucan, Director of PTC Mathcad Product Strategy. He answered numerous questions about PTC Mathcad Prime 3.0 functionality, and provided valuable insight that greatly improved this book. I thank him for his assistance.

Grateful acknowledgement is given to Dixie M. Griffin, Jr., PhD, former Professor of Civil Engineering at Louisiana Tech University. A significant number of the engineering examples in this book were adapted from worksheets provided by Dr. Griffin. He has accumulated thousands of PTC Mathcad worksheets over the years. Many of these worksheets are posted on his Mathcad webpage – http://pages.suddenlink.net/drgriffinsmathcad

Many engineers have helped to improve my knowledge and experience as an engineer. Indirectly, they have helped create this book. There are too many to mention by name, but I wish to express thanks to the engineers in the Structural Engineers Association of Utah. I also wish to thank my colleagues at work. It is a pleasure to work with them.

This list of thanks would not be complete without special thanks to my wife Cherie, who put up with the late nights, early mornings, and Saturdays spent working on this book. She has been a patient stalwart supporter of this effort. I could not have done it without her support. I also want to thank my five sons, two daughter-in-laws, and three grandchildren who are the joy of my life.

Brent Maxfield

PART

Building Your PTC Mathcad Toolbox

I

Your PTC Mathcad tools are stored in your PTC Mathcad toolbox. Your PTC Mathcad toolbox is the place where you will store your PTC Mathcad skills, which will be discussed in Parts II and III. You first build your PTC Mathcad toolbox by learning about the basics of the PTC Mathcad program and the PTC Mathcad worksheet. The chapters in Part I teach about variables, expression editing, user-defined functions, and units. These chapters create a foundation upon which to build. They create your PTC Mathcad Toolbox.

CHAPTER 1

An Introduction to PTC® Mathcad Prime® 3.0

> If you are a current Mathcad 15 user, glance quickly through this chapter to learn about some of the new features of PTC® Mathcad Prime® 3.0. Much of it will be review, but there are things to be learned. Chapter 2 will contain a much more in-depth review of PTC Mathcad Prime for current Mathcad 15 users.

This chapter is intended to quickly teach you some fundamental PTC Mathcad concepts; we will only touch the surface. In later chapters, we will get into more depth, and build on the concepts covered in this chapter. This chapter also teaches techniques to create and edit PTC Mathcad expressions.

Chapter 1 will:

- Show how to do simple math in PTC Mathcad.
- Teach how to assign and display variables.
- Explain how to create and edit math expressions.
- Demonstrate the grouping highlight for editing expressions.
- Discuss the use of operators.
- Briefly discuss the PTC Mathcad ribbon tabs.
- Introduce and define math and text regions.
- Introduce built-in and user-defined functions.
- Introduce units.
- Introduce arrays and subscripts.
- Discuss the variable ORIGIN.
- Describe the difference between literal and array subscripts.
- Introduce range variables.
- Introduce X-Y plots.
- Encourage completing several PTC Mathcad Tutorials.

Before You Begin

If you don't already have PTC Mathcad installed on your computer, take a few minutes to download and install it. This will allow you to follow along and practice the concepts discussed in this book. It will also give you access to the PTC Mathcad Help and the Getting Started ribbon, which includes some PTC Mathcad Tutorials.

Essential PTC® Mathcad Prime® 3.0.

Essential PTC Mathcad Prime 3.0 is based on the U.S. version of PTC Mathcad Prime 3.0. It is also based on the U.S. keyboard. There may be slight differences in PTC Mathcad versions sold outside of the United States.

It is suggested that you read and do the exercises in the PTC Mathcad Tutorial before or just after reading this chapter. You can open the PTC Mathcad Tutorial from the **Getting Started** tab in the **Get Help in PTC Mathcad** group **(Getting Started>Get Help in PTC Mathcad>Tutorial)**. This opens the Tutorials in the PTC Mathcad Help Center. In this window you will see a list of PTC Mathcad exercises. You may choose to do them all, but for the purpose of this chapter, focus on the following topics: Entering and Evaluating an Expression, Editing an Equation, Defining and Evaluating Variables, Ordering Regions and Applying Labels, and Defining and Evaluating Functions. This chapter cannot replace the experience gained by completing the PTC Mathcad Tutorials.

If the **Getting Started** tab does not appear on your Ribbon Bar, click the round button in the upper left of the PTC Mathcad window, select Options, and then select the **Options** tab at the bottom of the dialog box. If the **Enable Getting Started Tab** does not have a check, click the box to enable it.

PTC Mathcad Basics

Whenever you open PTC Mathcad, a blank worksheet appears. You can liken this worksheet to a clean sheet of calculation paper waiting for you to put information onto it. The PTC Mathcad default worksheet has graph paper-like gridlines enabled by default. The images in this book have the gridlines turned off. If you want to turn off the gridlines, select the **Document** tab and from the **Page** group, select the **Show Grid** control (**Document>Page>Show Grid**).

Let's begin with some simple math. Type **5+3** =. You should get the following:

$$5+3=8$$

Now type **2+3 Spacebar Spacebar *2=**. You should get the following:

$$(2+3)\cdot 2=10$$

You can also assign variable names to these equations. To assign a value to a variable, type the variable name and then type the colon **:** key. For example type **a1 : 5+3**.

$$a1:=5+3$$

Now type **a1=**. This evaluates and displays the value of variable "a1."

$a1 = 8$

Let's assign another variable. Type **b1 : 2+3 Spacebar Spacebar*2**.

$b1 := (2+3) \cdot 2$

Now type **b1=**. This displays the value of variable "b1."

$b1 = 10$

Now that values are assigned to variable "a1" and variable "b1," you can use these variables in equations. Type **c1 : a1+b1**.

$c1 := a1 + b1$

Now type **c1=**. You should get the following result:

$c1 = 18$

As you begin using variables, it is important to understand the following PTC Mathcad protocol. In order to use a previously defined variable, the variable must be defined above or to the left of where it is being used. In other words, PTC Mathcad calculates from left to right, top to bottom.

As you can see, PTC Mathcad does not require any programming language to perform simple operations. Simply type the equations as you would write them on paper.

Creating Simple Math Expressions

There are two ways to create a simple expression. The first way is to just type as you would say the expression. For example you say 2 plus 5, so you would type the following **2+5**. You say 2 to the 4th power, so you would type **2^4**. You say the square root of 100, so you type **\100**.

The second way to create a simple expression is to type an operator such as +, —, *, or /. This will create empty placeholders (gray boxes) that you can then click to fill in the numbers or operands. For example, if you press the + key anywhere in your worksheet, you will get the following:

$|+\blacksquare$

Click in the first placeholder and type **2**, then use the right arrow key twice to move to the second placeholder or click in the second placeholder and type **5**. Your expression should now look like this:

$2+5|$

In this example 2 and 5 are operands of the addition (+) operator.

You can use the above procedure with any operator. Let's try the ***exponentiation*** operator. Press **^** to create the exponentiation operator. You can also click x^y from the **Operators** control on the **Math** tab **(Math>Operators and Symbols>Operators>Algebra>x^y)**. You should have the following:

$|\blacksquare^{\blacksquare}$

Click in the lower placeholder and type **2**, then use the right arrow key twice to move to the upper placeholder or click in the upper placeholder and type **4**. Your expression should now look like this:

$2^{4|}$

These methods of creating expressions work very well for creating simple expressions. As your expressions become more complex, there are a few things that you must learn.

Grouping

Creating more complex math expressions is very easy once you learn the concept of the grouping of terms and the effect of cursor location. The cursor is the vertical line in the expression. Pressing the spacebar will begin to highlight and group a portion of the expression. The first press of the spacebar highlights the term adjacent to the cursor. Pressing the spacebar again will highlight and group more of the expression. For example, if you type **2+5 Spacebar Spacebar**, you get the following:

$$2+5$$

Whatever is highlighted or grouped becomes the operand for the next operator. So, if you type **2+5 Spacebar Spacebar ^ 3**, you get the following:

$$(2+5)^{3}$$

In this case (2+5) is the x operand for the operator x to the power of y. Notice how the cursor is located adjacent to the number 3. This means that if you type any operator, the number 3 is the operand for the operator. Thus, if you type + 4, then you get the following:

$$(2+5)^{3+4}$$

But, if you press the **Spacebar** twice, the grouping highlight expands to enclose the whole expression.

$$(2+5)^{3}$$

This group becomes the operand for the next operator. Thus, if you now type **+ 4**, you get the following:

$$(2+5)^{3}+4$$

The whole expression became the operand for the ***addition*** operator.

It is very important to understand this concept of using the grouping highlight to determine what the operand of your next operator will be. You can also use parentheses to set the operand for operators. The following example will help reinforce these concepts.

Let's create the following expression:

$$\frac{\left(\frac{1}{2}-\frac{1}{3}\right)^{2}}{\sqrt{\frac{4}{5}+\frac{2}{7}}}$$

To create this expression, use the following steps:

1. Type `1/2 Spacebar Spacebar`. The grouping highlight now holds the fraction 1/2. This becomes the operand for the ***subtraction*** operator.

2. Type `− 1/3 Spacebar Spacebar Spacebar`. The grouping highlight should now hold both fractions. This becomes the operand for the ***exponentiation*** operator.

$$\frac{1}{2}-\frac{1}{3}$$

3. Type `^2 Spacebar Spacebar`. The grouping highlight should now hold the entire numerator. This becomes the operand for the ***division*** operator.

$$\left(\frac{1}{2}-\frac{1}{3}\right)^{2}$$

4. Type `/ \4 /5 Spacebar Spacebar Spacebar`. This makes everything under the radical the operand for the ***addition*** operator.

$$\frac{\left(\frac{1}{2}-\frac{1}{3}\right)^{2}}{\sqrt{\frac{4}{5}}}$$

5. Type `+ 2 / 7`. This completes the example.

$$\frac{\left(\frac{1}{2}-\frac{1}{3}\right)^{2}}{\sqrt{\frac{4}{5}+\frac{2}{7}}}$$

Notice how during each step, the spacebar was used to enlarge the grouping highlight to include the operand for the following operator.

The PTC Mathcad Tutorial has additional examples that provide worthwhile practice.

Editing Expressions

Another important concept to know is how to edit existing expressions.

Selecting Characters

If you click anywhere in an expression and begin pressing the spacebar, the grouping highlight expands to include more and more of the expression. How the grouping highlight expands depends on where you begin and on what side of a term the cursor is on. The best way to understand how it works is to experiment and to follow the examples in the PTC Mathcad Tutorial.

The general rule is that as the grouping highlight expands and crosses an operator, the operand for that operator is then included within the grouping highlight.

Deleting Characters

You can delete characters in your expressions by moving the cursor adjacent to the character. If the cursor is to the left of the character, press the **DELETE** key. If the cursor is to the right of the character, press the **BACKSPACE** key.

To delete multiple characters, drag select the portion of the expression you want to delete. If the cursor is to the left of the highlighted area, press the **DELETE** key. If the cursor is to the right of the highlighted area, press the **BACKSPACE** key.

Deleting and Replacing Operators

When you click an operator, the cursor disappears, the operator turns blue and flashes, and the operands for the operator are grouped on both sides of the operator. This is very useful to clearly understand the operands for the operators. The following examples illustrate various operators of the above example.

$$\frac{\left(\frac{1}{2}-\frac{1}{3}\right)^2}{\sqrt{\frac{4}{5}+\frac{2}{7}}} \quad \frac{\left(\frac{1}{2}-\frac{1}{3}\right)^2}{\sqrt{\frac{4}{5}+\frac{2}{7}}} \quad \frac{\left(\frac{1}{2}-\frac{1}{3}\right)^2}{\sqrt{\frac{4}{5}+\frac{2}{7}}} \quad \frac{\left(\frac{1}{2}-\frac{1}{3}\right)^2}{\sqrt{\frac{4}{5}+\frac{2}{7}}}$$

To replace an operator, click the operator or move the cursor until the cursor crosses the operator and the operator changes to a blue color and flashes. Both operands should now be highlighted. Next press the **DELETE** key. This will delete the operator leaving a small dot where the operator used to be. Now, use the arrow key to cross over the dot. This will group the operands on both sides of the dot. You are now ready to type a new operator.

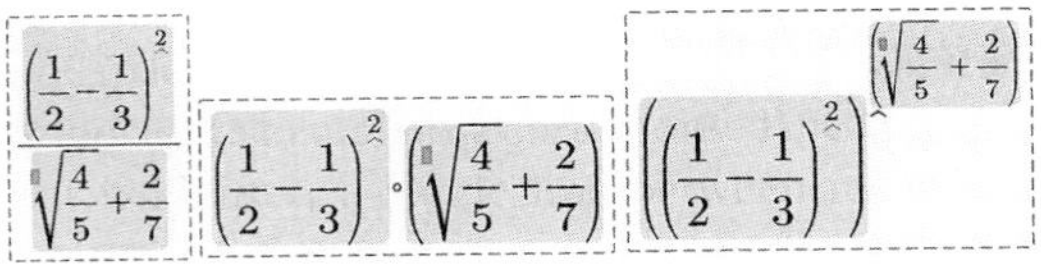

$$\frac{\left(\frac{1}{2}-\frac{1}{3}\right)^2}{\sqrt{\frac{4}{5}+\frac{2}{7}}} \quad \left(\frac{1}{2}-\frac{1}{3}\right)^2 \circ \left(\sqrt{\frac{4}{5}+\frac{2}{7}}\right) \quad \left(\left(\frac{1}{2}-\frac{1}{3}\right)^2\right)^{\left(\sqrt{\frac{4}{5}+\frac{2}{7}}\right)}$$

The best way to understand this concept is to experiment with it.

Modifying Expressions

When modifying expressions, it is best to use your cursor to select an element or drag select a grouping and then modify the selected elements. For example, let's change $(2x^2+3x-6)^2$ to $(2x^2+3x)^2-6^2$

Let's first select and delete the parentheses.

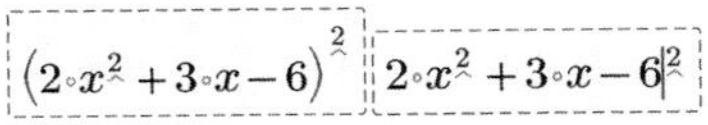

$$(2\circ x^2+3\circ x-6)^2 \quad 2\circ x^2+3\circ x-6^2$$

Next, select and group the $2x^2+3x$. If you drag select from left to right, the cursor will be on the right side of the grouping and will be ready to type the ***exponentiation*** operator **^2**.

$$2\circ x^2+3\circ x-6^2 \quad (2\circ x^2+3\circ x)^2-6^2$$

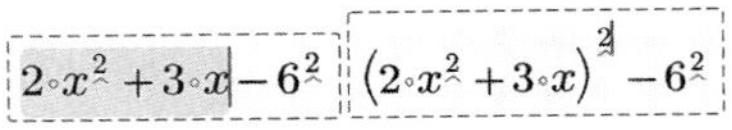

The parentheses are added automatically because the selected area was grouped prior to typing the ***exponentiation*** operator. Be aware that if you drag select from right to left, the cursor will be on the left side of the grouping. If you then type the ***exponentiation*** operator, the grouping will become the exponent of the ***exponentiation*** operator.

$$2\circ x^2+3\circ x-6^2 \quad \blacksquare^{2\circ x^2+3\circ x}-6^2$$

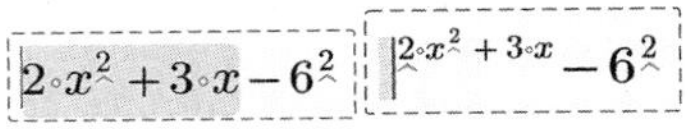

PTC Mathcad Workspace

Figure 1.1 shows the PTC Mathcad workspace. It consists of the PTC Mathcad worksheet and the following:

- PTC Mathcad Button
- Ribbon Bar
- Quick Access Toolbar
- Open Worksheets Bar
- Status Bar

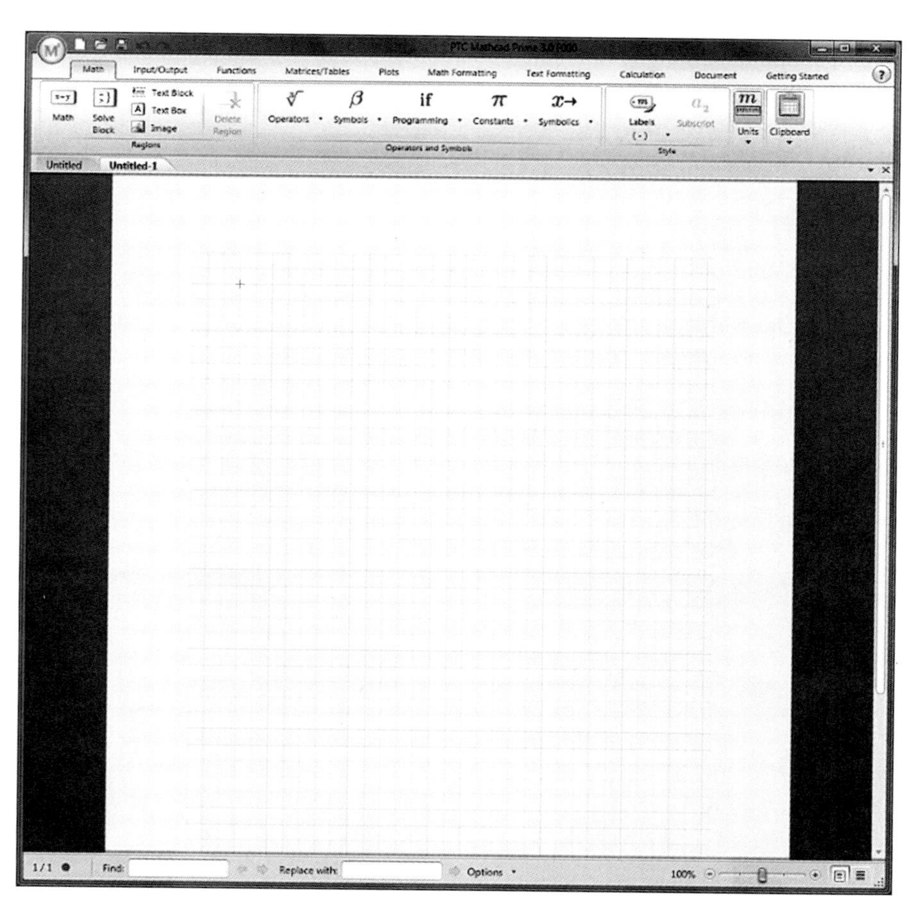

FIGURE 1.1

PTC Mathcad workspace

PTC Mathcad Button

This button is similar to the File tab used in many software programs. It allows you to open a new worksheet, open an existing worksheet, save a worksheet, save a template, print a worksheet, or close a worksheet. There is also an Options button where you can set PTC Mathcad global settings. This is where you can also manage your PTC Mathcad licenses.

Quick Access Toolbar

The Quick Access Toolbar is located at the top of the workspace. It contains commands that are used on a regular basis. You can add commands to the Quick Access toolbar by right clicking on a command and selecting, "Add to Quick Access Toolbar."

Open Worksheets Bar

The Open Worksheets Bar is located just below the Ribbon Bar and has a tab for each open worksheet. You can switch between worksheets by clicking on one of the tabs shown in the Open Worksheets Bar. To close the selected worksheet, click the "X" located on the right side of the bar.

Status Bar

The Status Bar is located at the bottom of the workspace. It contains page numbers, Find, Replace, and a zoom slider. This is also where you can switch between Page View and Draft View display modes. The circle in the Status Bar is green when the Auto Calculation mode is turned on.

Ribbon Bar

The Ribbon Bar is comprised of multiple tabs where you can easily access most of PTC Mathcad's features. Let's quickly explore some of the features contained in the Ribbon Bar.

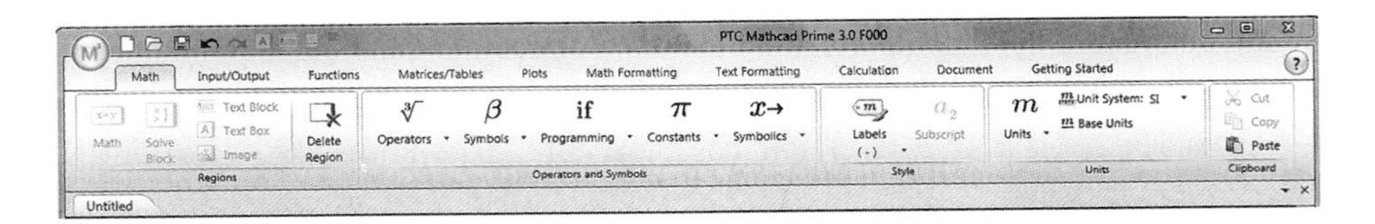

Math Tab

The **Math** tab allows you to quickly access many basic math operators and symbols. These are located in the **Operators and Symbols** group.

Operators

The **Operators** list contains numerous operators from algebra, calculus, Boolean comparison, and more.

Algebra

$\lvert x \rvert$	+	,	/	x^y
$x!$	()	÷	·	_
$\sqrt{x}$	%			

Calculus

⊛	*	d/dx	$\int dx$	lim
=	$\int'$	Π	Σ	

Comparison

∈	≡	⊕	>	≥
<	≤	∧	¬	∨
≠				

Definition and Evaluation

:=	=	→

Engineering

$\bar{z}$	°	∠	·

Vector and Matrix

×	$\lVert x \rVert$	[▯]	$M^{\langle\rangle}$	$M_{,}$
$M^{\rightleftharpoons}$	M^T	1..n	1,3..n	$\vec{\nabla}$

Symbols

The **Symbols** list contains lowercase and uppercase Greek letters. If you hover your mouse over a Greek letter its keyboard shortcut will appear. Refer to Appendix 3 for a complete list of keyboard shortcuts.

Lowercase Greek

α	β	γ	δ	ε
ζ	η	θ	ι	κ
λ	μ	ν	ξ	ο
π	ρ	σ	τ	υ
φ	ϕ	χ	ψ	ω

Uppercase Greek

Α	Β	Γ	Δ	Ε
Ζ	Η	Θ	Ι	Κ
Λ	Μ	Ν	Ξ	Ο
Π	Ρ	Σ	Τ	Υ
Φ	Χ	Ψ	Ω	

Math Symbols

ˆ	∐	∀	∃	∝
∡	∢	∥	∩	∪
∴	∵	⊆	⊇	°
î	ℂ	ℚ	ℝ	ℤ
∞				

Monetary Symbols

¢	£	¤	¥	€

Hebrew

א	ב	ג	ד

Tags

⁋	∐	©	®	†
※	∋			

Constants

The **Constants** list has symbols and variables for common math and physics constants.

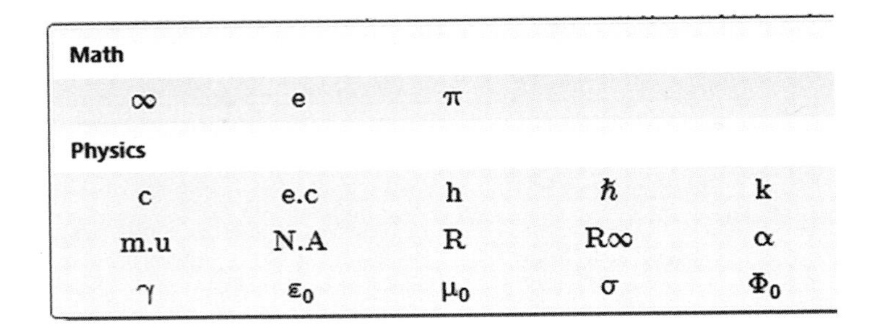

Summary of Equal Signs

There are four equal signs used in PTC Mathcad. It is important to understand the difference between them.

- The ***definition*** operator (:=) `COLON` is used to define variables, functions, or expressions.
- The ***evaluation*** operator (=) `EQUAL SIGN` is used to evaluate a variable, function, or expression numerically.
- The Boolean ***equal to*** operator (=) `CTRL+EQUAL SIGN` is used to evaluate the equality condition in a Boolean statement. It is also used for programming, solving, and in symbolic equations. It will be discussed in more detail in future chapters.
- The ***global definition*** operator (≡) `CTRL+TILDA~or CTRL+SHIFT+ACCENT` is used to assign a global variable. All global assignment definitions in the worksheet are scanned by PTC Mathcad prior to scanning for normal assignment definitions. This means that global assignments can be defined anywhere in the worksheet and still be recognized. Global assignments should be used with caution.

The ***definition*** operator, ***evaluation*** operator, and ***global definition*** operator are found on the **Operators** list under the **Definition and Evaluation** section. The Boolean ***equal to*** operator is found under the **Comparison** section.

Regions

A region is a location where information is stored on the worksheet. Your entire PTC Mathcad worksheet will be comprised of individual regions. There are many types of regions:

- Math
- Text (Text Block and Text Box)
- Plot
- Table
- Solve
- Image

We will discuss the Math, Text, and Plot regions in this chapter. The remaining regions will be covered later in the book.

Math Regions

Math regions contain variables, constants, expressions, functions, etc. These regions are basically anything except text regions and plot regions. These regions are created automatically whenever you create any expression or definition.

Text Regions

Text regions allow you to add notes, comments, titles, headings, and other items of interest to your calculation worksheet. There are two types of text regions: text block and text box. The text block extends from margin to margin. A text block will not overlap existing regions. As a text block expands, regions below the text block move down. When you insert a text block other regions are moved up or down to make room for the text block. It is useful for inserting long paragraphs of content on the page. A text box on the other hand is a text region that can be resized and moved around the worksheet. It can overlap other regions. A text box is useful for headings, captions, and short descriptions.

The insert **Text Block** and insert **Text Box** controls are contained on the **Math** tab in the **Regions** group (**Math>Regions>Text Box** and **Math>Regions>Text Block**). The keyboard shortcut to insert a text box is **CTRL+t**. The keyboard shortcut to insert a text block is **CTRL+SHIFT+T**. When you are finished typing the text, if you press the **ENTER** key, PTC Mathcad inserts a new paragraph in the same text region. In order to exit a text region, click outside the region or use the arrow keys to move outside the text region.

In Chapter 3, we will discuss text regions in much more depth.

Plot Regions

Plot regions will be covered later in this chapter.

Functions

Functions will be discussed briefly in Chapter 4 and built upon throughout the book. The following paragraphs will get you started.

Built-in Functions

PTC Mathcad has hundreds of built-in functions. You access these functions from the **Functions** tab on the ribbon.

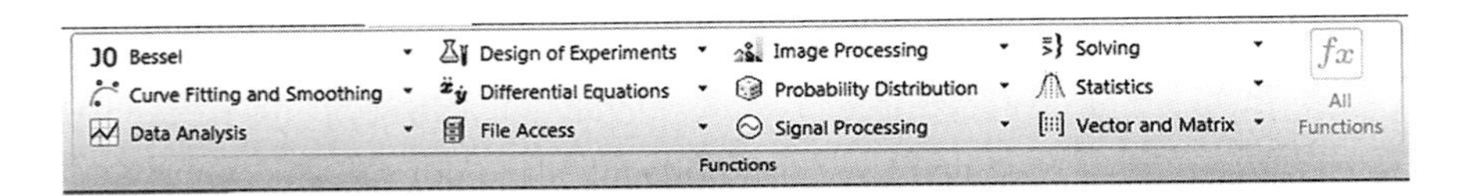

There are several default categories listed. The **All Functions** control places a list on the left side of the worksheet of all the PTC Mathcad functions listed by category. If you click the AZ↓ control, it lists the functions alphabetically.

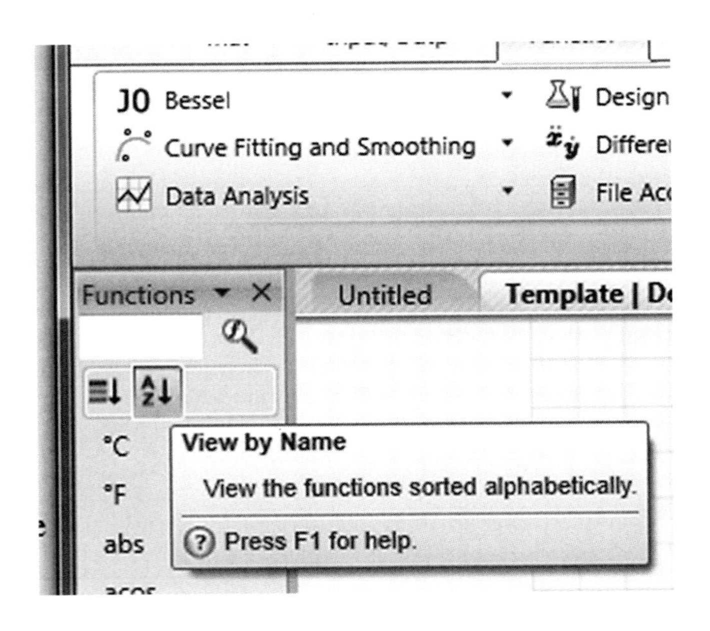

User-defined Functions

User-defined functions are very similar to built-in functions. They consist of a name, a list of arguments (in parentheses following the name), and a definition giving the relationship between the arguments. The name of the user-defined function is simply a variable name. See Figure 1.2 for an illustration of some user-defined functions.

$\mathrm{CircleArea}(r) := \pi \cdot r^2$ $\quad$ $\mathrm{CircleArea}(5) = 78.540$

$\mathrm{SquareArea}(L) := L^2$ $\quad$ $\mathrm{SquareArea}(4) = 16.000$

$\mathrm{RectangleArea}(L, H) := L \cdot H$ $\quad$ $\mathrm{RectangleArea}(4, 5) = 20.000$

$\mathrm{BoxVolume}(L, H, W) := L \cdot H \cdot W$ $\quad$ $\mathrm{BoxVolume}(4, 5, 10) = 200.000$

FIGURE 1.2

User-defined functions

Units

This section is only intended to get you started with units. Many experienced PTC Mathcad users still do not understand the significant benefits of using PTC Mathcad units, so it is important to read and study Chapter 5 – Units!

Once a unit is assigned to a variable, PTC Mathcad keeps track of it internally and displays the unit automatically. You will never need to remember the conversion factors for various units. You will never need to convert it from one unit system to another. PTC Mathcad does it all for you. All you need to do is tell PTC Mathcad how you want the unit displayed. For example, you can attach the unit of meters (m) to a variable. You will then be able to tell PTC Mathcad to display this variable in any unit of length such as: millimeters (mm), centimeters (cm), kilometers (km), inches (in), yards (yd), or miles (mi). PTC Mathcad does the conversion for you. If PTC Mathcad does not have the unit of measurement built in, you can define it, and then use it over and over.

Assigning Units to Numbers

To assign units to a number, simply multiply the number by the name of the unit. If you cannot remember the name of the unit, you can select from a list of over one hundred built-in PTC Mathcad units. These are found on the **Math** tab in the **Units** group on the **Units** list (**Math>Units>Units**).

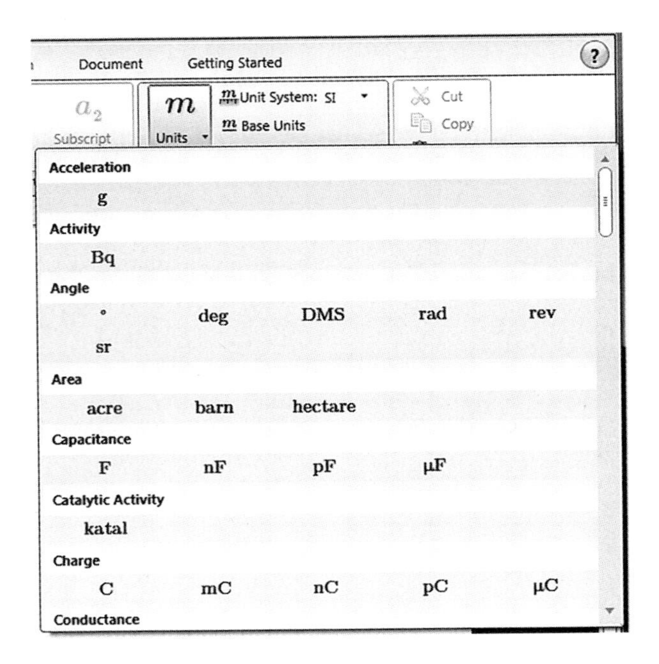

Mathcad displays units based on one of three "Unit Systems": **SI**, **US Customary System** (USCS), or **Centimeter-Gram-Second** (CGS). The worksheet system is shown on the **Math** tab in the **Units** group.

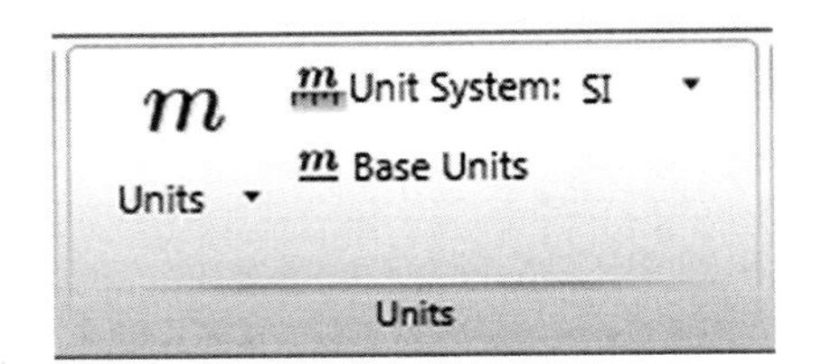

To change the system, select the drop-down box adjacent to the **Unit System** control.

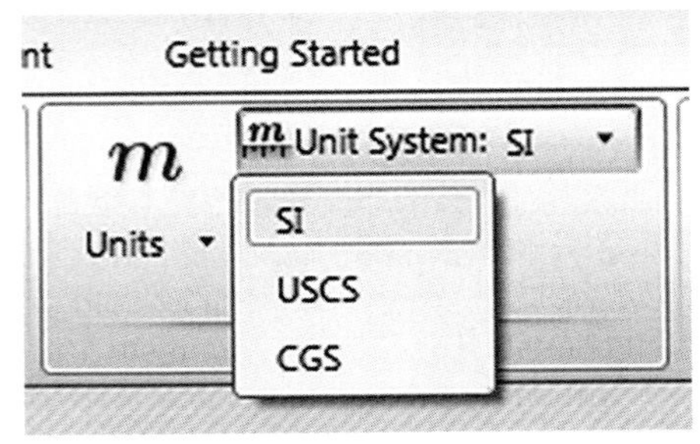

To assign units from the **Units** list, type a number, type the asterisk [*] and then select the desired unit from the "Units" drop-down list. Figure 1.3 shows some examples of units attached to numbers.

Examples of units attached to numbers and their default results. The PTC Mathcad default unit system is set to SI, so units display in the SI system.

If you do not know the unit names, then use the unit list on the Math tab. Type the number, type "*", and click the desired unit.

$1 \cdot ft = 0.305\ m$	Unit of length.
$180 \cdot deg = 3.142$	Unit of angle. The result is in radians.
$1 \cdot gal = 3.785\ L$	Unit of volume.
$1 \cdot min = 60.000\ s$	Unit of time.
$1 \cdot gm = 0.001\ kg$	Unit of mass.

FIGURE 1.3

Examples of units attached to numbers

PTC Mathcad does not require the ***multiplication*** operator between a value and the unit, but I recommend it as good practice.

Evaluating and Displaying Units

When you evaluate an expression with a unit attached, the unit that PTC Mathcad displays by default is based on the chosen default unit system (See Chapter 5). After evaluating an expression by pressing the [=] key, PTC Mathcad displays the default unit in a blue italic bold font label, and the cursor is placed to the right side of the unit label. Labels will be discussed later in the book, but they are a way for PTC Mathcad to differentiate between different type of variables. If you want PTC Mathcad to display a unit different than the default unit, use the backspace key to delete the unit, and then type the name of the unit you want displayed. You may also double click the unit to select it, and then type a new unit name, or choose a unit from the **Units** drop-down list on the **Math** tab. See Figure 1.4 for some examples of displaying units.

Chapters 15, 16, and 17 will show how to set and keep default unit systems for all new worksheets.

$\text{Length} := 3 \cdot ft + 1 \cdot m + 33 \cdot mm + 4\ yd$

$\text{Length} = 5.605\ m$

PTC Mathcad displays the default unit of length for the SI system. To display the result in inches, delete the "m" and type "in".

$\text{Length} = 220.669\ in$

To display the results in mm, delete the "m" and type "mm".

$\text{Length} = \left(5.605 \cdot 10^{3}\right)\ mm$

You can display the results in many different units.

$\text{Length} = 0.006\ km$

$\text{Length} = 0.003\ mi$

$\text{Length} = 0.028\ furlong$

$\text{Length} = 6.130\ yd$

$\text{Length} = 18.389\ ft$

FIGURE 1.4

Displaying results in different units

Arrays and Subscripts

An array is simply a vector or a matrix. A vector is a matrix with only a single column or single row. This section will briefly introduce the topic. Chapter 6 will have a much more in-depth discussion.

Creating Arrays

The **Insert Matrix** control is located on the **Matrices/Tables** tab in the **Matrices and Tables** group (**Matrices/Tables>Matrices and Tables>Insert Matrix**). The control allows you to select up to a 12x12 matrix.

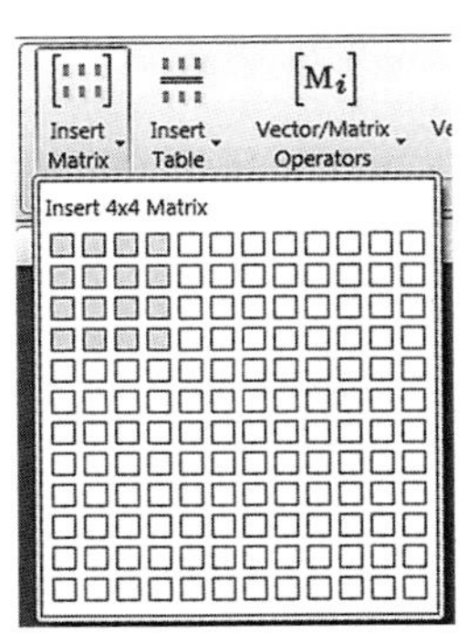

You can also type the keyboard shortcut **CTRL+m**. This command inserts a single placeholder within brackets. [ ] The **Rows and Columns** group on the **Matrices/Tables** tab has controls to add rows above or below the insertion point and to add columns to the left or right of the insertion point (**Matrices/Tables>Rows and Columns>Insert Right**).

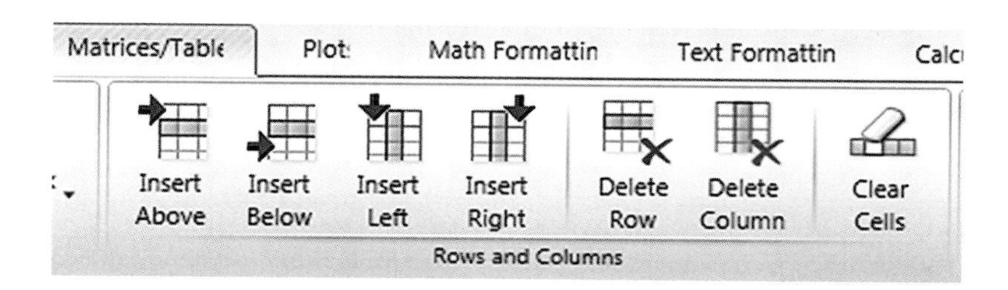

The keyboard shortcut **SHIFT+Spacebar** adds a column. The location is dependent on the location of the insertion point. If the insertion point is on the right of the placeholder, the column is inserted to the right. If the insertion point is on the left of the placeholder, the column is inserted to the left. The tab key will move the insertion point from placeholder to placeholder. When it reaches the end of available placeholders it will add a new row. You may also use the arrow keys to move between placeholders.

See Figure 1.5 for two sample matrix definitions using numbers and expressions.

ORIGIN

The value of the variable name ORIGIN tells PTC Mathcad the starting index of your array. The PTC Mathcad default for this variable is 0. This means that a vector or matrix begin indexing with zero. In other words, the first element is the 0th element. Thus, in Matrix_1 of Figure 1.5, the value of the 0th element of the matrix (Matrix_1(0,0)) would be 1.

I find it awkward to begin array numbering with 0. I like the first variable in an array to be labeled 1 rather than 0. I set the built-in variable ORIGIN to the value of 1.

For most scientific and engineering calculations, it is suggested that you change the value of ORIGIN from 0 to 1. With the value of ORIGIN set at 1, the first element of a matrix is the 1st element. Thus, in Matrix_1 of Figure 1.5, the value of the first

$$\text{Matrix\_1} := \begin{bmatrix} 1 & 2 & 3 & 4 \\ 5 & 6 & 7 & 8 \\ 9 & 10 & 11 & 12 \\ 13 & 14 & 15 & 16 \end{bmatrix}$$

$$\text{Matrix\_1} = \begin{bmatrix} 1.000 & 2.000 & 3.000 & 4.000 \\ 5.000 & 6.000 & 7.000 & 8.000 \\ 9.000 & 10.000 & 11.000 & 12.000 \\ 13.000 & 14.000 & 15.000 & 16.000 \end{bmatrix}$$

$$\text{Matrix\_2} := \begin{bmatrix} 1+2 & 2+3 & 3+4 & 4+5 \\ 5+6 & 6+7 & 7+8 & 8+9 \\ 9+10 & 10+11 & 11+12 & 12+13 \\ 13+14 & 14+15 & 15+16 & 16+17 \end{bmatrix}$$

$$\text{Matrix\_2} = \begin{bmatrix} 3.000 & 5.000 & 7.000 & 9.000 \\ 11.000 & 13.000 & 15.000 & 17.000 \\ 19.000 & 21.000 & 23.000 & 25.000 \\ 27.000 & 29.000 & 31.000 & 33.000 \end{bmatrix}$$

FIGURE 1.5

Sample matrix definitions

element of the matrix (Matrix_1(1,1)) would be 1. For the remainder of this book, the value of ORIGIN will be set at 1.

To change the value of ORIGIN, use the **ORIGIN** control on the **Calculation** tab in the **Worksheet Settings** group (**Calculation>Worksheet Settings>ORIGIN>1**).

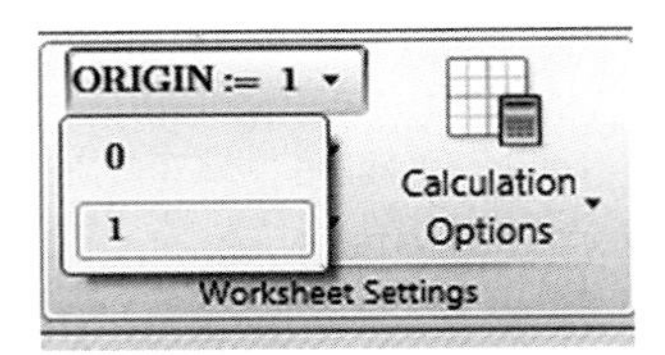

Subscripts

A discussion of arrays would not be complete without a discussion of subscripts. It is critical to understand the difference between two types of subscripts because they behave very differently. These two types of subscripts are called literal subscripts and array subscripts.

Literal Subscripts

Literal subscripts are part of a variable name. They allow you to have variable names such as F_s, $f_{y.}$, or H_2S0_4. The literal subscript in a variable name is controlled by the **Subscript** control on the **Math** tab in the **Style** group **(Math>Style>Subscript)**. It is a toggle switch that turns the subscript on and off.

To type a literal subscript, type the first part of the variable name, then activate the **Subscript** control and type the subscript portion of the variable name. You may then deactivate the **Subscript** control and continue typing the variable name. A variable name may have multiple literal subscripts. You cannot change a typed character to a literal subscript. You must delete the character and replace it with a subscript. See Figure 1.6 for an example of variable names using literal subscripts.

$$Example_1$$
$$Sample_2$$
$$f'_c$$
$$f_y$$
$$H_2O$$
$$H_2SO_4$$

FIGURE 1.6

Example of variable names using literal subscripts

Array Subscripts

An array subscript is not part of the variable name. An array subscript allows PTC Mathcad to display the value of a particular element in an array. It is used to refer to a single element in the array. The array subscript is created by using the ***matrix index*** operator, or by using the `[` key.

If you want PTC Mathcad to display the value of the first element in Matrix_1 in Figure 1.5 you would type: `Matrix_1[1,1=` and get a result of **$\mathbf{Matrix\_1_{1,1}=1.000}$**. Remember we changed ORIGIN from 0 to 1. If you want PTC Mathcad to display the value of the element in the 3rd row, 4th column, you would type `Matrix_1[3,4=` and get a result of **$\mathbf{Matrix\_1_{3,4}=12.000}$**.

In the above example, the variable name was Matrix_1. The variable contains a 4 row - 4 column matrix. The array subscript is not part of the variable name. It is only used to display an element of the array.

You can use an array subscript to assign an element in an array.

$$Matrix\_1_{1,1} := 20$$

$$Matrix\_1 = \begin{bmatrix} 20.000 & 2.000 & 3.000 & 4.000 \\ 5.000 & 6.000 & 7.000 & 8.000 \\ 9.000 & 10.000 & 11.000 & 12.000 \\ 13.000 & 14.000 & 15.000 & 16.000 \end{bmatrix}$$

FIGURE 1.7

Changing the value of a single array element

You can also use an array subscript to assign elements of an array. If you type `Matrix_1[1,1 : 20` then the value of the 1st element in Matrix_1 will be changed from 1 to 20. See Figure 1.7.

Figure 1.8 shows how to use array subscripts with a vector. Figure 1.9 shows how to use array subscripts to assign new values to vectors and arrays.

$$Vector\_1 := \begin{bmatrix} 2 \\ 22 \\ 222 \\ 444 \end{bmatrix} \qquad Vector\_1 = \begin{bmatrix} 2.000 \\ 22.000 \\ 222.000 \\ 444.000 \end{bmatrix}$$

$$Vector\_1_{1} = 2.000$$

$$Vector\_1_{2} = 22.000$$

$$Vector\_1_{3} = 222.000$$

$$Vector\_1_{4} = 444.000$$

With ORIGIN set to one, there is no zero element.

$$Vector\_1_{0} = ? \qquad Vector\_1_{0} = ?$$

This array index is invalid. The index must be an integer, not less than ORIGIN, and not greater than the last element.

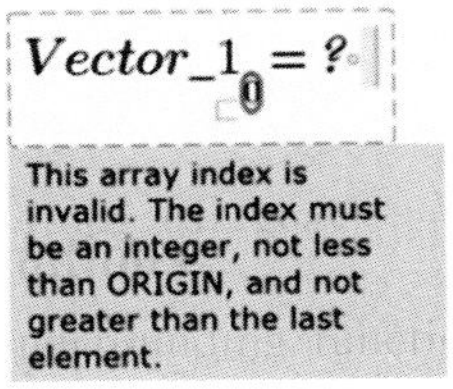

FIGURE 1.8

Using array subscripts with a column vector

You can add additional elements to an array by defining them with array subscripts.

$\mathbf{Vector\_1}_6 := 888$ $\mathbf{Matrix\_1}_{5,5} := 55$

$$\mathbf{Vector\_1} = \begin{bmatrix} 2.000 \\ 22.000 \\ 222.000 \\ 444.000 \\ 0.000 \\ 888.000 \end{bmatrix} \quad \mathbf{Matrix\_1} = \begin{bmatrix} 20.000 & 2.000 & 3.000 & 4.000 & 0.000 \\ 5.000 & 6.000 & 7.000 & 8.000 & 0.000 \\ 9.000 & 10.000 & 11.000 & 12.000 & 0.000 \\ 13.000 & 14.000 & 15.000 & 16.000 & 0.000 \\ 0.000 & 0.000 & 0.000 & 0.000 & 55.000 \end{bmatrix}$$

FIGURE 1.9

Using array subscripts

When you create or click an array subscript, you see a blue open "⊏" icon which represents the ***matrix index*** operator.

Figure 1.10 compares the differences between a literal subscript and an array subscript.

Literal Subscript:

Define the variable "var_2".

$\mathbf{var_2} := 40 \cdot cm$

Array Subscript:

Define the variable "var" as a column array.

$$\mathbf{var} := \begin{bmatrix} 1 \\ 3 \\ 5 \\ 7 \end{bmatrix}$$

Display the second element of the array. (Remember that ORIGIN is set to 1.)

$\mathbf{var}_2 = 3.000$

Notice how the two variable names below appear nearly the same. The subscript on the array subscript is slightly smaller and is lower than the subscript on the literal subscript.

$\mathbf{var_2} = 0.400\ m$ $\mathbf{var}_2 = 3.000$

This is how each region appears when you click inside it.

$\mathbf{var_2} = 0.4\ m$ $\mathbf{var}_{⊏2} = 3$

FIGURE 1.10

Comparing literal and array subscripts

Range Variables

Range variables will be used extensively in later chapters, and Chapter 6 has an entire section devoted to discussing the differences between range variables and vectors. This section will only introduce the concept.

A range variable is similar to a vector in that it takes on multiple values. It has a range of values. The range of values has a beginning value, an ending value and uniform incremental values between the beginning and ending values. Range variables can be used to iterate a calculation over a specific range of values, or to plot a function over a specific range of values. They are often used as integer subscripts for defining arrays. A range variable looks like this: **RangeVariableA := 1 , 1.5 .. 5**. This range variable begins with 1.0. The second number in the range variable sets the increment value. PTC Mathcad takes the difference between the first and second numbers and uses this as the incremental value. In the above case, the increment is 0.5. The last number in this range is 5.0. Thus, this range variable has the values: 1.0, 1.5, 2.0, 2.5, 3.0, 3.5, 4.0, 4.5, and 5.0.

To define a range variable, type the variable name followed by the colon `:`. This creates the variable definition. In the placeholder, type the beginning value, and then type a comma. This adds two placeholders separated by two periods.

$$rv := 1,\ \square\ ..\ \square$$

Now enter the incremental value first of the two placeholders. Now type `Period Period`. This moves the cursor to the last placeholder. Enter the ending value in the placeholder. If the second value is less than the beginning value, then the range variable will be decreasing, and the last value sets the lower limit to the range variable. See Figure 1.11 for sample range variables and their displayed results.

Plotting - X-Y Plots

Plots Tab

The **Plots** tab is shown below. The **Plots** tab allows you to quickly insert 2-dimensional X-Y plots, Polar plots, and 3-dimensional plots. Plotting will be discussed at length in Chapter 7, but let's take a quick look at how to create some simple plots.

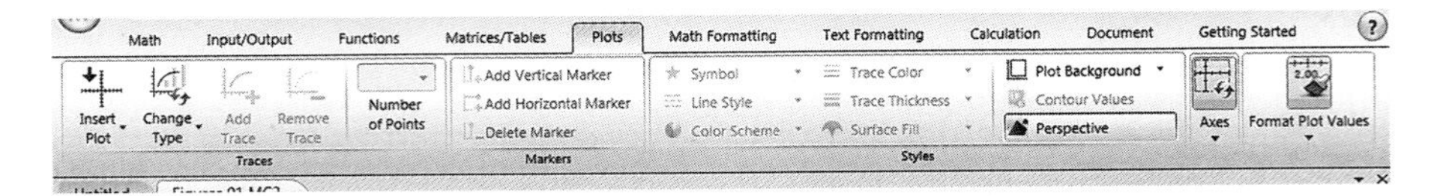

The second value sets the increment.

$$\text{RangeVariable\_A} := 5, 5.1 .. 6$$

$$\text{RangeVariable\_A} = \begin{bmatrix} 5.000 \\ 5.100 \\ 5.200 \\ 5.300 \\ 5.400 \\ 5.500 \\ 5.600 \\ 5.700 \\ 5.800 \\ 5.900 \\ 6.000 \end{bmatrix}$$

Range variables may also decrease.

$$\text{RangeVariable\_B} := 10, 8 .. -10$$

$$\text{RangeVariable\_B} = \begin{bmatrix} 10.000 \\ 8.000 \\ 6.000 \\ 4.000 \\ 2.000 \\ 0.000 \\ -2.000 \\ -4.000 \\ -6.000 \\ -8.000 \\ -10.000 \end{bmatrix}$$

If the second value is not given, then an increment of 1 is used.

$$\text{RangeVariable\_C} := 1 .. 10$$

$$\text{RangeVariable\_C} = \begin{bmatrix} 1.000 \\ 2.000 \\ 3.000 \\ 4.000 \\ 5.000 \\ 6.000 \\ 7.000 \\ 8.000 \\ 9.000 \\ 10.000 \end{bmatrix}$$

FIGURE 1.11

Sample range variables

To create a simple X-Y plot, click the **Insert Plot** control in the **Traces** group on the **Plots** tab, and then select **XY Plot** (**Plots>Traces>Insert Plot>XY Plot**).

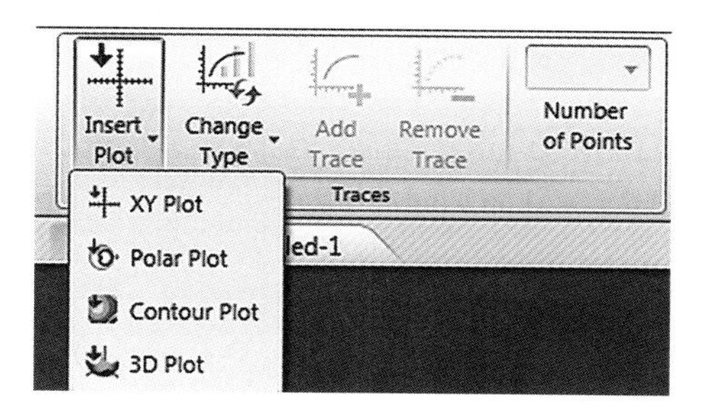

You may also type CTRL+2. This places a blank ***X-Y plot*** operator on the worksheet.

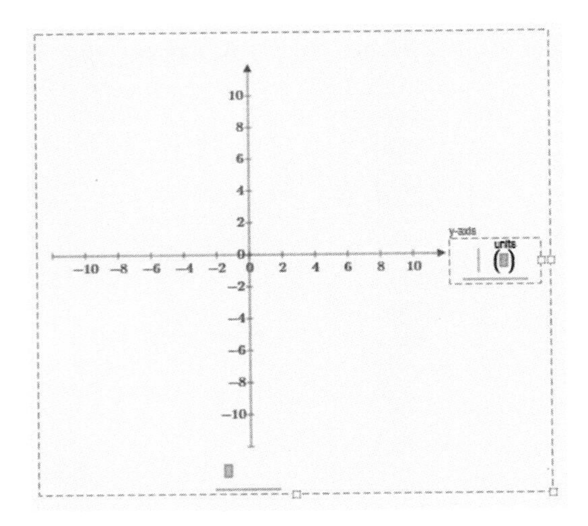

Click on the bottom middle placeholder. This is where you type the x-axis variable. Type the name of a previously undefined variable. The variable is allowed to be "x" but can be any PTC Mathcad variable name. Next, click on the middle right placeholder, and type an expression using the variable named on the x-axis. Click outside the operator to view the X-Y plot. PTC Mathcad automatically selects the range for both the x-axis and the y-axis. See Figure 1.12.

Another way to create a plot is to define a user-defined function prior to creating the plot. Open the ***X-Y Plot*** operator by typing CTRL+2. Click the bottom placeholder and type a variable name for the x-axis. This variable name does not need to be the same one used as the argument to define the function. On the right placeholder, type the name of the function. Use the variable name from the x-axis as the argument of the function. Here again, PTC Mathcad selects the range for both the x-axis and the y-axis. See Figure 1.13.

If you use a previously defined variable, then PTC Mathcad will not plot a graph over a range of values. It will only plot the value of the variable used. In some cases, this may only be a single point. For a plot, it is important to use only undefined variables. We will discuss the use of range variables in plots in Chapter 7. This is a case where a previously defined variable can be used.

Setting Plotting Ranges

PTC Mathcad automatically sets the x-axis domain range from −10 to 10. The y-axis values are automatically set to match the function being plotted. The x-axis and y-axis plot ranges may be changed by editing the first and last tick

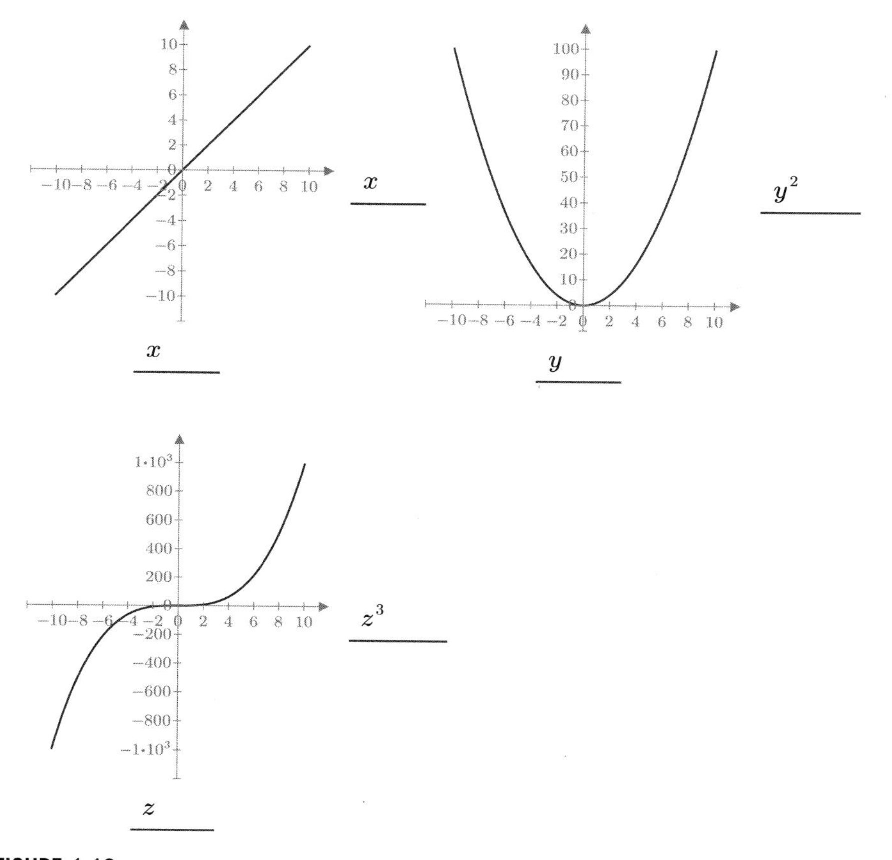

FIGURE 1.12

X-Y plot examples

marks for each axis. The interval may be changed by editing the second tick mark. Refer to Chapter 8 for a more detailed description of editing tick marks. See Figures 1.14 and 1.15.

Programming, Symbolic Calculations, Solving and Calculus

There are many wonderful PTC Mathcad features that we have not covered in this chapter, but this chapter is titled "An Introduction to PTC Mathcad." If we covered all the features, then we would need a book to discuss them. The rest of this book is about teaching you some of the essential features of PTC Mathcad.

Define three functions

$$\mathbf{ff}(\mathbf{h}) := \mathbf{h} \quad \mathbf{gg}(\mathbf{h}) := \mathbf{h}^2 \quad \mathbf{hh}(\mathbf{h}) := \mathbf{h}^3$$

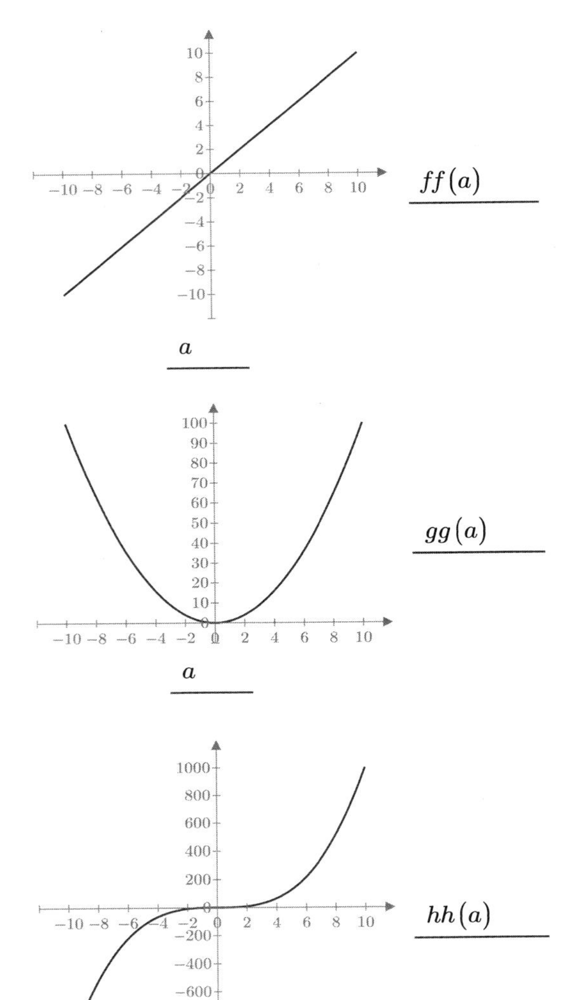

Note:
The variable used on the x axis must be a previously undefined variable.

FIGURE 1.13

X-Y plot of functions

$$D(t) := \frac{1}{2} \cdot g \cdot t^2$$

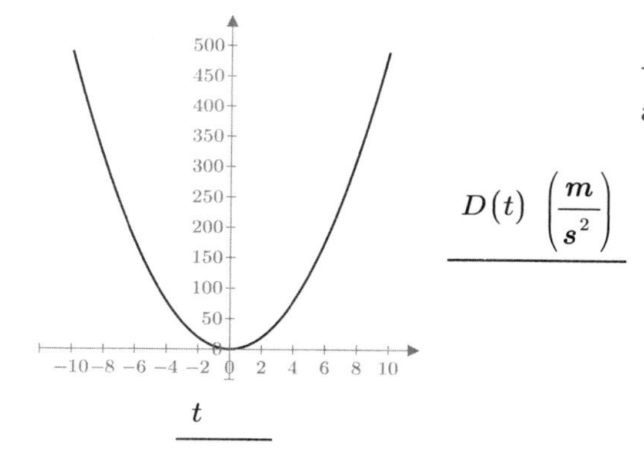

The PTC Mathcad default x-axis limits are -10 and 10.

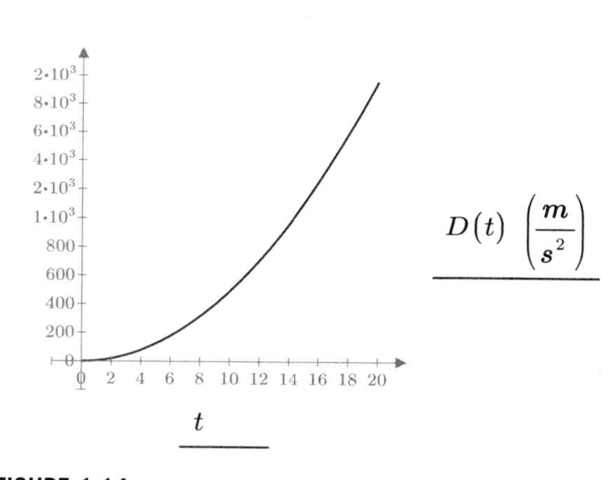

In this plot, the x-axis limits are changed to 0 and 20, with a tick mark interval of 2.

FIGURE 1.14

Setting plot range

In future chapters we will build on the concepts learned in this chapter. We will also discuss how to use PTC Mathcad programming to create useful and powerful functions. We will discuss the use of symbolic calculations to return algebraic results rather than numeric results. Chapter 10 will discuss some of PTC Mathcad's powerful solving features. In Chapter 13, we will demonstrate how PTC Mathcad can solve calculus and differential equation problems. In Chapter 21 we will discuss how to integrate MS Excel spreadsheets into your PTC Mathcad calculations. Part IV will discuss how to use PTC Mathcad to create and organize scientific and engineering calculations.

In this plot, the y-axis limits were set between 600 and 1200 with a tick mark interval of 100. Note how the function is truncated to only show the limits set by the y-axis.

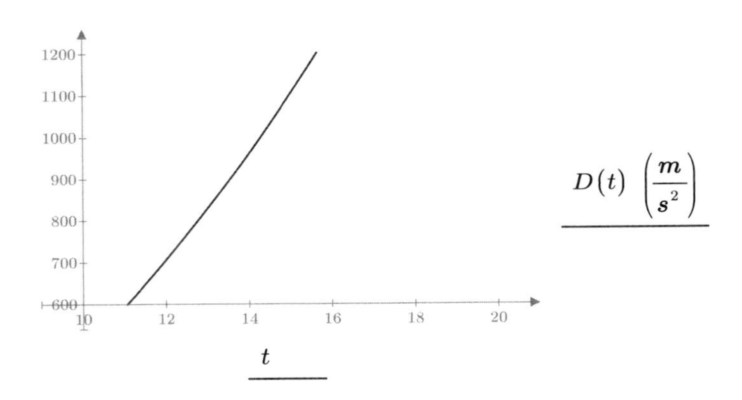

$$D(10) = 490.333 \frac{m}{s^2}$$

FIGURE 1.15

Setting plot range

Getting Started Tab

The **Getting Started** tab is your one-stop place to access PTC Mathcad information. From this tab, you can access PTC Mathcad Help, Tutorials, Engineering Resources, a list of keyboard shortcuts, and many online resources. This is a great place to just explore and learn more about PTC Mathcad.

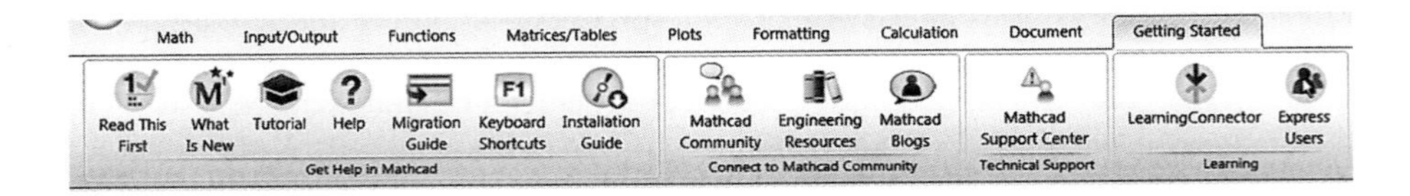

Summary

The intent of this chapter was to get you up and running with PTC Mathcad by introducing key PTC Mathcad features. It is also intended to whet your appetite for the information covered in future chapters. The best way to gain an understanding of the concepts introduced in this chapter is to practice. If you have not done so already, open the PTC Mathcad Tutorials and go through the tutorials mentioned at the beginning of this chapter.

In Chapter 1 we:

- Showed how to create PTC Mathcad expressions.
- Illustrated how to edit PTC Mathcad expressions using the grouping highlight.
- Described the PTC Mathcad Ribbon Bar.
- Differentiated between the different PTC Mathcad equal signs.
- Discussed regions.
- Introduced functions, units, arrays, and plotting.
- Introduced range variables.
- Emphasized the difference between literal subscripts and array subscripts.
- Described the variable ORIGIN.

Practice

The PTC Mathcad Prime 3.0 figures and examples used in this book are available for download from the book's website. The reader is encouraged to download the files and use them to practice the concepts learned. Additional examples and problems are also provided. To access this content go to http://store.elsevier.com/9780124104105 *and click on the Resources tab, and then click on the link for the Online Companion Materials.*

1. Enter the following equations into a PTC Mathcad worksheet.

$$\frac{-b+\sqrt{b^2-4\cdot a\cdot c}}{2\cdot a}$$

$$\frac{F_0}{\sqrt{m^2\cdot\left(\omega_0{}^2-\omega^2\right)^2+b^2\cdot\omega^2}}$$

$$\left(\frac{2}{3}-\frac{F_y\left(\frac{1}{r_T}\right)^2}{1530\mathrm{x}10^3\cdot C_b}\right)\cdot F_y$$

$$\frac{-1}{2} \cdot d_2 + \sqrt{\frac{2 \cdot v_2^{\,2} \cdot d_2}{g} + \frac{d_2^{\,2}}{4}}$$

2. Give each of the above equations a variable name. Assign variable names and a value to the variables used in the equations. These variable assignments will need to be made above the equation definition. Show the result. Change some of the input variable values and see the impact they have on the results.
3. Choose 10 equations from your field of study (or from a physics book) and enter them into a PTC Mathcad worksheet. Assign the variables that the equation needs prior to entering the equation. Select appropriate variable names. Don't select easy equations. Pick long complicated formulas that will give you some practice entering equations.
4. Choose some of the above equations and change some of the operators in the equation.
5. Chose some of the above equations and make the equation wrap at an ***addition*** or ***subtraction*** operator.

CHAPTER

PTC® Mathcad Prime® 3.0 for Current Mathcad 15 Users

2

So, you are a current Mathcad 15 user, and you are considering switching to PTC® Mathcad Prime® 3.0. Is it time to switch? The answer is, absolutely! With PTC Mathcad Prime 3.0, there are finally enough features to make the switch worthwhile for most people; however, it will not be without some frustration.

Let's admit it. PTC Mathcad Prime is different than Mathcad 15, and there will be a learning curve as you begin using it. Many familiar keystrokes will not work with PTC Mathcad Prime, and other familiar keystrokes cause different actions to occur. PTC Mathcad Prime 3.0 still does not have all of the features that are available in Mathcad 15. Many hoped-for features are not included.

So, with all these issues, why make the change? The reason is that PTC Mathcad Prime 3.0 has feature and function improvements over Mathcad 15. A few of these improvements will be discussed shortly, and more will be discussed in later chapters.

The purpose of this chapter is to help make a smoother transition from Mathcad 15 to PTC Mathcad Prime 3.0. It is assumed that you have glanced through Chapter 1 to get a feel for some of the new features in PTC Mathcad Prime.

Differences to Become Accustomed to

Let's first highlight some things that are different in PTC Mathcad Prime. Some of these changes will get you very frustrated as you begin using PTC Mathcad Prime 3.0. You will also realize that some of the differences are actually good things once you get familiar with them.

Tab Key

This will trip you up for a very long time. For some unfortunate reason, the tab key no longer moves the cursor to the next tab stop. In PTC Mathcad Prime 3.0, the tab key moves you down to the next region. If you are in the last region in the worksheet (which is often the case), the tab key moves you back up to the first region in the worksheet. This is very frustrating.

When this happens, remember: **SHIFT+TAB**. This will bring you back to where you were. You can then use the arrow keys to move to the right of the region.

Another frustrating thing is that PTC Mathcad Prime 3.0 does not have tab stops.

When you are in a region, the ENTER key will move your cursor down directly below the current region. This is a similar behavior to Mathcad 15.

Essential PTC® Mathcad Prime® 3.0.

Extra Spacebar

PTC Mathcad requires one extra press of the spacebar to select an expression. For example, in Mathcad 15, if you want to type the expression a:=x^2 + 3, you would type **a : x^2 Spacebar+3**. In PTC Mathcad Prime you need to type **a : x^2 Spacebar Spacebar+3**. In Mathcad 15, the 2 was already selected, so the spacebar expanded the cursor to include the x^2. In PTC Mathcad Prime, the 2 is not selected. The first spacebar selects the 2 and the second spacebar selects the x^2. This is not a bad thing. It will be frustrating for a while, but you will get used to it. The reason for the change has to do with the improved methods of editing expressions.

Editing Expressions

This is different, but much better because it allows you to actually select operators instead of being on the left or right side of the operators as in Mathcad 15. The new highlight and grouping editor was discussed in Chapter 1.

Creating Text Boxes

This is another one of the changes that will cause you frustration for a very long time. In Mathcad 15, when you typed a spacebar after entering some text, Mathcad 15 automatically converted the text and created a text box. In PTC Mathcad Prime 3.0, when you type some text and then type the spacebar it selects the text that you just typed. It does not create a text box.

In Mathcad 15, the keyboard shortcut for creating a text box is the double-quote key ". When you type the double-quote key " in PTC Mathcad Prime, you get matching double-quotes with a curser between.

This creates a string, not a text box. This is another thing that will cause you much frustration. The keyboard shortcut for a text box is **CTRL+t**. Refer to Chapter 3 for more information about text boxes and a new text region called a text block.

Literal Subscripts

There is an exciting change to literal subscripts with PTC Mathcad Prime that we will discuss shortly, but typing a period no longer works to create a literal subscript. The new keyboard shortcut is much more cumbersome to use. It is **CTRL+MINUS**.

The good news is that literal subscripts are now a toggle. You can turn them on and off, so that you can create a variable name such as H_2SO_4. Use the same keyboard shortcut to turn subscripts off.

You cannot create a subscript after the fact. You must first delete the character(s), then activate the subscript, and then retype the subscript. In Mathcad 15 you could just type a period to the left of the subscript, or delete the period if you wanted to remove the subscript.

Page Breaks

PTC Mathcad Prime 3.0 behaves much like a word processor. In Mathcad 15 you could scroll multiple pages below the last region in a worksheet. With PTC Mathcad Prime you can only scroll down to the bottom of the page. You must type **CTRL+ENTER** to add a new page. Even though this is similar to a word processor page, it still causes frustration if you want to type new regions on a new page. You will need to continually add new pages.

With PTC Mathcad Prime there is no visual clue provided when you insert a manual page break. Mathcad 15 added a horizontal page break line that could be moved up or down.

Regions to the Right of the Right Margin Do Not Print

In Mathcad 15 you had the option of printing or not printing regions to the right of the right margin. In PTC Mathcad Prime 3.0 you cannot print information to the right of the right margin. The regions to the right of the right margin are only visible in Draft View. When Page View is used, and when there are regions to the right of the page, PTC Mathcad Prime displays an arrow along the right margin. Clicking the arrow will switch to Draft View.

Units

In Mathcad 15, the unit placeholder was located adjacent to the displayed unit, and the cursor stayed on the left side of the **evaluation** operator: Unit = 1.524 m ▪. You needed to press the **TAB** key in order to move the cursor to the unit placeholder. In PTC Mathcad Prime, the unit placeholder is not adjacent to the unit; the unit occupies the unit placeholder, and the cursor is placed adjacent to the unit: Unit = 1.524·m|. To change the displayed unit, you need to delete the current unit and type a new unit in its place. The unit placeholder becomes visible once the displayed unit is deleted.

Unit = 1.524· |

Creating a Range Variable

To create a range variable in Mathcad 15 you would type `1, 2; 15`. In PTC Mathcad Prime you type `1, 2..15`. The new method is more logical because you type what you see. The semicolon no longer works. As soon as you type the comma, then PTC Mathcad Prime inserts the range variable placeholders. You can use the arrow keys to go to the placeholders or enter the incremental value and then type two periods to move to the ending value.

Single Quote Does Not Add Parentheses

In Mathcad 15, typing a single quote key created a pair of parentheses. In PTC Mathcad Prime 3.0 the single quote key adds an accent character '. This is a good thing because it now allows you to use the accent character in variable names such as f'_c. Another good thing in PTC Mathcad Prime is that when you type one parenthesis, two matching parentheses are created. The change from using a single quote key to using a parenthesis will be frustrating only for a short time, until you get used to typing "(" to create a pair of parentheses.

Highlighting Regions

In Mathcad 15 you could right-click a region and click a check box to highlight a region, and then choose the color. In PTC Mathcad Prime the highlight feature is on the **Math Formatting** tab. You need to select the region(s), and then select the highlight color from the **Math Font** group. You can remove the highlight by clicking the remove format button 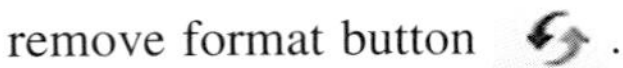 .

It is important to understand that you are changing the worksheet defaults if you change a highlight color (or any font characteristic such as font, font size, or font color) from the Math Font group (See the image below.) and you do not have a region selected. All current and future regions will be affected except those where you have already made similar region specific changes. Region specific changes are not affected by changing worksheet defaults. If your regions start looking different and you do not know why, you may have accidently modified the worksheet font settings.

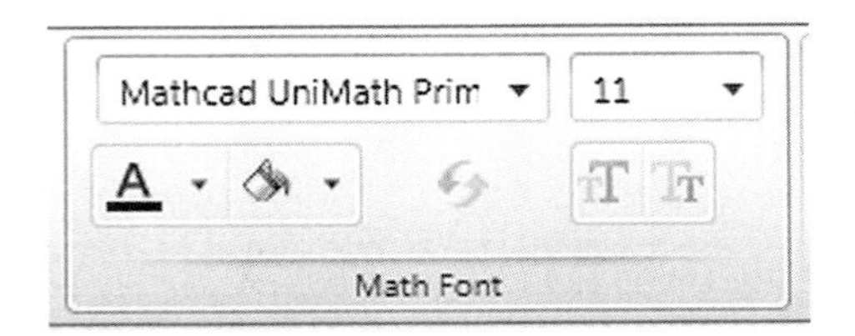

Exciting New Features in PTC Mathcad Prime

Once you understand the differences between PTC Mathcad Prime 3.0 and Mathcad 15, PTC Mathcad Prime 3.0 will begin to grow on you. You will begin to favor PTC Mathcad Prime 3.0 over Mathcad 15 because of some of its exciting new features.

Labels

Labels are very similar to Mathcad 15 styles, but are much more useful. They allow PTC Mathcad Prime 3.0 to automatically associate different elements such as variables, units, functions, and constants. This allows you to use "m" as a variable name and as a unit name. Chapter 16 has a lengthy discussion about the differences between Mathcad 15 styles and PTC Mathcad Prime labels.

Ribbon Bar

The ribbon bar allows quick access to most features within two- or in a few instances three-mouse clicks. It is a great addition to PTC Mathcad Prime. The location of some of the features is not extremely intuitive, so it takes a little while to get comfortable with it, but after the short learning curve the ribbon bar is easy to use.

When I first started using PTC Mathcad Prime, I was always looking for the "Insert" tab. This tab is not needed because most of the insert controls are available by clicking the right mouse button. The insert controls are split between several of the tabs. A majority of the insert controls can be found on the **Math** tab. The controls on this tab allow you to insert math regions, text boxes, text blocks, images, solve blocks, and symbols. Functions are inserted from the **Functions** tab. Matrices and tables are inserted from the **Matrices/Tables** tab, and plots are inserted from the **Plots** tab.

Another tab I was searching for was the "View" tab. Most of the view controls are found on the **Document** tab. The controls on this tab allow you to control Page View or Daft View, show or hide gridlines, add headers and footers, and set margins.

Open Worksheets Bar

Each open worksheet has a tab displayed just below the ribbon bar. This allows you to quickly see which documents are open in PTC Mathcad and also allows you to quickly switch between open documents.

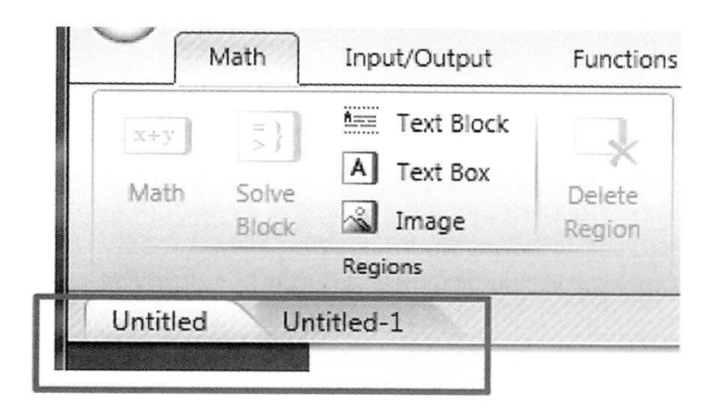

Status Bar

The status bar is located at the bottom of the PTC Mathcad workspace. It lists page numbers for the active worksheet, has the find and replace boxes, zoom slider, and icons to switch between Page View and Draft View. It also has a color coded circle that is green when **Auto Calculation** is turned on.

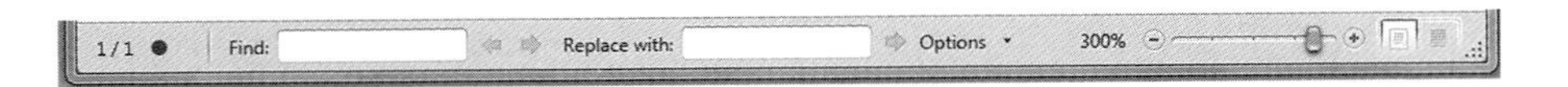

Ability to "pin" a Worksheet on the Recently Used Worksheets List

When you open a worksheet you will see open circles to the right of the worksheets listed in the Recently Used Worksheets list. Click the circle, and it becomes a solid circle, and the worksheet will remain at the top of the list.

Remove Empty Space

If you right click at the top of an empty space between regions and select "Remove Space," all empty space between regions will be removed.

Mixed Units in Arrays

In Mathcad 15 you could not mix unit dimensions in a single array. In PTC Mathcad Prime, you can use mixed units in arrays. See Chapter 5.

Operators

In Mathcad 15, there were many operators that had two forms such as a definite and indefinite integral, derivative and nth derivative, square root and nth root, etc. In PTC Mathcad Prime these operators are combined and there are optional placeholders that can be left empty.

Features from Mathcad 15 that Are Not in Mathcad Prime 3.0

This is a topic that really gets me frustrated, and I am not alone. It seems that by now all features available in Mathcad 15 should be available in PTC Mathcad Prime 3.0. This is an area where you can make a difference. Let PTC know of your frustration and tell them which features you want added into Mathcad Prime 4.0. I think in this case the squeaky wheel will get the grease. If they hear from enough of us, they may get the message.

The following is my list. Your list may include additional items. Please let PTC know of your list, so that we can finally get these features added into PTC Mathcad Prime 4.0.

I hope that Mathcad Prime 4.0 will have:

1. Text styles.
2. Customizable units.
3. Exponential threshold.
4. Ability to customize the function categories that are displayed on the **Functions** tab on the ribbon bar.
5. More text formatting features such as subscripts and superscripts.
6. More ability to individually control top, bottom, left and right page margins, and header and footer margins.
7. Align regions.
8. View regions.
9. Mixed fractions.
10. A right-click menu for an active or selected region that allows for cutting, copying, highlighting, and disabling the region.
11. More color choices for fonts and highlights.
12. A redefinition warning.
13. More plotting features.
14. Auto save.
15. Locked areas.
16. Worksheet protection.
17. Hyperlinks.
18. Ruler and tabs.

Summary

Once you get past the learning curve and become accustomed to PTC Mathcad Prime 3.0, you will begin to appreciate its new features and enhancements. YES, we all wish it had more of the features of Mathcad 15, but they are coming. This lack of features should not deter you from beginning to use and enjoy PTC Mathcad Prime 3.0.

> Throughout this book we will try to highlight the differences between Mathcad 15 and PTC Mathcad Prime. These differences will be highlighted like this paragraph. This will allow you to quickly scan the book and learn the new features of PTC Mathcad Prime 3.0.

CHAPTER 3

Variables and Regions

Variables

Variables are one of the most important features of PTC Mathcad. As in algebra, variables define constants and create relationships. As we saw in Chapter 1, your PTC Mathcad worksheet will be full of variables. It is therefore important to quickly gain a solid foundation in their use.

Chapter 3 will:

- Discuss types of variables.
- Give rules for naming variables.
- List characters that can be used in variable names.
- Introduce string variables.
- Tell how to move and resize regions.
- Discuss the use of Find and Replace.

Types of Variables

Variable definitions can consist of numbers or constants such as A := 1, or B := 67. They can consist of equations such as C := A + B, or D := A + 3. You can set one variable equal to another such as E := A. Variable definitions can also consist of strings of characters such as F := "This is an example of a string variable." Variable definitions can even have logic programs associated with them so that the value of the variable depends on the outcome of Boolean logic. As you go through this book you will see that variable definitions can be very simple or very complex. For the purpose of this chapter we will stay with simple examples. More detailed examples will follow in later chapters.

Rules for Naming Variables

Case and Font

> In PTC Mathcad 15 variable names were sensitive to case, font, font size, and style. PTC® Mathcad Prime® allows you to change font, font size, and font color without affecting the variable name definition. PTC Mathcad Prime uses labels which are a type of style. Variable names are sensitive to labels. Labels are discussed extensively in Chapter 16.

Essential PTC® Mathcad Prime® 3.0.

The first important thing to remember about variable names is that they are case sensitive. Thus the variable "ANT" is different than the variable "ant" (uppercase versus lowercase). Variable names are not sensitive to font, and font characteristics such as font size, or font color, thus the variable "Ant" is the same variable as "**Ant**." These font characteristics are controlled by the **Math Formatting** tab. This tab is shown below.

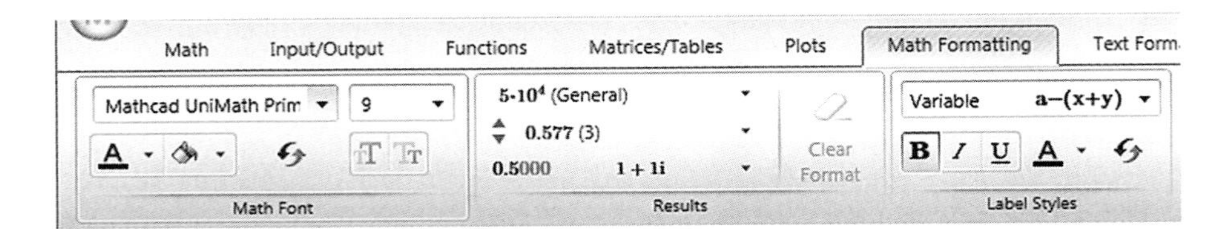

Variable names can be modified by any of the features in the Math Font group without affecting the variable name. See Figure 3.1.

If you do not have a specific region selected, then the changes made from the Math Font group will affect all current and future math regions in your worksheet.

The font characteristics in the **Label Styles** group affect all variable names. They cannot be made on an individual region. The use of labels is covered extensively in Chapter 16. For this discussion it is important to understand that PTC Mathcad

This example illustrates that variable names are not sensitive to the formatting features found in the **Math Font** group on the **Math Formatting** tab.

Ant := 1	Default font.
Ant := 2	Defined Ant with different font face.
Ant = 2	The last definition is recognized.
Ant := 3	Defined Ant with different color.
Ant = 3	The last definition is recognized.
Ant := 4	Defined Ant with larger font and with highlight.
Ant = 4	The last definition is recognized.

FIGURE 3.1

Effect of math font changes to variable names

automatically assigns labels for things such as variables, units, built-in functions, and built-in constants (such as e and π). Font changes made in the **Label Styles** group affect all variables assigned to that label. So, hold off on making changes from this group until we have had a discussion about labels. Also be aware that variable names are sensitive to labels.

Characters That Can Be Used in Variable Names

There are some rules for naming variables:

- Variable names can consist of uppercase and lowercase letters.
- The digits 0 through 9 can be used in a variable name, except that the leading character in a variable name cannot be a digit. PTC Mathcad interprets anything beginning with a digit to be a number and not a variable.
- Variable names may consist of Greek letters. The easiest way to insert Greek letters is to use the following Ribbon command: **Math>Operators and Symbols>Symbols**. Select the desired Greek letter from the list, and it will be inserted into your worksheet. If you pause briefly prior to clicking your desired Greek letter, a pop-up box will appear listing the name of the Greek letter and also indicating the keyboard shortcut for the letter. To use the keyboard shortcut, type the equivalent roman letter and then type **CTRL+G**. See Figure 3.2 for a table of equivalent Greek letters.
- A single dot (period) may be used anywhere including as the first character of a variable name.
- Variable names may also include most ASCII characters. To access the ASCII character in MS Windows, click **Start>All Programs>Accessories>System Tools>Character Map**.
- Variable names can include the symbols and constants listed on the **Math** tab in the **Operators and Symbols** group. **(Math>Operators and Symbols>Symbols** and **Math>Operators and Symbols>Constants)**.
- Variable names may not include the operators listed on the **Math** tab in the **Operators and Symbols** group in the **Operators** list. There is an exception to this. See Special Text Mode below.

Literal Subscripts

> In Mathcad 15 you typed a period to begin a literal subscript, and once you began a subscript, everything following the period was a subscript. In PTC Mathcad Prime, type **CTRL+ −** to begin and end a literal subscript, because in PTC Mathcad Prime, the subscript command is a toggle. This allows a variable name to have multiple subscripts such as H_2SO_4.

α	a	η	h	ο	o	ϖ	v
β	b	ι	i	π	p	ω	w
χ	c	φ	j	θ	q	ξ	x
δ	d	κ	k	ρ	r	ψ	y
ε	e	λ	l	σ	s	ζ	z
ϕ	f	μ	m	τ	t		
γ	g	ν	n	υ	u		
A	A	H	H	O	O	ς	V
B	B	I	I	Π	P	Ω	W
X	C	ϑ	J	Θ	Q	Ξ	X
Δ	D	K	K	P	R	Ψ	Y
E	E	Λ	L	Σ	S	Z	Z
Φ	F	M	M	T	T		
Γ	G	N	N	Y	U		

FIGURE 3.2

Table of equivalent Greek letters

Literal subscripts were discussed in Chapter 1. To type a subscript, type the first part of the variable name, and then use the ribbon control **Math>Style>Subscript**, or type **CTRL+−**. The insertion point will drop down half of a line. The literal subscript is a toggle allowing you to use a combination of letters and subscripts. Repeat the command to switch off the subscript allowing you to begin typing normal characters. To remove subscripted characters, delete the characters and switch the subscript toggle off. A variable name may begin with a subscript, but in order to do this you must first type a letter and use the arrow key to move the cursor to the left side of the letter, and then activate the subscript control and type the subscript letter.

Remember that a literal subscript looks similar to an array subscript, but it behaves much differently. Array subscripts were discussed in Chapter 1 and will be discussed further in Chapter 6.

Special Text Mode

As a general rule, keyboard symbols may be used in variable names, but PTC Mathcad operators may not. However, many keyboard symbols are also PTC Mathcad shortcuts that insert a PTC Mathcad operator or perform another PTC Mathcad function. (See the Appendix for a list of keyboard shortcuts.) This prevents you from

Note what happens when you type "A!B". Notice the dot between the ! and B. It is waiting for an operator to be added.

A!·B $A!\,B$

When you type "A+B" PTC Mathcad will not let you add the definition operator, because the + is recognized as the addition operator.

$A+B$ $A+B$

FIGURE 3.3

Operators may not be used in a variable name

using many keyboard symbols in your variable names. When you try to use a symbol that is also a PTC Mathcad shortcut, PTC Mathcad inserts the operator or executes the command referenced by the shortcut. For example, if you want to use a variable name A!B and type **A ! B**, then PTC Mathcad inserts the ***factorial*** operator, not the ! symbol. This is because the ! symbol is the keyboard shortcut for the ***factorial*** operator. If you want a variable name A+B and you type **A + B :**, PTC Mathcad will not recognize the variable name, and will not let you type the definition operator. See Figure 3.3.

There is an indirect way for PTC Mathcad to use both symbols and operators in variable names by providing a special text mode. To do this, create a string (see next section) by typing the double quotes **"** key. The string name should be the variable name including the operator in the name. After typing the name of the variable, delete the double quote at the end of the variable name. What remains is a variable name with an operator included. You can then assign or evaluate the variable just as with any other name. See Figure 3.4.

String Variables

A string is a sequence of characters between double quotes. It has no numeric value, but it can be defined as a variable. To create a string variable, type the variable name followed by pressing the colon **:** key. Type the double quotes key **"** in the placeholder. You will see an insertion line between a pair of double quotes. You can then type any combination of letters, numbers or other characters. When you are finished with the string, press **ENTER**.

"A+B"	Type a double quote " and then the variable name. This creates a string name.
A+B	Delete the double quote from the string name.
A+B := 5	Type the **definition** operator : and define the value.

The keystrokes for the above are " **A + B DELETE : 5**

A+B = 5	Repeat the same when using the **evaluation** operator. " **A + B DELETE =**

FIGURE 3.4

Using an operator in a variable name

String variables are useful to use as error messages. If you need a certain input to be a positive number, you can assign a string variable to have the value, "Input must be positive." If a number less than zero is entered, PTC Mathcad can display this string variable as an error message. String variables are also useful as a means of displaying whether or not a certain condition is met. You can assign one variable to have the value "Yes" and another variable to have the value "No." If a specific condition is met PTC Mathcad can display the string variable associated with "Yes." If the specific condition is not met PTC Mathcad can display the string variable associated with "No." These logic programs will be discussed in Chapter 9, "Simple Logic Programming." See Figure 3.5 for some examples of text strings.

TextString1 := "This is a text string"

TextString2 := "Yes"

TextString3 := "No"

TextString1 = "This is a text string"
TextString2 = "Yes" **TextString3** = "No"

FIGURE 3.5

Examples of text strings

Why Use Variables

Figure 3.6 shows three different ways of getting similar results. The first method shown is very direct. It is just like using a calculator. If you were not going to be saving the worksheet and you needed a quick answer, the first method works fine.

The following example shows three methods of getting the same results.

Use direct numbers.

$5+7=12$

$12 \cdot 3=36$

$36-8=28$

Use intermediate results.

$\mathbf{Answer1} := 5+7$ $\mathbf{Answer1} = 12$

$\mathbf{Answer2} := \mathbf{Answer1} \cdot 3$ $\mathbf{Answer2} = 36$

$\mathbf{Answer3} := \mathbf{Answer2} - 8$ $\mathbf{Answer3} = 28$

Assign variable names to input and output values.

$\mathbf{Input1} := 5$ $\mathbf{Input2} := 7$ $\mathbf{Input3} := 3$ $\mathbf{Input4} := 8$

$\mathbf{Answer4} := \mathbf{Input1} + \mathbf{Input2}$ $\mathbf{Answer4} = 12$

$\mathbf{Answer5} := \mathbf{Answer4} \cdot \mathbf{Input3}$ $\mathbf{Answer5} = 36$

$\mathbf{Answer6} := \mathbf{Answer5} - \mathbf{Input4}$ $\mathbf{Answer6} = 28$

FIGURE 3.6

Using variables in calculations

Just type in the numbers to get an answer. Use the result of the first equation, and type it into the second equation. Use the result of the second equation, and type it into the third equation.

The second method shown is to assign a variable name to the intermediate answers. The benefit of this method is that you will always have the result of each expression available to use in other expressions in your worksheet. The equations shown are very simple and basic, but in scientific or engineering calculations the equations or expressions can be very complex. Once the result of an expression is calculated by PTC Mathcad, you want to capture it for future use. You do this by assigning the result to a variable.

The third method shown is to assign all values to variable names. There are four input values and three output results. You may be saying to yourself, "Why would I type all those extra key strokes? It is much more time consuming to type Input1+Input2 than just typing 5+7." Well, let's assume that the numbers 5, 7, 3, and 8 represent some type of engineering input. You now use these numbers over and over in your calculations. If you keep using just the numbers 5, 7, 3, and 8 in your many different equations, then what happens if at some point the input value 7 is changed to 9? If you have used the variable Input2=7, then you just change the value of Input2 from 7 to 9 and you are done. PTC Mathcad does the rest. If you didn't use the input variable, then you will need to go through your worksheet and change (or attempt to change) every instance where the number 7 represented the input variable from 7 to 9. This could be an impossible task if you have a complex worksheet.

Remember that the ultimate goal of this book is to teach you how to use PTC Mathcad as a tool for creating technical calculations. Because of this, it is recommended that you get into the habit of using the third method illustrated in Figure 3.6. Most of the examples used in this book will use this method. We will assign the input values to variable names, assign a variable name to the expression, and then display the results of the expression.

Chapter 14 provides some useful naming guidelines for variables to be used in your scientific and engineering calculations.

Regions

In Chapter 1 we discussed how a PTC Mathcad worksheet is comprised of many different regions. This section will discuss how to manipulate and organize regions.

> Unfortunately, many of the region alignment features are not yet in PTC Mathcad Prime 3.0. Some of the missing features are: ruler, tabs, show regions, alignment gridlines, align across, and align down.

Understanding the Difference Between "activate" and "select"

You "activate" a region by clicking within a region. You "select" a region by drag selecting the region. You do this by clicking outside a region, holding your left mouse button, and dragging over the region. When a region is activated by clicking inside the region, a dashed blue line appears around the region.

$$5+3=8$$

When a region is selected by using drag select, the region becomes highlighted with a light blue color, and there is no dashed line.

$$5+3=8$$

Activate a region to modify elements within the region. Select a region when you want to delete the region.

Selecting and Moving Regions

You can select and move a single region or multiple regions. To move a single region, activate or select the region, and then place your cursor near the perimeter of the region until the single arrow changes to a four headed arrow. Now left click and hold the mouse button. Drag the region to where you want it.

To move multiple regions, first "drag select" the regions. To do this, click outside of a region, and then hold the left mouse button and drag it across several regions and release the mouse button. All regions within this area will now be selected, and each will be highlighted blue. To select non-adjacent regions, hold the **CTRL** key and click within each desired region. To move the selected regions, place the cursor near the edge of one of the regions, left click and hold the mouse button. Drag the regions to a new location. You may also use the arrow keys to move the selected regions.

Separating Regions

There may be times when you have regions that overlap. This can happen when you paste some regions into your worksheet. If this happens, you can undo the paste, or you can select the overlapping regions and use the **Separate Regions** control from the **Spacing** group on the **Document** tab **(Document>Spacing>Separate Regions)**. This control will allow you to separate the regions vertically or horizontally.

Text Regions

> Text regions do not begin when you type a spacebar following some text. They do not begin when you type the double quote key (**"**). The double quote key creates a text string. PTC Mathcad Prime has two types of text regions: A text block and a text box. See below for a description. The keyboard shortcut for adding a text box is **CTRL+t**. The keyboard shortcut for adding a text block is **CTRL+SHIFT+t** (**CTRL+T**). Remember CTRL + lowercase t for text box and uppercase T for text block. You may also use the controls found on the **Math** tab.

In Chapter 1 we learned that there are two types of text regions: a text block and a text box. The text block extends from margin to margin and as it expands the regions below are moved down. As you insert a text block, overlapping regions are moved up or down to make room for the text block. The text box may overlap other regions, and regions are not moved as the text box expands. You are able to control the width of the text box. The controls for adding text blocks and text boxes are found on the **Math** tab in the **Regions** group. The keyboard shortcut for adding a text box is **CTRL+t**. The keyboard shortcut for adding a text block is **CTRL+SHIFT+t(CTRL+T)**. Remember CTRL + lowercase t for text box and uppercase T for text block.

I like to add the Text Box and Text Block controls to my Quick Access Toolbar. To do this, go to the **Math** tab, right click the Text Box command, and select "Add to Quick Access Toolbar." Do the same for the Text Block command.

The following sections will focus on how to modify and edit text regions.

Changing Font Characteristics

Once you create a text region you can type text just as you would in a word processor. You can change font characteristics such as font type, font size or font color. You can also use such things as bold, italic, or underline. PTC Mathcad Prime 3.0 does not allow strikeout, subscript, or superscript. The commands for changing the font characteristics are found on the **Text Formatting** tab. To change the font characteristics while in a text region, highlight the text and change any of the font characteristics. See Figure 3.7.

To change the font characteristics for future text regions click outside of a text region and change any of the font characteristics. All new text regions will now have the new font characteristics. Existing text boxes will not be affected. You must have an existing text box selected before you can change its font characteristics.

I've had times when the text being typed in new text boxes was unexpectedly different from what previous text boxes had. I could not understand why. This was very frustrating, until I finally realized that I had inadvertently changed a font characteristic. This characteristic was changed for all future text boxes and text blocks but not the existing boxes and blocks. This condition happens when you want to change some existing text, but do not have the text region or the individual text selected. When you change the font characteristic, it changes all future text boxes. You may not realize this until later when you create a new text region.

To change the font characteristics of all your existing text regions, type **CTRL+A** to select all regions in your worksheet (text and math regions). Now, make any changes you would like to the font characteristics. These changes will be made to all text regions. The selected math regions will not be affected by the text region changes.

Text Box 1: This is the default text.

Text Box 2: All text within a single text box may be changed.

Text Box 3: <u>**Individual text**</u> within a text box may be changed.

FIGURE 3.7

Text formatting of text regions

Controlling the Width of a Text Box

When you create a text box, the width of the text box is fixed. Typed text will wrap to a new line when the edge of the text box is reached. You may use your mouse to extend or shrink the right edge of the text box by clicking on the handle located on the right side of the text box.

This is a text box. You may use your mouse to click on the right side of the box and make the box wider or narower.

Paragraph Properties

A text region is similar to a simple word processor. You are able to format the text region in much the same way as you would in a word processor. We discussed earlier how to change the font characteristics of text in a text region. You can also set a few paragraph characteristics such as: indents, bullets, automatic numbering, and alignment.

Areas

> You cannot lock an area in PTC Mathcad Prime 3.0.

Areas are collapsible regions. You can include text regions, math regions, and plot regions in an area and then collapse the area so that the regions are not visible. The control to insert an area is on the **Document** tab in the **Regions** group (**Document>Regions>Area**).

The regions inside the Area will not be visible once the Area is collapsed.

$a := 1$ $b := 2$ $c := a + b = 3.00$

$c = 3.00$

After an area is collapsed, you see a + symbol and a horizontal line. All of the regions inside the collapsed are still effective, you just cannot see them. When an Area is expanded, the + changes to a − .

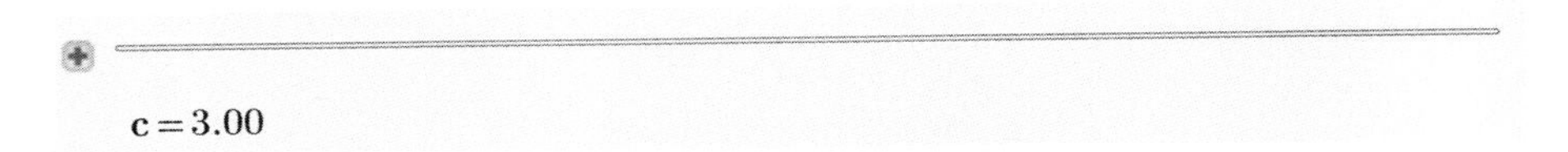

Additional Information About Math Regions

Math Regions in Text Regions

When you are in a text region, you may add a math region by clicking **Math** from the **Regions** group on the **Math** tab.

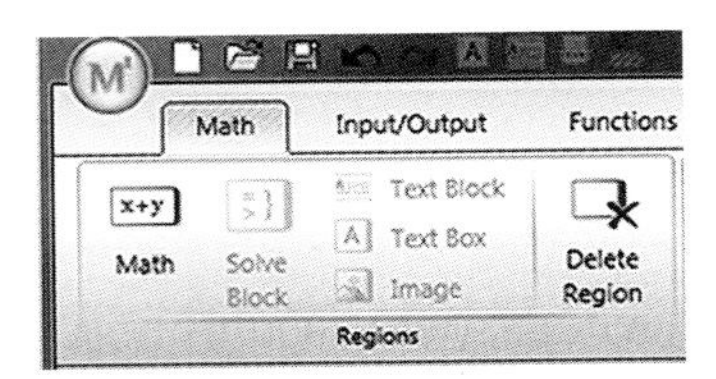

This places a blank placeholder in the text region, where you can type a math expression. After you are finished with the expression, use the right arrow to move you back into the text box, where you can continue typing text. See Figure 3.8.

Before inserting a math region, I like to add one or two spaces after the insertion point. This makes it easier to continue typing after I insert the math region. It is not necessary, but it makes it easier to see the cursor after leaving the math region.

Math Regions That Do Not Calculate

There will be many times when you want to display an equation prior to the point where the variables are defined. When you try to do this, PTC Mathcad will give you an error message.

You can include a math region in a text region. For example $\textbf{Example} := \dfrac{600\ N}{3\ m \cdot 2\ m}$ Use the **Math** control from the **Regions** group on the **Math** tab. $\textbf{Example} = 100\ Pa$.

FIGURE 3.8

Including math regions in a text region

$NoVariables_1 := a \cdot x^2 + b \cdot x + c$ — This region used the **Disable Region** control on the **Calculation** tab. The region is light gray.

$NoVariables_1 = ?$

$NoVariables_1 = ?$ — This variable does not exist because the above expression was disabled.

This variable is undefined. Check that the label is set correctly.

$NoVariables_2 = a \cdot x^2 + b \cdot x + c$ — This example uses a Boolean equal (CTRL + =).

$NoVariables_2 = ?$ — This variable does not exist because the Boolean equal does not define a value.

FIGURE 3.9

Displaying math regions without having variables defined

There are several ways to work around this:

- Type in the expression using variable names that have not yet been defined. Before clicking out of the PTC Mathcad region, select **Disable Region** from the **Controls** group on the **Calculation** tab (**Calculation>Controls>Disable Region**). This will cause the region to turn light gray and will prevent Mathcad from evaluating the expression.
- Use the Boolean equal to operator **CTRL+EQUAL SIGN** instead of the assignment operator **COLON**. This does not make a variable assignment, but it allows you to display what you want without an error.

See Figure 3.9 for an example of using these methods of displaying expressions.

Find and Replace

Find

The **Find** and **Replace** features can easily help you to either find or replace variables or text in your worksheets. These features are found in the status bar located at the bottom of the worksheet.

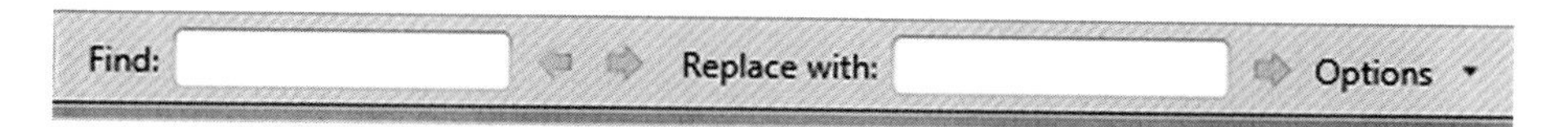

The **Options** list allows you to tell PTC Mathcad to look in only text regions, in only math regions or both. By default, PTC Mathcad looks in both math and text regions. The control also allows you to search parts of words, whole words, or to match case. If you want to search for a portion of a word, be sure that the "Whole Words Only" control is not activated.

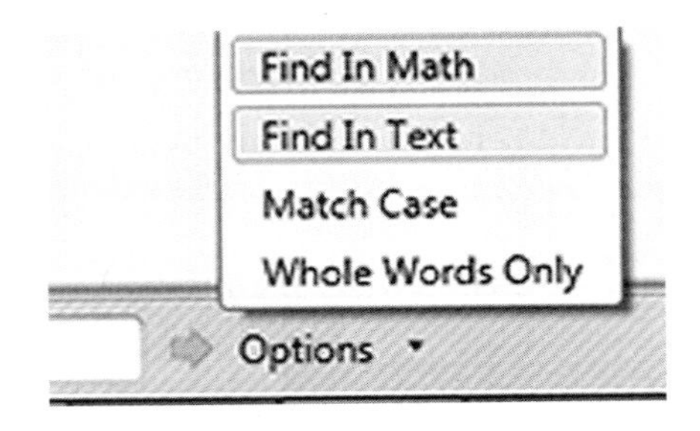

There is a bug in the **Find** and **Replace** feature in PTC Mathcad Prime 3.0. It has to do with searching for portions of a word in math regions. When PTC Mathcad finds a portion of a word in a math region, rather than selecting only the portion, it selects the entire word. Thus, when using **Replace**, instead of only replacing the portion, the entire word is replaced. This is a known issue and will be fixed in future versions of PTC Mathcad Prime. There is not an issue with finding and replacing parts of a word in text boxes.

If you only want to find something, then leave the "Replace" box empty. Type what you want to find in the "Find" box, and then use the right facing arrow to the right of the "Find" box. If you are searching for a Greek letter you will need to copy a Greek letter from the worksheet and paste it into the "Find" box.

Replace

To find and replace, fill in both boxes. Use the arrow next to the "Find" box to find the words, and the arrow next to the replace box to replace the words. Figures 3.10 and 3.11 illustrate a bug in PTC Mathcad Prime 3.0. When the "Whole Words Only" control is not activated, the intended behavior is to have PTC Mathcad be able to select only a portion of a word and then replace only the portion. The current incorrect behavior is that when PTC Mathcad finds a portion of a word in a math region, the entire word is selected. Figure 3.11 shows what happened after we replaced all instances of "Example" with "Test." The intended behavior should have been "Test1," "Test2," and "Test3."

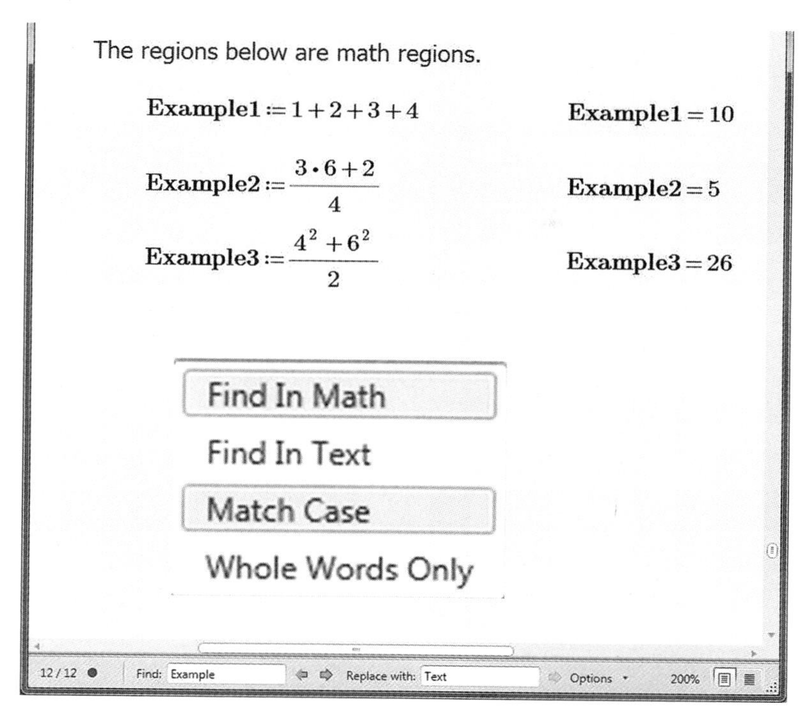

FIGURE 3.10

Using the Replace dialog box

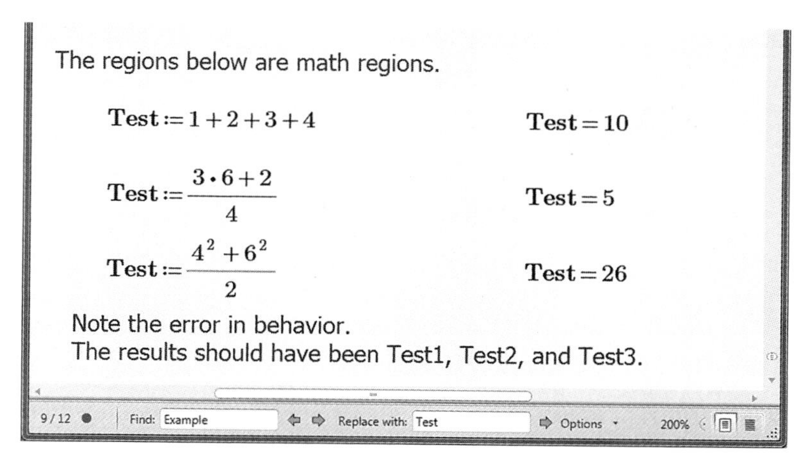

FIGURE 3.11

After using Replace

Inserting and Deleting Lines

PTC Mathcad Prime does not have the "Insert Lines" features found in Mathcad 15 where you could specify the number of lines to insert.

If you need more space between regions, just hold down the **ENTER** key until you have added the desired space between regions. PTC Mathcad also has an **Add Space** command that will insert a single line between regions. The space will be inserted below the location of your cursor. Access this command from the ribbon bar: **Document>Spacing>Add Space>Single Line**. You can also right click above the region you want to move and select "Add Space." The **Add Space** control on the ribbon also lets you choose to add a full page below your cursor. When pasting new information into PTC Mathcad, it is wise to insert blank spaces between the area where you are pasting the information. This will prevent regions from overlapping.

A great feature in PTC Mathcad Prime allows you to remove all empty space between regions. This is very handy if you have several pages of blank worksheets between regions. You access this command from the Ribbon bar: **Document>Spacing>Remove Space>Empty Space**. This control is also available by right clicking and selecting "Remove Space." You can also use the **DELETE** key to delete one blank line at a time.

I like to add the Remove Empty Space control to the Quick Access Toolbar. Do this by right clicking on the control and choosing "Add to Quick Access Toolbar."

Use caution when deleting spaces using the **DELETE** key. Once the spaces are removed PTC Mathcad begins deleting portions of your text or math regions. I have erased these on multiple occasions. Thankfully there is the undo button.

Page View and Page Breaks

The default PTC Mathcad worksheet is displayed in "Page View." This view shows exactly what will appear on the printed worksheet. It will display headers, footers, and all margins. It is similar to a page shown in a word processing program.

The **Document** tab has controls that allow you to determine how your worksheets are displayed on the screen.

View

> In PTC Mathcad Prime everything to the right of the right margin is not printable. This differs from Mathcad 15 where you had the option of printing regions to the right of the right margin.

The **View** control allows you to switch between the "Page View" and "Draft View."

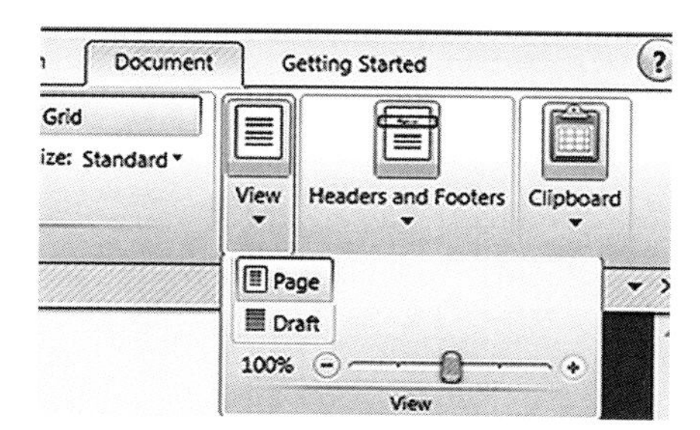

In Draft View, you are able to add and view items to the right of the right margin. This area is not printable, and everything added to this area will not show in Page View. If there are regions in this area and you are in Page View, PTC Mathcad displays a small arrow in the right margin indicating that there are regions that are not viewable. If you click the arrow, then the view switches to Draft View. Refer to Figure 3.12 and Figure 3.13

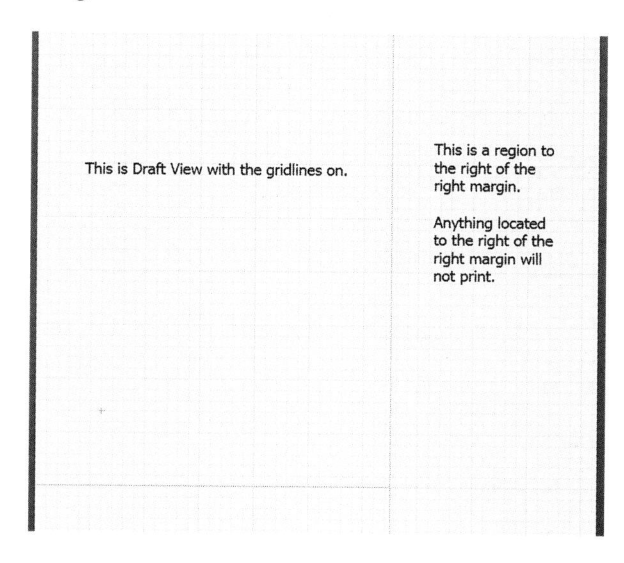

FIGURE 3.12

Draft View

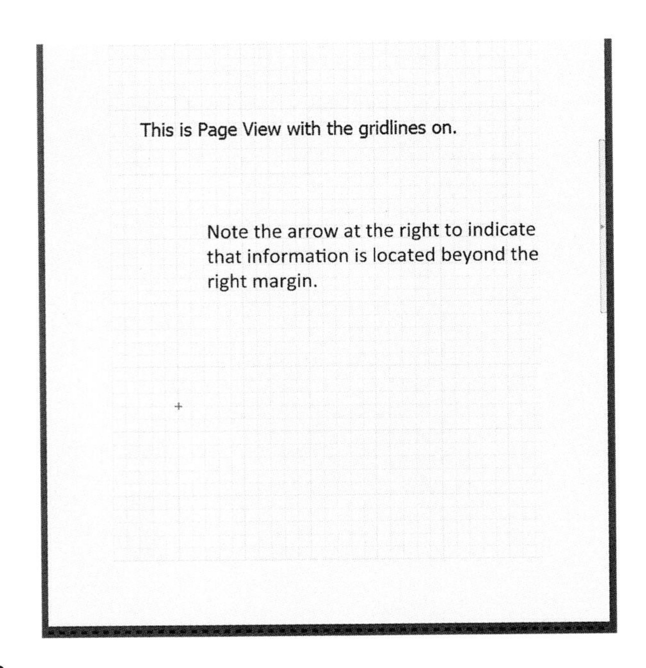

FIGURE 3.13

Page View

Gridlines

By default the worksheet will have gridlines displayed.

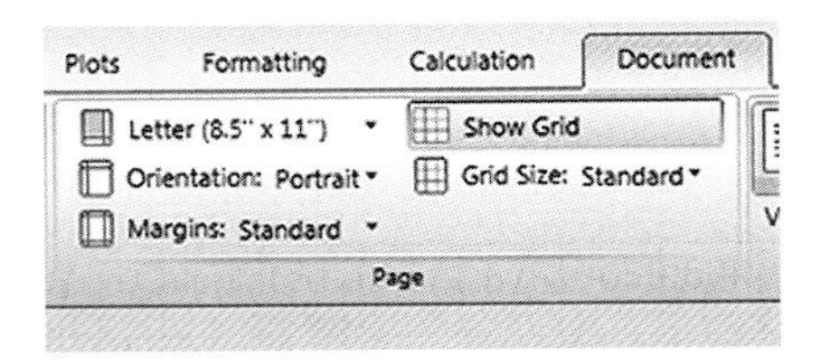

The regions added to the worksheet will snap to the gridlines. The **Page** group has a control to show or hide the gridlines and a control to change from the default "Standard" grid to "Fine" grid. Figure 3.14 shows the difference between the Standard and Fine grids.

Page Breaks

> Mathcad 15 used a horizontal line placed anywhere in the worksheet to create a page break. PTC Mathcad Prime does not have this feature. It behaves more like a word processor where complete pages are displayed, and a new page is added when a page break is used.

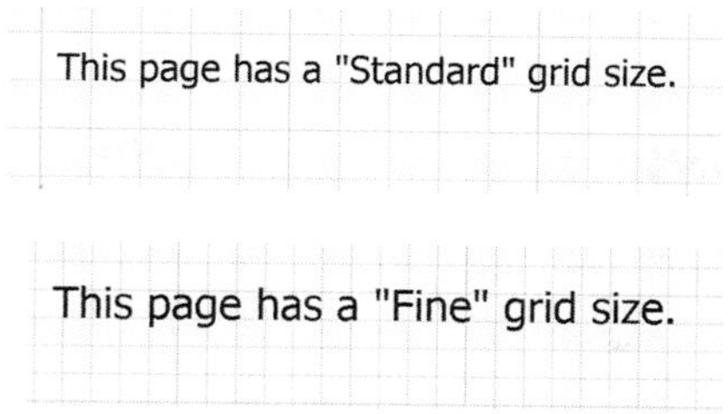

This page has a "Standard" grid size.

This page has a "Fine" grid size.

FIGURE 3.14

Gridline size

If you want to add a new page to your worksheet, press **CTRL+ENTER** or use the **Add Space** control in the **Spacing** group (**Document>Spacing>Add Space>Page**).

Summary

Variables and regions are an important part of technical calculations in PTC Mathcad. Learning how to use them effectively will make it much easier to create scientific and engineering calculations. Knowing how to format text regions can also make your calculation worksheets look better and be easier to follow.

In Chapter 3 we:

- Learned that variables are case sensitive, but not sensitive to font, font size, and font color.
- Discussed which characters can be used in variable names.
- Learned about the special text mode.
- Discussed string variables.
- Emphasized the importance of defining variables in PTC Mathcad calculations.
- Discussed moving regions.
- Described the attributes of text regions.
- Explained paragraph properties in text regions.
- Demonstrated the use of find and replace.
- Discussed Page View, Draft View, and gridlines.

Practice

The PTC Mathcad Prime 3.0 figures and examples used in this book are available for download from the book's website. The reader is encouraged to download the files and use them to practice the concepts learned. Additional examples and problems are also provided. To access this content go to http://store.elsevier.com/

9780124104105 *and click on the Resources tab, and then click on the link for the Online Companion Materials.*

1. Open a new PTC Mathcad worksheet.
2. Add 10 simple variable definitions from your field of study, and include some subscript names.
3. Add 10 variable definitions that are two words or more. Use two different methods for differentiating between words.
4. Add 10 variable definitions that include characters requiring the use of the special text mode.
5. Add 10 string variable definitions.
6. Create a text box and write two or three paragraphs about the things you learned in this chapter. After creating the paragraphs, change some of the paragraph characteristics and font characteristics.
7. Create a text block between two of your regions. As you add text, watch how the regions below the text block move downward as the text block expands.
8. Use the Find and Replace features to search for and replace certain text and math characters.

CHAPTER

Simple Functions

4

This chapter introduces PTC Mathcad's built-in functions and describes their basic use. It also introduces user-defined functions. We will hold off our in-depth discussion of the built-in PTC Mathcad functions until Chapter 7.

The power of user-defined functions is not realized even by many long-time PTC Mathcad users. In some instances, user-defined functions can be confusing and complicated, causing many users to ignore them. After briefly discussing built-in functions, this chapter will focus on simple user-defined functions. The goal of this chapter is to make you comfortable with the concept and use of simple user-defined functions.

Chapter 4 will:

- Introduce built-in functions.
- Discuss what an "argument" is.
- Introduce user-defined functions.
- Show different types of arguments.
- Give examples of different function names and argument names.
- Tell when to use a user-defined function.
- Describe how to use variables in user-defined functions.
- Give examples of user-defined functions in technical calculations.
- Provide warnings about the use of functions.

Built-in Functions

Built-in functions range from very simple to very complex. Some simple built-in PTC Mathcad functions are: ***sin()***, ***cos()***, ***ln()***, and ***max()***. Every PTC Mathcad function is set up in a similar way. The function name is given, followed by a pair of parentheses. The information that is typed within the parentheses is called the argument. Every function has a name and an argument. The function takes the information from the argument (contained within the parentheses), processes the information based on rules that are defined for the specific function, and then returns a result.

The **Functions** tab on the ribbon contains all of the built-in PTC Mathcad functions. The functions are organized by category. All of the categories do not appear on the **Functions** tab. To see a list of all the built-in functions that PTC Mathcad has, select the **ALL Functions** control.

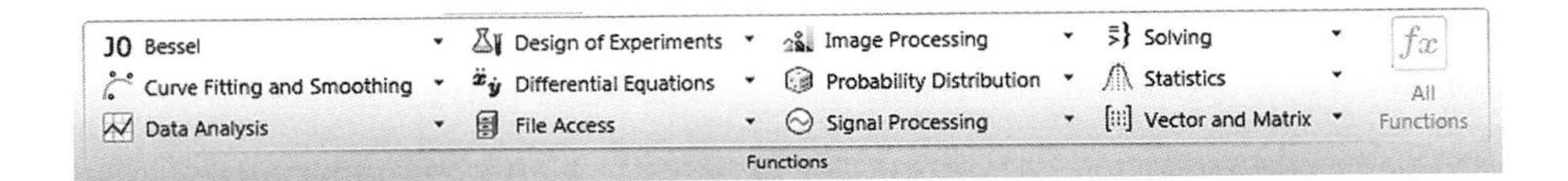

When the **All Functions** control is clicked, a list on the left hand side of the PTC Mathcad worksheet will appear that shows all of the built-in functions of PTC Mathcad. The default is to group the functions by category. See Figure 4.1. Select the triangle to the left of the category to see the functions in that category. If you want to see an alphabetical list of all PTC Mathcad functions by name, select the A↓Z button at the top of the function list. See Figure 4.2.

If you hover your cursor over the name of a function, an information box will appear listing the name of the function and its arguments in parentheses. It tells you what the function does, and lists any restrictions on the arguments (such as if the argument must be in radians, or whether it must be an integer).

Some functions expect only a single argument. Other functions expect two arguments. Some functions require multiple arguments or have optional arguments. There are some functions that can have a variable number of arguments. These will be indicated by three dots following the listed arguments. Figure 4.3 shows four functions, each with a different numbers of arguments.

Take a moment to scan the complete list of PTC Mathcad's built-in functions from the **Function** list. In later chapters, we will discuss selected functions. If you are interested in learning about a specific function, refer to PTC Mathcad Help. Figure 4.4 shows examples of using four different PTC Mathcad functions requiring various argument lengths.

A PTC Mathcad expression can have an unlimited number of functions included. To insert a function within an expression, simply select the location in the expression where you want the function to be, and then click on the function name from the **Functions** tab. If you are familiar with the name of the function and its arguments, you can simply type the name of the function and include the required arguments between parentheses. Remember that PTC Mathcad function names are case sensitive. If you type a function name and PTC Mathcad does not recognize it, use the **Function** list to see if the first letter is uppercase or lowercase. Figure 4.5 shows an expression using multiple built-in functions.

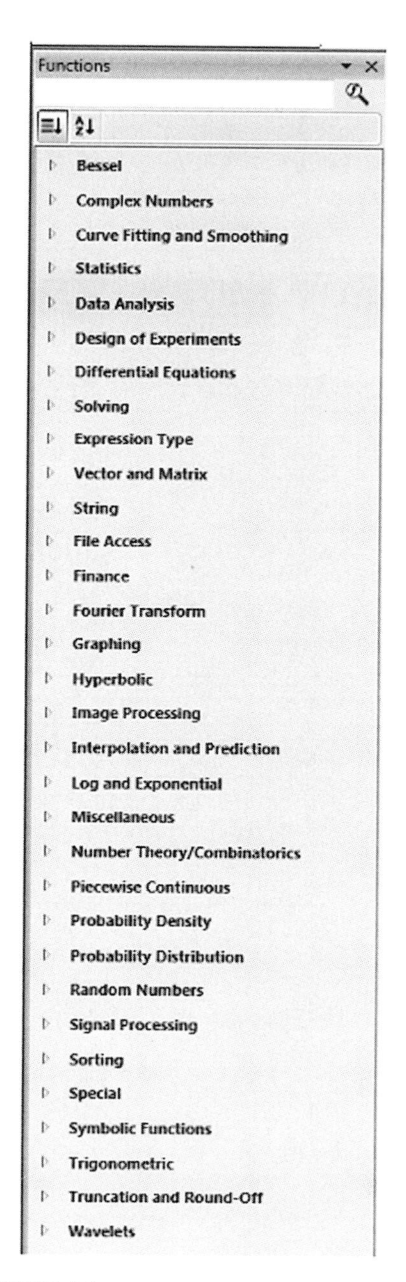

FIGURE 4.1

Functions by category

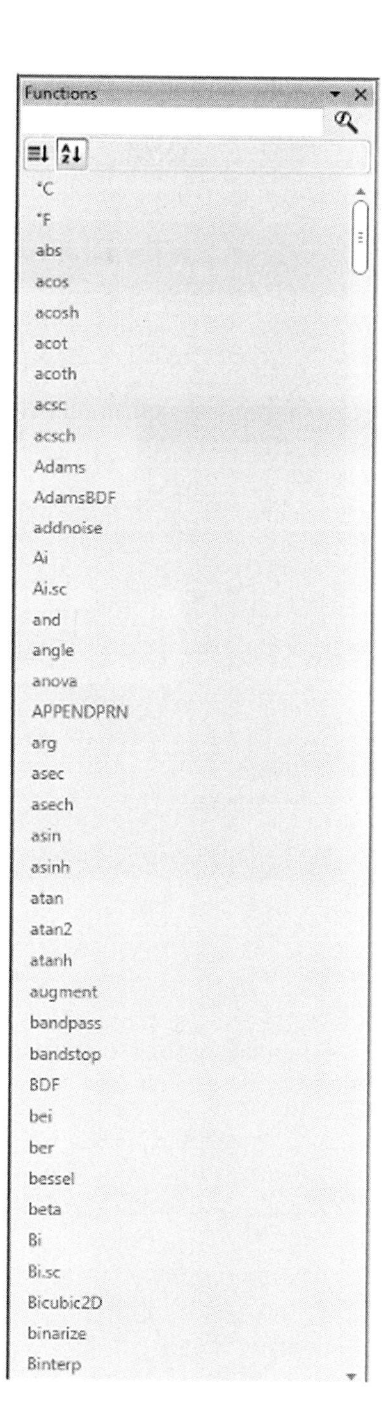

FIGURE 4.2

Functions by name

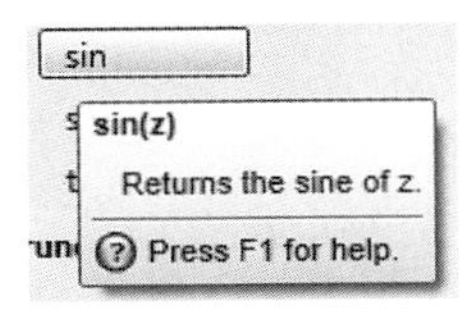

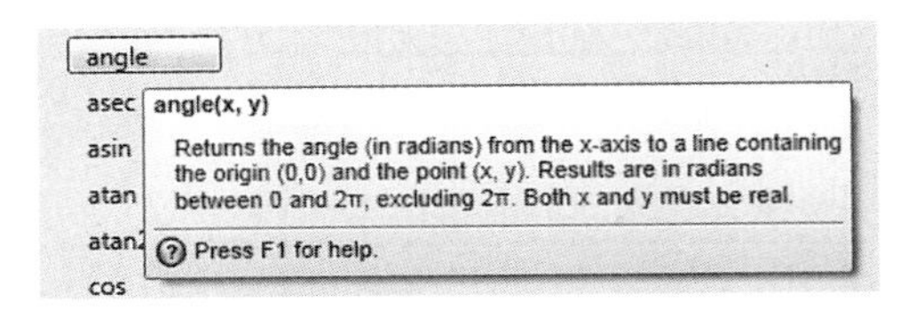

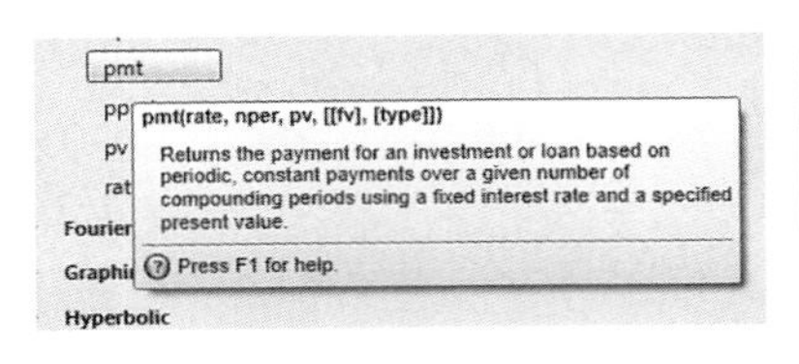

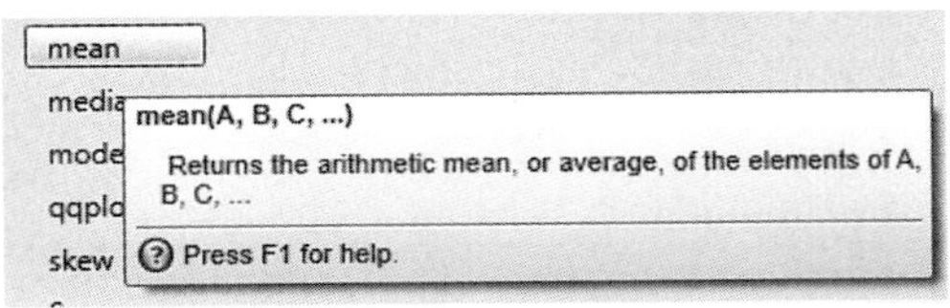

FIGURE 4.3

Arguments of functions

$$\text{Example}_{1a} := \underline{\sin}\left(\frac{\pi}{4}\right)$$

$$\text{Example}_{1a} = 0.707$$

$$\text{Example}_{1b} := \underline{\text{angle}}(3, 4)$$

In the second result, "deg" was typed in the unit placeholder.

$$\text{Example}_{1b} = 0.927 \qquad \text{Example}_{1b} = 53.13\ deg$$

$$\text{Example}_{1c} := \underline{\text{pmt}}\left(\frac{10\%}{12}, 36, 10000, 0, 0\right)$$

See PTC Mathcad Help for information about the *pmt* function.

$$\text{Example}_{1c} = -322.672$$

$$\text{Example}_{1d} := \underline{\text{mean}}(1, 3, 5, 7, 10)$$

The list of arguments for this function can be any length.

$$\text{Example}_{1d} = 5.2$$

FIGURE 4.4

Examples of using various types of arguments

$$\text{Example}_2 := 1 + \underline{\cos}\left(\frac{\pi}{2} + 7 \cdot \underline{\tan}\left(\frac{\pi}{4}\right)\right) + \underline{\sin}(\pi)$$

$$\text{Example}_2 = 0.343$$

This is the default result.

$$\text{Example}_2 = 0.343\ rad$$

Result with "rad" typed in placeholder.

$$\text{Example}_2 = 19.653\ deg$$

Result with "deg" typed in placeholder.

FIGURE 4.5

Example of using built-in functions within an expression

Labels

We have briefly mentioned labels in previous chapters. Chapter 16 has an in-depth discussion about labels, but we need to understand a few important concepts about labels as we discuss functions.

On the **Math** tab in the **Style** group there is a control called **Labels** (**Math>Style>Labels**). If you click the control you get the following list:

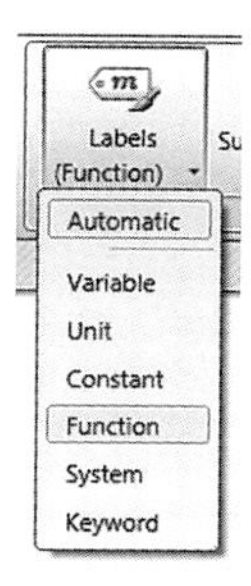

There are six types of labels. PTC Mathcad automatically assigns one of these label names to variables, functions, constants, units, etc. You normally do not need to worry about these labels because PTC Mathcad does this for you. Figure 4.6 illustrates some of the labels that are automatically assigned to various elements.

Notice the following labels in this figure:

1. The variable name has the "Variable" label.
2. The function *sin* has the "Function" label.
3. The units *cm*, *m* and *deg* have the "Unit" label.
4. The constant π has the "Constant" label.

These were automatically assigned by PTC Mathcad.

$\text{Variable} := 30\ cm$ $\quad\quad$ $\text{Variable} = 0.3\ m$

$\text{Variable2} := \underline{\text{sin}}\left(\frac{\pi}{2}\right)$ $\quad\quad$ $\text{Variable2} = 1$

$\text{Variable3} := \underline{\text{cos}}(45\ deg)$ $\quad\quad$ $\text{Variable3} = 0.707$

FIGURE 4.6

PTC Mathcad labels

For this book, the default label styles were modified from the PTC Mathcad default labels so that they would be easier to distinguish. This book uses a bold underline font for the Function label. Refer to Chapters 16 and 17 for a discussion about modifying label font characteristics and creating templates.

You can see what label PTC Mathcad assigns to a particular element by clicking on the element and looking at the **Labels** control in the **Style** group on the **Math** tab. In the below example, the "Function" label was assigned.

User-defined Functions

Remember from Chapter 1 that user-defined functions are very similar to built-in functions. User-defined functions consist of a variable name, a list of arguments, and a definition giving the relationship between the arguments. The same rules that apply to naming variables also apply to naming functions. Refer to Chapter 3 for a list of the naming rules. When creating a user-defined function the arguments listed in parentheses do not need to be defined previously in your worksheet. The function is simply telling PTC Mathcad what to do with the function arguments. However, when evaluating a user-defined function, the arguments must be defined. (An exception is when you are using symbolics. This will be discussed in Chapter 10)

This is a simple user-defined function: SampleFunction(x) := x^2. "SampleFunction" is the name of the function, and "x" is the argument. When you evaluate the function by typing: **SampleFunction(2) =**, PTC Mathcad takes the value of the argument (2), and applies it everywhere there is an occurrence of the argument in the definition. So here, PTC Mathcad replaces "x" with the number "2" and squares the number, returning the value of 4. See Figure 4.7 for examples of this function with various arguments. Notice that the argument may also be the result of an expression. Any expression is allowed as long as the result of the expression is a value that is expected by the function. The beauty of this is that you don't need to calculate the value of the function argument if it is the result of another equation.

Let us look at some sample user-defined functions. See Figure 4.8. Notice the different types of function names and function arguments. It doesn't matter what letter or combination of letters you use for the argument. You can even use characters as arguments. You may need to use the special text mode discussed in Chapter 3. See Figure 4.9.

Assigning the "Function" Label to User-defined Functions

When you create a user-defined function, PTC Mathcad automatically assigns a "Variable" label to the name of the function. This means that built-in functions will have a different appearance than user-defined functions by default. Figure 4.10

$\text{SampleFunction}_1(x) := x^2$	Function definition.
$\text{SampleFunction}_1(2) = 4$	Simple positive argument.
$\text{SampleFunction}_1(-4) = 16$	Simple negative argument.
$\text{SampleFunction}_1(\sqrt{16}) = 16$	Argument using the result of a built-in function.
$\text{SampleFunction}_1\left(\frac{1}{2}\right) = 0.25$	Argument using the result of an expression.
$\text{SampleFunction}_1(2 \cdot 4) = 64$	Argument using the result of an expression.
$\text{SampleFunction}_1\left(4 \cdot \sin\left(\frac{\pi}{4}\right)\right) = 8$	Argument using the result of an expression with a built-in function.

FIGURE 4.7

Sample Function with various arguments

$f(a) := a^{\frac{1}{3}}$ $\quad f(27) = 3$	Function names can be a single letter.
$F_2(a) := 2 \cdot a^2 + a + 3$ $F_2(3) = 24$	Note that the "a" in this function is totally independent of the "a" in the function above. Each argument applies only to the defined function.
$\text{ExampleFunction}_2(\text{Dog}) := 2^{\text{Dog}}$ $\text{ExampleFunction}_2(4) = 16$	Arguments are not limited to single letters.
$\text{AnotherFunction}(x2) := \sqrt{x2 + 8}$ $\text{AnotherFunction}(8) = 4$	Arguments may also consist of letters and numbers. This is not recommended as it appears that the argument is "x" multiplied by 2, rather than "x2".

FIGURE 4.8

Examples of various function names and arguments

$\text{SampleFunction}_3(@) := @^2 + 2$ $\text{SampleFunction}_3(10) = 102$	Most symbols may be used. See example below if the symbol is the shortcut for an operator.

The ! is the shortcut for the factorial operator.

$\text{SampleFunction}_4(!) := ! \cdot 3 + 4$ $\text{SampleFunction}_4(10) = 34$	To use the ! symbol you need to first use the double quotes to get a string, and then type !. After this, delete the double quotes leaving only the !.

FIGURE 4.9

Example of using a symbol as the argument

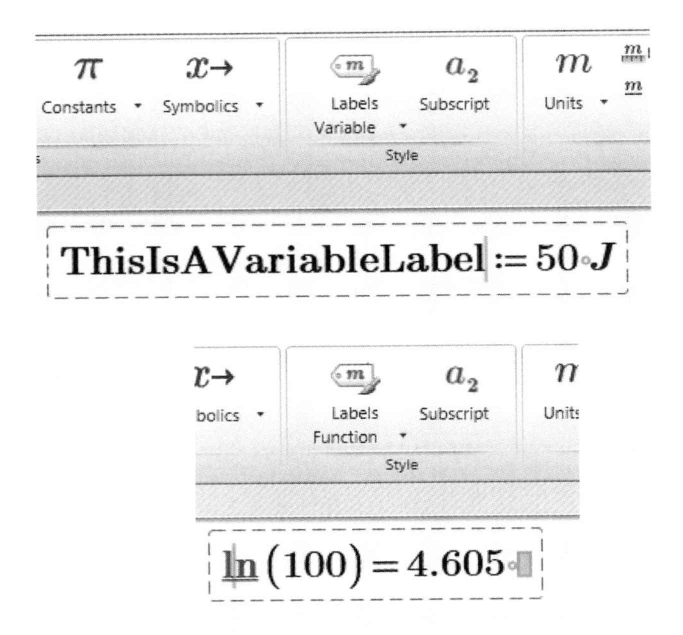

FIGURE 4.10

Labels

shows a "Variable" label on the top and a "Function" Variable on the bottom. Each of these were assigned by PTC Mathcad. If you would like all your functions to have the same appearance, then you can assign the "Function" label to your user-defined function. You do this by placing the cursor in the name of the function, clicking on the **Labels** control, and selecting Function **(Math>Style>Labels>Function)**. Once this is done, the user-defined function will be assigned with a "Function" label and have the same font characteristics as built-in functions. If there is no other variable with the same name as the user-defined function, then PTC Mathcad will automatically recognize the user-defined function as having the "Function" label. See Figure 4.11.

PTC Mathcad does a pretty good job of assuming the correct label. Occasionally, you will need to manually change the label so that PTC Mathcad will recognize a variable.

Why Use User-defined Functions?

Once a user-defined function is defined, it can be used over and over again. This makes user-defined functions very powerful. User-defined functions are helpful if

$\text{AreaOfCircle}(r) := \pi \cdot r^2$

User-defined functions have a "Variable" label by default.

$\underline{\text{VolumeOfSphere}}(r) := \frac{4}{3} \cdot \pi \cdot r^3$

You can assign a "Function" label for user-defined fuctions.

$\text{AreaOfCircle}(3\ m) = 28.274\ m^2$

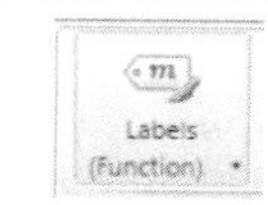

$\underline{\text{VolumeOfSphere}}(3\ m) = (1.131 \cdot 10^5)\ L$

If you type "VolumeOfSphere(3m=" This is the result that you get. Because there is no other variable named VolumeOfSphere PTC Mathcad assumes that you want to use the variable VolumeOfSphere with the label "Function".

If you look at the **Labels** control on the **Math** tab, it shows (Function). The parentheses means that this is an assumed, not an assigned label.

$\underline{\text{VolumeOfSphere}}(3\ m) = 113.097\ m^3$

If you assign the "Function" label to VolumeOfSphere, then the Label control does not show parentheses because it is an assigned label.

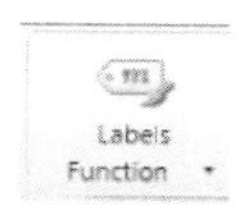

FIGURE 4.11

Assigning the "Function" label to user-defined functions

you are repeatedly doing the same steps over and over again in your calculations. This is illustrated in Figure 4.12.

Using Multiple Arguments

Until now, we have been using a single argument in the argument list. PTC Mathcad allows you to have several arguments in the argument list. These are illustrated in Figure 4.13.

Variables in User-defined Functions

You are allowed to use a variable in your function definition that is not a part of your argument list; however, this variable must be defined prior to using it in your user-defined function. The value of the variable—at the time you define your user-defined function—becomes a permanent part of the function. If you redefine the variable below this point in your worksheet, the function still uses the value at the time the function was defined. This case is illustrated in Figures 4.14 and 4.15.

When you evaluate a previously defined user-defined function, the values you put between the parentheses in the argument list may also be previously defined variables. This is illustrated in Figures 4.16–4.18.

Suppose that you need to find the result of $x^3 + 2 \cdot x^2 - x + 3$ for many different values of x. You can set up this expression: $y := x^3 + 2 \cdot x^2 - x + 3$. You can then define x:=1 and type y=. You could then redefine x to be x:=5, and then type y= to get another value of y. This method works, but it takes time and is not convenient because you must continuously redefine "x," and your results do not remain.

$x := 1$

$y := x^3 + 2 \cdot x^2 - x + 3$ $\quad y = 5$

You would need to redefine "x" to be another number in order to get a result for a different value of y. Your original answer using x:=1 will also be lost.

(In later chapters we will discuss range variables and arrays which will make the above scenario easier.)

Another way to solve the above situation is to define a function. After the function is defined, you can type the function using numerous values for the argument.

$y(x) := x^3 + 2 \cdot x^2 - x + 3$

$y(1) = 5$ $\quad y(10) = 1193$

$y(5) = 173$ $\quad y(100) = 1.02 \cdot 10^6$

FIGURE 4.12

Using a Function

$\mathbf{MultipleArgument_1(a, b) := a^2 + b^2}$

$\mathbf{MultipleArgument_1(2, 3) = 13}$

$\mathbf{MultipleArgument_2(dog, cat, goat) := \dfrac{\sqrt{dog} + 2^{cat}}{goat}}$

$\mathbf{MultipleArgument_2(4, 3, 2) = 5}$

FIGURE 4.13

Using multiple arguments

$\text{Input}_1 := 4$

$\text{VariableExample}(V) := \text{Input}_1 \cdot V$

$\text{VariableExample}(3) = 12$

The value of "Input1" in the above function remains the value at the point at which the function was defined (4).

$\text{Input}_1 := 8$ Redefine the value of "Input1".

$\text{VariableExample}(3) = 12$

In the above example, "Input1" was redefined from 4 to 8. This did not affect the result of the function. The value of "Input1" in the function depends on the value of "Input1" at the point at which the function was defined (4), not the new value of "Input1" (8).

FIGURE 4.14

Using variables in user-defined functions

$\text{Input}_1 := 4$

$\text{Input}_1 := 8$ Redefine the value of "Input1".

$\text{VariableExample}(V) := \text{Input}_1 \cdot V$

$\text{VariableExample}(3) = 24$

In this example, the value of "Input1" in the function IS changed. This is because the value of "Input1" was changed from 4 to 8 PRIOR to the definition of the user-defined function.

FIGURE 4.15

Changing the value of a variable in a user-defined function

$\text{Distance}(a, b) := \sqrt{a^2 + b^2}$

$\text{Distance}(3, 4) = 5$

$x_1 := 3$ $y_1 := 4$

$x_2 := 5$ $y_2 := 12$

$x_3 := 20$ $y_3 := 25$

$\text{Distance}(x_1, y_1) = 5$

$\text{Distance}(x_2, y_2) = 13$

$\text{Distance}(x_3, y_3) = 32.016$

The argument list can use previous defined variables.

FIGURE 4.16

Using variables in the argument list

$B := 10$	This defines variable "B".
$\mathbf{Function_1}(B) := B^2$	The argument "B" in this function is independent of the value of variable "B" above. In this case, "B" defines the argument of the function "Function1". It does not depend on the value of variable "B" above.
$\mathbf{Function_1}(3) = 9$	Uses 3 as the argument of "Function1". Result of "Function1" is 3^2.
$\mathbf{Function_1}(B) = 100$	Uses variable "B" as the argument of "Function1". Result of "Function1" is 10^2.
$B := 12$	Redefines variable "B".
$\mathbf{Function_1}(B) = 144$	Result of the function changes because the value of variable "B" has changed. Result of "Function1" is 12^2.

FIGURE 4.17

Note the difference between variable "B" and argument "B"

$C_1 := 2$	This defines variable "C1".
$\mathbf{Function_2}(D) := {C_1}^D$	In this function "C1" is a fixed variable from outside the function, and "D" is the argument of the function (taken from the argument list).
$\mathbf{Function_2}(3) = 8$	Result of "Function2" is 2^3. C1=2, Argument D=3.
$\mathbf{Function_2}(C_1) = 4$	Result of "Function2" is 2^2. The variable "C1" becomes the value of the argument in "Function2". C1=2, Argument D=2 (Value of variable "C1").
$C_1 := 4$	Redefine variable "C1".
$\mathbf{Function_2}(C_1) = 16$	Result of "Function2" is 2^4. C1=2 (for the function), Argument D=4 (New value of variable "C1"). Remember that for the function, "C1" remains the original value at the time that the function was defined, not the redefined value of "C1".

FIGURE 4.18

Note how "C1" is captured in the function and how it also becomes the argument "D"

Examples of User-defined Functions

Let's now look at two examples of how user-defined functions can be used in technical calculations. See Figures 4.19 and 4.20.

Write a function to calculate the area of a trapezoid.

$$Trap_{Area}(B1, B2, h) := \frac{1}{2} \cdot h \cdot (B1 + B2)$$

$Trap_{Area}(5, 10, 18) = 135$ $\quad Trap_{Area}(8, 0, 9) = 36$

Write a function to calculate the center of gravity above the base of a trapezoid.

$$Trap_{CG}(B1, B2, h) := \frac{h \cdot (2 \cdot B2 + B1)}{3 \cdot (B1 + B2)}$$

$Trap_{CG}(5, 10, 18) = 10$ $\quad Trap_{CG}(8, 0, 9) = 3$

In order to use these functions and results in technical calculations, you may want to assign the input values and results to variable names such as these:

Input values to be used in functions.

$T1_{B1} := 5$ $\quad T1_{B2} := 10$ $\quad T1_{Height} := 18$

$T2_{B1} := 8$ $\quad T2_{B2} := 0$ $\quad T2_{Height} := 9$

Results are assigned to variable names. These results can now be used later in the calculations.

$T1_{Area} := Trap_{Area}(T1_{B1}, T1_{B2}, T1_{Height}) = 135$

$T1_{CG} := Trap_{CG}(T1_{B1}, T1_{B2}, T1_{Height}) = 10$

$T2_{Area} := Trap_{Area}(T2_{B1}, T2_{B2}, T2_{Height}) = 36$

$T2_{CG} := Trap_{CG}(T2_{B1}, T2_{B2}, T2_{Height}) = 3$

FIGURE 4.19

Finding the area and center of gravity of a trapezoid

Write a function to calculate the surface area of a right circular cylinder.

$$\mathbf{Cylinder_{Area}}(\mathrm{r}, \mathrm{h}) := 2 \cdot \pi \cdot \mathrm{r} \cdot \mathrm{h} + 2 \cdot \pi \cdot \mathrm{r}^2$$

$$\mathbf{Cylinder_{Area}}(4\ ft, 12\ ft) = 402.124\ ft^2$$
$$\mathbf{Cylinder_{Area}}(3\ m, 10\ m) = 245.044\ m^2$$

Write a function to find the volume of a right circular cylinder.

$$\mathbf{Cylinder_{Volume}}(\mathrm{r}, \mathrm{h}) := \pi \cdot \mathrm{h} \cdot \mathrm{r}^2$$

$$\mathbf{Cylinder_{Volume}}(4\ ft, 12\ ft) = 603.186\ ft^3$$
$$\mathbf{Cylinder_{Volume}}(3\ m, 10\ m) = 282.743\ m^3$$

In order to use these functions and results in technical calculations, you may want to assign the input values and results to variable names such as these:

Input values to be used in functions.

$\mathrm{r}_1 := 4\ ft$ $\qquad$ $\mathrm{h}_1 := 12\ ft$

$\mathrm{r}_2 := 3\ m$ $\qquad$ $\mathrm{h}_2 := 10\ m$

Results are assigned to variable names. These results can now be used later in the calculations.

$$\mathbf{Area_1} := \mathbf{Cylinder_{Area}}(\mathrm{r}_1, \mathrm{h}_1) = 402.124\ ft^2$$

$$\mathbf{Volume_1} := \mathbf{Cylinder_{Volume}}(\mathrm{r}_1, \mathrm{h}_1) = 603.186\ ft^3$$

$$\mathbf{Area_2} := \mathbf{Cylinder_{Area}}(\mathrm{r}_2, \mathrm{h}_2) = 245.044\ m^2$$

$$\mathbf{Volume_2} := \mathbf{Cylinder_{Volume}}(\mathrm{r}_2, \mathrm{h}_2) = 282.743\ m^3$$

FIGURE 4.20

Finding the surface area and volume of a right circular cylinder

Passing a Function to a Function

It is possible to use a previous user-defined function in a new user-defined function. Figure 4.21 gives an example of a user-defined function "SectionModulus" that calculates the section modulus of a rectangular beam. The arguments for this function are b and d. The new function "Stress" calculates the stress in the beam for a given moment M. This function uses the "SectionModulus" function. Thus, the arguments for "Stress" must include all the arguments needed for both functions.

The second example in Figure 4.21 is a bit more complicated, but much more powerful. In this example, the user-defined function "Sample" uses a function name as an argument. The function used will not be defined until the user-defined function "Sample" is executed. This means that when you execute the user-defined function "Sample," you need to include a variable argument for "x," and also include a function argument for "H." The function can be a user-defined function or a PTC Mathcad function that uses a single argument.

$$\text{SectionModulus}(b,d) := \frac{1}{6} \cdot b \cdot d^2$$

This function uses the SectionModulus function, which gets the arguments from the Stress function.

$$\text{Stress}(b,d,M) := \frac{M}{\text{SectionModulus}(b,d)}$$

$$\text{Stress}(.5 \cdot m, 1.3 \cdot m, 800000 \cdot N \cdot m) = 5680.473\ kPa$$

The above example used a static function. The next example shows the included function to be variable.

$F_1(a) := a^2$ $G_1(b) := \frac{1}{b^2}$ Define two user-defined functions.

This user-defined function uses two arguments. They are "x" - a variable and "H" a function name to be defined later. The function actually used will be given as an argument of the user-defined function "Sample".

$$\text{Sample}(x,H) := \frac{3}{x} \cdot H(x)$$

This example uses the user-defined function F1.

$\text{Example}_1 := \text{Sample}(2, F_1) = 6$ $\frac{3}{2} \cdot 2^2 = 6$

This example uses the user-defined function G1.

$\text{Example}_2 := \text{Sample}(2, G_1) = 0.375$ $\frac{3}{2} \cdot \frac{1}{2^2} = 0.375$

This example uses the PTC Mathcad function sin().

$\text{Example}_3 := \text{Sample}(2, \underline{\sin}) = 1.364$ $\frac{3}{2} \cdot \underline{\sin}(2) = 1.364$

This example uses the PTC Mathcad function ln().

$\text{Example}_4 := \text{Sample}(2, \underline{\ln}) = 1.04$ $\frac{3}{2} \cdot \underline{\ln}(2) = 1.04$

FIGURE 4.21

Function in a function

Warnings

In Mathcad 15, you had to avoid creating variables with the same names as functions. In PTC® Mathcad Prime®, you can use labels to distinguish between variables and functions with the same names. Actually, in Mathcad 15 you could use styles to do the same thing, but labels are more automated than the styles of Mathcad 15.

$h(x) := x^2 + x + 1$	Defines function "h".
$h(2) = 7$	
$h := 3$	Redefines "h" as a variable.
$h(2) = ?$	Function "h" is no longer recognized, because it was redefined as variable "h".
$h(2) = ?$ This value must be a function.	

If h(x) uses the label "Function," it will avoid this issue.

$h(x) := x^2 + x + 1$	h(x) now has the "Function" label.	
$h(2) = 7$	PTC Mathcad assumes that h with the function label is wanted. (The label has parentheses around it.)	Labels (Function)
$h := 3$		
$h(2) = ?$	Typing h(2= now gives an error.	
$h(2) = ?$ This value must be a function.		
$h(2) = 7$	In order to get results, you must manually assign the "Function" label to "h". The "Function" label now does not have parentheses because it is not an assumed label.	Labels Function
$h = 3$	Typing h= gives this result. PTC Mathcad assumes that you want h with the variable label. (The label has parentheses around it.)	Labels (Variable)

FIGURE 4.22

Using labels to distinguish between variables and functions with the same name

Several warnings will help make the use of functions more effective:

- Be careful not to redefine a user-defined function. User-defined function names are similar to variables. If you want to have a variable with the same name as a user-defined function, then assign the "Function" label to the user defined function. This will allow you to use the same name for both the variable and the user-defined function. See Figure 4.22.
- Be careful not to redefine a built-in function. This only occurs when you assign a "Function" label to a variable. See Figure 4.23.

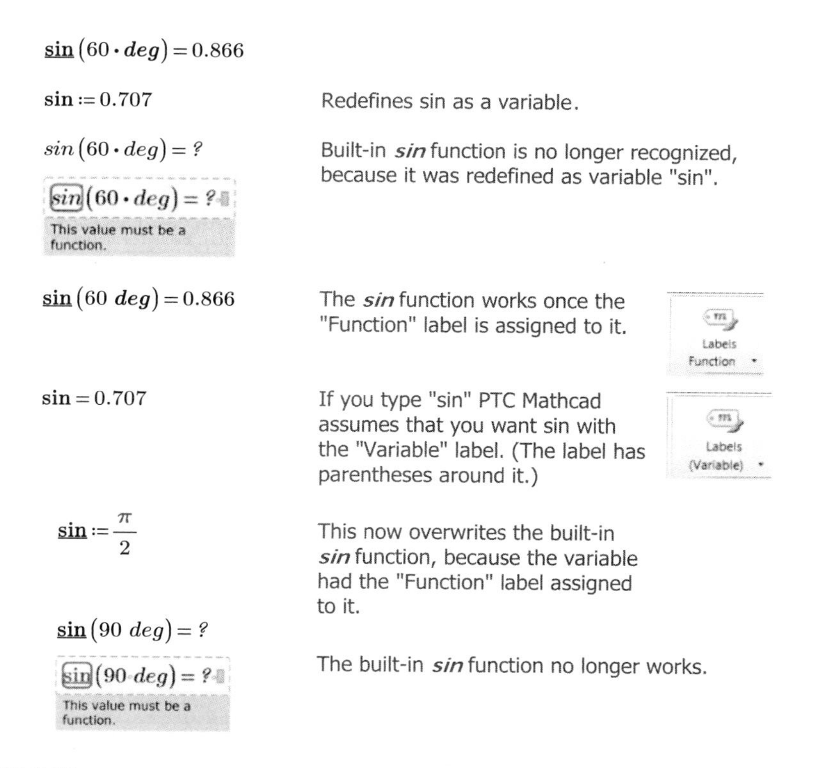

FIGURE 4.23

Be careful not to overwrite built-in functions

- Remember that if you use a variable in your function definition, the value of the variable does not change, even if you later rename the variable.
- Once you define your user-defined function, you will not be able to display the function definition again in your worksheet (unless you use symbolics that are discussed in Chapter 10). If you type the name of the function and an equal sign, PTC Mathcad does not display the function definition. See Figure 4.24. This may make it difficult to remember exactly what the function definition was. You must also remember what order the arguments go in. One option for displaying the user-defined function definition later in your worksheet is to copy the function definition and paste the definition where you want it displayed. This essentially redefines the function to the same definition, so you will need to disable the region from the **Calculation** tab. You may wonder, why worry about disabling a definition? You get the same answer as before because it is the same definition. The problem comes if you later redefine the original user-defined function. PTC Mathcad will use the value of the changed function until it comes to the point in your worksheet where you pasted the original

$g_1(x,y) := x^2 + \frac{y}{2}$ Function definition

PTC Mathcad does not display the original definition

$g_1 =$ "Function: g₊1₊{VARIABLE}"

In order to display the function definition at a later point in your worksheet, copy the original definition and paste it where you want it displayed. WARNING: BE SURE TO DISABLE THE PASTED FUNCTION! (**Calculation>Controls>Disable Region**) You do not want to have two definitions of the same function in your worksheet, because if you change the original function, you may forget to change the other copies of the function. This will result in incorrect results in your calculations.

$g_1(x,y) := x^2 + \frac{y}{2}$ This is a disabled definition. PTC Mathcad ignores the definition. If you want to display the function after it is defined be sure to disable the definition. WARNING: If the original function is changed, this display will be WRONG.

$g_1(3,4) = 11$ The result of this fuction is based on the original definition, not the disabled definition. If the original function is changed, this result will also change.

To display the original function using Symbolics (Chapter 10), you use the symbolic evaluation operator (**Math>Operators and Symbols>Symbolics>Explicit>explicit**). Follow these steps:

- Type the name of the function and its arguments.
- Click on "explicit" in the **Symbolics** list.
- Type a comma and type the name of the function (without the argument).
- Click outside of the region.

$g_1(x,y) \xrightarrow{explicit,\, g_1} x^2 + \frac{y}{2}$

$g_1(a,b) \xrightarrow{explicit,\, g_1} a^2 + \frac{b}{2}$

You may use your original arguments, or you may use new arguments.

FIGURE 4.24

How to display a function definition later in your calculations

function. From this point on, PTC Mathcad will use the old function definition. By disabling the function definition, PTC Mathcad will continue to use the new version of the function. However, PTC Mathcad will still display an old version of the function. This may lead to some confusion in your calculation, but not an incorrect answer.

- There is a way to display the original function using PTC Mathcad's symbolic processor. Symbolic calculations will be discussed in Chapter 10. Figure 4.24 shows how to use Symbolics to show a function after it has been defined.

Engineering Examples

Engineering Example 4.1: Column Buckling

The Euler column formula predicts the critical buckling load of a long column with pinned ends. The Euler formula is $P_{cr} = \frac{\pi^2 \cdot E \cdot I}{L^2}$ where E is the modulus of elasticity in (force/length2), I is the moment of inertia (length4), L is the length of the column.

Create a user-defined function to calculate the critical buckling load of a column.

$$P_{cr}(E, I, L) := \frac{\pi^2 \cdot E \cdot I}{L^2}$$

Calculate the critical buckling load for:

1. A steel column with E=29,000ksi, I=37 in4, and L=20ft.
2. A wood column with E=1,800,000 psi, I=5.36 in4, and L=10ft.

$P_1 := P_{cr}(29000 \cdot ksi, 37 \cdot in^4, 20 \cdot ft)$ $\quad P_1 = 183.856\ kip$

$P_2 := P_{cr}(1800000 \cdot psi, 5.36 \cdot in^4, 10 \cdot ft)$ $\quad P_2 = 6.61\ kip$

Engineering Example 4.2: Torsional Shear Stress

The torsional shear stress of a linear elastic homogeneous and isotropic shaft is given by the equation: $\tau = \frac{T \cdot p}{J}$, where T=torque, p=distance from the shaft center, and J=polar moment of of inertia with respect to the longitudinal axis of the shaft. For a solid circular shaft, J=π*D4/32.

Create a user-defined function to calculate the shear stress in a circular shaft.

$$J(D) := \frac{\pi \cdot D^4}{32} \qquad \tau(D, T) := \frac{T \cdot \frac{D}{2}}{J(D)}$$

$$\tau_1 := \tau(1.5 \cdot in, 8000 \cdot in \cdot lbf) \qquad \tau_1 = 12.072\ ksi$$

$$\tau_2 := \tau(4 \cdot in, 300 \cdot in \cdot kip) \qquad \tau_2 = 23.873\ ksi$$

The twist in a shaft can be calculated by the equation $\theta = \frac{T \cdot L}{J \cdot G}$, where L is the length of the shaft and G is the shear modulus of elasticity or modulus of rigidity (for steel G= 11,200 lbf).

Create a user-defined function to calculate the twist of a circular shaft.

$$G := 11200 \cdot ksi \qquad \theta(D, T, L) := \frac{T \cdot L}{J(D) \cdot G}$$

$$\theta_1 := \theta(1.5 \cdot in, 8000 \cdot in \cdot lbf, 50 \cdot in) \quad \theta_1 = 4.117\ deg \quad \theta_1 = 0.072\ rad$$

$$\theta_1 := \theta(4 \cdot in, 300 \cdot in \cdot kip, 80 \cdot in) \quad \theta_1 = 4.885\ deg \quad \theta_1 = 0.085\ rad$$

Summary

We have just scratched the surface of user-defined functions. The intent of this chapter was to introduce simple built-in functions, and to get you comfortable with the concept and use of user-defined functions. There is much more to learn. User-defined functions will be discussed in much more detail in later chapters of this book.

In Chapter 4 we:

- Discussed PTC Mathcad's built-in functions.
- Expanded the discussion of user-defined functions.
- Explained the benefits of using user-defined functions rather than expressions.
- Showed how to include multiple arguments and variables in user-defined functions.
- Issued several warnings about using built-in and user-defined functions.

Practice

The PTC Mathcad Prime 3.0 figures and examples used in this book are available for download from the book's website. The reader is encouraged to download the files and use them to practice the concepts learned. Additional examples and problems are also provided. To access this content go to http://store.elsevier.com/9780124104105 *and click on the Resources tab, and then click on the link for the Online Companion Materials.*

Note: Save your worksheet with the following user-defined functions for use with the practice exercises in Chapter 5.

1. Write user-defined functions to calculate the following. Choose your own descriptive function names. These functions will have a single argument. Evaluate each function for two different input arguments.
   a. The volume of a circular sphere with a radius of R1.
   b. The surface area of a sphere with a radius of R1.
   c. Converting degree Celsius to degree Fahrenheit.
   d. Converting degree Fahrenheit to degree Celsius.
   e. The surface area of a square box with length L1.
2. Write user-defined functions to calculate the following. Choose your own descriptive function names. These functions will have multiple arguments. Evaluate each function for two sets of input arguments.
   a. The inside area of a pipe with an outside diameter of D1 and thickness of T1.
   b. The material area of a pipe with an outside diameter of D1 and thickness of T1.
   c. The volume of a box with sides: L1, L2, and L3.
   d. The surface area of a box with sides: L1, L2, and L3.

e. The distance traveled by a free-falling object (neglecting air resistance) with an initial velocity, acceleration, and time: V0, a, and T.

3. From your area of study (or from a physics book) write 10 user-defined functions. At least five of these should be with multiple arguments. Evaluate each function for two sets of input arguments.
4. The bending moment (at any point x) of a simply supported beam with a uniformly distributed load is defined by the following formula: M_x = w*x*(L-x)/2, where w = force per unit of length, L = length of beam, and x is the distance from the left support. Write a user-defined function to calculate the bending moment at any point x. Use the arguments w, L, and x. When testing the formula be sure to input your values in constant units of length. (Unless you attach units to your input arguments, which will be discussed in the next chapter.)
5. The maximum deflection of a simply supported beam with a uniformly distributed load is defined by the following formula: Deflection = 5*w*L^4/(384*E*I). Write a user-defined function to calculate the maximum deflection of the beam. Use the arguments w, L, E, and I. When testing the formula, be sure to input your values in constant units of force and length. The units are as follows: w = force/unit of length; L = length; E = force/$length^2$; and I = $length^4$.
6. The moment of inertia I of a rectangle about its centroidal axis is defined by the following formula: I = b*d^3/12. Write a user-defined function to calculate the moment of inertia. Use the arguments b and d.

CHAPTER

5 Units!

One of the most powerful features of PTC Mathcad is its ability to attach units to numbers. PTC Mathcad units will be one of your best friends as a scientist or engineer. PTC Mathcad units will simplify your work, and they are one of the best means you have of catching mistakes. You may remember the 1999 Mars Climate Orbiter that ended in disaster by burning up in the atmosphere of Mars. Why? Because engineers failed to convert English measures of rocket thrust into metric measures. Using PTC Mathcad units will help prevent problems like this from occurring. Once units are attached to values, PTC Mathcad understands the unit dimensions and can display the units in any desired unit system. It does this easily and automatically. You are not left wondering what units the results represent. If you use units consistently, PTC Mathcad will alert you to problems in your expressions. For example, if you expected the results to be in lbf/ft^2, but they are in lbf/ft, then something is missing from your expression or from your input.

Chapter 5 will:

- Discuss unit dimensions, units, and the default unit systems.
- Show how to assign units to numbers.
- Discuss units of force and units of mass.
- Explain how to create custom units.
- Emphasize the importance of understanding labels.
- Describe how to use units in equations and functions.
- Present ways of dealing with units in empirical formulas.
- Discuss how to use units of temperature.
- Introduce the DMS, hhmmss, and FIF functions and how to use them as units.
- Illustrate the use of custom dimensionless units.
- Show how to use the unit placeholder as a scaling factor.

Essential PTC® Mathcad Prime® 3.0.

Introduction

Remember from Chapter 1 that to assign units to a number or variable you should multiply the number by the name of the unit. If you do not know the name of the unit, you may use the Ribbon Bar: **Math>Units>Units**, to insert the desired unit.

To change the displayed unit, delete the text of the displayed unit name and retype the desired unit name, or use the **Units** group from the **Math** tab to insert the desired unit name. When you delete a displayed unit name, a box with a vertical line appears following the result. This is referred to as the unit placeholder and is where you type or insert the desired unit name. $5 \cdot ft = 1.524 \cdot |$ So when we refer to the unit placeholder, remember that it appears once the displayed unit name is deleted.

Definitions

For our discussion of units, we will use the following terminology:

- Unit dimension: A physical quantity, such as mass, time, or pressure, which can be measured.
- Unit: A means of measuring the quantity of a unit dimension.
- Base unit dimension: One of eight basic unit dimensions: length, mass, time, temperature, luminous intensity, substance, current or charge, and money.
- Base unit: A default unit measuring one of the seven base unit dimensions, plus money. For example, the following are base units: meter (m), kilogram (kg), second (s), Kelvin (K), candela (cd), mole (mol), Ampere (A), and money (¤).
- Derived unit dimension: A unit dimension derived from a combination of any of the seven base unit dimensions. For example, the following are derived unit dimensions: area (length^2), pressure (mass/(length*time^2), energy (mass*length^2/time^2) and power (mass*length^2/time^3).
- Derived unit: A unit measuring a derived unit dimension. For example, the following are derived units: pounds per square inch (psi), Joules (J), British Thermal Unit (BTU), and Watts (W).
- Unit system: A group of units used to measure base unit dimensions and derived unit dimensions. There are three unit systems available in PTC Mathcad:
  - SI, with base units of meter (m), kilogram (kg), second (s), Kelvin (K), candela (cd), mole (mol), and Ampere (A).
  - U.S. Customary System, with base units of feet (ft), pound mass (lb), second (s), Kelvin (K), candela (cd), mole (mol), and Ampere (A).
  - CGS, with base units of centimeter (cm), gram (gm), second (s), Kelvin (K), candela (cd), mole (mol), and statampere (statamp).
- Default unit system: The unit system you tell PTC Mathcad to use. PTC Mathcad defaults to the SI unit system, unless you change it.

Changing the Default Unit System

When you open PTC Mathcad, the SI unit system is the default unit system. To change to another unit system in your current worksheet, use the **Unit System** control from the **Units** group on the **Math** tab (**Math>Units>Unit System**). See Figure 5.1.

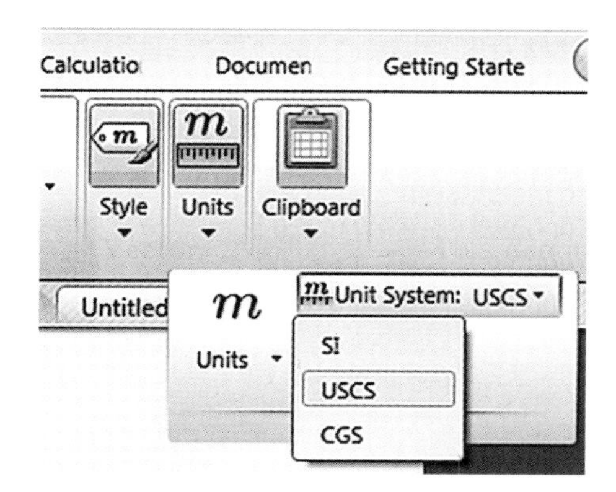

FIGURE 5.1

Setting default unit system

> Custom unit systems did not make it into PTC® Mathcad Prime® 3.0. ☹ They are promised for PTC Mathcad Prime 4.0.

Using and Displaying Units

To add a series of numbers with units attached, the units must all be from the same unit dimension. For example, you can add any units of length, or you can add any units of time, but you cannot add units of length to units of time. Figure 5.2 shows some examples of adding simple units. If you attempt to add different unit dimensions, PTC Mathcad warns that the unit dimensions do not match.

The units PTC Mathcad displays by default are based on the chosen unit system, but you can change the displayed units. Let's now look at how to display different units. See Figures 5.3 and 5.4.

Examples of unit addition.

Attached units can be from any unit system as long as the unit dimensions are the same (such as length, force, etc).

If you don't know the unit names then use the **Units** list from the **Math** tab.

$\text{Units}_A := 6 \cdot m + 4 \cdot mm = 6.004\ m$ Type: 6*m+4*mm=

$\text{Units}_B := 6 \cdot ft + 3 \cdot in = 1.905\ m$ Type 6*ft+3*in=

PTC Mathcad's default unit system is set to SI, so PTC Mathcad displays units of length in meters.

$\text{Units}_C := 6 \cdot N + 1 \cdot kip = \left(4.454 \cdot 10^3\right) N$

$\text{Units}_D := 1 \cdot day + 2 \cdot hr + 25 \cdot s = \left(9.363 \cdot 10^4\right) s$

$\text{Units}_E := 5 \cdot m + \boxed{3 \cdot N} = ?$

$\text{Units}_E := 5 \cdot m + \boxed{3 \cdot N} = ?$

These units are not compatible.

PTC Mathcad warns that the units are not compatible.

FIGURE 5.2

Unit addition

$\text{Units}_F := 5.605 \cdot m$

If you want to change the displayed units, then delete the displayed unit, and type the unit you want displayed. After deleting the displayed unit, you can also insert a new unit from the **Units** control on the **Math** tab.

$\text{Units}_F = 5.605\ m$

$\text{Units}_F = 5.605$ Delete the displayed unit. The unit placeholder appears.

$\text{Units}_F = 18.389\ ft$ Type the desired display unit in the placeholder.

FIGURE 5.3

Changing displayed units

You may also double click on the displayed unit to highlight the unit and then select a new unit from the **Units** control on the **Math** tab.

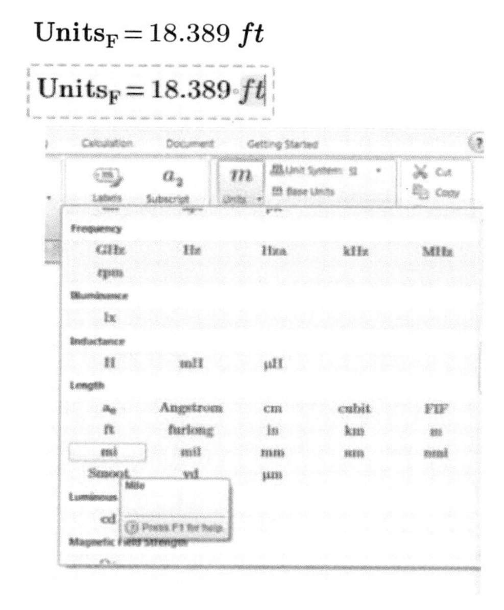

$\mathbf{Units_F} = 0.003483\ mi$

FIGURE 5.4

Changing displayed units

If you type inconsistent units in the unit placeholder, PTC Mathcad will add units that make the result consistent. See Figure 5.5.

PTC Mathcad can combine units. Look at the examples in Figure 5.6.

$\mathbf{Units_G} := 5 \cdot ft + 4 \cdot in$

$\mathbf{Units_G} = 1.626\ m$ — PTC Mathcad defaults to m in the SI default unit system.

$\mathbf{Units_G} = 1.626\ \frac{m}{s} \cdot s$ — If you type "s" in the unit placeholder, PTC Mathcad adds an "s" in the denominator in order for the final units to be units of length.

FIGURE 5.5

Balancing displayed units

Multiply units of length to get units of area.

$\mathbf{Units_H} := 40 \cdot m \cdot 59 \cdot m = \left(2.36 \cdot 10^3\right)\ m^2$

$\mathbf{Units_H} = \left(2.54 \cdot 10^4\right)\ ft^2$ — Hint: To get the superscript, use the "^" character above the number 6 key. Type mm^2.

$\mathbf{Units_H} = 0.236\ \textit{hectare}$ — Area displayed as hectare.

Length divided by time gives velocity.

$\mathbf{Units_I} := \frac{10 \cdot ft}{2 \cdot s} = 1.524\ \frac{m}{s}$ — Default display.

$\mathbf{Units_I} = 5\ \frac{ft}{s}$ — Velocity displayed as feet per second.

$\mathbf{Units_I} = 3.409\ \textit{mph}$ — Velocity displayed as miles per hour.

$\mathbf{Units_I} = 2.962\ \textit{knot}$ — Velocity displayed as knots.

FIGURE 5.6

Displaying combined units

To attach units of area or volume, use the "^" symbol (which is above the number 6 key) to raise the unit to a power. For area type `m^2` or `ft^2`. For volume type `m^3` or `ft^3`.

Derived Units

A derived unit dimension is derived from combinations of any of the seven base unit dimensions. A derived unit measures a derived unit dimension. Some examples of derived unit dimensions include: acceleration, area, conductance, permeability, permittivity, pressure, viscosity, etc. Some examples of derived unit names are: atmosphere, hectare, farad, joule, newton, watt, etc.

PTC Mathcad can display derived unit dimensions by the derived unit name or as a combination of the base unit names. The default is to use the derived unit names. To display derived unit dimensions as a combination of base unit names, turn on the **Base Units** control from the **Units** group on the **Math** tab (Math>Units>Base Units). See Figure 5.7. Figure 5.8 compares the display of units with and without Base Units turned on.

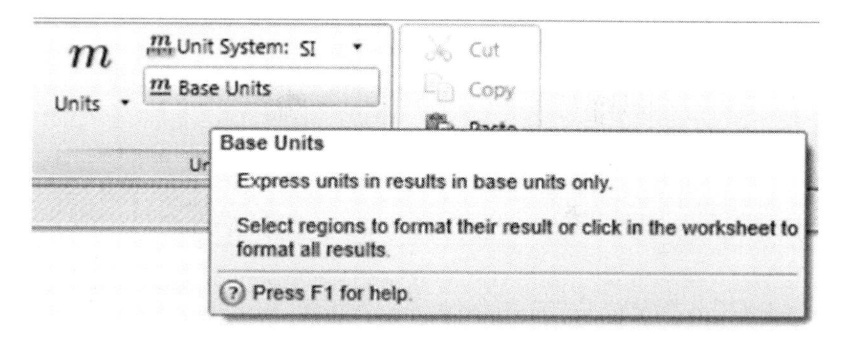

FIGURE 5.7

Unit display

Compare the display of derived units with and without **Base Units** on.

Define Variables

Pressure	Energy	Work	Volume
$\text{Press} := 1 \cdot Pa$	$\text{Energy} := 1 \cdot J$	$\text{Work} := 1 \cdot J$	$\text{Vol} := 1 \cdot m^3$

Base Units off

$\text{Press} = 1\ Pa$	$\text{Energy} = 1\ J$	$\text{Work} = 1\ J$	$\text{Vol} = 1000\ L$

Base Units on

$\text{Press} = 1\ \frac{kg}{m \cdot s^2}$	$\text{Energy} = 1\ \frac{kg \cdot m^2}{s^2}$	$\text{Work} = 1\ \frac{kg \cdot m^2}{s^2}$	$\text{Vol} = 1\ m^3$

FIGURE 5.8

Compare results of units formatting

Custom Default Unit System

PTC Mathcad Prime 3.0 does not have custom units. ☹

When you choose one of the three default unit systems, you get pre-selected base units and pre-selected derived units. In Mathcad 15 you were allowed to tell PTC Mathcad what base units to use and what derived units to use. For example, the default unit of volume in the SI unit system is liter, and the default unit of volume in the USCS is gal. In Mathcad 15 you could tell PTC Mathcad to display units of volume as m^3 or ft^3. This was a very useful feature. Please contact PTC and let them know of your desire to add a custom unit system to PTC Mathcad Prime 4.0.

Units of Force and Units of Mass

It is important to understand how PTC Mathcad considers units of force and units of mass. In the USCS unit system, "lbf" (pound force) is a unit of force, and "lb", "lbm" (pound mass), and "slug" are units of mass. Many people in the United States do not understand the difference between these units. In the SI unit system, "N" (Newton) and "kgf" (kilogram force) are units of force, and "kg" is a unit of mass. Figure 5.9 shows the relationship between various units of force and mass in the USCS unit system. Figure 5.10 shows the relationship between various units of force and mass in the SI unit system.

Note: For this example, the default unit system was changed to US.

$g = 32.174 \frac{ft}{s^2}$ — g is a built-in PTC Mathcad unit for the acceleration of gravity.

$\text{Units}_J := 1 \cdot lbm \cdot g$ — lbf (pound force) = lbm (mass) * acceleration of gravity.

$\text{Units}_J = 1\ lbf$

$1 \cdot lbf = 32.174 \frac{lbm \cdot ft}{sec^2}$ — Shows the relationship between lbf (pound force) and lbm (pound mass).

$1 \cdot slug = 32.174\ lb$ — 1 slug = 32.174 pound mass.

Note: In order to eliminate confusion as to whether lb mean mass or force, it is suggested to always use lbm in lieu of lb.

FIGURE 5.9

Relationship between various units of mass and force in the USCS system

$g = 9.807 \ \frac{m}{s^2}$ g is a built-in PTC Mathcad unit for the acceleration of gravity.

$\text{Units}_K := 1 \cdot kg \cdot g$ mass * acceleration of gravity = Force

$\text{Units}_K = 9.807 \ N$ This is the way PTC Mathcad defaults in the SI unit system.

$\text{Units}_K = 1 \ kgf$ Force displayed as kgf. 1 kgf =1 kg * g

$1 \cdot kgf = 9.807 \ N$ Shows relationship between kgf and N.

$1 \cdot N = 0.102 \ kgf$

Note: Sometimes it is easier to display force as kgf rather than N because it eliminates the 9.807 factor.

FIGURE 5.10

Relationship between various units of mass and force in the SI system

In order to avoid confusion in the USCS unit system, use "lbm" as a unit of mass (rather than "lb") and "lbf" as a unit of force.

Creating Custom Units

Even though PTC Mathcad has over one hundred built-in units, you will still need to create your own custom units from time to time. This is very easy to do. You define custom units the same way you define variables. For example, if you want to define a unit of "cfs" for cubic feet per second, then follow the steps in Figure 5.11. In order for your custom unit to be recognized, it must be assigned the **Unit** label. To do this, click within the variable name and select **Math>Style>Labels>Unit**.

In order to have the custom unit available anywhere in your worksheet, place the definition at the top of your worksheet. You may also use the ***global definition*** operator when creating custom units. The ***global definition*** operator appears as a triple equal sign. All global definitions in the worksheet are scanned by PTC Mathcad prior to scanning the normal definitions. This way the unit definition does not need to be at the top of your worksheet. To define a global custom unit, type the name of the unit, press the tilde key ~, and then type the definition. See Figure 5.12.

$cfs := \frac{ft^3}{sec}$

This defines "cfs" as a custom unit of ft^3 divided by second. Click cfs and then from the **Math** tab, select **Labels** > **Unit**.

$\mathbf{Units_L} := \frac{1000 \cdot ft^3}{1 \cdot min}$

$\mathbf{Units_L} = 0.472\ \frac{m^3}{s}$

PTC Mathcad default in SI units.

$\mathbf{Units_L} = 16.667\ cfs$

Type: Units CTRL+MINUS L=BACKSPACE BACKSPACE cfs [enter]
The custom unit "cfs" is attached to the result.

FIGURE 5.11

Creating custom units

Create a custom unit for Million Gallons per Day (MGD) using the ***global definition*** operator.

$\mathbf{Flow} := 400\ cfs = 11.327\ \frac{m^3}{s}$

$\mathbf{Flow} = 258.527\ mgd$

Note that mgd is used above the definition of mgd.

$\mathbf{mgd} \equiv \frac{1000000 \cdot gal}{day}$

This is a global definition. The unit will now be available anywhere in the worksheet rather than just below and to the right. Use the tilde key (~) to get the ***global definition*** operator. You may also choose the global definition equal sign from the **Operators** control on the **Math** tab.

FIGURE 5.12

Global definition of custom unit

Note: Be sure to assign the **Unit** label to custom unit definitions.

Custom Defined Unit

$cup := 8 \cdot fl\_oz$ $tbsp := \frac{1}{16} \cdot cup$ $tsp := \frac{tbsp}{3}$

$1\ cup = 16\ tbsp$ $1\ cup = 48\ tsp$ $1\ tsp = 4.929\ mL$

$\text{Custom}_1 := 1 \cdot gal$

PTC Mathcad Default Display (SI)

$\text{Custom}_1 = 3.79\ L$

Using Custom Unit Display

$\text{Custom}_1 = 16\ cup$

$\text{Custom}_1 = 256\ tbsp$

$\text{Custom}_1 = 768\ tsp$

FIGURE 5.13

Examples of custom units

Figure 5.13 gives some more examples of custom units. Notice how easily PTC Mathcad deals with the mixing of U.S. and SI units.

Units in Equations

Now that you understand the concept of units, let's explore the use of units in equations and functions. Units are almost always a part of any engineering equation. In order to take advantage of PTC Mathcad's amazing unit system, it is important to always attach units to your variables. Get in the habit of using units even for simple calculations. There are a few cases where this is not possible. These cases will be noted as they occur.

Figure 5.14 shows the formula for kinetic energy. Notice how the dimension of mass, length, and time combine to form a unit of energy (Joules).

Figure 5.15 shows the final velocity of an object based on initial velocity (v0), acceleration (a), and distance (s). For this example the default unit system was changed to U.S.

The formula for kinetic energy is: $\frac{Mass \cdot Velocity^2}{2}$

$$\text{Mass} := 2 \cdot kg \qquad \text{Velocity} := 200 \cdot \frac{cm}{s}$$

$$\text{KineticEnergy}_1 := \frac{\text{Mass} \cdot \text{Velocity}^2}{2}$$

$$\text{KineticEnergy}_1 = 4\ J$$

FIGURE 5.14

Example of units in an equation

Note: For this example, the default unit system was changed to U.S.

The formula for final velocity based on initial velocity (V0), acceleration (a), and distance (s) is: $\sqrt{v_0^{\ 2} + 2 \cdot a \cdot s}$

$$\text{InitialVelocity}_1 := 80 \cdot \frac{ft}{s} \qquad \text{Acceleration}_1 := 10 \cdot \frac{ft}{s^2} \qquad \text{Distance}_1 := 300 \cdot ft$$

$$\text{FinalVelocity}_1 := \sqrt{\text{InitialVelocity}_1^{\ 2} + 2 \cdot \text{Acceleration}_1 \cdot \text{Distance}_1}$$

$\text{FinalVelocity}_1 = 111.355\ \frac{ft}{s}$ PTC Mathcad defaults to ft/s in the USCS unit system.

$\text{FinalVelocity}_1 = 75.924\ mph$ Result after attaching mph to the unit placeholder.

$\text{FinalVelocity}_1 = 33.941\ \frac{m}{s}$ Result after attaching m/s to the unit placeholder.

FIGURE 5.15

Example of units in an equation

Note: For this example, the default unit system was changed to U.S.

The following input quantities are the same as in Figure 5.15. They are just input using different units.

$\text{InitialVelocity}_2 := 54.545 \cdot mph$ $\text{Acceleration}_2 := 3.048 \cdot \frac{m}{sec^2}$

$\text{Distance}_2 := 100 \cdot yd$

$\text{FinalVelocity}_2 := \sqrt{\text{InitialVelocity}_2{}^2 + 2 \cdot \text{Acceleration}_2 \cdot \text{Distance}_2}$

$\text{FinalVelocity}_2 = 111.355\ \frac{ft}{s}$ PTC Mathcad defaults to ft/s in the USCS unit system.

$\text{FinalVelocity}_2 = 75.924\ mph$ Result after attaching mph to the unit placeholder.

$\text{FinalVelocity}_2 = 33.941\ \frac{m}{s}$ Result after attaching m/s to the unit placeholder.

Notice how the input units of initial velocity do not need to be in ft/sec. The input units of acceleration do not need to be in ft/sec2, nor does the input distance need to be in feet. They can be in any units of length and time. PTC Mathcad does all the conversion for you! The result is exactly the same as in Figure 5.15, even though the input units were all different.

FIGURE 5.16

Same example as Figure 5.15, but using mixed units

Let's look at the same equation, but using units from different unit systems. Notice how PTC Mathcad does all the conversions for you. See Figure 5.16.

Using Labels to Distinguish Between Variables and Units

PTC Mathcad Prime uses labels to distinguish between variables with the same name.

If the same equation as used in Figure 5.14 is rewritten with only single letter variables, the variable "m" is the same name as the built-in unit of meter.

The formula for kinetic energy is: $\frac{m \cdot v^2}{2}$

$m := 2 \cdot kg$ $\qquad v := 200 \cdot \frac{cm}{s}$

$KineticEnergy_2 := \frac{m \cdot v^2}{2}$ $\qquad KineticEnergy_2 = 4\ J$

The above equation worked, because the "m" used above has a **Variable** label, which is different than "m" with a **Unit** label.

Because of the above definition, PTC Mathcad now has two definitions of "m" and there will be times when you need to tell PTC Mathcad which one you intend to use.

$mass := 5 \cdot m$ $\qquad mass = 10\ kg$ $\qquad$ PTC Mathcad defaults to using the m with the **Variable** label, so there is no problem with this definition.

If you want to use "m" as a unit of lenth, then you need to manually select the Unit label from the **Labels** control on the **Math** tab.

$Distance1 := 50\ m$ $\qquad Distance1 = 100\ kg$ $\qquad$ PTC Mathcad defaults to using the "m" with the **Variable** label, so this is an issue.

To solve the issue, select the "m" and then select the **Unit** label from the **Labels** control on the **Math** tab.

$Distance2 := 50\ m$ $\qquad Distance2 = 50\ m$ $\qquad$ This is now correct.

FIGURE 5.17

Using labels to distinguish between variables and units

In Mathcad 15, you needed to be careful not to overwrite built-in units. For example, if you used the variable "m" to represent mass, it overwrote the definition of "m" as a unit of length. PTC Mathcad Prime solves this issue by using labels. We touched briefly on this subject when we talked about custom units. Figure 5.17 shows how to use labels to distinguish between variables and units.

Units in User-defined Functions

Using units in user-defined functions is similar to using units in equations, except that you need to include the units in the arguments and not the function. Figure 5.18 uses a function similar to the equation used in Figure 5.15.

Creates a user-defined function based on initial velocity (v0), acceleration (a), and distance (s). Do not include units in the definition.

$$\mathbf{FinalVelocity}\left(v_0, a, s\right) := \sqrt{{v_0}^2 + 2 \cdot a \cdot s}$$

Units must be attached to each argument.

$$\mathbf{FinalVelocity}\left(80 \cdot \frac{ft}{s}, 10 \cdot \frac{ft}{sec^2}, 300 \cdot ft\right) = 111.355 \frac{ft}{s}$$

In the below example, the numbers used for the arguments are the same as above, but no units are attached to the numbers. The numeric result is the same as above, but no units are attached to the result. For engineering calculations you want units attached to all results. Therefore, make sure that units are attached to all the input information.

$$\mathbf{FinalVelocity}\left(80, 10, 300\right) = 111.355$$

Input arguments can be mixed units.

$$\mathbf{FinalVelocity}\left(54.545 \cdot mph, 3.048 \cdot \frac{m}{sec^2}, 100 \cdot yd\right) = 111.355 \frac{ft}{s}$$

If no units are attached to the arguments, and the units do not match, then the numeric result is incorrect. BE SURE TO ATTACH UNITS TO ALL THE INPUT ARGUMENTS.

$$\mathbf{FinalVelocity}\left(54.545, 3.048, 100\right) = 59.873$$

FIGURE 5.18

Units in user-defined functions

Units in Empirical Formulas

Many engineering equations have empirical formulas. There are times when units may not work with the empirical equations. This can occur when the empirical formula raises a number to a power that is not an integer such as $x^{1/2}$ or $x^{2/3}$. If units are attached to x then the units of the result will not be accurate. In order to resolve this problem, first divide the variable by the units expected of the equation, and then multiply the results by the same unit. For example, the shear strength of concrete is based on the square root of the concrete strength in psi (lbf/in^2). See Figures 5.19, 5.20, and 5.21 to see how to resolve the use of units in empirical formulas.

$\phi := 0.85 \qquad f'_c := 4000 \cdot psi$ Input variables: phi and strength of concrete.

$\mathbf{ShearStrength_1} := 2 \cdot \phi \cdot \sqrt{f'_c}$ Empirical formula for shear strength of concrete based on f'c. Result should be in psi.

$\mathbf{ShearStrength_1} = (7.318 \cdot 10^3) \dfrac{lb^{\frac{1}{2}}}{ft^{\frac{1}{2}} \cdot s}$ Incorrect result because of the units under the square root.

The result needs to be in psi. The empirical formula takes the square root of f'c (in psi) and expects the result in psi, but this did not happen in the above equation.

In order to resolve this problem, divide f'c by psi to make it unitless, and then multiply the result by psi.

$\mathbf{ShearStrength_2} := 2 \cdot \phi \cdot \sqrt{\dfrac{f'_c}{psi}} \cdot psi = 107.517\ psi$ Correct result.

FIGURE 5.19

Units in empirical formulas

If an empirical formula expects a number to be in a particular unit (in this case psi), then it is important to divide the number by psi, even if the variable was input in a different unit system. For example, if you are using a U.S. formula, and if f'c were input in MPa, you would still need to divide by psi, not MPa. The result can be displayed as MPa. See below for an example.

Change the input to MPa. (This step isn't really necessary. PTC Mathcad already knows that f'c was 27.579 MPa. It is done only to emphasize that f'c was input as SI.)

$f'_c := 27.579 \cdot MPa$

$f'_c = (4 \cdot 10^3)\ psi$ — Even though f'c was input in metric units, PTC Mathcad knows that it is the same as 4000 psi and the result will be the same.

$ShearStrength_3 := 2 \cdot \phi \cdot \sqrt{\frac{f'_c}{psi}} \cdot psi$ — Since the equation is written for U.S. units, divide by psi, not MPa.

$ShearStrength_3 = 107.517\ psi$ — The result is the same, even though f'c was input in metric units.

$ShearStrength_3 = 0.741\ MPa$ — The result can be displayed as MPa.

$ShearStrength_4 := 2 \cdot \phi \cdot \sqrt{\frac{f'_c}{MPa}} \cdot MPa$ — Result is incorrect if you try to divide by MPa. (Because the formula was written for U.S. units.)

$ShearStrength_4 = (1.295 \cdot 10^3)\ psi$

$ShearStrength_4 = 8.928\ MPa$ — Incorrect result.

FIGURE 5.20

Units in empirical formulas

- Don't stop using units if they appear to not work with your equation.
- Divide the affected variables by the unit in the system expected in the equation.
  - For example, if the equation expects feet, divide by feet.
- You may need to multiply again by the same units at some point in the equation.

Figure 5.22 illustrates another empirical equation using units.

SIUnitsOf(x)

I do not recommend using the ***SIUnitsOf(x)*** function to create a unitless number. The best way to create a unitless number is to divide by the unit that you want the number to represent. For example if you want the number to represent feet, divide by feet.

The same shear strength equation in SI form is: $0.166 \cdot \phi \cdot \sqrt{f'_c}$ where f`c is in MPa.

$f'_c = (4 \cdot 10^3)\ psi$ $\quad f'_c = 27.579\ MPa$

These values were input in the previous example.

$$SI\_ShearStrength := 0.166 \cdot \phi \cdot \sqrt{\frac{f'_c}{MPa}} \cdot MPa$$

This equation was written for SI units; therefore, divide by MPa.

$$SI\_ShearStrength = 107.472\ psi$$

$$SI\_ShearStrength = 0.741\ MPa$$

The results are the same as using the U.S. equation.

When using empirical formulas it is critical to know what units the equation was written for. It is then critical to divide by the units the equation was written for, not what the input units are.

The following illustrates why this is important. Notice how f'c divided by psi = 4000 and f'c divided by MPa = 27.58. These are the numeric numbers expected in the U.S. and SI shear equations respectively.

$$\frac{f'_c}{psi} = 4000 \qquad \frac{f'_c}{MPa} = 27.579$$

FIGURE 5.21

Same example as in Figure 5.20, but in SI form

PTC Mathcad has a function called ***SIUnitsOf(x)***. The PTC Mathcad definition of this function is, "The dimensions of x scaled to the default International System (SI) unit, regardless of your chosen unit system." If x has no units, the function returns 1. This means PTC Mathcad takes the unit dimension of x and returns the SI base unit dimension or derived unit dimension of x. If the unit system in your worksheet is set to SI, then this function is usually clear. If the unit system in your worksheet is set to USCS, remember that the function returns a SI unit dimension, but displays the result in USCS. Let's give an example. If you have chosen SI as your default unit system and you define "Length:=2 m," then ***SIUnitsOf(Length)*** returns 1 m, because meter is the default unit of length in the SI unit system. If you have chosen USCS as your default unit system and you define "length:=2 ft," then ***SIUnitsOf(Length)*** returns 3.281 ft. This is 1 m converted to 3.281 ft. Meter is the default SI base unit for the length unit dimension. The result displays 1 meter in the USCS system. The actual quantity of the dimension returned by the function is the same no matter what default unit system you have chosen; it is only the displayed unit value that changes. See Figure 5.23.

The Hazen-Williams equation for calculating the fluid velocity in a pipe system is $Velocity = 1.318 \cdot C_H \cdot R^{0.63} \cdot S^{0.54}$. Where C_H is a coefficient related to the pipe material, L is length of pipe in feet, R is the hydraulic radius (which for a pipe flowing full is the area in feet divided by the circumference in feet), and S is the slope of the energy line or the hydraulic gradient hf/L.

Calculate the velocity given the following:

24 inch diameter pipe with C_H of 110, and hydraulic gradient: 90 feet over 1000 feet.

$C_H := 110 \qquad D := 24 \cdot in \qquad h_f := 90 \cdot ft \qquad Length := 1000 \cdot ft$

The third term calculates the hydraulic radius (area/circumference), and the fourth term calculates the slope (rise/run).

$$Velocity := 1.318 \cdot C_H \cdot \left(\frac{\frac{\pi \cdot D^2}{4}}{\pi \cdot D} \right)^{0.63} \cdot \left(\frac{h_f}{Length} \right)^{0.54}$$

$$Velocity = 25.524\ ft^{\frac{63}{100}}$$

Result is not accurate because of the empirical formula and units.

Now divide the units out of the equation, and multiply by ft/s.

$$Velocity := 1.318 \cdot C_H \cdot \left(\frac{\frac{\pi \cdot \left(\frac{D}{ft}\right)^2}{4}}{\pi \cdot \frac{D}{ft}} \right)^{0.63} \cdot \left(\frac{h_f}{Length} \right)^{0.54} \cdot \frac{ft}{s}$$

$$Velocity = 25.52\ \frac{ft}{s}$$

Try with no units used.

$$1.318 \cdot 110 \cdot \left(\frac{\frac{\pi \cdot 2^2}{4}}{\pi \cdot 2} \right)^{0.63} \cdot \left(\frac{90}{1000} \right)^{0.54} = 25.524$$

Result is the same.

FIGURE 5.22

Example of another empirical formula

The function *SIUnitsOf(x)* returns the SI units of the variable "x".

When the SI is chosen as the default unit system, then the ***SIUnitsOf(x)*** function always returns the value of 1 times the default base unit. It does not matter what the magnitude of the unit is, the ***SIUnitOf(x)*** function always returns the value of 1 times the default base unit.

$\mathrm{Length}_1 := 5 \cdot cm$

$\underline{\mathrm{SIUnitsOf}}\left(\mathrm{Length}_1\right) = 1\ m$

$\mathrm{Length}_2 := 500 \cdot \mathrm{m}$

$\underline{\mathrm{SIUnitsOf}}\left(\mathrm{Length}_2\right) = 1\ m$

$\mathrm{Torque} := 25 \cdot N \cdot \mathrm{m}$

$\underline{\mathrm{SIUnitsOf}}\left(\mathrm{Torque}\right) = 1\ J$

$\mathrm{Pressure} := 25 \cdot Pa$

$\underline{\mathrm{SIUnitsOf}}\left(\mathrm{Pressure}\right) = 1\ Pa$

$\mathrm{Power} := 4 \cdot \dfrac{N \cdot \mathrm{m}}{s}$

$\underline{\mathrm{SIUnitsOf}}\left(\mathrm{Power}\right) = 1\ W$

The quantity of the result of ***SIUnitOf(x)*** is always the same; however, it may be displayed differently depending on the chosen default unit system. The following values are displayed when the U.S. default unit system is chosen. When converted back to the SI default units, these quantities are exactly the same as shown above.

$\underline{\mathrm{SIUnitsOf}}\left(\mathrm{Length}_1\right) = 3.281\ ft$

$\underline{\mathrm{SIUnitsOf}}\left(\mathrm{Length}_2\right) = 3.281\ ft$

$\underline{\mathrm{SIUnitsOf}}\left(\mathrm{Torque}\right) = 23.73\ \dfrac{1}{s^2} \cdot ft^2 \cdot lbm$

$\underline{\mathrm{SIUnitsOf}}\left(\mathrm{Pressure}\right) = 0.000145\ psi$

$\underline{\mathrm{SIUnitsOf}}\left(\mathrm{Power}\right) = 1\ W$

FIGURE 5.23

SIUnitsOf(x)

Unit Scaling Functions

Converting between some units is not multiplicative. For example, you cannot convert between degrees Fahrenheit and degrees Celsius by multiplying by a single number. The same is true about converting radians to degrees-minutes-seconds. For these types of conversions, PTC Mathcad uses unit scaling functions. They technically are not units, but they appear at the end of a number as a unit. PTC Mathcad must use a function to display the result. The following are unit scaling functions: °***F***, °***C***, ***DMS*** (Degrees, Minutes, Seconds), ***hhmmss*** (Hours, Minutes, Seconds), and ***FIF*** (Feet, Inch, Fractions).

Each of the above functions are found in both the **All Functions** list (**Function> Functions>All Functions>Miscellaneous**) and the **Units** list (**Math>Units>Units>Angle, or Length, or Temperature, or Time**). To use these as units at the end of a value, they must be inserted from the **Units** list and not the **All Functions** list. Remember that if you type these units rather than inserting them, you must use the **Unit** label.

Using unit scaling functions illustrates an interesting way to use many of the functions of PTC Mathcad. Rather than using the form of ***function***(argument) (such as log(150) = 2.18), you can write the function in the form of argument ***function*** where the argument is listed first without the parentheses (such as 150log = 2.18). Try it.

Let's look at a few different unit scaling functions.

Fahrenheit and Celsius

As discussed above, the °***F*** and °***C*** units are actually functions, but PTC Mathcad treats them as if they are units, and you can use them as if they are units. If you hold your cursor over one of these unit functions when inserting them from the **Units** list on the **Math** tab, PTC Mathcad informs you that these units are actually functions. See Figure 5.24.

How do the unit scaling functions work? Remember that to evaluate a function you type the function name and then list its arguments in parentheses. If you type °***F***(32) = you get 273.15 K. Use the Symbols section of the Ribbon Bar to get the degree character: (**Math>Symbols>Math Symbols**). If you delete the K and insert °***C***, (**Math>Units>Units>Temperature**), then you get the display °***F***(32)=0.00°***C***. Figure 5.25 shows the use of °***F*** and °***C*** as functions. Figure 5.26 shows the use of °***F*** and °***C*** as units.

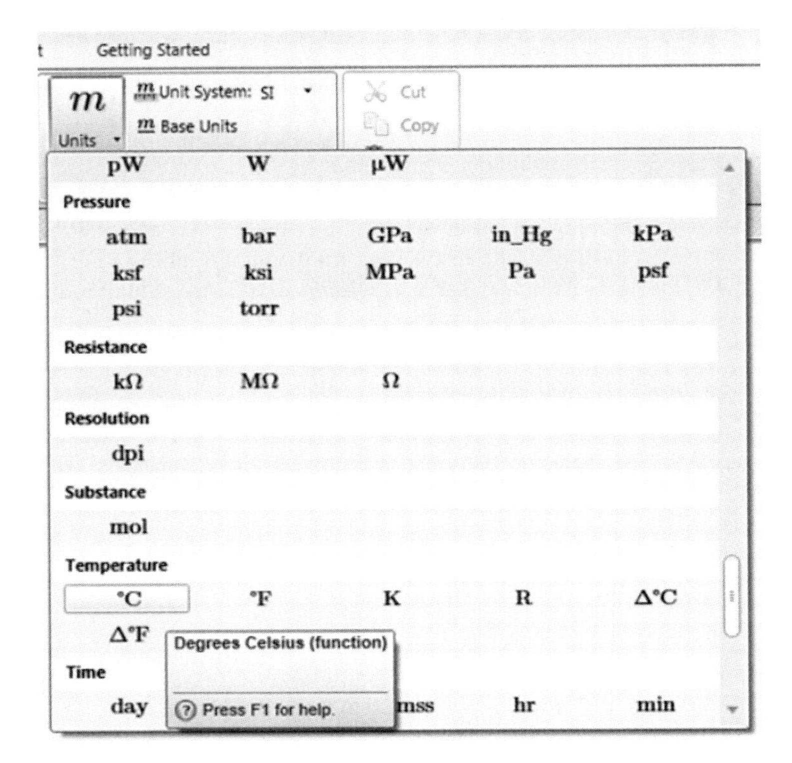

FIGURE 5.24

Degrees Celsius is actually a function

Change in Temperature

You may add or subtract Kelvin and Rankine temperatures as you would any units. For example 400K-300K=100K, or 200R+50R=250R.

When adding or subtracting Fahrenheit and Celsius temperatures, you must remember that PTC Mathcad converts to Kelvin before doing any addition or subtraction. Thus, 212°F-200°F is the same thing as taking 373.15K-366.48K=6.67K (−447.67°F) or 671.67R-659.67R=12R (-447.67°F). The answer you wish to obtain is 12°F. PTC Mathcad has solved this issue by creating a unit called Δ°F.

The following functions were inserted from the **All Functions** list (Miscellaneous category).

$°F(32) = 273.15\ K$	$°C(100) = 373.15\ K$
$°F(32) = 491.67\ R$	$°C(100) = 671.67\ R$
$°F(32) = 0\ °C$	$°C(100) = 100\ °C$
$°F(32) = 32\ °F$	$°C(100) = 212\ °F$
$250 \cdot K = 250\ K$	$300 \cdot R = 166.667\ K$
$250 \cdot K = 450\ R$	$300 \cdot R = 300\ R$
$250 \cdot K = -23.15\ °C$	$300 \cdot R = -106.483\ °C$
$250 \cdot K = -9.67\ °F$	$300 \cdot R = -159.67\ °F$

FIGURE 5.25

Fahrenheit and Celsius as functions

The following were inserted as units from the **Units** list.

$32\ °F = 273.15\ K$	$100\ °C = 373.15\ K$
$32\ °F = 491.67\ R$	$100\ °C = 671.67\ R$
$32\ °F = 0\ °C$	$100\ °C = 100\ °C$
$32\ °F = 32\ °F$	$100\ °C = 212\ °F$

FIGURE 5.26

Fahrenheit and Celsius as units

You may subtract Kelvin and Rankine temperatures as you would any units.The result is an absolute temperature. When subtracting Fahrenheit or Celsius temperatures you usually want to find the difference between the temperatures.

When you calculate the change in temperature you must display the result in terms of the change in degrees $\Delta°F$ and not the actual temperature. The $\Delta°F$ and $\Delta°C$ are available in the **Units** list on the **Math** tab.

$$\textbf{TempChange}_1 := °F(212) - °F(200)$$

$$\textbf{TempChange}_1 = 12\ \Delta°F$$

$$\textbf{TempChange}_1 = 6.667\ \Delta°C$$

If you do not display the result in $\Delta°F$ or $\Delta°C$, then PTC Mathcad will display the temperature difference as the difference between Kelvin or Rankine temperature. See below.

$°F(212) = 373.15\ K$ $\qquad$ $°F(212) = 671.67\ R$

$°F(200) = 366.483\ K$ $\qquad$ $°F(200) = 659.67\ R$

$\textbf{TempChange}_1 = 6.667\ K$ Default display

Calculated as 373.15K - 366.483K = 6.667 K = -447.67 $°F$.

$6.667\ K = -447.669\ °F$ $\qquad$ $6.667\ K = -266.483\ °C$

$$\textbf{TempChange}_1 = -447.67\ °F$$

$$\textbf{TempChange}_1 = -266.483\ °C$$

$\textbf{TempChange}_1 = 12\ R$ Calculated as 671.67R - 659.67R = 12R = -447.67 $°F$.

$12 \cdot R = -447.67\ °F$ $\qquad$ $12 \cdot R = -266.483\ °C$

FIGURE 5.27

Change in temperature

This simply means the change in temperature in degree Fahrenheit. Therefore, if you do a Fahrenheit subtraction you must display the result in Δ°F, not °F. If you do a Celsius subtraction, you must display the result in Δ°C, not °C. The Δ°F and Δ°C are available from the **Units** list on the **Math** tab (**Math>Units>Units>Temperature**). See Figure 5.27.

You may add Kelvin and Rankine temperatures just as you would any unit.

$200 \cdot K + 50 \cdot K = 250\ K$

$200 \cdot R + 50 \cdot R = 138.889\ K$

$200 \cdot R + 50 \cdot R = 250\ R$

If you are adding Fahrenheit or Celsius temperatures, you must use $\Delta°F$ and $\Delta°C$.

$\text{TempChange}_2 := 50\ °F + 30\ °F$

$\text{TempChange}_2 = 555.189\ K$

$\text{TempChange}_2 = 539.67\ °F$

Adding $50\ °F$ and $30\ °F$ is the same as adding 283.15K and 272.04K, which is equal to 555.19K or $539.67\ °F$.

$50\ °F = 283.15\ K$

$30\ °F = 272.039\ K$

$283.15 \cdot K + 272.039 \cdot K = 555.189\ K$

$555.189\ K = 539.67\ °F$

$\text{TempChange}_3 := °F(50) + 30 \cdot \Delta°F$

$\text{TempChange}_3 = 80\ °F$

$\text{TempChange}_3 = 299.817\ K$

$80\ °F = 299.817\ K$

To be accurate, you must take $50\ °F$ and add a change of $30\ °F$.
This is the same thing as taking $50\ °F$ and adding 30R or 30*(5/9)K.
In other words $1\ \Delta°F$ =1R and $1\ \Delta°C$ =1K.

$$283.15 + 30 \cdot \frac{5}{9} = 299.817$$

FIGURE 5.28

Adding temperatures

When adding Fahrenheit and Celsius temperatures, you actually want to add a change in temperature, not add an absolute Kelvin or Rankine temperature. This is illustrated in Figure 5.28.

Figure 5.29 gives an example of using Fahrenheit in engineering calculations.

Degrees Minutes Seconds (DMS)

Another unit scaling function is the ***DMS*** function. This function converts degrees, minutes and seconds to decimal degrees. It will also display radians or decimal degrees as degrees, minutes and seconds. For the ***DMS*** function, you need a 3 row, 1 column vector with degrees in the top element, minutes in the middle element, and seconds in the bottom element.

Calculate the temperature of mixed air from two sources.

$\mathrm{Temp_1} := 85\ °F$

$\mathrm{Temp_2} := 20\ °F$

$\mathrm{CFM_1} := 200 \cdot cfm$

$\mathrm{CFM_2} := 50 \cdot cfm$

$$\mathrm{Temp_{Final}} := \frac{\mathrm{Temp_1} \cdot \mathrm{CFM_1} + \mathrm{Temp_2} \cdot \mathrm{CFM_2}}{\mathrm{CFM_1} + \mathrm{CFM_2}} \qquad \mathrm{Temp_{Final}} = 295.372\ K$$

$\mathrm{Temp_{Final}} = 72\ °F$

FIGURE 5.29

Example of using Fahrenheit in engineering calculations

The ***DMS*** function may also use the form ***DMS***(Deg,Min,Sec). I find this much easier to input than using the column vector. There is no way to have PTC Mathcad display the results in this format.

The format and use of the ***DMS*** function is illustrated in Figure 5.30.

Hours Minutes Seconds (*hhmmss*)

The ***hhmmss*** function converts hours, minutes and seconds into decimal time—either seconds, minutes, hours, days, etc. It can also be used in the units placeholder to display decimal units of time into hours, minutes and seconds. The display of ***hhmmss*** differs from the ***DMS*** function. Instead of using a vector, the ***hhmmss*** function uses a text string with the hours, minutes and seconds separated by colons. Thus 2 hours, 32 minutes, and 14 seconds would be: "2:32:14." The input can use a text string within parentheses with the form ***hhmmss***("hh:mm:ss.sss") or it can use values separated by commas in the form ***hhmmss***(hh,mm,ss.sss). Figure 5.31 gives some examples.

Using *DMS* as a function

$DMS(\blacksquare)$ Insert the DMS function from the **All Functions** list in the **Miscellaneous** category.

$DMS([\blacksquare])$ Click on the placeholder and type CTRL+m to insert the matrix operators.

$$DMS\left(\begin{bmatrix}29\\30\\15\end{bmatrix}\right)$$

Add degrees to the placeholder and type TAB.
Add minutes to the 2nd placeholder and type TAB.
Add seconds to the bottom placeholder.

$$DMS\left(\begin{bmatrix}29\\30\\15\end{bmatrix}\right)=0.515$$

The result default display is in radians.

$$DMS\left(\begin{bmatrix}29\\30\\15\end{bmatrix}\right)=29.50416667\ \mathbf{deg}$$

Click on the units placeholder and type "deg" to get decimal degrees.

Using *DMS* as a unit.

$$45\cdot \mathbf{deg}=0.785$$

The default display is in radians.

$$45\ \mathbf{deg}=\begin{bmatrix}45\\0\\0\end{bmatrix}\ \mathbf{DMS}$$

Type DMS in the units placeholder to get degrees, minutes, and seconds. The top placeholder is degrees, the middle placeholder is minutes, and the bottom placeholder is seconds.

$$29.504167\cdot \mathbf{deg}=\begin{bmatrix}29\\30\\15\end{bmatrix}\ \mathbf{DMS}$$

The format for the argument of **DMS** can be a column vector (as shown above) or it can be values separated by commas (as shown below).

$$DMS(29,30,15)=0.515 \qquad DMS(29,30,15)=\begin{bmatrix}29\\30\\15\end{bmatrix}\ \mathbf{DMS}$$

FIGURE 5.30

Degrees minutes seconds *(DMS)*

Using *hhmmss* as a function.

$\mathbf{Time_1} := \boldsymbol{hhmmss}(\text{“12:32:15”})$

The hours, minutes, and seconds must be input between quote marks " and must be separated by colons.

$\mathbf{Time_1} = 45135.00\ \boldsymbol{s}$

$\mathbf{Time_1} = 752.25\ \boldsymbol{min}$

$\mathbf{Time_1} = 12.538\ \boldsymbol{hr}$

$\mathbf{Time_1} = 0.522\ \boldsymbol{day}$

$\mathbf{Time_1} = \text{“12:32:15”}\ \boldsymbol{hhmmss}$ Delete the unit and type hhmmss.

Using *hhmmss* as a unit.

$\mathbf{Time_2} := 40000 \cdot \boldsymbol{s}$

$\mathbf{Time_2} = \text{“11:6:40”}\ \boldsymbol{hhmmss}$ Delete the unit and type hhmmss.

$54.25 \cdot \boldsymbol{min} = \text{“0:54:15”}\ \boldsymbol{hhmmss}$

$4.7525 \cdot \boldsymbol{day} = \text{“114:3:36”}\ \boldsymbol{hhmmss}$

$0.125 \cdot \boldsymbol{yr} = \text{“1095:43:35.747”}\ \boldsymbol{hhmmss}$

The format for the argument of **hhmmss** can be a string separated by colons (as shown above) or it can be values separated by commas (as shown below).

$\boldsymbol{hhmmss}(12, 32, 15) = 45135.00\ \boldsymbol{s}$

$\boldsymbol{hhmmss}(12, 32, 15) = \text{“12:32:15”}\ \boldsymbol{hhmmss}$

FIGURE 5.31

Hours minutes seconds (*hhmmss*)

The ***hhmmss*** function may also use the form ***hhmmss***(hh,mm,ss.sss). I find this much easier to input than using the text string separated by colons. There is no way to have PTC Mathcad display the results in this format.

$\text{FIF\_Length} := FIF(\text{“1' 2–1/2''”})$

$\text{FIF\_Length} = 0.368\ m$

$\text{FIF\_Length} = 14.5\ in$

$\text{FIF\_Length} = 1.208\ ft$

The feet, inches, and fractions must be input between quote marks. The format is: FIF, left parenthesis, double quote, feet, single quote, Spacebar, inches, dash, numerator/denominator, double quote, double quote, right parenthesis. FIF("feet' inches-numerator/denominator"").

Using *FIF* in the units placeholder.

$18.5 \cdot ft = \text{“18' 6''”}\ FIF$ Delete the unit and type FIF.

$25.5 \cdot in = \text{“2' 1–1/2''”}\ FIF$

$157.75 \cdot in = \text{“13' 1–3/4''”}\ FIF$

$2.575 \cdot ft = \text{“2' 6–9/10''”}\ FIF$

$1 \cdot \mathrm{m} = \text{“3' 3–47/127''”}\ FIF$

The format for the argument of **FIF** can be a string (as shown above) or it can be values separated by commas (as shown below). **FIF**(feet,inch,fraction).

$FIF\left(2, 2, \frac{1}{2}\right) = 0.673\ m$ $FIF\left(2, 2, \frac{1}{2}\right) = 26.5\ in$

$FIF\left(2, 2, \frac{1}{2}\right) = \text{“2' 2–1/2''”}\ FIF$

FIGURE 5.32

Feet inch fraction (*FIF*)

Feet Inch Fraction (*FIF*)

The ***FIF*** function is similar to the ***hhmmss*** function. It uses a string as the argument for the function, and when used as a unit, it also displays a string. The format uses a string with a single quote for feet and double quote for inches and fractions of inches. A dimension of 2 feet $3^1/_2$ inches would be input as a text string like this: ***FIF*** ("2' 3-1/2"). Notice the (2) double quotes at the end. One to indicate inches, the other to close the text string. The keystrokes are: **FIF, left parenthesis, double quote, 2, single quote, Spacebar, 3, dash, 1/2, double quote, Right Arrow, EQUAL.** Note that the right hand double quote and right hand parenthesis were automatically entered. The Right Arrow is required to move outside of the string. FIF ("feet'inches-numerator/denominator"). The spacebar after the single quote and the dash after the inches are optional. See Figure 5.32.

The ***FIF*** function may also use the form FIF(feet,inches,fraction). I find this much easier to input than the text string. There is no way to have PTC Mathcad display the results in this format.

Money

PTC Mathcad allows you to use money as a unit. PTC Mathcad uses a base currency symbol ¤, which you can insert from the **Units** list control on the **Math** tab (**Math>Units>Units>Money>¤**).

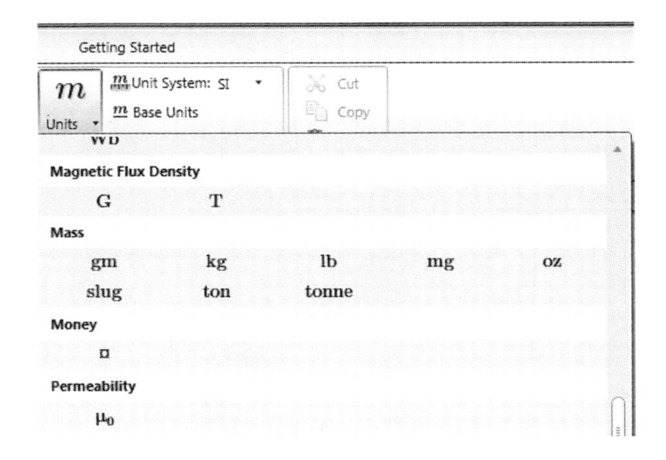

You can use this currency symbol as is to represent a generic currency, but a better way is to assign a currency to be equal to the ¤ symbol. For example, you can assign $:=¤. You can then create relationships between $ and other currencies. Be sure to assign the **Unit** label to each currency definition. See Figure 5.33.

Dimensionless Units

There may be times when you want to attach a unit to a number that is not one of the eight basic dimensions of length, mass, time, temperature, luminous intensity, substance, current or charge, and money. In order to do this, type the name of your unitless dimension and define it as the number 1. This means that you can attach the unit to a number and not affect its value. Once you have defined a dimensionless unit, you can create other units that have relationships with this unit. Be sure to assign the **Unit** label to each unitless dimension you create. Figure 5.34 gives two examples of dimensionless units. You can create dimensionless units for just about anything.

Use the base currency unit.

$\text{ProductCost} := \frac{20\ ¤}{kg}$ $\qquad$ $\text{Quantity} := 10\ kg$

$\text{TotalCost} := \text{ProductCost} \cdot \text{Quantity} = 200\ ¤$

$\text{TotalCost} = ?\ \$$ $\qquad$ PTC Mathcad does not understand $.

Now define a relationship of $ to the base currency unit.

$\$:= ¤$

Set dollars to be equal to the base currency unit.
Be sure to set the label on **$** to **Units**.

$\text{TotalCost} = 200\ ¤$ $\qquad$ The default display is the base currency unit.

$\text{TotalCost} = 200\ \$$ $\qquad$ Delete the base currency unit and type $.

Create a relationship between **$** and **€** and *£*.

$€ := 1.31\ \$$ $\qquad$ The Euro and GBP are found at **Math>Operators and Symbols>Symbols>Monetary Symbols>** **€** or *£*.

$£ := 1.53\ \$$ $\qquad$ Be sure to set the label on **€** and *£* to **Units**.

$\text{TotalCost} = 200\ ¤$

$\text{TotalCost} = 200\ \$$

$\text{TotalCost} = 152.672\ €$

$\text{TotalCost} = 130.719\ £$

$\text{ProductB\_Cost} := 30\ \frac{\$}{gal}$ $\qquad$ $\text{Volume} := 10\ gal$

$\text{Total} := \text{ProductB\_Cost} \cdot \text{Volume} = 300\ ¤$

$\text{Total} = 300\ \$$ $\qquad$ $\text{Total} = 229.008\ €$ $\qquad$ $\text{Total} = 196.078\ £$

FIGURE 5.33

Money

Here are a few things to consider when using dimensionless units.

- PTC Mathcad will not automatically attach these types of units.
- You must attach them yourself.
- PTC Mathcad does not do a consistency check and warn you if you have attached inconsistent units.
- Inconsistent units can create wrong results. See Figure 5.34.

$Bottles := 1$

Be sure to set the label to **Units**.

$Case := 24 \cdot Bottles$

$Pallet := 16\ Case$

How many bottles are on 12 pallets? How many cases?

$12\ Pallet = 4608$

$12\ Pallet = 4608\ Bottles$

$12\ Pallet = 192\ Case$

Nonsense Example

$Widget := 1$ $\quad \text{Product}_1 := 4 \cdot Widget$

$Gadget := 2 \cdot Widget$ $\quad \text{Product}_2 := 4 \cdot Gadget$

$Watzut := 3 \cdot Gadget$ $\quad \text{Product}_3 := 4 \cdot Watzut$

$\text{Product}_1 = 4$ $\quad \text{Product}_2 = 8$ $\quad \text{Product}_3 = 24$

$\text{Product}_1 = 4\ Widget$ $\quad \text{Product}_2 = 8\ Widget$ $\quad \text{Product}_3 = 24\ Widget$

$\text{Product}_1 = 2\ Gadget$ $\quad \text{Product}_2 = 4\ Gadget$ $\quad \text{Product}_3 = 12\ Gadget$

$\text{Product}_1 = 0.667\ Watzut$ $\quad \text{Product}_2 = 1.333\ Watzut$ $\quad \text{Product}_3 = 4\ Watzut$

Be careful to not attach the wrong unit.

$\text{Product}_1 = 4\ Bottles$

$\text{Product}_1 = 3.053\ \frac{1}{¤} \cdot €$

PTC Mathcad will not warn you if you use inconsistent units. The unit **Bottles** has a value of 1. Attaching **Bottles** to "Product1" did not change its value, but an incorrect unit was attached and PTC Mathcad did not warn of an incorrect unit.

FIGURE 5.34

Dimensionless units

Using the Unit Placeholder for Scaling

In addition to being used to display units, the unit placeholder can be used for scaling. Let's say that you want to calculate the value of sin(x) at 8 points around a circle. If you create a range variable, it can display as decimal radians or as degrees. Is there a way to display the values in terms of π? Yes! You can put π in the unit placeholder (**p+CTRL g**) and the values will then be scaled to π. Note that when π is inserted, it will have the **Units** label assigned to it. You must select π and then change the label to **Constant** in order for it to work. See Figure 5.35.

Calculate the value of sin(x) at 8 points around a circle.

$\text{Values} := 0, \frac{\pi}{4} .. 2\pi$ Create a range variable.

$$\text{Values} = \begin{bmatrix} 0 \\ 0.785 \\ 1.571 \\ 2.356 \\ 3.142 \\ 3.927 \\ 4.712 \\ 5.498 \\ 6.283 \end{bmatrix} \qquad \underline{\sin}(\text{Values}) = \begin{bmatrix} 0 \\ 0.707 \\ 1 \\ 0.707 \\ 0 \\ -0.707 \\ -1 \\ -0.707 \\ 0 \end{bmatrix}$$

$$\text{Values} = \begin{bmatrix} 0 \\ 45 \\ 90 \\ 135 \\ 180 \\ 225 \\ 270 \\ 315 \\ 360 \end{bmatrix} deg$$

You can display values as degrees by typing deg in the unit placeholer.

Use π in the unit placeholder so the values will be scaled to π.

$\text{Values} = ?\ \pi$

$\text{Values} = ?\ \pi$

This variable is undefined. Check that the label is set correctly.

If you type π in the units placeholder PTC Mathcad assumes that it is a unit and assigns the **Unit** lablel to π.

You need to select π and then select the **Constant** label for π.

$$\text{Values} = \begin{bmatrix} 0 \\ 0.25 \\ 0.5 \\ 0.75 \\ 1 \\ 1.25 \\ 1.5 \\ 1.75 \\ 2 \end{bmatrix} \pi$$

Now PTC Mathcad displays the values scaled to π.

FIGURE 5.35

Using the unit placeholder for scaling

Summary

Units are essential when using PTC Mathcad for technical calculations. If you do not use them, you will be missing out on one of the most useful features of PTC Mathcad.

In Chapter 5 we:

- Showed how to attach units to numbers by multiplying the number by the unit.
- Explained that PTC Mathcad will keep track of all units in variables with similar unit dimensions no matter what units are displayed.
- Illustrated that units in equations do not need to match as long as the unit dimensions are the same. PTC Mathcad does the conversion for you!
- Discussed how results can be displayed in any unit system.
- Showed that if PTC Mathcad does not have the unit you want, you can define it yourself.
- Emphasized the importance of understanding the **Unit** label.
- Encouraged the use of "lbm" instead of "lb" to avoid confusion of mass and force in the U.S. unit system.
- Emphasized that in user-defined functions the units are attached to the arguments and not in the definition of the function.
- Learned that when using empirical formulas, divide the numbers (which the formula expects to be unitless) by the units expected in the equation, then multiply by the expected units.
- Suggested to avoid using the function ***SIUnitOf(x)*** to create a unitless number.
- Explained the use of custom scaling functions such as °F, °C, ***DMS***, ***hhmmss***, and ***FIF***.
- Introduced the concept of dimensionless units.
- Showed how to use the unit placeholder as a scaling factor.

Practice

The PTC Mathcad Prime 3.0 figures and examples used in this book are available for download from the book's website. The reader is encouraged to download the files and use them to practice the concepts learned. Additional examples and problems are also provided. To access this content go to http://store.elsevier.com/9780124104105 *and click on the Resources tab, and then click on the link for the Online Companion Materials.*

1. On a PTC Mathcad worksheet type `Unit_1 : 10*m`. Now display this variable with the following units: Meters, centimeters, millimeters, kilometers, feet, inches, yards, miles, furlongs, and Angstrom.
2. In order to get a feel for the PTC Mathcad default unit system, open three different PTC Mathcad worksheets and set each worksheet to a different default

unit system (SI, CGS, and USCS). Type the following in each of the three different default unit systems.

a. g = (Gravity)
b. 25ft^2 = (Area)
c. 25F = (Capacitance)
d. 25C = (Charge)
e. 25A = (Current)
f. 25IBTU = (Energy)
g. 25N*m = (Energy)
h. 25ft^3[Spacebar Spacebar]/min = (Flowrate)
i. 25lbf = (Force)
j. 25N/m = (Force per length)
k. 25km = (Length)
l. 25cd = (Luminous intensity)
m. 25 G = (Magnetic flux density)
n. 25kg = (Mass)
o. 251bf/ft^3 = (Force density)
p. 25V = (Potential)
q. 25kW = (Power)
r. 25N/m^2 = (Pressure)
s. 25 K = (Temperature)
t. 25m/s = (Velocity)
u. 25ft^3 = (Volume)

3. Go through each of the displayed units from the previous exercise and change the displayed units to a different unit (if it is available).

4. From your field of study (or another field), try to find five (or more) units (or combination of units) that are not defined by PTC Mathcad. Create custom units for these units. Use the global definition. (For example, `cfs[CTRL+SHIFT+~]1*ft^3[Spacebar Spacebar]/s`.) Assign the "Unit" label to the definition.

5. Go back to the user-defined functions you created in the Chapter 4 practice exercises. Attach units to the input arguments. Make sure that the result is in the unit dimension you expect to see. Then change the displayed units to other units of your choice.

6. Create four dimensionless units that are all related, and perform arithmetic operations using them.

7. John did not have a tape measure to measure a room, so he used his shoes. Using his shoes end to end, the room measured 20 shoes wide by 25.5 shoes long. When he returned, he measured his shoes. They measured 280 mm long. Create a new unit "JohnShoes." Define it as 280mm. Define the width and length of the room in units of "JohnShoes." Display the width and length in meters and feet. Calculate the area of the room and display it in JohnShoes^2.

PART

Hand Tools for Your PTC Mathcad Toolbox

You have now built your PTC Mathcad toolbox, and it is time to start filling it. Part II will introduce some simple PTC Mathcad tools. The goal of Part II is to get you comfortable with PTC Mathcad, and you will soon see that it is an easy-to-learn, yet powerful resource.

The chapters in this part will focus on essential features such as vectors, matrices, simple PTC Mathcad functions, plotting, and simple logic programming. These are important to understand before the more complex and powerful tools are discussed in Part III.

CHAPTER

Arrays, Vectors, and Matrices 6

An understanding of vectors and matrices can make engineering calculations much more effective. PTC Mathcad performs many complex vector and matrix operations. This chapter will not delve into all of these operations. The purpose of this chapter is to illustrate the benefits that can be had by using simple vectors and matrices in technical calculations.

Chapter 6 will:

- Review how to define arrays.
- Review the ORIGIN variable and recommend changing its default value.
- Review the concept of array subscripts.
- Review the concept of range variables and tell how to define them and use them.
- Compare the difference between range variables and vectors.
- Tell how to convert a range variable into a vector.
- Show how to display vector and matrix output.
- Illustrate how to attach units to vectors and matrices.
- Describe how to use simple math operators with vectors and matrices.
- Provide several examples of how vectors and matrices can be used in technical calculations.
- Demonstrate how to use vectors to evaluate the same equation or function for various input values.

Review of Chapter 1

Chapter 1 introduced and defined arrays. It also described how to create and modify arrays. Let's review a few of the important concepts covered in Chapter 1.

A column vector is a single-column array. A row vector is a single-row array. A matrix is an array with multiple rows and columns.

Use the **Matrices/Tables** tab to insert a vector or matrix (**Matrices/Tables>Matrices and Tables>Insert Matrix**). This will allow you to use your mouse to visually select the size of the inserted array. You may also type **CTRL + M** to insert a 1x1 matrix, then type **SHIFT + Spacebar** to add a new column, or use the **Tab** key to insert a new row.

The easiest way to add or delete rows and columns is from the **Matrices/Tables** tab (**Matrices/Tables>Rows and Columns>Insert Above or Insert Below**).

Essential PTC® Mathcad Prime® 3.0.

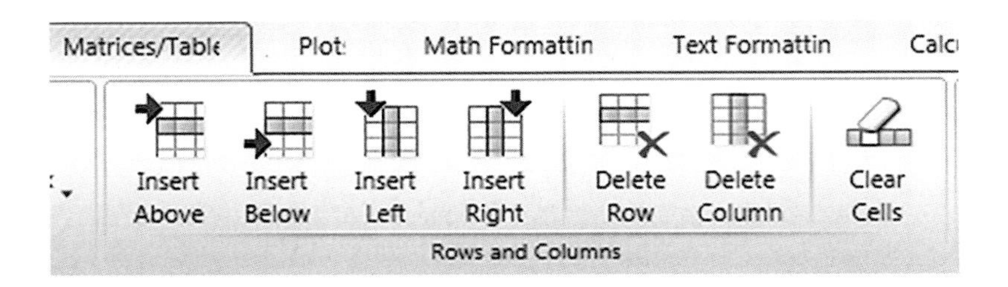

The keyboard shortcut to add a row is **SHIFT + ENTER**. The row will be added below the current row unless the cursor is located at the leftmost entry point of the leftmost element in a row, in which case the row will be added above the current row. The keyboard shortcut to add a column is **SHIFT + Spacebar**. If the cursor is located on the left side of an element, then the column will be added to the left of the element. If the cursor is located on the right side of an element, then the column will be added to the right of the element.

The built-in variable ORIGIN tells PTC Mathcad the starting index of your array. This book recommends changing the value of ORIGIN to 1, and examples are given with ORIGIN set at 1. The location to change ORIGIN from the default value of 0 to 1 is in the **Worksheet Settings** group on the **Calculation** tab (**Calculation>Worksheet Settings>ORIGIN>1**).

Chapter 1 distinguished between literal subscripts and array subscripts. Array subscripts are not a part of the variable name (as are literal subscripts). Array subscripts are a means of displaying or defining the value of a particular element in an array. The array subscript is created by using the **[** key. When you create or click an array subscript, you see a blue open "⊏" icon, which represents the matrix index operator.

Tables

> Tables are new to PTC® Mathcad Prime®. They are used for inputting vertical data sets. When you use an input table, you are essentially creating column vectors.

We have discussed different methods of creating vectors and matrices. This section will introduce tables as a new way of inputting information into an array.

A table is a region where you define vertical data sets. It could more appropriately be named an "input" table. It is comprised of columns and rows like in an array, but it also has some unique features. Each column in a table comprises one named data set. A table is essentially creating a series of column vectors. The top row in a table is where you define the variable name for each data set (column vector). The second row is where you define the units for the data set. The third row of a table is where you begin adding data. It becomes the first row of your column vector.

To insert a table, use the **Insert Table** control in the **Matrices and Tables** group of the **Matrices/Tables** tab (**Matrices/Tables>Matrices and Tables>Insert Table**). The keyboard shortcut is **CTRL + 6**.

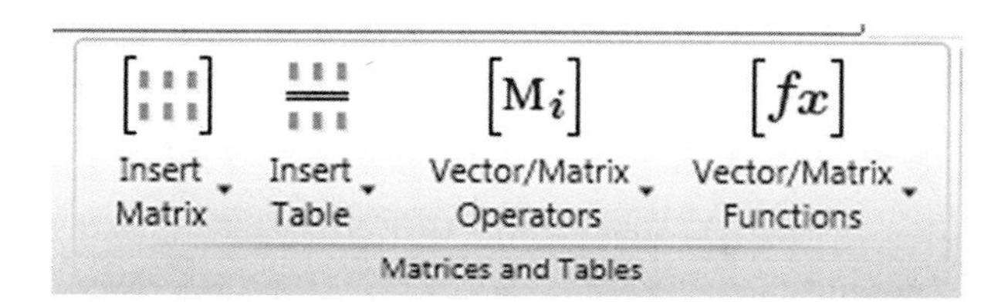

The **Insert Table** control works just like the **Insert Matrix** control. You can use your mouse to select the size of the table you want to insert. The number of rows you select does not include the top two rows of the table. So if you select 5 rows, then the table will have 5 data rows in addition of the name definition row, and the unit row. You input information into a table the same as you do for an array. Once you reach the end of the rows, you can use the tab key to add additional rows.

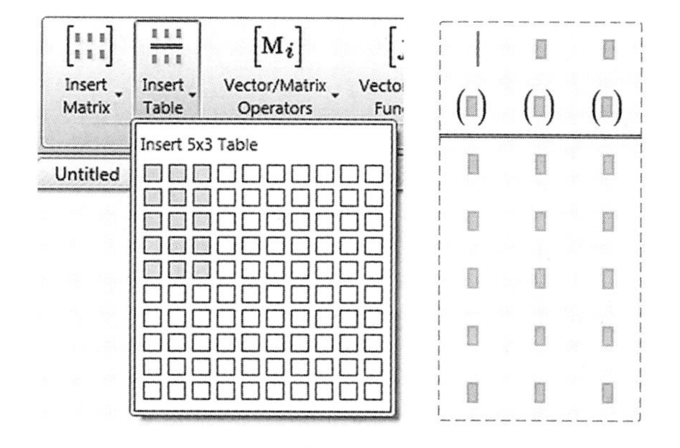

Figure 6.1 provides an example of inputting data using a table, and then plotting the data as points on a plot.

The following table has three data sets. The first data set does not have any units attached. The second column has units of meters, and the third column has units of Newtons. The third column has one more row of data than the second column.

Element	Length (m)	Force (N)
1	.5	2
2	1	2
3	1.5	2
4	2	3
5		3

$$\text{Element} = \begin{bmatrix} 1.000 \\ 2.000 \\ 3.000 \\ 4.000 \\ 5.000 \end{bmatrix} \quad \text{Force} = \begin{bmatrix} 2.000 \\ 2.000 \\ 2.000 \\ 3.000 \\ 3.000 \end{bmatrix} N$$

$$\text{Length} = \begin{bmatrix} 0.500 \\ 1.000 \\ 1.500 \\ 2.000 \end{bmatrix} m$$

After data has been input into a table, each data set is now an individual column vector. You do not refer to a table as a single variable.

You can plot the data as points on a plot. (Chapter 7 will discuss plotting in more detail.)

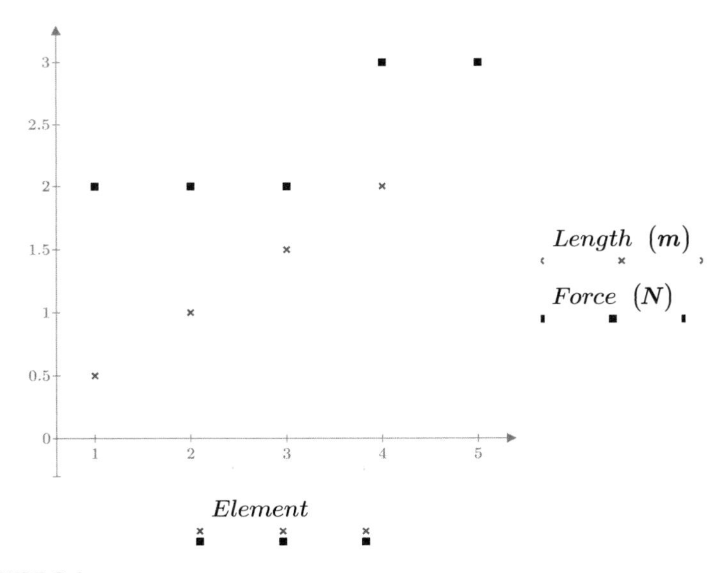

FIGURE 6.1

Using a table to input data

Range Variables

Range variables were introduced in Chapter 1. Remember that range variables have a beginning value, a second value that establishes an incremental value, and an ending value. To define a range variable, type the variable name followed by the colon `:`. In the placeholder, type the beginning value followed by a comma. This inserts two new placeholders, one following the comma and the other following two periods. In the first placeholder type the second value of the series. This establishes the incremental value. In the second placeholder enter the ending value. You can get to the second placeholder by using the right arrow key, but it is easier to just type two periods. If the incremental value is one, you can eliminate the second value by typing the beginning value, then typing two periods and the ending value. To get the following, type `a : 1,3..21`.

$$a := 1,3..21$$

Range Variables vs. Vectors

You can only view and cannot access individual elements of a range variable. This differs from a vector, where each element is accessible. Because range variables and vectors are similar, it is important to understand the difference between them. Table 6.1 is a comparison between range variables and vectors:

Table 6.1 Comparing Range Variables and Vectors

Range Variables	Vectors
Range variables must increment (up or down) in uniform steps.	Vectors may have numbers in any order.
Range variables must be real.	Vectors may use real or complex numbers.
You cannot access individual elements of range variables.	Each element of a vector can be accessed by using array subscripts.
When using range variables in calculations, the results are displayed, but the individual results are not accessible.	When using vectors in calculations, the results are also displayed, but each individual result is accessible.
Range variables can be used to iterate calculations over a range of values. The calculation is performed once for each value in the range.	Vectors can also be used as arguments for calculations. The calculation is performed once for each value in the vector.
Range variables are often used as subscripts to write or access data in vectors and matrices.	Range variables (starting at ORIGIN and incrementing by 1) can be used to create a vector of values.
Range variables begin at the defined beginning value.	Vectors use ORIGIN as the first element.

Range variables are best used to increment expressions, iterate calculations, and to set plotting limits. When you use range variables to iterate a calculation, it is important to understand that PTC Mathcad begins at the beginning value and iterates every value in the range. You cannot tell PTC Mathcad to only use part of the range variable. If you use a range variable as an argument for a function, the result is another range variable, which means that the result is displayed, but you cannot access individual elements of the result. You cannot assign the result to a variable. Even though it is possible to use a range variable as an argument for a function, it is best to use a vector, so that each element of the result can be assigned and accessed. Figures 6.2 and 6.3 illustrate this. In the next section we show how to convert a range variable to a vector, so that it can be used as an argument for a function.

Converting a Range Variable to a Vector

I have been using Mathcad for almost 20 years. I just recently discovered this tip. To convert a range variable to a vector, first define the range variable, and then type the ***evaluation*** operator (=) immediately following the definition. See Figure 6.4.

To create the vector in Figure 6.4, the keystrokes are: **`VectorA : 1,2..10=`**.

The above tip illustrates a very important feature of PTC Mathcad. When you define a variable and place the ***evaluation*** operator in conjunction with the ***definition*** operator, the variable becomes defined as the result of the evaluation. In other words, if A:=[a range variable definition]=[vector display], then PTC Mathcad interprets this as A:=[vector]. This can be expressed more generally as A:=[expression]=[result] means that A:=[result]. This may seem obvious, and in most cases it does not have any special effect. However, there are times when this is important, such as in creating a vector from a range variable. For another example, see Figure 13.3.

Using Range Variables to Create Arrays

Figures 6.2 and 6.3 show how PTC Mathcad does not allow range variables to be assigned to a variable. There is an exception to this if you use the same range variable on both sides of the definition, and if the range variable begins with (or is greater than) ORIGIN and uses positive integer increments. The range variable is used for the array subscripts for defining the array, and can also be used to define

This example compares the difference between using a range variable and a vector as input into a function.

1. Using a Range Variable.

$o := 0.5, 1..3$ Define the range variable "o".

$\mathrm{CircleArea}(r) := \pi \cdot r^2$

$$\mathrm{CircleArea}(o \cdot cm) = \begin{bmatrix} 0.785 \\ 3.142 \\ 7.069 \\ 12.566 \\ 19.635 \\ 28.274 \end{bmatrix} cm^2$$

$\mathrm{CircleOutput} := CircleArea(o \cdot cm)$ $\mathrm{CircleOutput} := CircleArea(o \cdot cm)$

This value must be a scalar or a matrix.

2. Using a Vector.

In order to make each result accessible, you need to assign the results to a vector.

$$v := \begin{bmatrix} 0.5 \\ 1 \\ 1.5 \\ 2 \\ 2.5 \\ 3 \end{bmatrix} \cdot cm$$

Create a vector "v".

$$\mathrm{Area} := \mathrm{CircleArea}(v) = \begin{bmatrix} 0.785 \\ 3.142 \\ 7.069 \\ 12.566 \\ 19.635 \\ 28.274 \end{bmatrix} cm^2$$

By using a vector, each element of the results is now accessible.

$\mathrm{Area}_1 = 0.785\ cm^2$

$\mathrm{Area}_6 = 28.274\ cm^2$

FIGURE 6.2

Range variable vs. vector in user-defined functions

1. Using a range variable.

Use the range variable "o" from Figure 6.2. Use an expression rather than a function, which was used in Figure 6.2.

$$\pi \cdot (o \cdot cm)^2 = \begin{bmatrix} 0.785 \\ 3.142 \\ 7.069 \\ 12.566 \\ 19.635 \\ 28.274 \end{bmatrix} cm^2 \qquad o = \begin{bmatrix} 0.500 \\ 1.000 \\ 1.500 \\ 2.000 \\ 2.500 \\ 3.000 \end{bmatrix}$$

Now try to assign the same expression to a variable. It will not work.

$$\text{CircleOutput} := \pi \cdot (o \cdot cm)^2 = ?$$

$$\text{CircleOutput} := \pi \cdot (o \cdot cm)^2 = ?$$

This value must be a scalar or a matrix.

2. Using a vector.

Use the vector "v" from Figure 6.2. Assign the expression to a variable.

$$\text{CircleArea\_1} := \pi \cdot v^2 = \begin{bmatrix} 0.785 \\ 3.142 \\ 7.069 \\ 12.566 \\ 19.635 \\ 28.274 \end{bmatrix} cm^2 \qquad v = \begin{bmatrix} 0.500 \\ 1.000 \\ 1.500 \\ 2.000 \\ 2.500 \\ 3.000 \end{bmatrix} cm$$

This time it works and each element of the results is now accessible.

$$\text{CircleArea\_1}_1 = 0.785\ cm^2 \qquad \text{CircleArea\_1}_4 = 12.566\ cm^2$$

FIGURE 6.3

Range variable vs. vector in expressions

To convert a range variable to a vector, simply type the evaluation operator following the definition.

$$\mathbf{VectorA} := 1, 2 .. 10 = \begin{bmatrix} 1.000 \\ 2.000 \\ 3.000 \\ 4.000 \\ 5.000 \\ 6.000 \\ 7.000 \\ 8.000 \\ 9.000 \\ 10.000 \end{bmatrix}$$

The definition operator created a range variable.

The evaluation operator displays the range variable as a column vector. It actually does more than just display the column vector.

Placing the evaluation operator with the definition operator has the effect of assigning the column vector to the variable.

In other words, VectorA is no longer just a range variable. It is now a column vector.

Compare

$$\mathbf{RangeA} := 1, 2 .. 10$$

$$\mathbf{VectorA} = \begin{bmatrix} 1.000 \\ 2.000 \\ 3.000 \\ 4.000 \\ 5.000 \\ 6.000 \\ 7.000 \\ 8.000 \\ 9.000 \\ 10.000 \end{bmatrix} \qquad \mathbf{RangeA} = \begin{bmatrix} 1.000 \\ 2.000 \\ 3.000 \\ 4.000 \\ 5.000 \\ 6.000 \\ 7.000 \\ 8.000 \\ 9.000 \\ 10.000 \end{bmatrix}$$

They both look the same, but VectorA is now a vector and RangeA is still a range variable.

$\mathbf{VectorA}_1 = 1.000$ $\qquad$ $\boxed{RangeA}_1 = ?$ $\qquad$ $\boxed{RangeA}_1 = ?$

This value must be a vector.

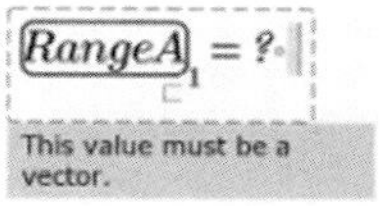

FIGURE 6.4

Converting a range variable to a vector

In the following example, the range variable "a" is used to create the vector "Sample." The subscript "a" in the definition is an array subscript created using the [key. The definition tells PTC Mathcad to create a vector called "Sample." Each value of the range variable will be used to create the vector.

$a := 1, 2..6$

$Sample_a := a^2 \cdot 2$

Note: The a in the variable name "Sample" is an array subscript.

PTC Mathcad uses the procedure listed below to calculate the values of the vector "Sample." It uses every value in the range variable from 1 to 6. The element value is the same as the range value.

$$Sample = \begin{bmatrix} 2.000 \\ 8.000 \\ 18.000 \\ 32.000 \\ 50.000 \\ 72.000 \end{bmatrix}$$

Element	Range Variable	Value
1	1	1^2*2=2
2	2	2^2*2=8
3	3	3^2*2=18
4	4	4^2*2=32
5	5	5^2*2=50
6	6	6^2*2=72

Use the range variable "b" to control the output displayed.

$b := 3, 4..6$

$$Sample_b = \begin{bmatrix} 18.000 \\ 32.000 \\ 50.000 \\ 72.000 \end{bmatrix}$$

FIGURE 6.5

Using a range variable to create a vector

the value of each element. Figures 6.5 and 6.6 give examples of how you can use range variables to create vectors. In both of these figures, the value of each created element is based on the values of the range variable. Figure 6.7 gives an example of using two range variables to create a matrix.

Using Units in Range Variables

It is possible to add units to the range variable definition, but it is generally discouraged. Range variables with units should be used primarily for plotting.

To add units, you simply attach units to the beginning value, second value, and ending value. You are required to input a second value when using units. The second and ending values do not need to use the same units, but the units used must be from the same unit dimension. For example, your beginning value can be in feet, the

Let's look at the same equation as in Figure 6.5, but using different range variables.

Values less than ORIGIN

$d := 0, 1 .. 10$

The range variable "d" does not work because ORIGIN is set to 1 for this worksheet, and there is not a zero element in the vector.

$Sample\_2_d = ?\ d^2 \cdot 2$

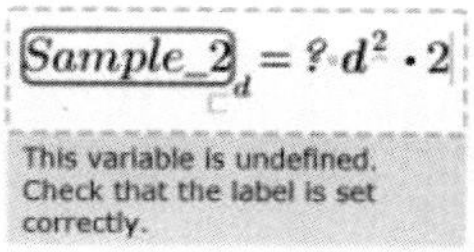

$Sample\_2_d = ?\ d^2 \cdot 2$

This variable is undefined. Check that the label is set correctly.

Non-consecutive numbers

$f := 2, 4 .. 9$

$Sample\_3_f := f^2 \cdot 2$

$$Sample\_3 = \begin{bmatrix} 0.000 \\ 8.000 \\ 0.000 \\ 32.000 \\ 0.000 \\ 72.000 \\ 0.000 \\ 128.000 \end{bmatrix}$$

Element	Range Variable	Value
1		0
2	2	2^2*2=8
3		0
4	4	4^2*2=32
5		0
6	6	6^2*2=72
7		0
8	8	8^2*2=128
9		Not created

Non-integers

$h := 1, 1.5 .. 5$

To create a vector with a range variable, the values in the range variable must be integers to correspond with the element numbers in the vector.

$Sample\_4_h := h^2 \cdot 2 = ?$

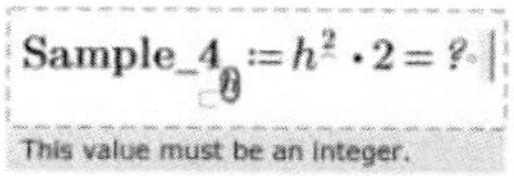

$Sample\_4_h := h^2 \cdot 2 = ?$

This value must be an integer.

FIGURE 6.6

Using range variables to create a vector

$k := 1..4$ For this example, use the range variable k as the index variable.

$n := 1..3$

$Sample\_5_{k,n} := k + 2 \cdot n$

$$Sample\_5 = \begin{bmatrix} 3.000 & 5.000 & 7.000 \\ 4.000 & 6.000 & 8.000 \\ 5.000 & 7.000 & 9.000 \\ 6.000 & 8.000 & 10.000 \end{bmatrix}$$

		Column 1	Column2	Column 3
		n=1	n=2	n=3
Row 1	k=1	1+2*1=3	1+2*2=5	1+2*3=7
Row 2	k=2	2+2*1=4	2+2*2=6	2+2*3=8
Row 3	k=3	3+2*1=5	3+2*2=7	3+2*3=9
Row 4	k=4	4+2*1=6	4+2*2=8	4+2*3=10

FIGURE 6.7

Using range variables to create a matrix

second value can be in inches, and the ending value can be in feet. This way the increment will be in inches. See Figure 6.8.

Calculating Increments from the Beginning and Ending Values

If the increments do not allow the range variable to stop exactly on the ending value, PTC Mathcad will stop the range variable short of the last value. See Figure 6.9. This could cause some unexpected results in your calculations. If it is difficult to calculate an increment that will stop exactly at the last value, you can enter a formula into the second placeholder. The formula is (Last value — First value)/(Number of increments) + First value. See Figure 6.10.

Displaying Arrays

> Mathcad 15 had much more control over the display of arrays. You could display an array in matrix form or in table form. The table option is not available in PTC Mathcad Prime 3.0. It is possible to display the column and row index numbers of an array.

If you move your cursor to the right side or the bottom side of the matrix a gray resizing bar will appear, and the cursor changes to a two-headed arrow. If you click on either of these resizing bars and drag the mouse, you can change the size of the displayed matrix. If you move the cursor to the bottom right corner, you can change both the row and column size at the same time. If you make the displayed size smaller than the matrix size, PTC Mathcad displays three dots in the lower right

Using consistent units

$q := 1 \cdot ft, 2 \cdot ft .. 6\ ft$

$$q = \begin{bmatrix} 0.305 \\ 0.610 \\ 0.914 \\ 1.219 \\ 1.524 \\ 1.829 \end{bmatrix} m$$

$$q = \begin{bmatrix} 1.000 \\ 2.000 \\ 3.000 \\ 4.000 \\ 5.000 \\ 6.000 \end{bmatrix} ft$$

Using units of feet and inches

$r := 1\ ft, 13\ in .. 2\ ft$

$$r = \begin{bmatrix} 0.305 \\ 0.330 \\ 0.356 \\ 0.381 \\ 0.406 \\ 0.432 \\ 0.457 \\ 0.483 \\ 0.508 \\ 0.533 \\ 0.559 \\ 0.584 \\ 0.610 \end{bmatrix} m \quad r = \begin{bmatrix} 12.000 \\ 13.000 \\ 14.000 \\ 15.000 \\ 16.000 \\ 17.000 \\ 18.000 \\ 19.000 \\ 20.000 \\ 21.000 \\ 22.000 \\ 23.000 \\ 24.000 \end{bmatrix} in \quad r = \begin{bmatrix} 1.000 \\ 1.083 \\ 1.167 \\ 1.250 \\ 1.333 \\ 1.417 \\ 1.500 \\ 1.583 \\ 1.667 \\ 1.750 \\ 1.833 \\ 1.917 \\ 2.000 \end{bmatrix} ft$$

Using Functions

$t := °F(0), °F(10) .. °F(100)$

$$t = \begin{bmatrix} 255.37 \\ 260.93 \\ 266.48 \\ 272.04 \\ 277.59 \\ 283.15 \\ 288.71 \\ 294.26 \\ 299.82 \\ 305.37 \\ 310.93 \end{bmatrix} K \quad t = \begin{bmatrix} 0 \\ 10 \\ 20 \\ 30 \\ 40 \\ 50 \\ 60 \\ 70 \\ 80 \\ 90 \\ 100 \end{bmatrix} °F \quad t = \begin{bmatrix} -17.78 \\ -12.22 \\ -6.67 \\ -1.11 \\ 4.44 \\ 10 \\ 15.56 \\ 21.11 \\ 26.67 \\ 32.22 \\ 37.78 \end{bmatrix} °C$$

Using units of minutes and seconds

$u := 1\ min, 61\ s .. 2\ min$

$$u = \begin{bmatrix} 60 \\ 61 \\ 62 \\ 63 \\ 64 \\ 65 \\ 66 \\ \vdots \end{bmatrix} s \quad u = \begin{bmatrix} 1 \\ 1.02 \\ 1.03 \\ 1.05 \\ 1.07 \\ 1.08 \\ 1.1 \\ 1.12 \\ \vdots \end{bmatrix} min$$

FIGURE 6.8

Range variables with units

$$\text{RangeVariable\_D} := 0, 0.4..3$$

$$\text{RangeVariable\_D} = \begin{bmatrix} 0.000 \\ 0.400 \\ 0.800 \\ 1.200 \\ 1.600 \\ 2.000 \\ 2.400 \\ 2.800 \end{bmatrix}$$

The last value of RangeVariable_D does not end at the ending value in the definition. This could cause unexpected results in your calculations.

FIGURE 6.9

Range variable where increment does not stop at ending value

$$\text{RangeVariable\_E} := 1, 1.9..10 \qquad \text{RangeVariable\_F} := 1, \frac{10-1}{10} + 1..10$$

Formula = ((Last value-First value)/Number of increments) + First value

$$\text{RangeVariable\_E} = \begin{bmatrix} 1.000 \\ 1.900 \\ 2.800 \\ 3.700 \\ 4.600 \\ 5.500 \\ 6.400 \\ 7.300 \\ 8.200 \\ 9.100 \\ 10.000 \end{bmatrix} \qquad \text{RangeVariable\_F} = \begin{bmatrix} 1.000 \\ 1.900 \\ 2.800 \\ 3.700 \\ 4.600 \\ 5.500 \\ 6.400 \\ 7.300 \\ 8.200 \\ 9.100 \\ 10.000 \end{bmatrix}$$

The values for RangeVariable_E and RangeVariable_F are the same. RangeVariable_F uses a formula to get 10 increments between the first and last values. The increment for RangeVariable_E needed to be calculated prior to entering the increment.

FIGURE 6.10

Calculating increments

corner. These dots are called the ***Matrix Navigator*** and indicate that the matrix is larger than what is being displayed.

$$a = \begin{bmatrix} 1 & 2 & 3 & 4 \\ 2 & 3 & 4 & 5 \\ 3 & 4 & 5 & 6 \\ 4 & 5 & 6 & 7 \end{bmatrix} \qquad a = \begin{bmatrix} 1 & 2 & 3 & \\ 2 & 3 & 4 & \\ 3 & 4 & 5 & \\ & & & \ddots \end{bmatrix}$$

Displaying and Resizing a Large Matrix

PTC Mathcad will display up to a 12x12 matrix by default. When a matrix has more than 12 rows or columns, PTC Mathcad will by default only display 12 rows and 12 columns and then display the ***Matrix Navigator*** in the lower right corner. When you click in this region, PTC Mathcad displays the index numbers for the rows and columns. See Figure 6.11.

The ***Matrix Navigator*** allows you to view hidden parts of the matrix. When you click on the ***Matrix Navigator*** a small ***Matrix Navigator*** window appears. The window has a light gray window that represents the full size of the matrix. A dark gray window represents the displayed portion of the matrix relative to the full size of the matrix. Your cursor changes to a four-headed arrow when it is placed over the dark gray window. If you click and hold your mouse, you can move the display window within the light gray window and different portions of the matrix will be displayed. The light gray display window also shows two numbers that represent the beginning row and column of the displayed matrix. This is illustrated in the figure below.

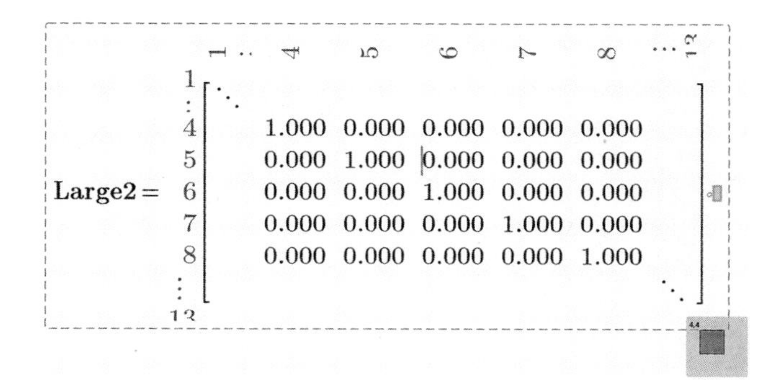

You can also use the gray resizing bars to expand the display to be larger than the default matrix display. The expanded displayed matrix may be larger than the width of the printed page. If this is the case, you can reduce the math font of the region. To do this, select the region, then from the **Math Formatting** tab, in the **Math Font** group select a smaller font from the **Font Size** control (**Math Formatting>Math Font>Font Size**).

Large1 := identity(12)   **Large2** := identity(13)

By default PTC Mathcad displays up to a 12x12 matrix.

$$\textbf{Large1} = \begin{bmatrix}
1.000 & 0.000 & 0.000 & 0.000 & 0.000 & 0.000 & 0.000 & 0.000 & 0.000 & 0.000 & 0.000 & 0.000 \\
0.000 & 1.000 & 0.000 & 0.000 & 0.000 & 0.000 & 0.000 & 0.000 & 0.000 & 0.000 & 0.000 & 0.000 \\
0.000 & 0.000 & 1.000 & 0.000 & 0.000 & 0.000 & 0.000 & 0.000 & 0.000 & 0.000 & 0.000 & 0.000 \\
0.000 & 0.000 & 0.000 & 1.000 & 0.000 & 0.000 & 0.000 & 0.000 & 0.000 & 0.000 & 0.000 & 0.000 \\
0.000 & 0.000 & 0.000 & 0.000 & 1.000 & 0.000 & 0.000 & 0.000 & 0.000 & 0.000 & 0.000 & 0.000 \\
0.000 & 0.000 & 0.000 & 0.000 & 0.000 & 1.000 & 0.000 & 0.000 & 0.000 & 0.000 & 0.000 & 0.000 \\
0.000 & 0.000 & 0.000 & 0.000 & 0.000 & 0.000 & 1.000 & 0.000 & 0.000 & 0.000 & 0.000 & 0.000 \\
0.000 & 0.000 & 0.000 & 0.000 & 0.000 & 0.000 & 0.000 & 1.000 & 0.000 & 0.000 & 0.000 & 0.000 \\
0.000 & 0.000 & 0.000 & 0.000 & 0.000 & 0.000 & 0.000 & 0.000 & 1.000 & 0.000 & 0.000 & 0.000 \\
0.000 & 0.000 & 0.000 & 0.000 & 0.000 & 0.000 & 0.000 & 0.000 & 0.000 & 1.000 & 0.000 & 0.000 \\
0.000 & 0.000 & 0.000 & 0.000 & 0.000 & 0.000 & 0.000 & 0.000 & 0.000 & 0.000 & 1.000 & 0.000 \\
0.000 & 0.000 & 0.000 & 0.000 & 0.000 & 0.000 & 0.000 & 0.000 & 0.000 & 0.000 & 0.000 & 1.000
\end{bmatrix}$$

When a matrix size exceeds 12x12, by default only 12 rows and columns are displayed and 3 dots in the lower right.

$$\textbf{Large2} = \begin{bmatrix}
1.000 & 0.000 & 0.000 & 0.000 & 0.000 & 0.000 & 0.000 & 0.000 & 0.000 & 0.000 & 0.000 & 0.000 & \\
0.000 & 1.000 & 0.000 & 0.000 & 0.000 & 0.000 & 0.000 & 0.000 & 0.000 & 0.000 & 0.000 & 0.000 & \\
0.000 & 0.000 & 1.000 & 0.000 & 0.000 & 0.000 & 0.000 & 0.000 & 0.000 & 0.000 & 0.000 & 0.000 & \\
0.000 & 0.000 & 0.000 & 1.000 & 0.000 & 0.000 & 0.000 & 0.000 & 0.000 & 0.000 & 0.000 & 0.000 & \\
0.000 & 0.000 & 0.000 & 0.000 & 1.000 & 0.000 & 0.000 & 0.000 & 0.000 & 0.000 & 0.000 & 0.000 & \\
0.000 & 0.000 & 0.000 & 0.000 & 0.000 & 1.000 & 0.000 & 0.000 & 0.000 & 0.000 & 0.000 & 0.000 & \\
0.000 & 0.000 & 0.000 & 0.000 & 0.000 & 0.000 & 1.000 & 0.000 & 0.000 & 0.000 & 0.000 & 0.000 & \\
0.000 & 0.000 & 0.000 & 0.000 & 0.000 & 0.000 & 0.000 & 1.000 & 0.000 & 0.000 & 0.000 & 0.000 & \\
0.000 & 0.000 & 0.000 & 0.000 & 0.000 & 0.000 & 0.000 & 0.000 & 1.000 & 0.000 & 0.000 & 0.000 & \\
0.000 & 0.000 & 0.000 & 0.000 & 0.000 & 0.000 & 0.000 & 0.000 & 0.000 & 1.000 & 0.000 & 0.000 & \\
0.000 & 0.000 & 0.000 & 0.000 & 0.000 & 0.000 & 0.000 & 0.000 & 0.000 & 0.000 & 1.000 & 0.000 & \\
0.000 & 0.000 & 0.000 & 0.000 & 0.000 & 0.000 & 0.000 & 0.000 & 0.000 & 0.000 & 0.000 & 1.000 & \\
 & & & & & & & & & & & & \ddots
\end{bmatrix}$$

When you click within the region, index numbers are displayed. If you move your cursor to the right side or bottom of the matrix, gray resizing bars appear.

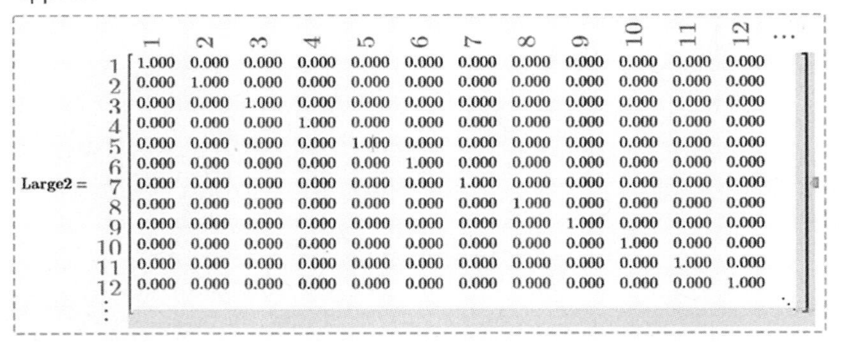

FIGURE 6.11

Display of large matrices

Use the gray resizing bars on the right and bottom of the matrix to expand the display to be larger than the default matrix display.

You may need to reduce the math font in order to display the entire matrix on a printed page.

$$\text{Large2} = \begin{bmatrix} 1.000 & 0.000 & 0.000 & 0.000 & 0.000 & 0.000 & 0.000 & 0.000 & 0.000 & 0.000 & 0.000 & 0.000 & 0.000 \\ 0.000 & 1.000 & 0.000 & 0.000 & 0.000 & 0.000 & 0.000 & 0.000 & 0.000 & 0.000 & 0.000 & 0.000 & 0.000 \\ 0.000 & 0.000 & 1.000 & 0.000 & 0.000 & 0.000 & 0.000 & 0.000 & 0.000 & 0.000 & 0.000 & 0.000 & 0.000 \\ 0.000 & 0.000 & 0.000 & 1.000 & 0.000 & 0.000 & 0.000 & 0.000 & 0.000 & 0.000 & 0.000 & 0.000 & 0.000 \\ 0.000 & 0.000 & 0.000 & 0.000 & 1.000 & 0.000 & 0.000 & 0.000 & 0.000 & 0.000 & 0.000 & 0.000 & 0.000 \\ 0.000 & 0.000 & 0.000 & 0.000 & 0.000 & 1.000 & 0.000 & 0.000 & 0.000 & 0.000 & 0.000 & 0.000 & 0.000 \\ 0.000 & 0.000 & 0.000 & 0.000 & 0.000 & 0.000 & 1.000 & 0.000 & 0.000 & 0.000 & 0.000 & 0.000 & 0.000 \\ 0.000 & 0.000 & 0.000 & 0.000 & 0.000 & 0.000 & 0.000 & 1.000 & 0.000 & 0.000 & 0.000 & 0.000 & 0.000 \\ 0.000 & 0.000 & 0.000 & 0.000 & 0.000 & 0.000 & 0.000 & 0.000 & 1.000 & 0.000 & 0.000 & 0.000 & 0.000 \\ 0.000 & 0.000 & 0.000 & 0.000 & 0.000 & 0.000 & 0.000 & 0.000 & 0.000 & 1.000 & 0.000 & 0.000 & 0.000 \\ 0.000 & 0.000 & 0.000 & 0.000 & 0.000 & 0.000 & 0.000 & 0.000 & 0.000 & 0.000 & 1.000 & 0.000 & 0.000 \\ 0.000 & 0.000 & 0.000 & 0.000 & 0.000 & 0.000 & 0.000 & 0.000 & 0.000 & 0.000 & 0.000 & 1.000 & 0.000 \\ 0.000 & 0.000 & 0.000 & 0.000 & 0.000 & 0.000 & 0.000 & 0.000 & 0.000 & 0.000 & 0.000 & 0.000 & 1.000 \end{bmatrix}$$

FIGURE 6.12

Display of large matrices

You can select a preset font size, or you can type a font size into the control. See Figure 6.12.

> The above was not possible in Mathcad 15 because you could not change the font size.

Show/Hide Indices

There may be times when you want to see the index number of the rows and columns in an array. The display of index numbers is controlled by the **Show Indices** control in the **Result Format** group on the **Matrices/Tables** tab (**Matrices/Tables>Result Format>Show Indices**). This is an on/off toggle control. The PTC Mathcad default is off. The control can be applied to all regions in the worksheet, but you can also select and apply the control to an individual region. Once the control is applied to an individual region, it overrides the worksheet setting.

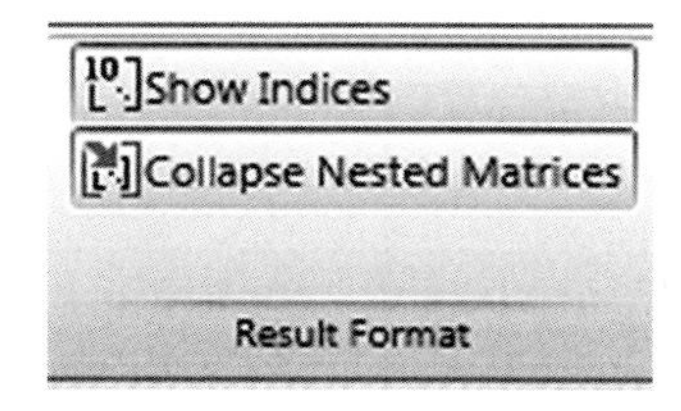

This control in PTC Mathcad Prime 3.0 has some unexpected behavior. You would expect that if the control is turned on you would always see indices, and when it is turned off you would not see indices. The first thing to understand is that this control turns on or off the display of indices for when the region is not activated. See Figure 6.13.

The second thing to understand is that indices are never displayed when the full matrix is displayed. This applies to regions that are activated or not activated, and no matter how large the matrix is.

The third thing to understand is that when the full matrix is not displayed, index numbers always appear when the region is activated, even if the **Show Indices** control is turned off. See Figure 6.14.

This region has the Show Indices control turned off.

Large2 =

1.000	0.000	0.000	0.000	0.000	0.000	0.000	0.000	0.000	0.000	0.000	0.000	
0.000	1.000	0.000	0.000	0.000	0.000	0.000	0.000	0.000	0.000	0.000	0.000	
0.000	0.000	1.000	0.000	0.000	0.000	0.000	0.000	0.000	0.000	0.000	0.000	
0.000	0.000	0.000	1.000	0.000	0.000	0.000	0.000	0.000	0.000	0.000	0.000	
0.000	0.000	0.000	0.000	1.000	0.000	0.000	0.000	0.000	0.000	0.000	0.000	
0.000	0.000	0.000	0.000	0.000	1.000	0.000	0.000	0.000	0.000	0.000	0.000	
0.000	0.000	0.000	0.000	0.000	0.000	1.000	0.000	0.000	0.000	0.000	0.000	
0.000	0.000	0.000	0.000	0.000	0.000	0.000	1.000	0.000	0.000	0.000	0.000	
0.000	0.000	0.000	0.000	0.000	0.000	0.000	0.000	1.000	0.000	0.000	0.000	
0.000	0.000	0.000	0.000	0.000	0.000	0.000	0.000	0.000	1.000	0.000	0.000	
0.000	0.000	0.000	0.000	0.000	0.000	0.000	0.000	0.000	0.000	1.000	0.000	
0.000	0.000	0.000	0.000	0.000	0.000	0.000	0.000	0.000	0.000	0.000	1.000	
												⋱

This region has the Show Indices control turned on.

Large2 =

	0	1	2	3	4	5	6	7	8	9	10	11	...
0	1.000	0.000	0.000	0.000	0.000	0.000	0.000	0.000	0.000	0.000	0.000	0.000	
1	0.000	1.000	0.000	0.000	0.000	0.000	0.000	0.000	0.000	0.000	0.000	0.000	
2	0.000	0.000	1.000	0.000	0.000	0.000	0.000	0.000	0.000	0.000	0.000	0.000	
3	0.000	0.000	0.000	1.000	0.000	0.000	0.000	0.000	0.000	0.000	0.000	0.000	
4	0.000	0.000	0.000	0.000	1.000	0.000	0.000	0.000	0.000	0.000	0.000	0.000	
5	0.000	0.000	0.000	0.000	0.000	1.000	0.000	0.000	0.000	0.000	0.000	0.000	
6	0.000	0.000	0.000	0.000	0.000	0.000	1.000	0.000	0.000	0.000	0.000	0.000	
7	0.000	0.000	0.000	0.000	0.000	0.000	0.000	1.000	0.000	0.000	0.000	0.000	
8	0.000	0.000	0.000	0.000	0.000	0.000	0.000	0.000	1.000	0.000	0.000	0.000	
9	0.000	0.000	0.000	0.000	0.000	0.000	0.000	0.000	0.000	1.000	0.000	0.000	
10	0.000	0.000	0.000	0.000	0.000	0.000	0.000	0.000	0.000	0.000	1.000	0.000	
11	0.000	0.000	0.000	0.000	0.000	0.000	0.000	0.000	0.000	0.000	0.000	1.000	
⋮													⋱

FIGURE 6.13

Show or hide indices

$\mathbf{Matrix} := \underline{\mathbf{identity}}(6)$

If the full matrix is displayed, then index numbers do not display when the region is selected.

$$\mathbf{Matrix} = \begin{bmatrix} 1.000 & 0.000 & 0.000 & 0.000 & 0.000 & 0.000 \\ 0.000 & 1.000 & 0.000 & 0.000 & 0.000 & 0.000 \\ 0.000 & 0.000 & 1.000 & 0.000 & 0.000 & 0.000 \\ 0.000 & 0.000 & 0.000 & 1.000 & 0.000 & 0.000 \\ 0.000 & 0.000 & 0.000 & 0.000 & 1.000 & 0.000 \\ 0.000 & 0.000 & 0.000 & 0.000 & 0.000 & 1.000 \end{bmatrix}$$

If the full matrix is not displayed, then index numbers are displayed when the region is selected.

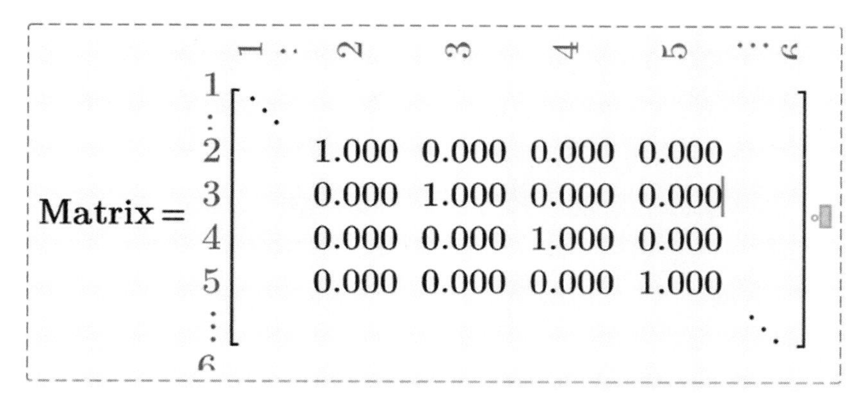

FIGURE 6.14

Display of indices when a region is selected

Using Units with Arrays

> PTC Mathcad Prime now allows mixed unit dimensions within a vector and matrix. ☺

Units are just as important with vectors and matrices as they are with other PTC Mathcad variables. Units are essential in technical calculations. PTC Mathcad allows you to now use mixed unit dimensions within vectors and matrices. This was not possible with previous versions of PTC Mathcad. This is a great improvement. If all the values in the array have the same units, then you can multiply the entire matrix by the unit rather than attaching the unit to each value. If the input values have different units, then multiply each individual value by the unit. See Figure 6.15. By default, PTC Mathcad does not allow you to control the display of individual results. In other words, there is no unit placeholder inside of the displayed array to change the displayed units. The unit placeholder is outside of the array, and you are stuck with the default display. See Figure 6.16.

Even though there is no unit placeholder inside of the displayed array, there is a way for you to display the values in the units you desire. You can add an array of units in the unit placeholder. See Figure 6.17.

$$\text{Matrix\_2} := \begin{bmatrix} 1 & 2 & 3 & 4 \\ 2 & 3 & 4 & 5 \\ 3 & 4 & 5 & 6 \\ 4 & 5 & 6 & 7 \end{bmatrix} \cdot m = \begin{bmatrix} 1.000 & 2.000 & 3.000 & 4.000 \\ 2.000 & 3.000 & 4.000 & 5.000 \\ 3.000 & 4.000 & 5.000 & 6.000 \\ 4.000 & 5.000 & 6.000 & 7.000 \end{bmatrix} m$$

$$\text{Matrix\_3} := \begin{bmatrix} 1\ m & 2\ cm & 3\ km & 4\ mm \\ 2\ ft & 3\ in & 4\ yd & 5\ mi \\ 3\ furlong & 4\ nmi & 5\ Angstrom & 6\ bohr \\ 4\ cubit & 5\ micron & 6\ \mu m & 7\ mil \end{bmatrix}$$

If all of the units are of the same unit dimension, then the displayed units will all be displayed with the same unit. You cannot change the display of individual elements.

$$\text{Matrix\_3} = \begin{bmatrix} 1.000 & 0.020 & 3.000 \cdot 10^{3} & 0.004 \\ 0.610 & 0.076 & 3.658 & 8.047 \cdot 10^{3} \\ 603.504 & 7.408 \cdot 10^{3} & 5.000 \cdot 10^{-10} & 3.175 \cdot 10^{-10} \\ 1.829 & 5.000 \cdot 10^{-6} & 6.000 \cdot 10^{-6} & 1.778 \cdot 10^{-4} \end{bmatrix} m$$

$$\text{Matrix\_3} = \begin{bmatrix} 3.281 & 0.066 & 9.843 \cdot 10^{3} & 0.013 \\ 2.000 & 0.250 & 12.000 & 2.640 \cdot 10^{4} \\ 1.980 \cdot 10^{3} & 2.430 \cdot 10^{4} & 1.640 \cdot 10^{-9} & 1.042 \cdot 10^{-9} \\ 6.000 & 1.640 \cdot 10^{-5} & 1.969 \cdot 10^{-5} & 5.833 \cdot 10^{-4} \end{bmatrix} ft$$

You can use mixed units inside of the array.

$$\text{Matrix\_4} := \begin{bmatrix} 2\ m \\ 3\ kg \\ 4\ N \\ 5\ W \end{bmatrix} = \begin{bmatrix} 2.000\ m \\ 3.000\ kg \\ 4.000\ N \\ 5.000\ W \end{bmatrix}$$

FIGURE 6.15

Using units with arrays

When using mixed units, you are stuck with the PTC Mathcad default units. You cannot modify the units that are displayed. In other words, you cannot change the display of the first element from m to cm.

$$\mathbf{Matrix\_5} := \begin{bmatrix} 2\ cm \\ 3\ gm \\ 4\ kN \\ 5\ MW \end{bmatrix} = \begin{bmatrix} 0.020\ m \\ 0.003\ kg \\ (4.000 \cdot 10^{3})\ N \\ (5.000 \cdot 10^{6})\ W \end{bmatrix}$$

$$\mathbf{Matrix\_5} = \begin{bmatrix} 0.020\ m \\ 0.003\ kg \\ (4.000 \cdot 10^{3})\ N \\ (5.000 \cdot 10^{6})\ W \end{bmatrix}$$

You can display individual elements with different units.

$$\mathbf{Matrix\_5}_1 = 20.000\ mm$$

$$\mathbf{Matrix\_5}_2 = 3.000\ gm$$

$$\mathbf{Matrix\_5}_3 = 4.000\ kN$$

$$\mathbf{Matrix\_5}_4 = 5.000\ MW$$

FIGURE 6.16

Display of mixed units in arrays

The PTC Mathcad default display:

$$\mathbf{Matrix\_5} = \begin{bmatrix} 0.020\ m \\ 0.003\ kg \\ (4.000 \cdot 10^{3})\ N \\ (5.000 \cdot 10^{6})\ W \end{bmatrix}$$

Add an array in the units placeholder with the units you want displayed.

$$\mathbf{Matrix\_5} = \begin{bmatrix} 2.000 \\ 3.000 \\ 4.000 \\ 5.000 \end{bmatrix} \begin{bmatrix} cm \\ gm \\ kN \\ MW \end{bmatrix} \qquad \mathbf{Matrix\_5} = \begin{bmatrix} 0.787 \\ 0.106 \\ 899.236 \\ 509.710 \end{bmatrix} \begin{bmatrix} in \\ oz \\ lbf \\ bhp \end{bmatrix}$$

FIGURE 6.17

Controlling the display of mixed units with an array in the unit placeholder

Calculating with Arrays

PTC Mathcad allows you to use many math operators with arrays just as you do with scalar variables, but you need to make sure that you follow the basic rules for matrix math. For example, you can add and subtract arrays just as you would other variables as long as the arrays are the same size. If they are different size arrays, PTC Mathcad will give you an error warning. You can multiply a scalar and an array. The result will be each element of the array multiplied by the scalar. If you multiply two array variables, PTC Mathcad assumes that you want a matrix dot product and gives a result (assuming that the two arrays are compatible with the dot product rules). There is a way to tell PTC Mathcad that you want the matrix cross product. There is also a way to tell PTC Mathcad that you want to multiply two arrays on an element-by-element basis and return a similar size array. These will be discussed shortly.

Addition and Subtraction

If vectors and matrices are exactly the same size, you may add or subtract them as you would any PTC Mathcad variable. The addition and subtraction is on an element-by-element basis. See Figure 6.18.

Multiplication

We will now discuss several different ways to multiply arrays.

Scalar Multiplication

You can multiply any vector or matrix by a scalar number. PTC Mathcad multiplies each element in the array by the scalar and returns the results. The scalar can be before or after the array. See Figure 6.19.

Dot Product Multiplication

When you multiply two arrays, PTC Mathcad assumes that you want the matrix dot product of the two arrays. (The dot product is calculated by multiplying each element of the first vector by the corresponding element of the complex conjugate of the second vector, and then summing the result. Refer to a text on matrix math for a discussion of the matrix dot product.) In order for the dot product to work, the number of columns in the first matrix must match the number of rows in the second matrix. In other words, the matrices must be of the size $m^x n$ and $n^x p$. The result is an $m^x p$ matrix. See Figure 6.20.

For addition and subtraction, the arrays must be the same size.

$\textbf{Matrix\_6} := \begin{bmatrix} 11 & 22 \\ 33 & 44 \end{bmatrix}$ $\textbf{Matrix\_7} := \begin{bmatrix} 2 & 3 \\ 4 & 5 \end{bmatrix}$ $\textbf{Matrix\_8} := \begin{bmatrix} 1 & 2 & 3 \\ 2 & 3 & 4 \end{bmatrix}$

$\textbf{AA} := \begin{bmatrix} 1 & 2 \end{bmatrix}$

$\textbf{Matrix\_9} := \textbf{Matrix\_6} + \textbf{Matrix\_7}$ The addition operator adds element-by-element.

$\textbf{Matrix\_9} = \begin{bmatrix} 13.000 & 25.000 \\ 37.000 & 49.000 \end{bmatrix}$

$\textbf{Matrix\_10} := \textbf{Matrix\_6} - \textbf{Matrix\_7}$ The subtraction operator subtracts element-by-element.

$\textbf{Matrix\_10} = \begin{bmatrix} 9.000 & 19.000 \\ 29.000 & 39.000 \end{bmatrix}$

$\textbf{Matrix\_11} := \boxed{Matrix\_6 + Matrix\_8}$ This does not work because the arrays are different sizes.

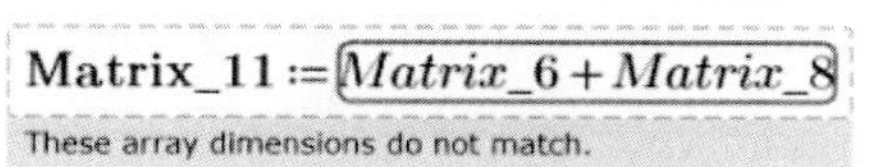

FIGURE 6.18

Addition and subtraction

You can multiply any vector or matrix by a scalar.

Refer to Figure 6.18 for the definition of Matrix_6, Matrix_7, and Matrix_8.

$$\textbf{Matrix\_12} := 2 \cdot \textbf{Matrix\_6} = \begin{bmatrix} 22.000 & 44.000 \\ 66.000 & 88.000 \end{bmatrix}$$

$$\textbf{Matrix\_13} := 3 \cdot \textbf{Matrix\_7} = \begin{bmatrix} 6.000 & 9.000 \\ 12.000 & 15.000 \end{bmatrix}$$

$$\textbf{Matrix\_14} := \textbf{Matrix\_8} \cdot 4 = \begin{bmatrix} 4.000 & 8.000 & 12.000 \\ 8.000 & 12.000 & 16.000 \end{bmatrix}$$

FIGURE 6.19

Scalar multiplication

The multiplication operator returns the matrix dot product of the arrays. The arrays must be of the size m*n and n*p. The number of columns in the first matrix must match the number of rows in the second matrix. The result is an m*p matrix.

$$\text{Matrix\_6} = \begin{bmatrix} 11 & 22 \\ 33 & 44 \end{bmatrix} \qquad \text{Matrix\_7a} := \begin{bmatrix} 2 & 3 \\ 4 & 5 \\ 6 & 7 \end{bmatrix}$$

$$\text{Matrix\_8} = \begin{bmatrix} 1 & 2 & 3 \\ 2 & 3 & 4 \end{bmatrix}$$

$$\text{Matrix\_15} := \text{Matrix\_7a} \cdot \text{Matrix\_6} = \begin{bmatrix} 121 & 176 \\ 209 & 308 \\ 297 & 440 \end{bmatrix}$$

$$\text{Matrix\_16} := \text{Matrix\_7a} \cdot \text{Matrix\_8} = \begin{bmatrix} 8 & 13 & 18 \\ 14 & 23 & 32 \\ 20 & 33 & 46 \end{bmatrix}$$

The below does not work because there is one column in the first matrix and two rows in the second.

$$\text{Matrix\_17} := \begin{bmatrix} 1 \\ 2 \end{bmatrix} \cdot Matrix\_7a = ?$$

$$\text{Matrix\_17} := \begin{bmatrix} 1 \\ 2 \end{bmatrix} \cdot Matrix\_7a = ?$$

These array dimensions do not match.

The below works because the first matrix has two columns and the second matrix has two rows.

$$\text{Matrix\_17} := \text{Matrix\_7a} \cdot \begin{bmatrix} 1 \\ 2 \end{bmatrix} = \begin{bmatrix} 8.000 \\ 14.000 \\ 20.000 \end{bmatrix}$$

FIGURE 6.20

Array dot product multiplication

Vector Cross Product Multiplication

PTC Mathcad can perform a vector cross product on two column vectors. Each vector must have three elements. The result is a vector perpendicular to the plane of the first two vectors. The direction is according to the right-hand rule. (Refer to a text on matrix math for a discussion of the vector cross product.) Use the **Operators** control list on the **Math** tab (**Math>Operators and Symbols>Operators>Vector and Matrix>x**) to insert the ***vector cross product*** operator, or type **CTRL + 8**. See Figure 6.21.

$$\text{Vector\_2} := \begin{bmatrix} 1 \\ 1 \\ 50 \end{bmatrix} \qquad \text{Vector\_3} := \begin{bmatrix} 2 \\ 3 \\ 60 \end{bmatrix}$$

$$\text{Vector\_4} := \text{Vector\_2} \times \text{Vector\_3} = \begin{bmatrix} -90.000 \\ 40.000 \\ 1.000 \end{bmatrix}$$

FIGURE 6.21

Vector cross product

Element-By-Element Multiplication (Vectorize)

The shortcut for the ***vectorize*** operator has been changed from CTRL + MINUS SIGN to CTRL + SHIFT + ^.

In order to do an element-by-element multiplication of same size arrays, you need to use the ***vectorize*** operator. This will tell PTC Mathcad to ignore the normal matrix rules and perform the operation on each corresponding element. The ***vectorize*** operator is an arrow above the expression pointing to the right.

$$\overrightarrow{A \cdot B}$$

To vectorize the multiplication operation, select the entire expression, and then from the Ribbon select **Matrices/Tables>Matrices and Tables>Vector/Matrix Operators>$\vec{V}$(Vectorization)**. You may also use the keyboard shortcut **CTRL+SHIFT + ^**. The ***vectorize*** operator is also found in the **Operators** list on the **Math** tab (**Math>Operators and symbols>Operator>Vector and Matrix**).

This places an arrow above the expression and tells PTC Mathcad to perform the operation on an element-by-element basis. When using a dot product multiplication, the number of columns in the first array must match the number of rows in the second array. When using the ***vectorize*** operator, the arrays must be exactly the same size because the multiplication is being done on a corresponding element-by-element

The ***vectorize*** operator is used to multiply arrays on an element-by-element basis. Compare this to Figure 6.20 - dot product multiplication.

$$\text{Matrix\_6} = \begin{bmatrix} 11 & 22 \\ 33 & 44 \end{bmatrix} \quad \text{Matrix\_7} = \begin{bmatrix} 2 & 3 \\ 4 & 5 \end{bmatrix} \quad \text{Matrix\_7a} := \begin{bmatrix} 2 & 3 \\ 4 & 5 \\ 6 & 7 \end{bmatrix}$$

$$\text{Matrix\_18} := \overrightarrow{\text{Matrix\_6} \cdot \text{Matrix\_7}} = \begin{bmatrix} 22 & 66 \\ 132 & 220 \end{bmatrix}$$

Keystrokes for the above: Matrix_6*Matrix_7 Spacebar Spacebar CTRL+SHIFT +^=.

The size of the arrays must match in order to use the ***vectorize*** operator.

$$\text{Matrix\_19} := \overrightarrow{Matrix\_6 \cdot Matrix\_7a} = ?$$

$$\text{Matrix\_19} := \overrightarrow{Matrix\_6 \cdot Matrix\_7a} = ?$$

Unknown error: unequal_matrix_dimensions.

FIGURE 6.22

Array element-by-element multiplication

basis (similar to addition and subtraction). See Figure 6.22 for examples of using the ***vectorize*** operator.

Division

For the case of X/Y, the result is dependent on whether X or Y are scalars or arrays. The result is also dependant on whether Y is a square matrix. Here are the rules:

- If Y is a square matrix and the number of columns matches the number of columns in X, then the result is the dot product of $X^* Y^{-1}$, where Y^{-1} is the inverse of the square matrix Y.
- If either X or Y is a scalar, then the division is done element-by-element.
- If both X and Y are arrays, then both arrays must be of the same size. If they are not square matrices, then PTC Mathcad does an element-by-element division. If they are square matrices, then the result is the dot product $X^*Y^{-1.}$

$$Matrix\_6 = \begin{bmatrix} 11 & 22 \\ 33 & 44 \end{bmatrix} \quad Matrix\_7 = \begin{bmatrix} 2 & 3 \\ 4 & 5 \end{bmatrix} \quad Matrix\_8 = \begin{bmatrix} 1 & 2 & 3 \\ 2 & 3 & 4 \end{bmatrix}$$

$$Matrix\_20 := \frac{Matrix\_6}{2} = \begin{bmatrix} 5.5 & 11 \\ 16.5 & 22 \end{bmatrix}$$

Each element of Matrix_6 is divided by the scalar.

$$Matrix\_21a := \frac{2}{Matrix\_7} = \begin{bmatrix} -5 & 3 \\ 4 & -2 \end{bmatrix}$$

Because Matrix_7 is a square matrix, the result is equal to 2*Matrix_7^-1.

$$Matrix\_7^{-1} = \begin{bmatrix} -2.5 & 1.5 \\ 2 & -1 \end{bmatrix} \qquad 2 \cdot Matrix\_7^{-1} = \begin{bmatrix} -5 & 3 \\ 4 & -2 \end{bmatrix}$$

$$Matrix\_21b := \frac{2}{Matrix\_8} = \begin{bmatrix} 2 & 1 & 0.667 \\ 1 & 0.667 & 0.5 \end{bmatrix}$$

Because Matrix_8 is not a square matrix, the result is element-by-element division.

FIGURE 6.23

Array division

- If both X and Y are square matrices, then in order to do an element-by-element division, you must use the ***vectorize*** operator. To do this, select the expression and type **CTRL + SHIFT + ^**.

  See Figures 6.23—6.26 for examples of array division.

Array Functions

PTC Mathcad has many array functions which add and extract data from arrays. This section will discuss a few of these functions. Remember that the starting index value relates to the value of ORIGIN. The PTC Mathcad default for ORIGIN is zero. This book recommends changing the value of ORIGIN to 1 (**Calculation>Worksheet Settings>ORIGIN>1**).

Creating Array Functions

- ***augment***(A,B,C,…): Returns an array formed by placing A, B, C, … left to right. A, B, C, … are scalars or single-row vectors, or they are column vectors or arrays having the same number of rows.

$$\text{Matrix\_6} = \begin{bmatrix} 11 & 22 \\ 33 & 44 \end{bmatrix} \quad \text{Matrix\_7} = \begin{bmatrix} 2 & 3 \\ 4 & 5 \end{bmatrix} \quad \text{Matrix\_8} = \begin{bmatrix} 1 & 2 & 3 \\ 2 & 3 & 4 \end{bmatrix}$$

$$\text{Matrix\_22} := \frac{\text{Matrix\_6}}{\text{Matrix\_7}} = \begin{bmatrix} 16.5 & -5.5 \\ 5.5 & 5.5 \end{bmatrix}$$

Because Matrix_7 is a square matrix, the result is equal to Matrix_6*Matrix_7^-1.

$$\text{Matrix\_7}^{-1} = \begin{bmatrix} -2.5 & 1.5 \\ 2 & -1 \end{bmatrix} \quad \text{Matrix\_6} \cdot \text{Matrix\_7}^{-1} = \begin{bmatrix} 16.5 & -5.5 \\ 5.5 & 5.5 \end{bmatrix}$$

$$\text{Matrix\_23} := \frac{\text{Matrix\_8}}{\begin{bmatrix} 2 & 4 & 6 \\ 4 & 6 & 8 \end{bmatrix}} = \begin{bmatrix} 0.5 & 0.5 & 0.5 \\ 0.5 & 0.5 & 0.5 \end{bmatrix}$$

Both matrices must be of the same size. Because the bottom matrix is not a square matrix, the result is element-by-element division.

$$\text{Matrix\_8a} := \begin{bmatrix} 1 & 2 & 3 \\ 2 & 3 & 4 \\ 5 & 6 & 7 \end{bmatrix}$$

Because Matrix_8a is a square matrix, the result is equal to Matrix_8*Matrix_8a^-1.

$$\text{Matrix\_23a} := \frac{\text{Matrix\_8}}{\text{Matrix\_8a}} = \begin{bmatrix} 1.000 & 0.000 & -0.750 \\ 1.000 & 1.000 & -1.000 \end{bmatrix}$$

$$\text{Matrix\_8a}^{-1} = \begin{bmatrix} -1.689 \cdot 10^{15} & 2.252 \cdot 10^{15} & -5.629 \cdot 10^{14} \\ 3.378 \cdot 10^{15} & -4.504 \cdot 10^{15} & 1.126 \cdot 10^{15} \\ -1.689 \cdot 10^{15} & 2.252 \cdot 10^{15} & -5.629 \cdot 10^{14} \end{bmatrix}$$

$$\text{Matrix\_8} \cdot \text{Matrix\_8a}^{-1} = \begin{bmatrix} 1.000 & 0.000 & -0.750 \\ 1.000 & 1.000 & -1.000 \end{bmatrix}$$

FIGURE 6.24

Array division

$$\frac{Matrix\_7}{Matrix\_8} = ?$$

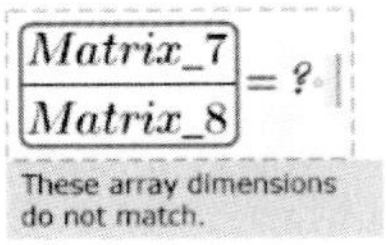

$$\frac{Matrix\_7}{Matrix\_8} = ?$$

These array dimensions do not match.

The division will not work because the matrices are different sizes.

FIGURE 6.25

Array division

$$\text{Matrix\_6} = \begin{bmatrix} 11 & 22 \\ 33 & 44 \end{bmatrix} \quad \text{Matrix\_7} = \begin{bmatrix} 2 & 3 \\ 4 & 5 \end{bmatrix} \quad \text{Matrix\_7}^{-1} = \begin{bmatrix} -2.500 & 1.500 \\ 2.000 & -1.000 \end{bmatrix}$$

$$\text{Matrix\_24a} := \overrightarrow{\frac{\text{Matrix\_6}}{\text{Matrix\_7}}} = \begin{bmatrix} 5.500 & 7.333 \\ 8.250 & 8.800 \end{bmatrix}$$

To apply the ***vectorize*** operator, select the expression and type CTRL+SHIFT+^. This allows an element-by-element division to occur.

This is the same as:

$$\text{Matrix\_24b} := \overrightarrow{\text{Matrix\_6} \cdot \text{Matrix\_7}^{-1}} = \begin{bmatrix} 5.500 & 7.333 \\ 8.250 & 8.800 \end{bmatrix}$$

FIGURE 6.26

Array division

- *stack*(A,B,C,…): Returns an array formed by placing A, B, C, … top to bottom. A, B, C, … are scalars and single-column vectors, or they are row vectors or arrays having the same number of columns.

Size Functions

- *cols*(A): Returns the number of columns in A.
- *rows*(A): Returns the number of rows in A.
- *length(v):* Returns the number of elements in column vector v.
- *last(v):* Returns the index of the last element in column vector v. This value relates to the value of ORIGIN.

Lookup Functions

> PTC Mathcad 15 had lowercase and uppercase lookup functions, where the uppercase functions allowed for the use of a modifier. PTC Mathcad Prime eliminates the uppercase lookup functions. The modifier is now an optional argument of the lowercase lookup functions.

Each lookup function includes an optional text string modifier that can be used to set the conditions of the comparison. The default value of the text string is "eq" meaning "equal." The possible values for the text string are as follows.

Modifier Value	Meaning
"near"	Returns the value closest to scalar z.
"gt"	Matches everything greater than scalar z.
"lt"	Matches everything less than scalar z.
"geq"	Matches everything greater than or equal to scalar z.
"leq"	Matches everything less than or equal to scalar z.
"not"	Matches everything not equal to scalar z.
"range"	Matches everything in the given range specified in a two-element vector z.
f	Matches everything that meets the conditions set by this user-defined comparison function. f is a user-defined function of two arguments. It returns either a 0 for false, or any other number for true. If f is not supplied, then the default will be equality.

- ***lookup***(z,A,B,"Modifier"): Looks in a vector or matrix A, for a given value z, subject to the conditions of the optional modifier, and returns the value(s) in the same position(s) (i.e., with the same row and column numbers) in another matrix, B. When multiple values are returned, they appear in a column vector.
- ***match***(z,A,"Modifier"): Looks in a vector or matrix A for a given value z, subject to the conditions of the optional modifier, and returns the index (indices) for each matching value. When the value (s) are returned, they appear in a column vector in a nested array.
- ***hlookup***(z,A,r, "Modifier"): Looks in the first row of matrix A for a given value z, subject to the conditions of the optional modifier, and returns the value(s) in the same columns(s) in the row specified, r. When multiple values are returned, they appear in a vector.
- ***vlookup***(z,A,c): Looks in the first column of matrix A for a given value z, subject to the conditions of the optional modifier, and returns the value(s) in the same row(s) in the column specified, c. When multiple values are returned, they appear in a column vector.
- ***vhlookup***(z1,z2,A,"Modifier"): Looks in the first column of matrix A for a given value z1 and in the first row of matrix A for a given value z2, subject to the conditions of the optional modifier, and returns the values(s) at the intersection.

The following example is similar to the one found in the Help Center, except with units attached.

$$\text{Matrix\_A} := \begin{bmatrix} \text{"Time"} & \text{"Degrees"} & \text{"Volts"} \\ 1\ s & 96.5\ °F & 2.6\ V \\ 2\ s & 97.1\ °F & 3.1\ V \\ 3\ s & 97.5\ °F & 3.3\ V \\ 4\ s & 97.1\ °F & 3.5\ V \\ 5\ s & 98.3\ °F & 3.8\ V \\ 6\ s & 98.5\ °F & 3.9\ V \end{bmatrix}$$

$z := \text{"Degrees"}$ $\qquad r := 3$

$$\underline{\text{hlookup}}(z, \text{Matrix\_A}, r) = [97.100]\ °F$$

$z := 4\ s$ $\qquad c := 2$

$$\underline{\text{vlookup}}(z, \text{Matrix\_A}, c) = [97.100]\ °F$$

$z := 6\ s$ $\qquad c := 3$

$$\underline{\text{vlookup}}(z, \text{Matrix\_A}, c) = [3.900]\ V$$

$z1 := 3\ s$ $\qquad z2 := \text{"Volts"}$

$$\underline{\text{vhlookup}}(z1, z2, \text{Matrix\_A}) = [3.300]\ V$$

$z := 97.1\ °F$

$$\underline{\text{match}}(z, \text{Matrix\_A}) = \begin{bmatrix} \begin{bmatrix} 3.000 \\ 2.000 \end{bmatrix} \\ \begin{bmatrix} 5.000 \\ 2.000 \end{bmatrix} \end{bmatrix}$$

Note: To display the nested array use:
Matrices/Tables>ResultFormat>Collapse Nested Matrices>off

FIGURE 6.27

Lookup functions with units

The matrix A can have units and the units may be mixed units. If mixed units are used, a match is found only when the units are of the same unit dimension. See Figures 6.27 and 6.28.

The degree of precision to which the comparisons adhere is determined by the TOL setting (**Calculation>Worksheet Setting>TOL**) in the worksheet. Reducing TOL makes the match more strict; increasing TOL makes the match less strict. See Figure 6.29.

Refer to Matrix_A from Figure 6.27.

In this example, PTC Mathcad returns all values in row 3 except the column "Degrees".

z := "Degrees"   r := 3   **modifier** := "not"

$$\underline{\text{hlookup}}(z, \text{Matrix\_A}, r, \text{modifier}) = \begin{bmatrix} 2.000\ s \\ 3.100\ V \end{bmatrix}$$

In this example, PTC Mathcad looks in the first column and matches everything less than 4 seconds, and returns values from column 3.

z := 4 *s*   c := 3   **modifier** := "lt"

$$\underline{\text{vlookup}}(z, \text{Matrix\_A}, c, \text{modifier}) = \begin{bmatrix} 2.600 \\ 3.100 \\ 3.300 \end{bmatrix} V$$

Add a corresponding matrix in order to use the lookup function.

$$\text{Matrix\_B} := \begin{bmatrix} \text{"Group"} & \text{"Number"} & \text{"Grade"} \\ \text{"a"} & 10 & 71 \\ \text{"b"} & 20 & 85 \\ \text{"c"} & 30 & 63 \\ \text{"d"} & 40 & 90 \\ \text{"e"} & 50 & 81 \\ \text{"f"} & 60 & 69 \end{bmatrix}$$

z1 := 3.1 *V*   **modifier** := "gt"

$$\underline{\text{lookup}}(z1, \text{Matrix\_A}, \text{Matrix\_B}, \text{modifier}) = \begin{bmatrix} 63.000 \\ 90.000 \\ 81.000 \\ 69.000 \end{bmatrix}$$

z1 := 97.5 *°F*   **modifier** := "lt"

$$\underline{\text{lookup}}(z1, \text{Matrix\_A}, \text{Matrix\_B}, \text{modifier}) = \begin{bmatrix} 10.000 \\ 20.000 \\ 40.000 \end{bmatrix}$$

FIGURE 6.28

Lookup functions with modifiers

Extracting Functions and Operators

- ***max***(A,B,C,...): Returns the largest value from A, B, C. This function will be discussed in Chapter 7. It works well for extracting the maximum value of a matrix.
- ***min***(A,B,C,...): Returns the smallest value in A, B, C.
- ***submatrix***(A,ir,jr,ic,jc): Returns the submatrix of array A consisting of elements in rows ir through jr and columns ic through jc of A. Remember "i" is the beginning and "j" is the ending index of the row "r" and column "c."

- ***Matrix Column*** operator $M^{\langle n \rangle}$. (**Matrices/Tables>Matrices and Tables>Vector/Matrix Operators>** $M^{\langle \rangle}$) It is also inserted by typing **CTRL + SHIFT + C** (CTRL + uppercase C-for column). Returns the nth column of matrix M.

$TOL = 0.001$ Original worksheet value of TOL.

$z := 4\ s \quad c := 2$

$\underline{vlookup}(z, Matrix\_A, c) = [97.100]\ °F$

$z := 1.4\ s$

$\underline{vlookup}(z, Matrix\_A, c, \text{“near”}) = [96.500]\ °F$

Increasing TOL lowers the degree of precision.

$TOL := .201$

$$\underline{vlookup}(z, Matrix\_A, c, \text{“near”}) = \begin{bmatrix} 96.500 \\ 97.100 \end{bmatrix} °F$$

In the above example, z=1.4 s was 0.4 s greater than 1 s. Changing TOL to 0.201 modified the range of lookup values to be between 1.199 s and 1.601 s. This now puts 2.0 s within 0.399 s of the lookup values.
Because this is less than 0.4 s, it is now included.

Increase TOL so that more values are included.

$TOL := 3$

$$\underline{vlookup}(z, Matrix\_A, c, \text{“near”}) = \begin{bmatrix} 96.500 \\ 97.100 \\ 97.500 \\ 97.100 \end{bmatrix} °F$$

Restore the value of TOL.

$TOL := 0.001$

FIGURE 6.29

Effect of TOL on lookup functions

> PTC Mathcad Prime has a ***Matrix Row*** operator $M^{\langle \rangle}$. You no longer have to invert a matrix in order to extract a row. ☺

- *Matrix Row* operator $M^{\langle\,\rangle}$ (**Matrices/Tables>Matrices and Tables>Vector/Matrix Operators>** $M^{\langle\,\rangle}$) It is also inserted by typing **CTRL+SHIFT+r** (CTRL+ uppercase R-for row). It returns the nth row of matrix M.

Sorting Functions

- *sort*(v): Returns a vector with the values from v sorted in ascending order.
- *reverse* (A): Reverses the order of elements in a vector, or the order of rows in matrix A.
- *csort* (A,n): Returns an array formed by rearranging rows of matrix A until column n is in ascending order.
- *rsort* (A,n): Returns an array formed by rearranging columns of matrix A until row n is in ascending order.
- *reverse* (*sort*(v)): Sorts in descending order.

PTC Mathcad Calculation Summary

It is not the intent of this chapter to provide a complete discussion of all the many different ways that PTC Mathcad can be used to process and manipulate vectors and matrices. PTC Mathcad has some very useful and powerful matrix features such as transpose, inverse, determinant, and statistical functions. An excellent discussion of vectors and matrices can be found in PTC Mathcad Help. Two sections can be referenced. The first section is titled "Vectors, Matrices, and Tables." The second section is titled "Matrix Operators" contained under Operators. This section discusses the many different operators that can be used with vectors and matrices. The PTC Mathcad Tutorial also has an excellent discussion of arrays.

Engineering Examples

We have spent several pages introducing vectors and matrices. Let's now give some examples of how to use arrays in your engineering equations.

Engineering Example 6.1 calculates fluid pressure and shows how vectors can be used in a user-defined function.

Engineering Example 6.2 shows how vectors can be used in expressions for many different input cases.

Engineering Example 6.3 shows the use of element-by-element multiplication when using matrices as input values.

Engineering Example 6.4 calculates the charge on a capacitor and compares the use of a range variable and a vector for multiple results.

Engineering Example 6.1: Using Vectors in a User-defined Function

The density of glycerine is 1249 kg/m^3. Calculate the pressure in a tank of glycerine at depths of 2, 4, 8, 10, and 12 meters.

$$\text{Density} := 1259 \, \frac{kg}{m^3}$$

$$\text{NetPressure}(\text{d}) := \text{Density} \cdot g \cdot \text{d}$$

Define a user-defined function to calculate the pressure based on depth.

$$g = 9.807 \, \frac{m}{s^2}$$

g is the acceleration of gravity - a built-in constant.

$$\text{Depths} := (2\,m, 4\,m .. 12\,m) = \begin{bmatrix} 2.000 \\ 4.000 \\ 6.000 \\ 8.000 \\ 10.000 \\ 12.000 \end{bmatrix} m$$

Use the evaluation operator "=" in the same line to convert the range variable to a vector.

Note: Earlier in the worksheet, the variable m was defined, so in order for "m" to represent meters, "m" must be assigned the "unit" label (**Math>Style>Labels>Unit**), or m must be inserted using the Units list (**Math>Units>Units>Length>m**).

The vector "Depths" is used as an input vector to the function "NetPressure".

The result is another vector.

$$\text{Pressures} := \text{NetPressure}(\text{Depths}) = \begin{bmatrix} 24.693 \\ 49.386 \\ 74.079 \\ 98.773 \\ 123.466 \\ 148.159 \end{bmatrix} kPa$$

$$\text{Pressures}_1 = 24.693 \, kPa$$

$$\text{Pressures}_4 = 98.773 \, kPa$$

When you use a range variable, you cannot assign the results.

$$\text{RV} := 2\,m, 4\,m .. 12\,m$$

$$\text{Test} := \text{NetPressure}(RV) = ?$$

Test := NetPressure(RV) = ?

This value must be a scalar or a matrix.

Engineering Example 6.2: Using Vectors in Expressions

Find the force caused by four different mass elements of 5 kg, 3 gm, 6 lbm, and 2 oz. and four different accelerations of 3 m/s^2, 2 m/s^2, 4 ft/s^2, and 5 ft/s^2.

It is not very efficient to use defined scalar values.

$\text{Mass} := 5\ kg \quad \text{Acceleration} := 3\ \frac{m}{s^2} \quad \text{F1} := \text{Mass} \cdot \text{Acceleration} = 15.000\ N$

$\text{Mass} := 3\ gm \quad \text{Acceleration} := 2\ \frac{m}{s^2} \quad \text{F2} := \text{Mass} \cdot \text{Acceleration} = 0.006\ N$

$\text{Mass} := 6\ lbm \quad \text{Acceleration} := 4\ \frac{ft}{s^2} \quad \text{F3} := \text{Mass} \cdot \text{Acceleration} = 0.746\ lbf$

$\text{Mass} := 2 \cdot oz \quad \text{Acceleration} := 5\ \frac{ft}{s^2} \quad \text{F4} := \text{Mass} \cdot \text{Acceleration} = 0.019\ lbf$

An easier way is to create a vector for the mass values and a vector for the acceleration values. The same equation now gives output for the four input conditions.

$$\text{Mass} := \begin{bmatrix} 5 \cdot kg \\ 3\ gm \\ 6\ lbm \\ 2\ oz \end{bmatrix} \qquad \text{Acceleration} := \begin{bmatrix} 3\ \frac{m}{s^2} \\ 2\ \frac{m}{s^2} \\ 4\ \frac{ft}{s^2} \\ 5\ \frac{ft}{s^2} \end{bmatrix}$$

$$\text{Force} := \text{Mass} \cdot \text{Acceleration} = 18.411\ N$$

$$\text{Force} := \overrightarrow{\text{Mass} \cdot \text{Acceleration}} = \begin{bmatrix} 15.000 \\ 0.006 \\ 3.318 \\ 0.086 \end{bmatrix} N$$

In order to do an element-by-element multiplication you must use the ***vectorize*** operator **CTRL+SHIFT+^**.

$$\text{Force} = \begin{bmatrix} 15.000 \\ 0.006 \\ 0.746 \\ 0.019 \end{bmatrix} \begin{bmatrix} N \\ N \\ lbf \\ lbf \end{bmatrix}$$

Add a matrix in the unit placeholder to display mixed units.

Engineering Example 6.3: Using Matrices in Functions and Expressions

Use a 2x2 mass matrix and a 2x2 acceleration matrix to calculate the force on various elements.

$$\text{Mass} := \begin{bmatrix} 5 & 7 \\ 6 & 8 \end{bmatrix} \cdot kg \qquad \text{Acceleration} := \begin{bmatrix} 3 & 7 \\ 5 & 9 \end{bmatrix} \cdot \frac{m}{sec^2}$$

1) Use a function.

$$\text{ForceFunction}(m, a) := m \cdot a$$

$$\text{Force1} := \text{ForceFunction}(\text{Mass}, \text{Acceleration}) = \begin{bmatrix} 50.000 & 98.000 \\ 58.000 & 114.000 \end{bmatrix} N$$

The above function used a dot product solution. You must use the *vectorize* operator to get an element-by-element multiplication.

$$\text{Force2} := \overrightarrow{\text{ForceFunction}(\text{Mass}, \text{Acceleration})} = \begin{bmatrix} 15.000 & 49.000 \\ 30.000 & 72.000 \end{bmatrix} N$$

2) Use an expression.

$$\text{Force3} := \text{Mass} \cdot \text{Acceleration} = \begin{bmatrix} 50.000 & 98.000 \\ 58.000 & 114.000 \end{bmatrix} N$$

The above expression used a dot product solution. You must use the ***vectorize*** operator to get an element-by-element multiplication.

$$\text{Force4} := \overrightarrow{\text{Mass} \cdot \text{Acceleration}} = \begin{bmatrix} 15.000 & 49.000 \\ 30.000 & 72.000 \end{bmatrix} N$$

Engineering Example 6.4: Comparison of Using Range Variables and Vectors

A charge Q (Coulomb) builds up on a capacitor when a resistor, capacitor, and battery are connected in series. The formula to define the charge is:

$$Q(t) = C \cdot V \left(1 - e^{\frac{-t}{R \cdot C}}\right)$$

Calculate the results for the first 4 seconds at half-second intervals, given R=4 Ohm, C=1 F, and V=9 Volts.

$$Q(t, R, C, V) := C \cdot V \cdot \left(1 - e^{\frac{-(t)}{R \cdot C}}\right)$$

$$\text{Resistance} := 4\ \Omega \qquad \text{Capacitor} := 1 \cdot F \qquad \text{Voltage} := 9 \cdot V$$

1) Use a range variable. Create a range variable from 0 to 4.

$$t := 0\ s, 0.5\ s .. 4\ s \qquad Q(t, \text{Resistance}, \text{Capacitor}, \text{Voltage}) = \begin{bmatrix} 0.000 \\ 1.058 \\ 1.991 \\ 2.814 \\ 3.541 \\ 4.183 \\ 4.749 \\ 5.248 \\ 5.689 \end{bmatrix} C$$

$$\text{Charge1} := Q(t, Resistance, Capacitor, Voltage)$$

$$\text{Charge1} := Q(t, Resistance, Capacitor, Voltage)$$

This value must be a scalar or a matrix.

Range variables can be used to display, but not assign results.

2) Use a vector.

Remember that the ***evaluation*** operator must be on the same line as the ***definition*** operator in order to convert it to a vector.

$$\text{Time} := 0\ s, 0.5\ s .. 4\ s = \begin{bmatrix} 0.000 \\ 0.500 \\ 1.000 \\ 1.500 \\ 2.000 \\ 2.500 \\ 3.000 \\ 3.500 \\ 4.000 \end{bmatrix} s$$

$$\text{Charge2} := Q(\text{Time}, \text{Resistance}, \text{Capacitor}, \text{Voltage})$$

$$\text{Time} = \begin{bmatrix} 0.000 \\ 0.500 \\ 1.000 \\ 1.500 \\ 2.000 \\ 2.500 \\ 3.000 \\ 3.500 \\ 4.000 \end{bmatrix} s \quad \text{Charge2} = \begin{bmatrix} 0.000 \\ 1.058 \\ 1.991 \\ 2.814 \\ 3.541 \\ 4.183 \\ 4.749 \\ 5.248 \\ 5.689 \end{bmatrix} C$$

Summary

Arrays are useful tools in engineering calculations. PTC Mathcad can do very advanced matrix computations. This chapter focused mostly on using arrays to perform multiple iterations of engineering expressions.

In Chapter 6 we:

- Reviewed how to create and modify the size of vectors and matrices.
- Set the ORIGIN built-in variable to 1.
- Reiterated the differences between range variables and vectors.
- Learned how to attach units to vectors and matrices.
- Learned how to reduce font size so that more information can be displayed on a single page.
- Learned how to add, subtract, multiply, and divide arrays.
- Illustrated how arrays can be used in technical calculations.

Practice

The PTC Mathcad Prime 3.0 figures and examples used in this book are available for download from the book's website. The reader is encouraged to download the files and use them to practice the concepts learned. Additional examples and problems are also provided. To access this content go to http://store.elsevier.com/9780124104105 *and click on the Resources tab, and then click on the link for the Online Companion Materials.*

1. Use the **Insert Matrix** control on the **Matrices/Tables** tab to create a 4 x 5 matrix. Fill in the matrix with some numbers.
2. In the matrix created above, insert a new column between the 3rd and 4th columns. Insert a new row between the 2nd and 3rd rows. Fill in new data.
3. In the above matrix, delete the 3rd row. Delete the 5th column.
4. Create 10 different arrays. Practice inserting and deleting rows and columns in the different arrays. Have at least two arrays larger than 20 x 20.
5. Assign variable names to the above arrays. Display each array without indices showing and with indices showing.
6. Select 20 elements from the above variables, and use the ***subscript*** operator to display the elements. Change the value of ORIGIN and see how the results change.
7. Create 10 range variables. Use various increments, some using fractions, decimals, and negative numbers. Use some formulas in the range variables.
8. Create four arrays of the following sizes: 1 x 3, 2 x 2, 2 x 3, and 3 x 3. Attach units of length to each matrix. Assign each array a variable name.

9. Using the arrays created above, create expressions to perform the following calculations. Practice using different forms for displaying results.
   a. Six addition expressions.
   b. Six subtraction expressions.
   c. Six dot product expressions.
   d. Six element-by-element multiplication expressions.
   e. 12 division expressions. For square matrices, create a dot product solution and an element-by-element solution.
10. Define a simple function with only one argument. Create a range variable that has at least six elements. Create an expression that uses the function with the range variable as the argument. Use the ***subscript*** operator to display at least three individual elements of the output vector.
11. Define another simple function with two arguments. Create two input vectors. Create an expression that uses the function and the input vectors (see Engineering Examples 5.1 and 5.2). Use the ***subscript*** operator to display at least three individual elements of the output vector.
12. Use the function and expression created above, but instead of using input vectors, use input matrices.
13. Verify that the results in the above three practice exercises are correct. Do you need to use the ***vectorize*** operator to get the correct results?

CHAPTER 7

Selected PTC Mathcad Functions

By now, you should be familiar with many PTC Mathcad functions. There have been many functions discussed in this book. Chapter 4 discussed how to use several different functions.

There are hundreds of PTC Mathcad functions. The following is a partial list of the different categories that have functions: Bessel Functions, Complex Numbers, Curve Fitting, Data Analysis, Differential Equations, Finance, Fourier Transforms, Graphing, Hyperbolic Functions, Image Processing, Interpolation, Logs, Number Theory, Probability, Solving, Sorting, Statistics, Trigonometry, Vectors, and Waves.

There are many functions that many of us have never heard of, and will never use. Several books have been written discussing the many PTC Mathcad functions. As explained earlier, the purpose of this book is to teach essential PTC Mathcad skills and the application of PTC Mathcad to technical calculations; therefore, a great deal of time will not be spent discussing specific PTC Mathcad functions. The PTC Mathcad Help Center is an excellent resource to learn the details of specific functions.

Of all the many PTC Mathcad functions, which ones are most important for technical calculations? The answer to this question depends upon your own perspective. A function that is important for one person may not be important for another.

With that said, this chapter will introduce and discuss several functions that (in the author's opinion) will be beneficial to many scientists and engineers doing technical calculations. These functions were chosen because they are easy to learn, and add power to technical calculations.

Chapter 7 will:

- Review the basic concept of built-in functions.
- Discuss the Calculation Toolbar.
- Introduce the following functions:
  - ***max***
  - ***min***
  - ***mean***
  - ***median***
  - ***floor*** and ***Floor***
  - ***ceil*** and ***Ceil***
  - ***trunc*** and ***Trunc***
  - ***round*** and ***Round***
  - ***Vector Sum***
  - ***Summation***

Essential PTC® Mathcad Prime® 3.0.

  - ***Range Sum***
  - ***if***
  - ***linterp***
- Discuss various categories of functions:
  - Curve fitting functions
  - String functions
  - Picture functions
  - Mapping functions
  - Polar notation
  - Angle functions
  - Reading and writing functions

Review of Built-in Functions

In Chapters 1 and 4, we learned that every built-in PTC Mathcad function is setup in a similar way. The name of the function is given, followed by a pair of parentheses The information required within the parentheses is called the argument. PTC Mathcad processes the argument(s) based on rules that are defined for the specific function.

The easiest way to insert a function into a worksheet is by using the **Functions** tab and then selecting the desired function. Remember to click the "**All Functions**" button to see a list of all the PTC Mathcad functions. Use the AZ button in the list to see the functions in alphabetical order. You may also type the name of the function followed by matching parentheses.

Refer to Chapter 4 for a more detailed description of functions.

Selected Functions

max and *min* Functions

The ***max*** and ***min*** functions are useful to select the maximum or minimum values from a list of values. The ***max*** function takes the form ***max***(A,B,C,...). PTC Mathcad returns the largest value from A, B, C, etc. The ***min*** function takes the same form, ***min***(A,B,C,...). In its simplest form, you can just type the list of values in the function. See Figure 7.1.

A, B, C, etc. can also be variable names. See Figure 7.2.

A, B, C, etc. can also be a list of arrays. See Figure 7.3.

Units can also be attached to A, B, C, etc. as long as they are of the same unit dimension. See Figure 7.4.

If A, B, C, etc. include a complex number, then the ***max*** function returns a complex number with the largest real part of any value, and i times the largest imaginary part of any value. For the ***min*** function, PTC Mathcad returns the

The max function selects the maximum value from a list of arguments. The min function selects the minumum value.

$\mathbf{Max1} := \underline{\mathbf{max}}(1,3,5,7,9,8,6,4,2,-4) = 9.00$

$\mathbf{Min1} := \underline{\mathbf{min}}(1,3,5,7,9,8,6,4,2,-4) = -4.00$

Note:
"min" is a function name and is also a unit of 60 seconds. Because of this PTC Mathcad does not automatically assign labels. This must be done manually (**Math>Style>Labels>Unit** or **Math>Style>Labels>Function**).

Before assigning labels	After assigning "Unit" and "Function" labels
$min = 60.00\ s$	$\boldsymbol{min} = 60.00\ s$
$min(1,2,3) = 1.00$	$\underline{\mathbf{min}}(1,2,3) = 1.00$

FIGURE 7.1

max and *min* functions with a list

The max and min functions can include a list of variable names.

$\mathbf{Var1} := 3 \qquad \mathbf{Var2} := 5 \qquad \mathbf{Var3} := [7\ \ 9\ \ 10\ \ 6]$

$\mathbf{Max3} := \underline{\mathbf{max}}(\mathbf{Var1}, \mathbf{Var2}, \mathbf{Var3}) = 10.00$

$\mathbf{Min3} := \underline{\mathbf{min}}(\mathbf{Var1}, \mathbf{Var2}, \mathbf{Var3}) = 3.00$

FIGURE 7.2

max and *min* functions with variables

The ***max*** and ***min*** functions can also be used to select the maximum and minimum values from vectors and matrices.

$$\mathbf{Matrix6.2} := \begin{bmatrix} 1 & 2 & 3 \\ 4 & 5 & 6 \\ 7 & 8 & 9 \end{bmatrix} \qquad \mathbf{Vector6.2} := \begin{bmatrix} 13 \\ 16 \\ 19 \end{bmatrix}$$

$\mathbf{Max3} := \underline{\mathbf{max}}(\mathbf{Matrix6.2}) = 9.00 \qquad \mathbf{Max4} := \underline{\mathbf{max}}(\mathbf{Vector6.2}) = 19.00$

$\mathbf{Min3} := \underline{\mathbf{min}}(\mathbf{Matrix6.2}) = 1.00 \qquad \mathbf{Min4} := \underline{\mathbf{min}}(\mathbf{Vector6.2}) = 13.00$

max and ***min*** can also select from multiple arrays.

$\mathbf{Max5} := \underline{\mathbf{max}}(\mathbf{Matrix6.2}, \mathbf{Vector6.2}) = 19.00$

$\mathbf{Min5} := \underline{\mathbf{min}}(\mathbf{Matrix6.2}, \mathbf{Vector6.2}) = 1.00$

FIGURE 7.3

max and *min* functions with arrays

The *max* and *min* functions may also include units from the same unit dimension. PTC Mathcad converts all units to consistent SI units, and then selects the maximum or minimum values. The result can be displayed in any unit.

$$\text{Max6} := \underline{\text{max}}(1\ m, 3\ m, 5\ m, 7\ m, 9\ cm, 8\ m, 6\ m, 4\ m, 25\ ft) = 8.00\ m$$

$$\text{Max6} = 26.25\ ft$$

$$\text{Min6} := \underline{\text{min}}(1\ m, 3\ m, 5\ m, 7\ m, 9\ cm, 8\ m, 6\ m, 4\ m, 25\ ft) = 0.09\ m$$

$$\text{Min6} = 0.30\ ft$$

$$\text{Max7} := \underline{\text{max}}(1\ min, 30\ s, 0.5\ day, 5\ hr) = 43200.00\ s$$

$$\text{Max7} = 12.00\ hr$$

$$\text{Min7} := \underline{\text{min}}(1\ min, 30\ s, 0.5\ day, 5\ hr) = 30.00\ s$$

$$\text{Matrix6.4} := \begin{bmatrix} 3\ ft & 5\ in & 10\ mm \\ 1\ m & 56\ cm & 1\ yd \\ 0.0005\ mile & 5000\ mil & 0.0005\ furlong \end{bmatrix}$$

$$\text{Matrix6.4} = \begin{bmatrix} 0.91 & 0.13 & 0.01 \\ 1.00 & 0.56 & 0.91 \\ 0.80 & 0.13 & 0.10 \end{bmatrix} m$$

$$\text{Max8} := \underline{\text{max}}(\text{Matrix6.4}) = 1.00\ m \qquad \text{Max8} = 3.28\ ft$$

$$\text{Min8} := \underline{\text{min}}(\text{Matrix6.4}) = 0.01\ m \qquad \text{Min8} = 10.00\ mm$$

FIGURE 7.4

max and *min* functions with units

smallest real part of any value, and i times the smallest imaginary part of any value. See Figure 7.5.

A, B, C, etc. can even be strings. For strings, "z" is larger than "a," thus the string "cat" is larger than the string "alligator." You cannot mix strings and numbers. See Figure 7.6.

mean and *median* Functions

PTC Mathcad has many statistical and data analysis functions. We will discuss only two of these functions. The ***mean*** function is useful for calculating averages of a list of values. The ***median*** function returns the value above and below which there are an equal number of values.

The ***mean*** function takes the form ***mean***(A,B,C,...). The arguments A, B, C, etc. can be scalars, arrays or complex numbers. The arguments can also have units

If the argument list for the ***max*** and ***min*** functions include complex numbers, then PTC Mathcad selects the largest or smallest real values and the largest or smallest imaginary part, even though they may be from different elements.

$$\mathbf{Vector6.5a} := \begin{bmatrix} 1+9i \\ 8 \\ 3+3i \\ 4+4i \end{bmatrix} \qquad \mathbf{Vector6.5b} := \begin{bmatrix} 1+9i \\ 8 \\ 3-3i \\ 4+4i \end{bmatrix}$$

$$\mathbf{Max9} := \underline{\mathbf{max}}(\mathbf{Vector6.5a}) = 8.00 + 9.00i$$

$$\mathbf{Min9} := \underline{\mathbf{min}}(\mathbf{Vector6.5a}) = 1.00$$

In this case 0*i is the smallest imaginary part.

$$\mathbf{Max10} := \underline{\mathbf{max}}(\mathbf{Vector6.5b}) = 8.00 + 9.00i$$

$$\mathbf{Min10} := \underline{\mathbf{min}}(\mathbf{Vector6.5b}) = 1.00 - 3.00i$$

FIGURE 7.5

max and *min* functions with complex numbers

The max and min functions can also select from a list of string variables.

For string variables "b" > "a".

AlphaMax := <u>**max**</u>(“cat”, “Alligator”) = “cat”

AlphaMin := *min*(“cat”, “Alligator”) = “Alligator”

FIGURE 7.6

max and *min* functions with string variables

attached. The ***mean*** function sums all the elements in the argument list and divides by the number of elements. If one or all of the arguments are arrays, PTC Mathcad sums all the elements in the arrays and counts all the elements in the arrays. If any of the arguments are complex numbers, PTC Mathcad returns i times the sum of the imaginary parts divided by the total number of all elements (not just the complex numbers). When using units, PTC Mathcad converts all units to SI units before taking the average. It then displays the average in the desired unit system. See Figure 7.7.

The ***median*** function returns the median of all the elements in the argument list. The median is the value above and below which there are an equal number of values. If there are an even number of values, the median is the arithmetic mean of the two central values. The arguments A, B, C, etc. can be scalars or arrays, but not complex numbers. The arguments can also have units attached. PTC Mathcad first sorts the arguments from lowest to highest prior to taking the median. See Figure 7.8.

The mean function calculates the average of a list of variables.

$$\mathbf{Matrix6.7a} := \begin{bmatrix} 3 & 5 & 7 \\ 9 & 11 & 13 \\ 15 & 17 & 19 \end{bmatrix} \qquad \mathbf{Vector6.7a} := \begin{bmatrix} 5 \\ 15 \\ 30 \end{bmatrix}$$

$$\mathbf{Mean1} := \underline{\mathbf{mean}}(\mathbf{Matrix6.7a}) = 11.00$$

$$\mathbf{Mean2} := \underline{\mathbf{mean}}(\mathbf{Vector6.7a}) = 16.67$$

$$\mathbf{Mean3} := \underline{\mathbf{mean}}(\mathbf{Matrix6.7a}, \mathbf{Vector6.7a}) = 12.42$$

The arguments may include units. PTC Mathcad converts all units to consistent SI units, and then takes the mean.

$$\mathbf{Matrix6.7b} := \begin{bmatrix} 3 \cdot ft & 5 \cdot in & 10 \cdot mm \\ 1 \cdot m & 56 \cdot cm & 1 \cdot yd \\ 0.0005 \cdot mile & 5000 \cdot mil & 0.0005 \cdot furlong \end{bmatrix}$$

$$\mathbf{Matrix6.7b} = \begin{bmatrix} 0.91 & 0.13 & 0.01 \\ 1.00 & 0.56 & 0.91 \\ 0.80 & 0.13 & 0.10 \end{bmatrix} m$$

$$\mathbf{Mean4} := \underline{\mathbf{mean}}(\mathbf{Matrix6.7b}) = 0.51\ m \qquad \mathbf{Mean4} = 1.66\ ft$$

If any of the arguments are complex numbers, PTC Mathcad takes the average of the real parts and the average of the imaginary parts (including the numbers with 0i).

$$\mathbf{Vector6.7c} := \begin{bmatrix} 1+9i \\ 8 \\ 3+3i \\ 4+4i \end{bmatrix} \qquad \mathbf{Vector6.7d} := \begin{bmatrix} 5+9i \\ 8 \\ 3-3i \\ 4+4i \end{bmatrix}$$

$$\mathbf{Mean5} := \underline{\mathbf{mean}}(\mathbf{Vector6.7c}) = 4.00 + 4.00i$$

$$\mathbf{Mean6} := \underline{\mathbf{mean}}(\mathbf{Vector6.7d}) = 5.00 + 2.50i$$

$$\mathbf{Mean7} := \underline{\mathbf{mean}}(\mathbf{Vector6.7c}, \mathbf{Vector6.7d}) = 4.50 + 3.25i$$

FIGURE 7.7

mean function

Truncation and Rounding Functions

We will discuss four truncation and rounding functions (***floor***, ***ceil***, ***trunc***, and ***round***). Each of these functions has two forms. One form is the lowercase form, and the other form is the uppercase form. We will discuss the lowercase forms first.

The ***median*** function returns the value above and below which there are an equal number of values.

$\text{Test\_median}_1 := \underline{\text{median}}(3,5,7,9,8,6,4) = 6.00$

$\text{Test\_median}_2 := \underline{\text{median}}(5 \cdot m, 8 \cdot m, 6 \cdot m, 9 \cdot m) = 7.00\ m$

$\text{Test\_median}_3 := \underline{\text{median}}(5 \cdot m, 8 \cdot m, 6 \cdot cm, 9 \cdot mm) = 2.53\ m$

$\underline{\text{median}}(3,4,5,6,7,8,9) = 6.00$

$\underline{\text{median}}(5,6,8,9) = 7.00$

$\underline{\text{median}}(0.009, 0.06, 5, 8) = 2.53$

$\frac{0.06+5}{2} = 2.53$

If there are an even number of values, then PTC Mathcad takes the arithmetic mean of the two central values.

$\text{Matrix6.7a} = \begin{bmatrix} 3.00 & 5.00 & 7.00 \\ 9.00 & 11.00 & 13.00 \\ 15.00 & 17.00 & 19.00 \end{bmatrix} \quad \text{Vector6.7a} = \begin{bmatrix} 5.00 \\ 15.00 \\ 30.00 \end{bmatrix}$

$\text{Test\_median}_4 := \underline{\text{median}}(\text{Matrix6.7a}) = 11.00$

$\text{Test\_median}_5 := \underline{\text{median}}(\text{Vector6.7a}) = 15.00$

$\text{Test\_median}_6 := \underline{\text{median}}(\text{Matrix6.7a}, \text{Vector6.7a}) = 12.00$

Check

$\underline{\text{median}}(3,5,5,7,9,11,13,15,15,17,19,30) = 12.00$

$\text{Matrix6.7b} = \begin{bmatrix} 0.91 & 0.13 & 0.01 \\ 1.00 & 0.56 & 0.91 \\ 0.80 & 0.13 & 0.10 \end{bmatrix} m$

$\text{Test\_median}_7 := \underline{\text{median}}(\text{Matrix6.7b}) = 0.56\ m$

Check

$\underline{\text{median}}(0.01, 0.10, 0.13, 0.13, 0.56, 0.80, 0.91, 0.91, 1.00) = 0.56$

FIGURE 7.8

median function

The function ***floor***(z) returns the greatest integer less than z.

The function ***ceil***(z) returns the smallest integer greater than z.

The function ***trunc***(z) returns the integer part of z by removing the fractional part. If z is greater than zero, then this function is identical to ***floor***(z). If z is less than zero, then this function is identical to ***ceil***(z).

The function ***round***(z,[n]) returns z rounded to n decimal places. The argument n must be an integer. If n is omitted (or equal to zero), it returns z rounded to the nearest integer. If n is less than zero, it returns z rounded to n places to the left of the decimal point.

The argument z can be a real or complex scalar or vector. It cannot be an array. See Figures 7.9 and 7.10 for examples of the lowercase truncate and round functions. The argument z in these lowercase functions may not have units attached, but there is still a way to use units with these functions. If your list of values has units attached, divide the list of values by the unit you want to use, and then multiply the function by the same unit. See Figure 7.11 for an example of using units with the lowercase truncate and round functions.

Use the following four values to examine the results of the following four functions: *floor*, *ceil*, *trunc*, and *round*.

$Var_{6.9a} := 3.49 \quad Var_{6.9b} := 3.51 \quad Var_{6.9c} := -3.49 \quad Var_{6.9d} := -3.51$

$Test\_floor_1 := \underline{floor}(Var_{6.9a}) = 3.00 \quad Test\_floor_2 := \underline{floor}(Var_{6.9b}) = 3.00$

$Test\_floor_3 := \underline{floor}(Var_{6.9c}) = -4.00 \quad Test\_floor_4 := \underline{floor}(Var_{6.9d}) = -4.00$

$Test\_ceil_1 := \underline{ceil}(Var_{6.9a}) = 4.00 \quad Test\_ceil_2 := \underline{ceil}(Var_{6.9b}) = 4.00$

$Test\_ceil_3 := \underline{ceil}(Var_{6.9c}) = -3.00 \quad Test\_ceil_4 := \underline{ceil}(Var_{6.9d}) = -3.00$

$Test\_trunc_1 := \underline{trunc}(Var_{6.9a}) = 3.00 \quad Test\_trunc_2 := \underline{trunc}(Var_{6.9b}) = 3.00$

$Test\_trunc_3 := \underline{trunc}(Var_{6.9c}) = -3.00 \quad Test\_trunc_4 := \underline{trunc}(Var_{6.9d}) = -3.00$

$Test\_round_1 := \underline{round}(Var_{6.9a}) = 3.00 \quad Test\_round_2 := \underline{round}(Var_{6.9b}) = 4.00$

$Test\_round_3 := \underline{round}(Var_{6.9c}) = -3.00 \quad Test\_round_4 := \underline{round}(Var_{6.9d}) = -4.00$

FIGURE 7.9

Lowercase truncate and round functions

The lowercase truncate and round functions may have vectors as arguments, but not matrices.

$$\mathbf{Var}_{6.10a} := \begin{bmatrix} 4.49 \\ 49.50 \\ -150.01 \\ -1499.99 \\ 5000.01 \end{bmatrix}$$

$$\mathbf{Test\_floor}_5 := \underline{\mathbf{floor}}\left(\mathbf{Var}_{6.10a}\right) = \begin{bmatrix} 4.00 \\ 49.00 \\ -151.00 \\ -1500.00 \\ 5000.00 \end{bmatrix}$$

$$\mathbf{Test\_ceil}_5 := \underline{\mathbf{ceil}}\left(\mathbf{Var}_{6.10a}\right) = \begin{bmatrix} 5.00 \\ 50.00 \\ -150.00 \\ -1499.00 \\ 5001.00 \end{bmatrix}$$

$$\mathbf{Test\_trunc}_5 := \underline{\mathbf{trunc}}\left(\mathbf{Var}_{6.10a}\right) = \begin{bmatrix} 4.00 \\ 49.00 \\ -150.00 \\ -1499.00 \\ 5000.00 \end{bmatrix}$$

$$\mathbf{Test\_round}_5 := \underline{\mathbf{round}}\left(\mathbf{Var}_{6.10a}\right) = \begin{bmatrix} 4.00 \\ 50.00 \\ -150.00 \\ -1500.00 \\ 5000.00 \end{bmatrix}$$

The *round* function has the form *round*(z,[n]), where n is an integer telling PTC Mathcad at which decimal place to round. If n is less than zero, PTC Mathcad rounds to n places to the left of the decimal point. If n is left blank (zero), PTC Mathcad rounds to an integer.

$$\mathbf{Test\_round}_6 := \underline{\mathbf{round}}\left(\mathbf{Var}_{6.10a}, 1\right) = \begin{bmatrix} 4.50 \\ 49.50 \\ -150.00 \\ -1500.00 \\ 5000.00 \end{bmatrix}$$

n=1 rounds to 1 decimal place.

$$\mathbf{Test\_round}_7 := \underline{\mathbf{round}}\left(\mathbf{Var}_{6.10a}, -1\right) = \begin{bmatrix} 0.00 \\ 50.00 \\ -150.00 \\ -1500.00 \\ 5000.00 \end{bmatrix}$$

n=-1 rounds to the 10's.

$$\mathbf{Test\_round}_8 := \underline{\mathbf{round}}\left(\mathbf{Var}_{6.10a}, -2\right) = \begin{bmatrix} 0.00 \\ 0.00 \\ -200.00 \\ -1500.00 \\ 5000.00 \end{bmatrix}$$

n=-2 rounds to the 100's.

FIGURE 7.10

Lowercase truncate and round functions with vectors

This figure is the same as Figure 7.10 except that units are used.

Even though the lowercase truncate and round functions do not allow the use of units, you can still use units with these functions. To do this, divide the list of units by the unit you want to truncate or round to, then multiply the function by the same unit.

$$\mathbf{Var}_{7.11} := \begin{bmatrix} 4.49 \\ 49.50 \\ -150.01 \\ -1499.99 \\ 5000.01 \end{bmatrix} \cdot m$$

$$\mathbf{Var}_{7.11} = \begin{bmatrix} 4.49 \\ 49.50 \\ -150.01 \\ -1499.99 \\ 5000.01 \end{bmatrix} m \qquad \mathbf{Var}_{7.11} = \begin{bmatrix} 14.73 \\ 162.40 \\ -492.16 \\ -4921.23 \\ 16404.23 \end{bmatrix} ft$$

$\mathbf{F7.11a} := floor\left(Var_{7.11}\right) = ?$ Units cannot be used with lowercase functions.

$\mathbf{F7.11a} := floor\left(Var_{7.11}\right) = ?$

These units are not compatible.

$$\mathbf{F7.11a} := \underline{\text{floor}}\left(\frac{\mathbf{Var}_{7.11}}{m}\right) \cdot m = \begin{bmatrix} 4.00 \\ 49.00 \\ -151.00 \\ -1500.00 \\ 5000.00 \end{bmatrix} m$$

$$\mathbf{F7.11b} := \underline{\text{floor}}\left(\frac{\mathbf{Var}_{7.11}}{ft}\right) \cdot ft = \begin{bmatrix} 14.00 \\ 162.00 \\ -493.00 \\ -4922.00 \\ 16404.00 \end{bmatrix} ft$$

Note that the rounding unit may be different than the input unit. You can use feet, even though the input was in meters.

$$\mathbf{F7.11c} := \underline{\text{trunc}}\left(\frac{\mathbf{Var}_{7.11}}{m}\right) \cdot m = \begin{bmatrix} 4.00 \\ 49.00 \\ -150.00 \\ -1499.00 \\ 5000.00 \end{bmatrix} m$$

FIGURE 7.11 *(Continued on next page)*

Lowercase truncate and round functions with units

$$F7.11d := \underline{ceil}\left(\frac{\mathbf{Var}_{7.11}}{m}\right)\cdot m = \begin{bmatrix} 5.00 \\ 50.00 \\ -150.00 \\ -1499.00 \\ 5001.00 \end{bmatrix} m$$

$$F7.11e := \underline{round}\left(\frac{\mathbf{Var}_{7.11}}{m}\right)\cdot m = \begin{bmatrix} 4.00 \\ 50.00 \\ -150.00 \\ -1500.00 \\ 5000.00 \end{bmatrix} m$$

$$F7.11f := \underline{round}\left(\frac{\mathbf{Var}_{7.11}}{ft}\right)\cdot ft = \begin{bmatrix} 15.00 \\ 162.00 \\ -492.00 \\ -4921.00 \\ 16404.00 \end{bmatrix} ft$$

The round function has the form round(z,[n]), where n is an integer telling PTC Mathcad at which decimal place to round. If n is less than zero, PTC Mathcad rounds to n places to the left of the decimal point. In n is left blank (zero), PTC Mathcad rounds to an integer.

$$F7.11g := \underline{round}\left(\frac{\mathbf{Var}_{7.11}}{m}, 1\right)\cdot m = \begin{bmatrix} 4.50 \\ 49.50 \\ -150.00 \\ -1500.00 \\ 5000.00 \end{bmatrix} m$$

$$F7.11h := \underline{round}\left(\frac{\mathbf{Var}_{7.11}}{m}, -1\right)\cdot m = \begin{bmatrix} 0.00 \\ 50.00 \\ -150.00 \\ -1500.00 \\ 5000.00 \end{bmatrix} m$$

$$F7.11i := \underline{round}\left(\frac{\mathbf{Var}_{7.11}}{ft}, -2\right)\cdot ft = \begin{bmatrix} 0.00 \\ 200.00 \\ -500.00 \\ -4900.00 \\ 16400.00 \end{bmatrix} ft$$

FIGURE 7.11

(Continued)

The uppercase forms of these equations are a bit more complicated. The uppercase functions introduce an additional argument "y." This argument must be a real, nonzero scalar or vector. The argument "y" tells PTC Mathcad to truncate or round to a multiple of y. The uppercase functions are specifically used for values with units. The lowercase functions are equivalent to the uppercase functions with y equal to one. (The n in ***round***(z,[n]) being zero.) If y is equal to three, then PTC Mathcad truncates or rounds the argument z to a multiple of three. Let's look at a few examples. See Figures 7.12, 7.13 (without units) and Figure 7.14 (with units).

The uppercase truncate and round functions introduce a second argument y. The functions have the form ***Floor***(z,y). The argument y tells PTC Mathcad to truncate or round to a multiple of y. This is useful for even numbers.

$\mathbf{Var}_{6.9a} = 3.49 \qquad \mathbf{Var}_{6.9b} = 3.51 \qquad \mathbf{Var}_{6.9c} = -3.49$

Using y=2 means that all results will be a multiple of 2 (or an even number).

$\mathbf{Test\_Floor}_9 := \underline{\mathbf{Floor}}\left(\mathbf{Var}_{6.9a}, 2\right) = 2.00$

$\mathbf{Test\_Floor}_{10} := \underline{\mathbf{Floor}}\left(\mathbf{Var}_{6.9b}, 2\right) = 2.00$

$\mathbf{Test\_Floor}_{11} := \underline{\mathbf{Floor}}\left(\mathbf{Var}_{6.9c}, 2\right) = -4.00$

$\mathbf{Test\_Ceil}_9 := \underline{\mathbf{Ceil}}\left(\mathbf{Var}_{6.9a}, 2\right) = 4.00$

$\mathbf{Test\_Ceil}_{10} := \underline{\mathbf{Ceil}}\left(\mathbf{Var}_{6.9b}, 2\right) = 4.00$

$\mathbf{Test\_Ceil}_{11} := \underline{\mathbf{Ceil}}\left(\mathbf{Var}_{6.9c}, 2\right) = -2.00$

$\mathbf{Test\_Trunc}_9 := \underline{\mathbf{Trunc}}\left(\mathbf{Var}_{6.9a}, 2\right) = 2.00$

$\mathbf{Test\_Trunc}_{10} := \underline{\mathbf{Trunc}}\left(\mathbf{Var}_{6.9b}, 2\right) = 2.00$

$\mathbf{Test\_Trunc}_{11} := \underline{\mathbf{Trunc}}\left(\mathbf{Var}_{6.9c}, 2\right) = -2.00$

$\mathbf{Test\_Round}_9 := \underline{\mathbf{Round}}\left(\mathbf{Var}_{6.9a}, 2\right) = 4.00$

$\mathbf{Test\_Round}_{10} := \underline{\mathbf{Round}}\left(\mathbf{Var}_{6.9b}, 2\right) = 4.00$

$\mathbf{Test\_Round}_{11} := \underline{\mathbf{Round}}\left(\mathbf{Var}_{6.9c}, 2\right) = -4.00$

FIGURE 7.12 *(Continued on next page)*

Uppercase truncate and round functions

Using y=0.1 means that all results will be a multiple of 0.1, or to the first decimal place.

$\text{Test\_Floor}_{12} := \underline{\text{Floor}}\left(\text{Var}_{6.9a}, 0.1\right) = 3.40$

$\text{Test\_Floor}_{13} := \underline{\text{Floor}}\left(\text{Var}_{6.9b}, 0.1\right) = 3.50$

$\text{Test\_Floor}_{14} := \underline{\text{Floor}}\left(\text{Var}_{6.9c}, 0.2\right) = -3.60$

$\text{Test\_Ceil}_{12} := \underline{\text{Ceil}}\left(\text{Var}_{6.9a}, 0.1\right) = 3.50$

$\text{Test\_Ceil}_{13} := \underline{\text{Ceil}}\left(\text{Var}_{6.9b}, 0.1\right) = 3.60$

$\text{Test\_Ceil}_{14} := \underline{\text{Ceil}}\left(\text{Var}_{6.9c}, 0.1\right) = -3.40$

$\text{Test\_Trunc}_{12} := \underline{\text{Trunc}}\left(\text{Var}_{6.9a}, 0.1\right) = 3.40$

$\text{Test\_Trunc}_{13} := \underline{\text{Trunc}}\left(\text{Var}_{6.9b}, 0.1\right)$

$\text{Test\_Trunc}_{14} := \underline{\text{Trunc}}\left(\text{Var}_{6.9c}, 0.1\right) = -3.40$

$\text{Test\_Round}_{12} := \underline{\text{Round}}\left(\text{Var}_{6.9a}, 0.1\right) = 3.50$

$\text{Test\_Round}_{13} := \underline{\text{Round}}\left(\text{Var}_{6.9b}, 0.1\right) = 3.50$

$\text{Test\_Round}_{14} := \underline{\text{Round}}\left(\text{Var}_{6.9c}, 0.1\right) = -3.50$

FIGURE 7.12

(Continued)

The uppercase truncate and round functions may have vectors as arguments, but not matrices. Let's look at the same variables from Figure 7.10 and use 3 values of y: y=2, y=0.1, and y=0.01.

$$\mathbf{Var}_{6.10a} = \begin{bmatrix} 4.49 \\ 49.50 \\ -150.01 \\ -1499.99 \\ 5000.01 \end{bmatrix}$$

Note:
In order to fill all examples in this figure, the results are not attached to a variable. This is not good engineering practice, because the results will not be available for later use.

$$\underline{\text{Floor}}\left(\mathbf{Var}_{6.10a}, 2\right) = \begin{bmatrix} 4.00 \\ 48.00 \\ -152.00 \\ -1500.00 \\ 5000.00 \end{bmatrix} \quad \underline{\text{Ceil}}\left(\mathbf{Var}_{6.10a}, 2\right) = \begin{bmatrix} 6.00 \\ 50.00 \\ -150.00 \\ -1498.00 \\ 5002.00 \end{bmatrix}$$

$$\underline{\text{Floor}}\left(\mathbf{Var}_{6.10a}, 0.1\right) = \begin{bmatrix} 4.40 \\ 49.50 \\ -150.10 \\ -1500.00 \\ 5000.00 \end{bmatrix} \quad \underline{\text{Ceil}}\left(\mathbf{Var}_{6.10a}, 0.1\right) = \begin{bmatrix} 4.50 \\ 49.50 \\ -150.00 \\ -1499.90 \\ 5000.10 \end{bmatrix}$$

$$\underline{\text{Floor}}\left(\mathbf{Var}_{6.10a}, 0.01\right) = \begin{bmatrix} 4.49 \\ 49.50 \\ -150.01 \\ -1499.99 \\ 5000.01 \end{bmatrix} \quad \underline{\text{Ceil}}\left(\mathbf{Var}_{6.10a}, 0.01\right) = \begin{bmatrix} 4.49 \\ 49.50 \\ -150.01 \\ -1499.99 \\ 5000.01 \end{bmatrix}$$

$$\underline{\text{Trunc}}\left(\mathbf{Var}_{6.10a}, 2\right) = \begin{bmatrix} 4.00 \\ 48.00 \\ -150.00 \\ -1498.00 \\ 5000.00 \end{bmatrix} \quad \underline{\text{Round}}\left(\mathbf{Var}_{6.10a}, 2\right) = \begin{bmatrix} 4.00 \\ 50.00 \\ -150.00 \\ -1500.00 \\ 5000.00 \end{bmatrix}$$

$$\underline{\text{Trunc}}\left(\mathbf{Var}_{6.10a}, 0.1\right) = \begin{bmatrix} 4.40 \\ 49.50 \\ -150.00 \\ -1499.90 \\ 5000.00 \end{bmatrix} \quad \underline{\text{Round}}\left(\mathbf{Var}_{6.10a}, 0.1\right) = \begin{bmatrix} 4.50 \\ 49.50 \\ -150.00 \\ -1500.00 \\ 5000.00 \end{bmatrix}$$

$$\underline{\text{Trunc}}\left(\mathbf{Var}_{6.10a}, 0.01\right) = \begin{bmatrix} 4.49 \\ 49.50 \\ -150.01 \\ -1499.99 \\ 5000.01 \end{bmatrix} \quad \underline{\text{Round}}\left(\mathbf{Var}_{6.10a}, 0.01\right) = \begin{bmatrix} 4.49 \\ 49.50 \\ -150.01 \\ -1499.99 \\ 5000.01 \end{bmatrix}$$

FIGURE 7.13

Uppercase truncate and round functions with vectors

The uppercase truncate and round functions are ideal for use with units. You do not need to divide the units out as you do with the lowercase functions.

$$\text{Var}_{6.14} := \begin{bmatrix} 4.49 \\ 49.50 \\ 150.01 \\ 1499.99 \\ 5000.01 \end{bmatrix} \cdot m \qquad \text{Var}_{6.14} = \begin{bmatrix} 14.73 \\ 162.40 \\ 492.16 \\ 4921.23 \\ 16404.23 \end{bmatrix} ft$$

$$\text{Test\_Floor}_{15} := \underline{\text{Floor}}\left(\text{Var}_{6.14}, 1 \cdot m\right) \quad \text{Test\_Round}_{15} := \underline{\text{Round}}\left(\text{Var}_{6.14}, 1 \cdot m\right)$$

$$\text{Test\_Floor}_{15} = \begin{bmatrix} 4.00 \\ 49.00 \\ 150.00 \\ 1499.00 \\ 5000.00 \end{bmatrix} m \qquad \text{Test\_Round}_{15} = \begin{bmatrix} 4.00 \\ 50.00 \\ 150.00 \\ 1500.00 \\ 5000.00 \end{bmatrix} m$$

$$\text{Test\_Floor}_{16} := \underline{\text{Floor}}\left(\text{Var}_{6.14}, 1 \cdot ft\right) \quad \text{Test\_Round}_{16} := \underline{\text{Round}}\left(\text{Var}_{6.14}, 1 \cdot ft\right)$$

Floor and ***Round*** to the nearest foot, but displayed in meters.

$$\text{Test\_Floor}_{16} = \begin{bmatrix} 4.27 \\ 49.38 \\ 149.96 \\ 1499.92 \\ 4999.94 \end{bmatrix} m \qquad \text{Test\_Round}_{16} = \begin{bmatrix} 4.57 \\ 49.38 \\ 149.96 \\ 1499.92 \\ 4999.94 \end{bmatrix} m$$

Floor and ***Round*** to the nearest foot, but displayed in feet.

$$\text{Test\_Floor}_{16} = \begin{bmatrix} 14.00 \\ 162.00 \\ 492.00 \\ 4921.00 \\ 16404.00 \end{bmatrix} ft \qquad \text{Test\_Round}_{16} = \begin{bmatrix} 15.00 \\ 162.00 \\ 492.00 \\ 4921.00 \\ 16404.00 \end{bmatrix} ft$$

FIGURE 7.14

Uppercase truncate and round functions with units

Summation Operator

PTC® Mathcad Prime® uses a single ***summation*** operator with optional placeholders in place of the three ***summation*** operators used in PTC Mathcad 15. $\sum$ $\sum$ $\sum$ Its shortcut is CTRL+$.

The ***summation*** operator is technically not a function, but it is useful in technical calculations, so it is included in this chapter. PTC Mathcad uses a single ***summation*** operator with three placeholders $\sum$. The top and bottom placeholders are optional. There are three ways of using the ***summation*** operator. We will refer to these as the ***vector sum*** operator, the ***range sum*** operator and the ***summation*** operator. The location of the ***summation*** operator is **Math>Operators and Symbols>Operators>Calculus>$\sum$**. The shortcut is **CTRL+$, or CTRL+SHIFT+4**.

The simplest use is the ***vector sum*** operator. This operator sums all the elements in a vector. It is useful when you want to add a variable series of numbers. If you include in a vector all the numbers you want to add, this function will give the sum of all the elements. See Figure 7.15 for some examples.

A second use of the ***summation*** operator uses all three placeholders, and is just referred to as the ***summation*** operator. It allows you to sum an expression or a function over a range of values. The bottom placeholder contains the index of the summation and the beginning limit. This index is independent of any variable name outside of the operator. It can be any variable name, but since it is independent to the operator, it is best to keep it a single letter. In the bottom placeholder type the index followed by the equal sign, followed by the value for the beginning limit. The upper placeholder contains the ending limit. The placeholder to the right is for an expression or function that typically contains the index, but it is not necessary. Let's look at some examples. See Figures 7.16 and 7.17.

Type CTRL+SHIFT+4 or use the **Operators** list on the **Math** tab to get the ***Summation*** operator. Type the name of the vector in the right placeholder.

$$\mathbf{Vector}_{6.15} := \begin{bmatrix} 5 \\ 4 \\ 3 \\ 2 \\ 1 \end{bmatrix} \qquad \sum \mathbf{Vector}_{6.15} = 15.00$$

Suppose you want to input various data, and then sum them.

$$\mathbf{Input}_1 := 50 \cdot \boldsymbol{Pa}$$

$$\mathbf{Input}_2 := 100 \cdot \boldsymbol{Pa}$$

$$\mathbf{Input}_3 := 40 \cdot \boldsymbol{Pa}$$

$$\mathbf{Input}_4 := 120 \cdot \boldsymbol{Pa}$$

Note that these variables are not literal subscripts (typed with the period key). They are array subscripts typed with the [key.

$$\mathbf{Input} = \begin{bmatrix} 50.00 \\ 100.00 \\ 40.00 \\ 120.00 \end{bmatrix} \boldsymbol{Pa}$$

Give the sum of the input variables.

$$\sum \mathbf{Input} = 310.00\ \boldsymbol{Pa}$$

Try the same example, but use a table for input.

In
(Pa)
50
100
40
120

$$\mathbf{In} = \begin{bmatrix} 50.00 \\ 100.00 \\ 40.00 \\ 120.00 \end{bmatrix} \boldsymbol{Pa} \qquad \sum \mathbf{In} = 310.00\ \boldsymbol{Pa}$$

FIGURE 7.15

Vector sum operator

Type CTRL+SHIFT+4 or use the **Operators** list on the **Math** tab to get the ***Summation*** operator. Type the name of the vector in the right placeholder.

Define the index, the beginning limit, the ending limit, and the expression.

$\sum_{i=1}^{4} 2^{i} = 30.00$ This is equivalent to: $2^1 + 2^2 + 2^3 + 2^4 = 30.00$

$f(x) := 2^x$ Define a function to use in the summation.

$\text{BeginningLimit} := 1$ The beginning limit and ending limit may be variables.

$\text{EndingLimit} := 4$

$\sum_{j=\text{BeginningLimit}}^{\text{EndingLimit}} f(j) = 30.00$ This is the same as above, except the function and limits were defined previously.

The summation expression does not need to contain the index, as shown below.

$\sum_{k=1}^{5} 2 = 10.00$ This is equivalent to: $2+2+2+2+2 = 10.00$

FIGURE 7.16

Summation operator

You can also use vectors and arrays with the *Summation* operator.

$$\mathbf{Vector}_{6.17} := \begin{bmatrix} 5 \\ 4 \\ 3 \\ 2 \\ 1 \end{bmatrix} \qquad \mathbf{Matrix}_{6.17} := \begin{bmatrix} 1 & 2 \\ 3 & 4 \end{bmatrix}$$

$$\mathbf{Summation\_1} := \sum_{l=1}^{3} \left(\mathbf{Vector}_{6.17} \cdot l\right)$$

This is equivalent to:

$$\mathbf{Summation\_1} = \begin{bmatrix} 30.00 \\ 24.00 \\ 18.00 \\ 12.00 \\ 6.00 \end{bmatrix} \qquad \begin{bmatrix} 5 \cdot 1 + 5 \cdot 2 + 5 \cdot 3 \\ 4 \cdot 1 + 4 \cdot 2 + 4 \cdot 3 \\ 3 \cdot 1 + 3 \cdot 2 + 3 \cdot 3 \\ 2 \cdot 1 + 2 \cdot 2 + 2 \cdot 3 \\ 1 \cdot 1 + 1 \cdot 2 + 1 \cdot 3 \end{bmatrix} = \begin{bmatrix} 30.00 \\ 24.00 \\ 18.00 \\ 12.00 \\ 6.00 \end{bmatrix}$$

$$\mathbf{Summation\_2} := \sum_{m=1}^{3} \left(\mathbf{Matrix}_{6.17} \cdot m\right)$$

$$\mathbf{Summation\_2} = \begin{bmatrix} 6.00 & 12.00 \\ 18.00 & 24.00 \end{bmatrix}$$

This is equivalent to:

$$\begin{bmatrix} 1 \cdot 1 + 1 \cdot 2 + 1 \cdot 3 & 2 \cdot 1 + 2 \cdot 2 + 2 \cdot 3 \\ 3 \cdot 1 + 3 \cdot 2 + 3 \cdot 3 & 4 \cdot 1 + 4 \cdot 2 + 4 \cdot 3 \end{bmatrix} = \begin{bmatrix} 6.00 & 12.00 \\ 18.00 & 24.00 \end{bmatrix}$$

FIGURE 7.17

Summation operator with vectors

The third use of the ***summation*** operator uses just two placeholders, and is referred to as the ***range sum*** operator. It is similar to the ***summation*** operator, except you must have a range variable defined prior to using it. It is used to sum an expression or a function over a range of variables. The bottom placeholder contains the name of the range variable. The placeholder to the right is for an expression or function that typically contains the name of the range variable, but it is not necessary. See Figures 7.18 and 7.19.

Figure 7.20 shows the use of the ***Range sum*** and ***Summation*** operators in an engineering example.

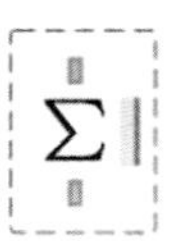

Type CTRL+SHIFT+4 or use the **Operators** list on the **Math** tab to get the ***Summation*** operator. Type the name of the vector in the right placeholder.

These examples are exactly the same as in Figure 7.16, except the Range sum operator is used.

In order to use the Range sum operator, the range must be defined previous to using the operator.

$i := 1, 2 .. 4$ Define a range variable.

$\sum_{i} 2^{i} = 30.00$ This is the same as in Figure 7.16, except the limits are defined by the range variable i.

$gg(x) := 2^{x}$ Define a function to use in the summation.

$BeginningCounter := 1$ Define limits for the range variable.

$EndingCounter := 4$

Define a range variable using previously defined variables.

$j := BeginningCounter .. EndingCounter$

$\sum_{j} gg(j) = 30.00$ This is the same as above, except the function was defined previously.

The summation expression does not need to contain the range variable.

$k := 1 .. 5$

$\sum_{k} 2 = 10.00$ This is equivalent to: $2+2+2+2+2 = 10.00$

FIGURE 7.18

Range sum operator

These examples are exactly the same as in Figure 7.17, except the Range sum operator is used.

You can also use vectors and arrays with the Range sum operator.

$$\mathbf{Vector}_{6.17} = \begin{bmatrix} 5.00 \\ 4.00 \\ 3.00 \\ 2.00 \\ 1.00 \end{bmatrix} \qquad \mathbf{Matrix}_{6.17} = \begin{bmatrix} 1.00 & 2.00 \\ 3.00 & 4.00 \end{bmatrix}$$

$$\mathbf{l} := 1..3$$

$$\mathbf{Summation\_3} := \sum_{l} \left(\mathbf{Vector}_{6.17} \cdot l\right)$$

This is equivalent to:

$$\mathbf{Summation\_3} = \begin{bmatrix} 30.00 \\ 24.00 \\ 18.00 \\ 12.00 \\ 6.00 \end{bmatrix}$$

$$\begin{bmatrix} 5 \cdot 1 + 5 \cdot 2 + 5 \cdot 3 \\ 4 \cdot 1 + 4 \cdot 2 + 4 \cdot 3 \\ 3 \cdot 1 + 3 \cdot 2 + 3 \cdot 3 \\ 2 \cdot 1 + 2 \cdot 2 + 2 \cdot 3 \\ 1 \cdot 1 + 1 \cdot 2 + 1 \cdot 3 \end{bmatrix} = \begin{bmatrix} 30.00 \\ 24.00 \\ 18.00 \\ 12.00 \\ 6.00 \end{bmatrix}$$

$$\mathbf{M} := 1..3$$

$$\mathbf{Summation\_4} := \sum_{M} \left(\mathbf{Matrix}_{6.17} \cdot M\right)$$

$$\mathbf{Summation\_4} = \begin{bmatrix} 6.00 & 12.00 \\ 18.00 & 24.00 \end{bmatrix}$$

This is equivalent to:

$$\begin{bmatrix} 1 \cdot 1 + 1 \cdot 2 + 1 \cdot 3 & 2 \cdot 1 + 2 \cdot 2 + 2 \cdot 3 \\ 3 \cdot 1 + 3 \cdot 2 + 3 \cdot 3 & 4 \cdot 1 + 4 \cdot 2 + 4 \cdot 3 \end{bmatrix} = \begin{bmatrix} 6.00 & 12.00 \\ 18.00 & 24.00 \end{bmatrix}$$

FIGURE 7.19

Range sum operator with vectors

Use the ***summation*** operator to calculate the total floor mass of a structure.

$\mathbf{TopFloor} := 4$ Define the top floor.

$\mathbf{Counter} := 1..\mathbf{TopFloor}$ Create a range variable for each floor.

$$\mathbf{Area} := \begin{bmatrix} 1400 \\ 1200 \\ 1200 \\ 1000 \end{bmatrix} \cdot m^2 \qquad \mathbf{Mass} := \begin{bmatrix} 700 \\ 700 \\ 700 \\ 500 \end{bmatrix} \cdot \frac{kg}{m^2}$$

Create two vectors with the area and mass/m^2 of each floor.

$$\mathbf{TotalMass} := \sum_{Counter} \left(\mathbf{Area}_{Counter} \cdot \mathbf{Mass}_{Counter}\right)$$

This equation takes the first element of Area and the first element of Mass and multiplies them. The counter then increments and the 2nd elements are multiplied. The process continues for the 3rd and 4th elements. It then takes the sum of the results.

$\mathbf{TotalMass} = 3160000.00\ kg$

Check

Note: Use CTRL+SHIFT+^ to get the vectorize operator.

$$\mathbf{test} := \overrightarrow{\left(\left(\begin{bmatrix} 1400 \\ 1200 \\ 1200 \\ 1000 \end{bmatrix} \cdot m^2 \cdot \begin{bmatrix} 700 \\ 700 \\ 700 \\ 500 \end{bmatrix} \cdot \frac{kg}{m^2}\right)\right)} \qquad \mathbf{test} = \begin{bmatrix} 980000.00 \\ 840000.00 \\ 840000.00 \\ 500000.00 \end{bmatrix} kg$$

$\sum \mathbf{test} = 3160000.00\ kg$

FIGURE 7.20

Range sum operator—engineering example

if Function

The *if* function allows PTC Mathcad to make a determination between two or more choices.

The *if* function is similar to the *if* function in Microsoft Excel. It takes the form *if*(Cond,x,y). Cond is an expression, typically involving a logical or Boolean operator. The function returns x if Cond is true, and y otherwise.

Let's look at some engineering examples. See Figures 7.21 and 7.22.

The *if* function takes the form *if*(cond,x,y). If cond is true then x. If cond is false then y.

Assume that some PTC Mathcad expressions returned two results.

$$\mathbf{Result_1} := 10 \cdot \boldsymbol{Hz}$$

$$\mathbf{Result_2} := 0 \cdot \boldsymbol{Hz}$$

Create two strings to use in the *if* function.

$\mathbf{IfTrue}$:= "Result1 is greater than Result2"

$\mathbf{IfFalse}$:= "Result1 is less than Result 2"

Is Result1 greater than Result2?

$$\mathbf{Result_3} := \text{if}\left(\mathbf{Result_1} > \mathbf{Result_2}, \mathbf{IfTrue}, \mathbf{IfFalse}\right)$$

$\mathbf{Result_3}$ = "Result1 is greater than Result2"

$$\mathbf{Result_4} := \text{if}\left(\mathbf{Result_2} = 0, \text{"Division by Zero"}, \frac{\mathbf{Result_1}}{\mathbf{Result_2}}\right)$$

$\mathbf{Result_4}$ = "Division by Zero"

FIGURE 7.21

if function

In this example, there is an input length. A specific formula needs to use the input length, but the length cannot be less than 25 ft.

$\mathbf{InputLength} := 20 \cdot ft$

$\mathbf{Length\_1} := \text{if}(\mathbf{InputLength} < 25 \cdot ft, 25 \cdot ft, \mathbf{InputLength})$

$\mathbf{Length\_1} = 25.00\ ft$

$\mathbf{InputFunction}(\mathbf{L}) := \text{if}(\mathbf{L} < 25 \cdot ft, 25 \cdot ft, \mathbf{InputLength})$

$\mathbf{Length\_2} := \mathbf{InputFunction}(\mathbf{InputLength})$

$\mathbf{Length\_2} = 25.00\ ft$

FIGURE 7.22

if function

linterp Function

PTC Mathcad has several interpolation and regression functions. The ***linterp*** function allows straight-line interpolation between points. It is a straight-line interpolation, and is the easiest to use. You may have a specific need to use some of the more advanced functions, but for our discussion, we will use the liner interpolation function.

The ***linterp*** function has the form ***linterp***(vx,vy,x). The value vx is a vector of real data values in ascending order. The value vy is a vector of real data values having the same number of elements as vector vx. The value x is the value of the independent variable at which to interpolate the value. It is best if the value x is contained within the data range of vx. If x is below the first value of vx, then PTC Mathcad extrapolates a straight line between the first two data points. If x is above the last value of vx, PTC Mathcad extrapolates a straight line between the last two data points.

The *linterp* function takes the form linterp(vx,vy,x), where vx and vy are vectors in ascending order, and x is the independent variable.

$$\mathbf{vx} := \begin{bmatrix} 1 \\ 2 \end{bmatrix} \qquad \mathbf{vy} := \begin{bmatrix} 4 \\ 8 \end{bmatrix}$$

The vectors vx and vy are the same length and have corresponding data.

$$\underline{\mathbf{linterp}}(\mathbf{vx}, \mathbf{vy}, 1.5) = 6.00$$

PTC Mathcad calculates the value of y for x=1.5.

FIGURE 7.23

linterp function

The ***linterp*** function draws a straight line between each data point and uses straight-line interpolation between the pairs of points. The ***linterp*** function is very useful if you have a table or graph of data and need to interpolate between data points. If your data is scattered, you may want to consider using a regression function instead.

Let's first look at a simple example. See Figure 7.23.

In this next example, there is a longer list of data values. PTC Mathcad uses linear interpolation between each pair of data points. See Figure 7.24.

Figure 7.25 is a bit more complicated engineering example. It uses interpolation to calculate pressures at different heights above the ground. It then uses some of the array information discussed in Chapter 6 to calculate force and overturning moments.

Time Velocity

The vectors Time and Velocity must be the same length and have corresponding data. PTC Mathcad uses linear interpolation between each data point.

$$\text{Time} := \begin{bmatrix} 1 \\ 2 \\ 3 \\ 4 \\ 5 \end{bmatrix} \cdot sec \qquad \text{Velocity} := \begin{bmatrix} 2.1 \\ 3.9 \\ 5.9 \\ 7.8 \\ 10.1 \end{bmatrix} \cdot \frac{m}{s}$$

$$\text{linterp}(\text{Time}, \text{Velocity}, 0.5 \cdot s) = 1.20 \, \frac{m}{s}$$

Use the *linterp* function to interpolate the velocity at different moments in time.

$$\text{linterp}(\text{Time}, \text{Velocity}, 1.5 \cdot s) = 3.00 \, \frac{m}{s}$$

$$\text{linterp}(\text{Time}, \text{Velocity}, 2.2 \cdot s) = 4.30 \, \frac{m}{s}$$

$$\text{linterp}(\text{Time}, \text{Velocity}, 6 \cdot s) = 12.40 \, \frac{m}{s}$$

Assign a vector of values to the x argument to calculate all values at the same time.

$$\text{xValues} := \begin{bmatrix} 0.5 \\ 1.5 \\ 2.2 \\ 6 \end{bmatrix} \cdot s \qquad \text{linterp}(\text{Time}, \text{Velocity}, \text{xValues}) = \begin{bmatrix} 1.20 \\ 3.00 \\ 4.30 \\ 12.40 \end{bmatrix} \frac{m}{s}$$

In order to reuse the interpolated values, assign the output to a variable.

$$\text{InterpolatedResults} := \text{linterp}(\text{Time}, \text{Velocity}, \text{xValues})$$

$$\text{InterpolatedResults} = \begin{bmatrix} 1.20 \\ 3.00 \\ 4.30 \\ 12.40 \end{bmatrix} \frac{m}{s} \qquad \text{xValues} = \begin{bmatrix} 0.50 \\ 1.50 \\ 2.20 \\ 6.00 \end{bmatrix} s$$

You can now access individual results from the vector "Interpolated Results."

$$\text{InterpolatedResults}_1 = 1.20 \, \frac{m}{s} \qquad \text{InterpolatedResults}_2 = 3.00 \, \frac{m}{s}$$

FIGURE 7.24

linterp function

This example uses linear interpolation to calculate the pressure on areas located at various heights above the ground. It uses pressure data taken from a table that lists pressures at various heights. The vectors HtData and PressureData are the same length and have corresponding data.

HtData and PressureData are taken from tables.

Area is the surface area of an object. Ht is the height above the ground to the centroid of the corresponding area.

$$\mathbf{Ht_{Data}} := \begin{bmatrix} 0 \\ 10 \\ 20 \\ 30 \\ 40 \\ 50 \\ 60 \\ 70 \end{bmatrix} \cdot ft \quad \mathbf{Pressure_{Data}} := \begin{bmatrix} 24 \\ 24 \\ 25 \\ 27 \\ 30 \\ 34 \\ 39 \\ 45 \end{bmatrix} \cdot psf \quad \mathbf{Area} := \begin{bmatrix} 10 \\ 12 \\ 9 \\ 11 \\ 20 \end{bmatrix} \cdot ft^2 \quad \mathbf{Ht} := \begin{bmatrix} 12 \\ 22 \\ 27 \\ 33 \\ 45 \end{bmatrix} \cdot ft$$

$$\mathbf{Pressure} := \underline{\mathbf{linterp}}\left(\mathbf{Ht_{Data}}, \mathbf{Pressure_{Data}}, \mathbf{Ht}\right)$$

Each value of Ht is interpolated between values of HtData and PressureData.

$$\mathbf{Pressure} = \begin{bmatrix} 24.20 \\ 25.40 \\ 26.40 \\ 27.90 \\ 32.00 \end{bmatrix} psf$$

Pressure is a vector of the interpolated pressures at the heights in the vector Ht.

The following part of this example uses vector information discussed in Chapter 6.
Calculate the force on each area, which is the area times the pressure.

$$\mathbf{Force} := \overrightarrow{(\mathbf{Area} \cdot \mathbf{Pressure})}$$

Use the ***vectorize*** operator to ensure that there is an element-by-element multiplication.

$$\mathbf{Force} = \begin{bmatrix} 242.00 \\ 304.80 \\ 237.60 \\ 306.90 \\ 640.00 \end{bmatrix} lbf$$

Calculate the overturning moment, which is the force times the height.

$$\mathbf{OTM} := \overrightarrow{(\mathbf{Force} \cdot \mathbf{Ht})} \qquad \mathbf{OTM} = \begin{bmatrix} 2904.00 \\ \vdots \end{bmatrix} ft \cdot lbf$$

Calculate the sum of the overturning moments. Use the Vector sum operator contained in the **Operators** list on the **Math** tab.

$$\mathbf{SumOTM} := \sum \mathbf{OTM} \qquad \mathbf{SumOTM} = 54952.50\ ft \cdot lbf$$

FIGURE 7.25

linterp function—engineering example

Miscellaneous Categories of Functions

Curve Fitting, Regression, and Data Analysis

PTC Mathcad has numerous curve fitting and data analysis functions. It has functions for linear regression, generalized regression, polynomial regression, specialized regression, and data analysis. The topic of data analysis alone can fill chapters. These topics are beyond the scope of this book. For additional information, refer to PTC Mathcad Help and Tutorials. There are also many resources available at www.ptc.com.

Error Function

The ***error*** function allows you to create your own custom error messages. This is useful when you are writing user-defined functions, or if you are creating PTC Mathcad programs. See Figure 7.26 for examples.

Using the *Error* function within a function.

$$\mathbf{ErrorTest}(\mathbf{a}) := \text{if}\left(\mathbf{a} > 0, \frac{1}{\mathbf{a}}, \underline{\mathbf{error}}(\text{“Argument must be greater than zero”})\right)$$

$\mathbf{ErrorTest}(2) = 0.50$

A custom error message appears when you use a negative number as an argument.

$\mathbf{ErrorTest}(3) = 0.33$

$ErrorTest(-3) = ?$

$ErrorTest(-3) = ?$

Argument must be greater than zero

Try another example, using a string variable.

$\mathbf{ErrorMessage} := \text{“Cannot divide by zero”}$

$$\mathbf{ErrorExample}(\mathbf{a}) := \text{if}\left(\mathbf{a} \neq 0, \frac{1}{\mathbf{a}}, \underline{\mathbf{error}}(\mathbf{ErrorMessage})\right)$$

$\mathbf{ErrorExample}(2) = 0.50$

$\mathbf{ErrorExample}(-2) = -0.50$

$ErrorExample(0) = ?$

The error message appears when you use 0 as an argument.

$ErrorExample(0) = ?$

Cannot divide by zero

FIGURE 7.26

Error function

String Functions

PTC Mathcad has many string functions that can add to or manipulate strings. These will not be described in any detail, but are presented to inform you of PTC Mathcad's ability to work with strings. For further information, refer to PTC Mathcad Help. The following can be done with strings:

- Add strings together to form one string using ***concat***().
- Return only a portion of a string using ***substr***().
- Tell where a certain phrase begins in a string using ***search***().
- Return how many characters are in a string using ***strlen***().
- Convert a scalar into a string using ***num2str***().
- Convert a number to a string using ***str2num***().
- Convert a string to a vector of ANSI codes using ***str2vec***().
- Convert a vector of integer ANSI codes to a string using ***vec2str***().
- Ask PTC Mathcad if a variable is a string using ***IsString***().

Picture Functions and Image Processing

> PTC Mathcad Prime 3.0 does not have the Picture Toolbar.

You can insert an image into PTC Mathcad by using the **Image** control in the **Regions** group on the **Math** tab (**Math>Regions>Image**). The keyboard shortcut is **CTRL+4**. This will bring up a small icon saying "Browse for Image." When you click this icon, an Open dialog box appears that allows you to browse your computer for an image file.

There are dozens of PTC Mathcad functions that can be used with picture images. Refer to PTC Mathcad Help for additional information.

Complex Numbers, Polar Coordinates, and Mapping Functions

PTC Mathcad recognizes either i or j to represent the imaginary portion of a complex number. When entering complex numbers, you must use a number in front of the i or j. For example type 1 + 1i or 1+1j, do not type 1+i or 1+j. Also, do not type 1+1 *i or 1+1 *j. If you type either i or j by itself, or if you use a multiplication, then PTC Mathcad will look for a variable i or j. You can choose to display complex results with either i or j. This is controlled by the **Complex Values** control located in the **Results** group on the **Math Formatting** tab. The Complex Values control has "1+1i" on the icon.

PTC Mathcad has several functions for working with complex numbers. The easy to remember ones are: ***Re***(Z) returns the real part of Z and ***Im***(Z) returns the imaginary part of Z.

Mapping Functions

PTC Mathcad has several mapping functions that are used in 2D and 3D plotting. These functions convert from rectangular coordinates to polar coordinates [***xy2pol***(x,y)], spherical coordinates [***xyz2sph***(x,y,z)], and to cylindrical coordinates [***xyz2cyl***(x,y,z)]. These functions need unitless numbers, and the results are returned as a vector with unitless radius and angles in radians. The inverse to these functions are ***pol2xy***(r,theta), ***sph2xy***(r,theta,phi), and ***cyl2xy***(r,q,f).

Polar Notation

It is possible to use complex numbers to convert between rectangular and polar coordinates, and to add rectangular coordinates. The following procedure was featured in the February 2004 PTC Mathcad Advisor Newsletter.

Let the real part of the complex number represent the x-coordinate and the imaginary part represent the y-coordinate. You can now add or subtract a series of x- and y-coordinates by adding and subtracting the complex numbers.

Now let's see how to convert to polar coordinates. Let the complex number be represented by "z." The radius for polar coordinates is the absolute value of "z." The angle is calculated using the function ***arg***(z). The angle is measured from π to $-\pi$. The result can also be displayed in degrees, if you add deg to the unit placeholder. You can create a user-defined function to convert from polar coordinates back to polar notation. See Figure 7.27 for the procedure.

Angle Functions

It may be useful to compare the results of PTC Mathcad's various angle functions: ***angle***(x,y), ***atan***(y/x), ***atan2***(x,y), and ***arg***(x+yi). These functions return the angle from the x-axis to a line going through the origin and the point (x,y).

angle(x,y): Returns the angle in radians between 0 and 2π, excluding 2π.

atan(y/x): Returns the angle in radians between $-\pi/2$ and $\pi/2$.

atan2(x,y): Returns the angle in radians between $-\pi$ and π, excluding $-\pi$.

arg(x+yi): Returns the angle in radians between $-\pi$ and π, excluding $-\pi$. The only difference between this function and atan2(x,y) is the way the arguments are input.

Reading from and Writing to Files

PTC Mathcad has numerous read and write functions. These are located in the Function Category "File Access" from the **Functions** list on the **Functions** tab. These functions are executed each time the worksheet is recalculated. Some key functions highlighted below. In these functions, the optional arguments may be omitted from right to left only. You cannot omit the first optional argument and expect the second optional argument to be recognized. You can read more about reading and writing files in the PTC Mathcad Help under the Function section in the Reading and Writing Files group.

READPRN("file"): This function returns a matrix formed from a structured ASCII text file. The argument is a text string with the path and file name to the

Let the x-coordinate be represented by the real part and the y-coordinate be represented by the imaginary part.

$\mathbf{PolarExample_1} := 3 + 3\mathrm{i}$

$\mathbf{Radius_1} := |\mathbf{PolarExample_1}|$ $\mathbf{Radius_1} = 4.24$

$\mathbf{Angle_1} := \underline{\mathbf{arg}}(\mathbf{PolarExample_1})$ $\mathbf{Angle_1} = 0.25\ \pi$ $\mathbf{Angle_1} = 45.00\ deg$

$\mathbf{PolarExample_2} := -3 - 3\mathrm{i}$

$\mathbf{Radius_2} := |\mathbf{PolarExample_2}|$ $\mathbf{Radius_2} = 4.24$

$\mathbf{Angle_2} := \underline{\mathbf{arg}}(\mathbf{PolarExample_2})$ $\mathbf{Angle_2} = -0.75\ \pi \mathbf{Angle_2} = -135.00\ deg$

This polar notation allows you to add x and y coordinates and then get a final radius and angle.

$\mathbf{PolarExample_3} := (3 + 3\mathrm{i}) + (4 + 12\mathrm{i}) + (3 - 5\mathrm{i}) = 10.00 + 10.00\mathrm{i}$

$\mathbf{Radius_3} := |\mathbf{PolarExample_3}|$ $\mathbf{Radius_3} = 14.14$

$\mathbf{Angle_3} := \underline{\mathbf{arg}}(\mathbf{PolarExample_3})$ $\mathbf{Angle_3} = 0.25\ \pi$ $\mathbf{Angle_3} = 45.00\ deg$

Create a user-defined function to convert from polar coordinates to a complex number.

$\mathrm{i} := 1\mathrm{i}$ Reset i from the range variable used in Figure 7.18.

$\mathbf{P2i}(\mathbf{mag}, \mathbf{angle}) := |\mathbf{mag}| \cdot (\underline{\mathrm{cos}}(\mathbf{angle} \cdot deg) + \underline{\mathrm{sin}}(\mathbf{angle} \cdot deg) \cdot \mathrm{i})$

$\mathbf{P2i}(2, 45) = 1.41 + 1.41\mathrm{i}$

$\mathbf{P2i}(10 \cdot \sqrt{2}, -135) = -10.00 - 10.00\mathrm{i}$

Assign a variable so that results can be reused.

$\mathbf{PolarExample_4} := \mathbf{P2i}(20, 145)$ $\mathbf{PolarExample_4} = -16.38 + 11.47\mathrm{i}$

FIGURE 7.27

Polar notation

data file. If a full path is not given, then the file is relative to the current working directory. The current working directory can be shown by typing **CWD=**. The file should be ASCII text only, with data arranged in rows and columns separated by spaces or tabs. A text header is allowed; however, once READPRN encounters a number, it assumes data has begun, so headers should contain no numbers.

WRITEPRN("file",M,[rows,[cols,["decsymb"]]]): This function writes the contents of M to a delimited ASCII file. The arguments are as follows:

- file: A text string with the path and file name of the data file. The path must be a full path or relative to the current working directory.
- M: The ***WRITEPRN*** function will write the contents of this variable. It can be an array or scalar. This function may be used on both sides of the ***definition*** operator. For example, two ways to use this function are: ***WRITEPRN***("TestFile",M)=, or NewVariable:=***WRITEPRN***("TestFile",M). In the second case, the variable "NewVariable" is assigned the contents of the file.

For rows, cols, and decsymb see above.

READFILE("file",["type", colwidths, [rows,[cols,[emptyfill,["decsymb"]]]]]): This function returns an array from the contents of a file of a specified type (delimited, fixed-width, or Excel). The arguments are as follows:

- file: A text string with the path and file name of the data file. The path must be a full path or relative to the current working directory.
- type: A text string with one of the following words: "delimited," "fixed," or "Excel."
- colwidths: This is required for a "fixed-width" type, and is omitted for other types. It is a vector. Each row specifies the number of characters in each fixed-width column in the data file.
- rows: This argument is optional. If omitted, PTC Mathcad reads every row of the data file. This argument can be a scalar or a two row vector. A scalar tells PTC Mathcad which row to start at. A two row vector tells PTC Mathcad a beginning row and an ending row. Row numbering begins with 1. ORIGIN does not affect the numbering.
- cols: This argument is optional; however, if used, rows should also be used. If omitted, PTC Mathcad reads every column of the data file. This argument can be a scalar or a two row vector. A scalar tells PTC Mathcad which column to start at. A two row vector tells PTC Mathcad a beginning column and an ending column. Column numbering begins with 1. ORIGIN does not affect the numbering.
- emptyfill: This argument tells PTC Mathcad what to do with missing entries in the data file. It can be a text string, or a scalar. The default is NaN. The built-in PTC Mathcad constant NaN represents a missing value. It stands for "not a number." Once input into a matrix it can be detected by the ***IsNaN*** function.

- decsymb (optional) is the decimal symbol to use. You can specify either "," (comma) or "." (dot). The default is "." (dot).

WRITEFILE("file",M): This function writes an Excel or tab delimited text file from the matrix M. The file type depends on the file extension in the "file" string (.xls, .xlsx, .txt, or .dat).

APPENDPRN("file",[M]): This function is similar to the ***WRITEPRN*** function, except that it adds the contents of the array M to the end of the file. The number of columns in the array M must match the number of columns in the existing file.

The PTC Mathcad default for the functions ***WRITEPRN*** and ***APPENDPRN*** is to append four significant digits in columns eight digits wide. This is controlled by two system variables: ***PRNPRECISION*** with a default value of 4 and ***PRNCOLWIDTH*** with a default value of 8. There is not a worksheet control for these system variables. If you want to change these values, you must redefine these variables in your worksheet.

READEXCEL("file",["range",[emptyfill,[blankrows]]]: This function returns a matrix from a defined range in an Excel file.

WRITEEXCEL("file",M,[rows,[cols]],["range"]): This function writes a matrix M to the defined range within the Excel "file". The "file" string must include the .xls or .xlsx file extension.

PTC Mathcad can read and write from many more file types such as: text files, binary data files, comma separated variable (CSV) data files, image files, and sound files. The description for these files types are included in the PTC Mathcad Help. Search for "Functions for reading and writing files."

Summary

We have reviewed only a handful of the hundreds of PTC Mathcad functions. Hopefully the functions discussed in this chapter will be helpful to your technical calculations. As you become more familiar with PTC Mathcad, you will add other functions to your PTC Mathcad toolbox. We will discuss additional functions in later chapters. In the meantime, open the **Functions** tab on the Ribbon Bar and look for functions that appear interesting to you. Use the PTC Mathcad Help to learn about these functions.

In Chapter 7 we:

- Reviewed the basics of PTC Mathcad functions.
- Learned about the following functions:
  - ***max*** and ***min*** functions
  - ***mean*** and ***median*** functions
  - truncation and rounding functions
  - ***summation*** operators

- *if* function
- interpolation functions
- Introduced the following categories of functions:
  - Curve fitting
  - String functions
  - Picture functions
  - Mapping functions
  - Polar notation
  - Angle functions
  - Reading and writing functions
- Encouraged you to search for and learn about additional functions that you can add to your PTC Mathcad toolbox.

Practice

The PTC Mathcad Prime 3.0 figures and examples used in this book are available for download from the book's website. The reader is encouraged to download the files and use them to practice the concepts learned. Additional examples and problems are also provided. To access this content go to http://store.elsevier.com/9780124104105 *and click on the Resources tab, and then click on the link for the Online Companion Materials.*

1. Use the ***max*** and ***min*** functions to find the maximum and minimum values of the following:
   - **a.** 10 m, 1 km, 3000 mm, 1 mi, 2000 yd, 10 furlong, 1 nmi.
   - **b.** 100 s, 1.25 min, 0.15 day, 0.05 week, 0.005 year. Create a custom unit for week.
   - **c.** 1000 Pa, 1 psi, 1 atm, 10 torr, 0.002 ksi, 0.002 MPa, 1in_Hg.
   - **d.** 1 bhp, 10 ehp, 10.1 kW, 19 mhp, 10 hpUK, 10 hhp, 10000 W.
2. Use the ***mean*** and ***median*** functions to determine the mean and median values from the list of values used in practice exercise 1.
3. Place the list of values from practice exercise 1 into four different vectors.
4. Use the ***max*** and ***min*** functions and the ***mean*** and ***median*** functions with the vectors created in practice exercise 3.
5. Use the ***floor***, ***ceil***, and ***trunc*** functions with the vectors created in practice exercise 3. Select two different units to use for each vector. For example, truncate to units of meters and feet for the first vector.
6. Use the ***round*** function with the vectors created in practice exercise 3. Round to the following: 2 decimal places, 1 decimal place, 0 decimal places, 10's, and 100's. Select two units for each case. For example, round to units of meters and feet for the first vector.
7. Use the ***Vector sum*** operator to calculate the sum of the vectors created in practice exercise 3. Display the result in three different units.

8. Use the ***Summation*** operator and the ***Range sum*** operator to calculate the total area of ten squares with sides incrementing from 1 m to 10 m.
9. Use the following variables for this practice exercise: R1 = 3Amp, R2 = 4Amp, R3= 10Volt, R4 = 20Volt. Create four variables using the *if* function to meet the following conditions:

Condition	First	Operator	Second	If true	If false
1	R1	<	R2	R3	R4
2	R1	=	R2	R3	R4
3	R2	=		"Division by zero"	R1/R2
4	R3	>=	R4	R1	R2

   a. Vary the values of the variables, to see how the results are affected.

10. Use the ***linterp*** function to interpolate the following values:

Time	Distance	Interpolate the distance for this time
0 sec	0.0 m	0.5 sec
1 sec	2.5 m	1.2 sec
2 sec	10.0 m	2.4 sec
3 sec	22.5 m	3.3 sec
4 sec	40.0 m	4.6 sec
5 sec	62.5 m	5.1 sec

CHAPTER

Plotting

8

Plots are an important part of engineering calculations because they allow visualization of data and equations. They are an important part of equation solving because they can help you select an initial guess for a solution. Plots also allow you to visualize trends in engineering data. This chapter will focus on using plots as a tool for visualization and solving of equations.

Chapter 8 will:

- Introduce the **Plots** tab.
- Review how to create simple 2D XY plots and polar plots.
- Show how to set plot ranges.
- Instruct how to graph multiple functions in the same plot.
- Discuss the use of range variables to control plots.
- Tell how to plot data points.
- Describe the steps necessary to format plots, including the use of log scale, scaling, numbering, and setting defaults.
- Show how the use of plots can help find the solutions to various engineering problems.
- Discuss plotting over a log scale.
- Introduce 3D plotting.
- Use engineering examples to illustrate the concepts.

Plots Tab

The **Plots** tab is shown below. The number of controls that show on the tab is dependent on how wide your PTC Mathcad workspace is. This tab is the nerve center for all things related to plotting. We will discuss each of the groups and controls on the **Plots** tab throughout the chapter. Many of the controls are dimmed until a plot region becomes active by clicking in the plot region.

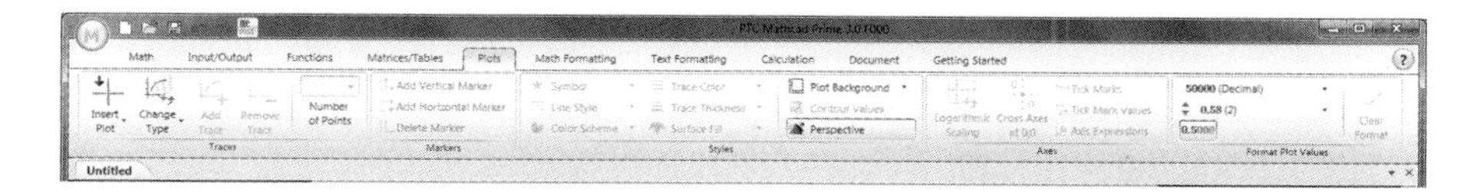

Essential PTC® Mathcad Prime® 3.0.

Let's first discuss the **Insert Plot** control from the **Traces** group. This control allows you to insert one of four plot types: **XY Plot**, **Polar Plot**, **Contour Plot**, and **3D Plot**.

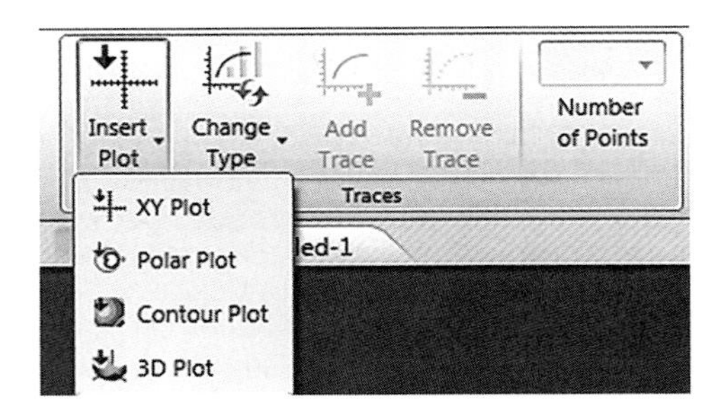

Creating a Simple XY Plot

> In Mathcad 15 the keyboard shortcut to insert an XY plot was **@** or **SHIFT+2**. In PTC® Mathcad Prime® the shortcut is **CTRL+2**. The shortcut for inserting a 3D plot was changed to **CTRL+3**.

In Chapter 1 we showed how to create a simple XY plot by typing **CTRL+2**, or by selecting **XY Plot** from the **Insert Plot** control on the **Plots** tab (**Plots>Traces>Insert Plot>XY Plot**). Remember that the x-axis variable in the bottom placeholder must be a previously undefined variable. If you use a previously defined scalar variable PTC Mathcad will only plot a single point instead of a range of points. The y-axis placeholder is where you place an expression or function using the variable on the x-axis.

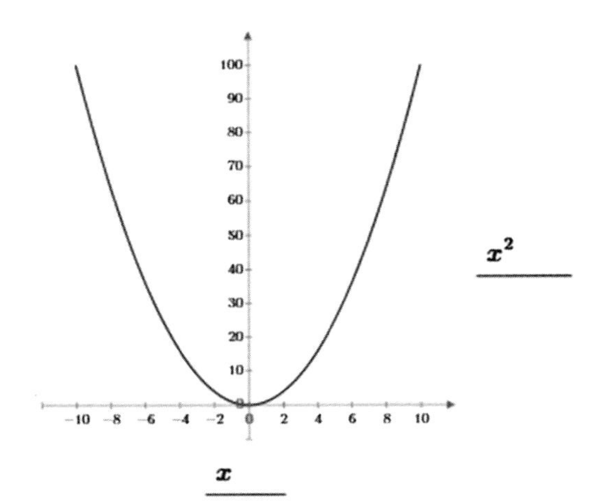

Creating a Simple Polar Plot

Creating a simple polar plot is similar to creating a simple XY plot. Type **CTRL+7**, or select **Polar Plot** from the **Insert Plot** control on the **Plots** tab (**Plots> Traces>Insert Plot> Polar Plot**).

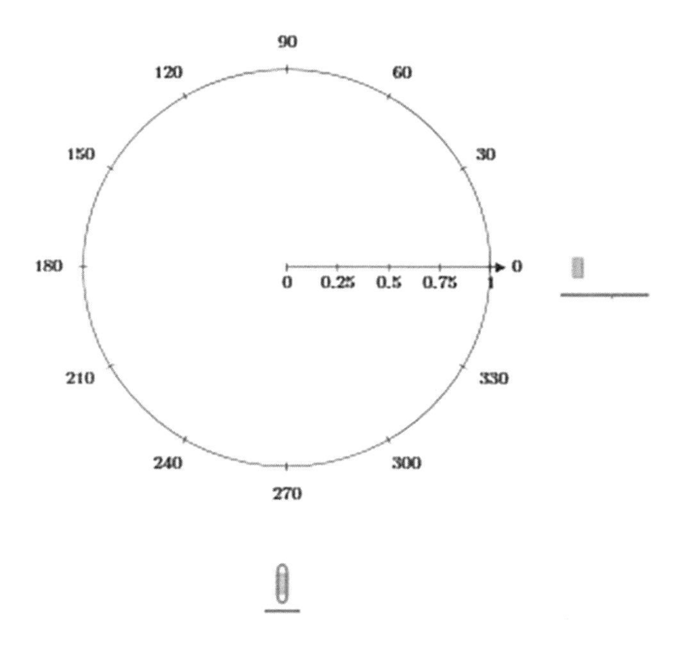

Click on the bottom (angular axis) placeholder. This is where you type the angular variable. Unless you specify otherwise, PTC Mathcad assumes the variable to be in radians. Type the name of a previously undefined variable. The variable can be any PTC Mathcad variable name. Next, click on the right side radial-axis placeholder and type an expression using the angular variable defined on the angular-axis. This sets the properties of the radial axis. Click outside of the plot region to view the plot. For every angle from 0 to 2π, PTC Mathcad plots a radial value. PTC Mathcad automatically selects the radial range. See Figure 8.1.

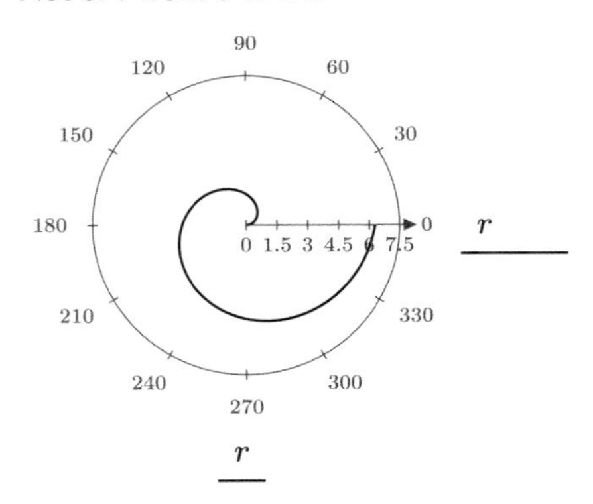

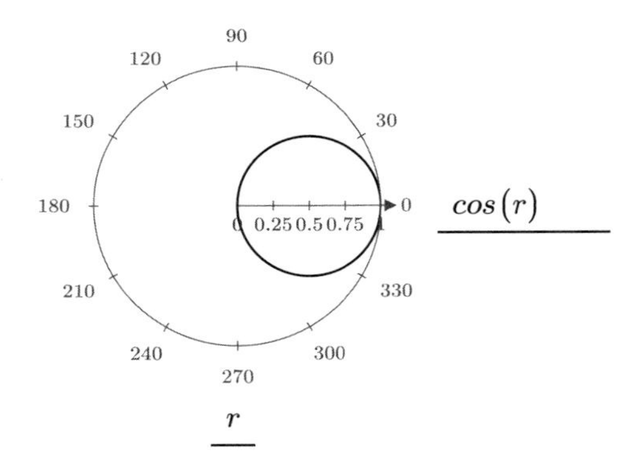

FIGURE 8.1

Polar plot of functions

You may also use a previously defined function in a simple polar plot. Begin the polar plot region by typing **CTRL+7**. Type the name of the angular variable in the bottom placeholder. Type the name of the function on the right placeholder using the angular variable name as the argument of the function. See Figure 8.2.

$f1(r) := r$ $\quad$ $f2(r) := \cos(6 \cdot r)$ $\quad$ Define functions

Plot of f1 from r=0 to 2 π

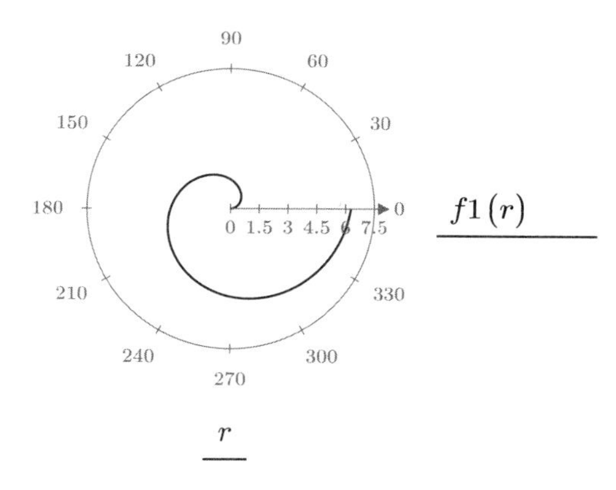

Plot of f2 from r=0 to 2 π

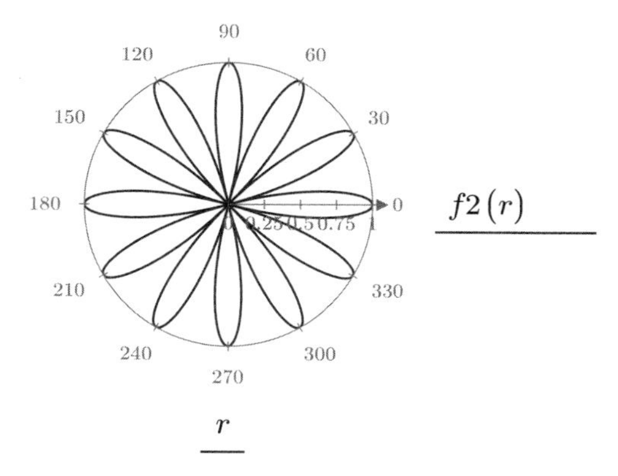

FIGURE 8.2

Polar plot of functions

XY Plot Range and Tick Marks

When you first create an XY plot, the default x-axis range is from −10 to 10. The x-axis tick mark interval defaults to 2. The range and interval are controlled by three tick marks: the Lower Limit Tick Mark, the Upper Limit Tick Mark, and the Interval Tick Mark. These tick marks become visible as you run your cursor along the x-axis. The image below shows the cursor over the Lower Limit Tick Mark.

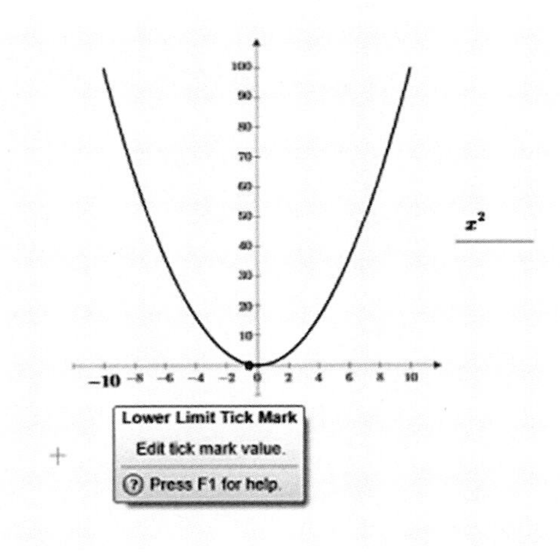

To change the Lower Limit and Upper Limit Tick Mark values, simply highlight the value and enter a new value. As you adjust the Lower Limit and Upper Limit Tick Mark values, PTC Mathcad adjusts the default interval.

The image below shows the Interval Tick Mark. It is located adjacent to the Lower Limit Tick Mark.

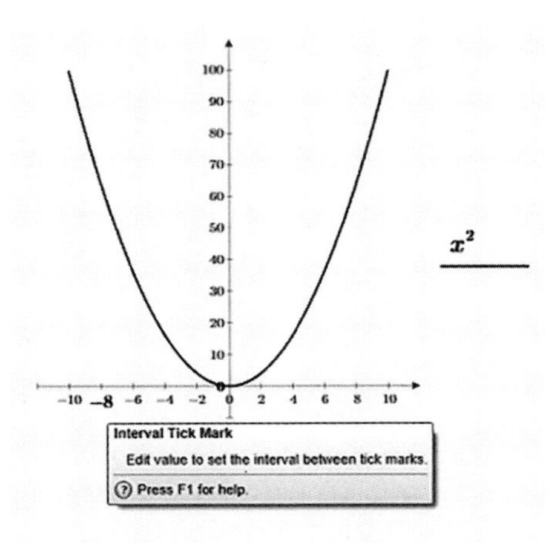

The interval value works similar to a range variable. The difference in values between the Interval Tick Mark and the Lower Limit Tick Mark sets the spacing of the tick marks along the x-axis. See Figure 8.3 for examples of changing tick marks.

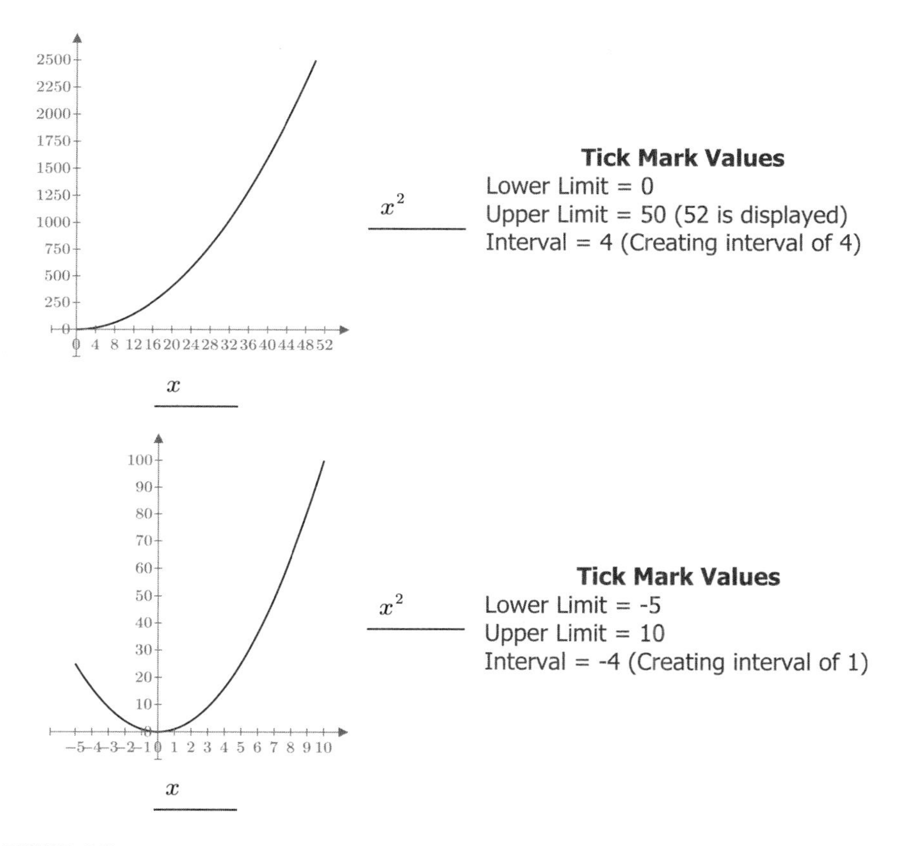

FIGURE 8.3

Changing tick marks

You can toggle the display of both the tick marks and the tick mark values of each axes independently. In order for the changes to take effect, you must have the desired axis selected by either selecting the x-axis or y-axis placeholder, or by clicking on one of the limit tick mark values. Once you have the axis selected, click the **Tick Mark** controls in the **Axes** group. The controls are dimmed until a plot region is made active by clicking in it. The **Tick Marks** and the **Tick Mark Values** are on by default. Clicking them will turn them off. The figure below has both the tick marks and the tick mark values turned off.

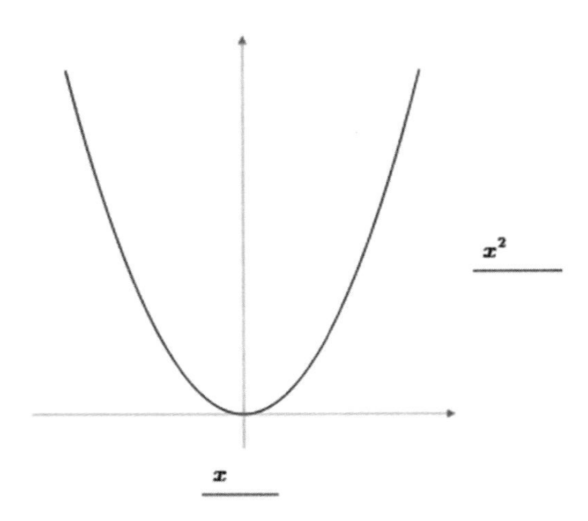

You can also turn off the display of what is in the x-axis and y-axis placeholders. To do this, click within the plot region, and select **Axis Expressions** from the **Axes** group. You will not see a change until after you click outside of the plot region because axis expressions (what is in the x-axis and y-axis placeholders) are always visible when a plot region is active. The figure below has the **Axis Expressions** turned off, in addition to the tick marks and tick mark values being turned off.

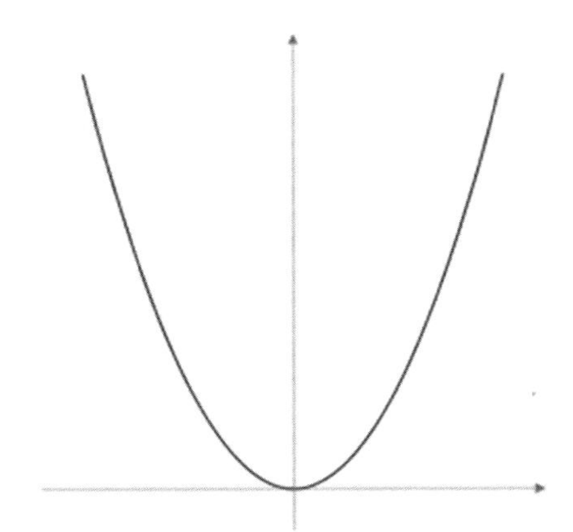

Number of Points Plotted

When PTC Mathcad graphs a function, it actually calculates multiple points and then draws a line (trace) between the points. The default value of points plotted is 500. You may change this number to be between 100 and 3000 for an XY Plot. The **Number of Points** control is in the **Traces** group. The control is dimmed until a plot region is made active. Figure 8.4 shows how changing the number of points can affect how the plot appears.

Number of Points set to 100

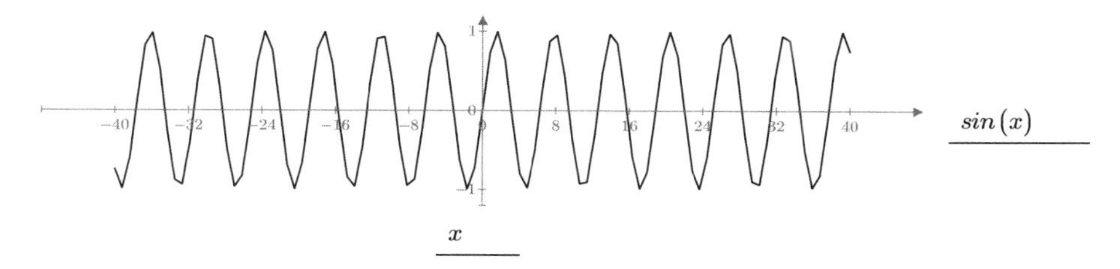

Number of Points set to 500

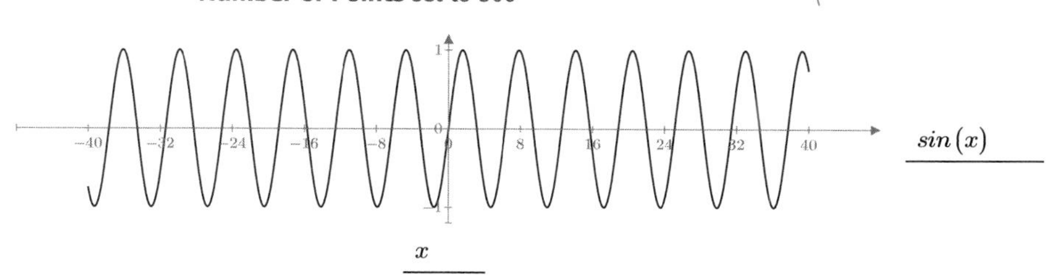

FIGURE 8.4

Number of Points

Showing Only Points

To illustrate the number of points plotted let's turn off the line and display the points, or symbols as they are referred to in PTC Mathcad. Let's first display the points. To do this, click in the y-axis placeholder. From the **Styles** group, click the **Symbol** control and select a solid circle (**Plots>Styles>Symbol>•**).

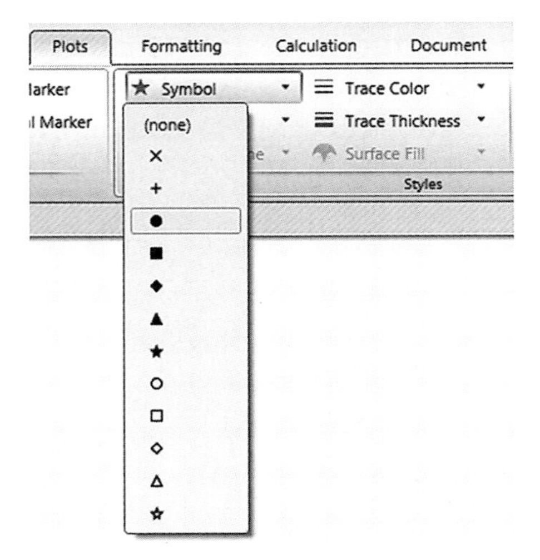

Next, click the **Line Style** control and select **(none)** (**Plots>Styles>Line Style>(none)**). Figure 8.5 shows the difference between **Number of Points** set to 100 and **Number of Points** set to 500.

Number of Points set to 100

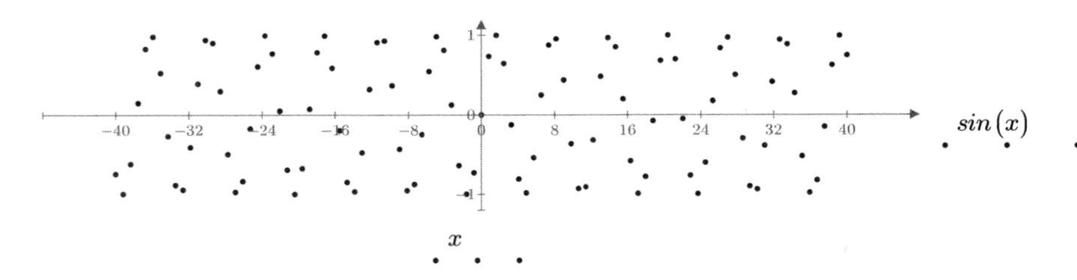

Number of Points set to 500

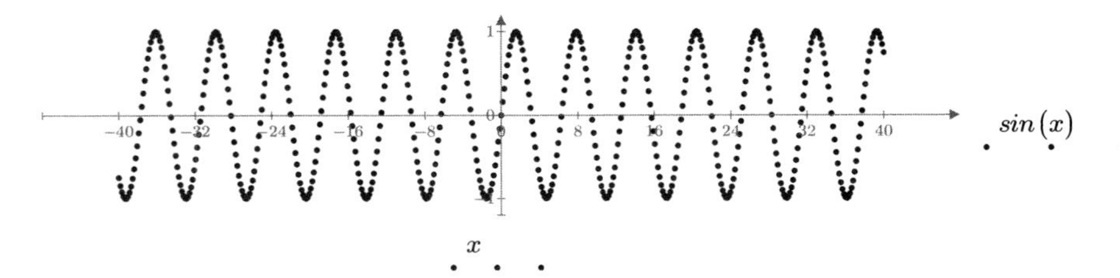

FIGURE 8.5

Number of Points without trace

Using Range Variables to Set Plot Domain

We have just seen how changing the Lower Limit and Upper Limit Tick Marks control the x-axis range. There is another way to control both the x-axis range and the number of points plotted. This is by using range variables. To plot using a range

$D(t) := \frac{1}{2} \cdot g \cdot t^2$ Define function.

$t_1 := 0, 1 .. 22$ Define range variables.

$t_2 := 0, 1 .. 100$

A simple XY plot with default range values

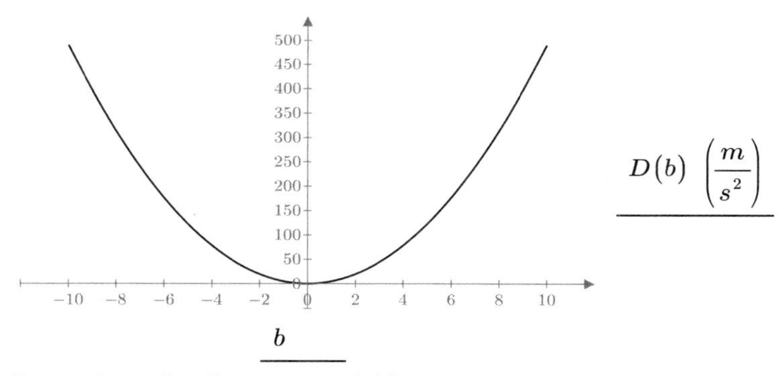

Expression using the range variable t_1

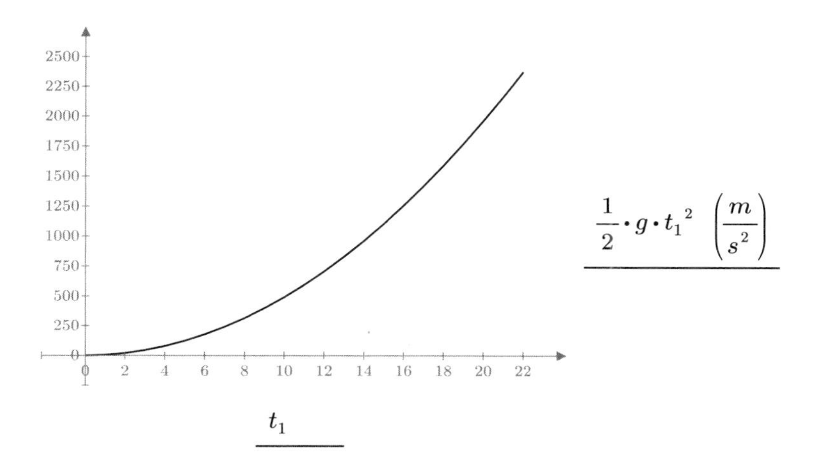

FIGURE 8.6 *(Continued on next page)*

Using range variables to set plot range

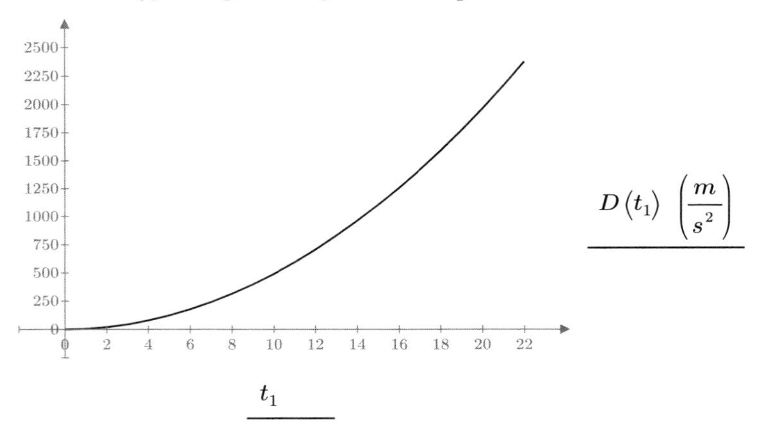

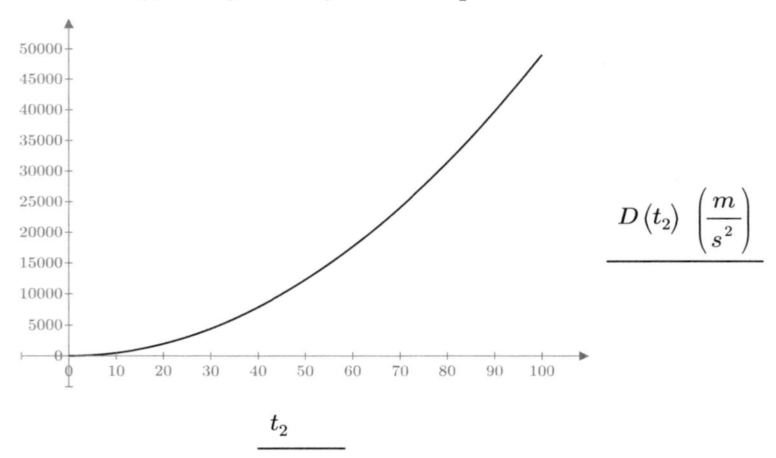

FIGURE 8.6

(continued)

variable, create the range variable before creating the plot region. Next, create the plot region, and type the name of the range variable in the x-axis placeholder. In the y-axis placeholder, type an expression or the name of a function using the range variable. See Figures 8.6 and 8.7. When using range variables, you are actually telling PTC Mathcad to plot each point in the range variable and to draw a line between the points. Because the number of points is defined by the range variable, the **Number of Points** control is dimmed and cannot be modified.

Create range variables and functions.

$\theta_1 := 0, \frac{2 \cdot \pi}{100} .. 1.8 \cdot \pi$ $\qquad$ $\theta_2 := 0, \frac{\pi}{50} .. \pi$

$f_3(\alpha) := \sin(\alpha)^2$ $\qquad$ $f_4(\alpha) := \sin(\alpha)^2 + \cos(\alpha)$

Function f_3 using defaults

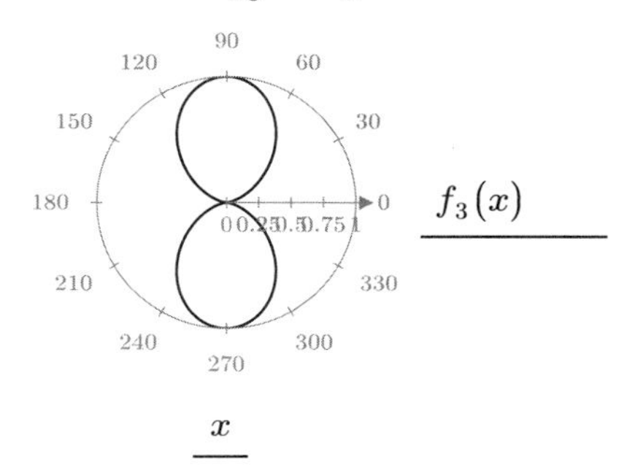

Function f_4 using defaults

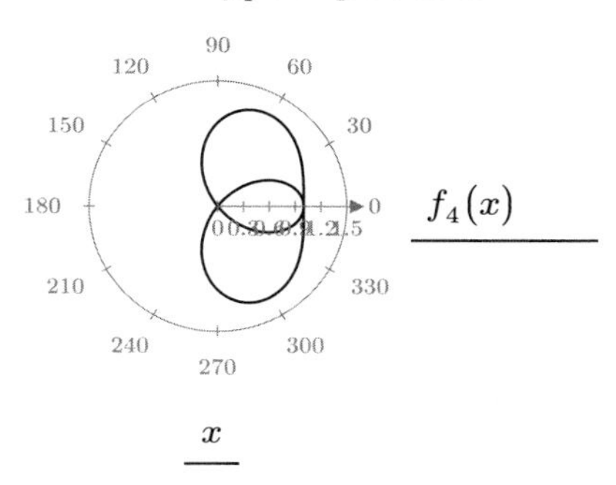

Function f_3 using range variable θ_1

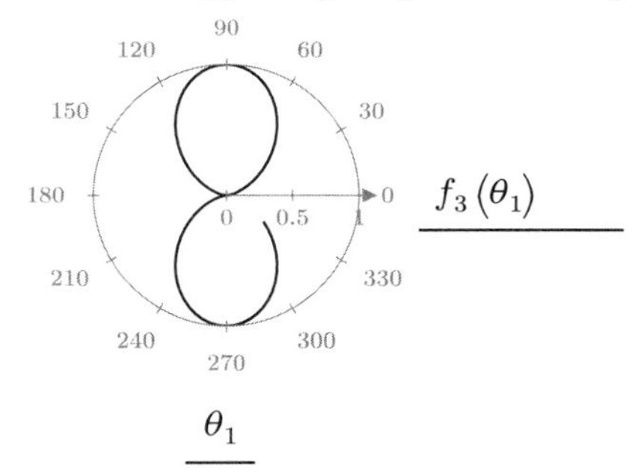

Function f_4 using range variable θ_1

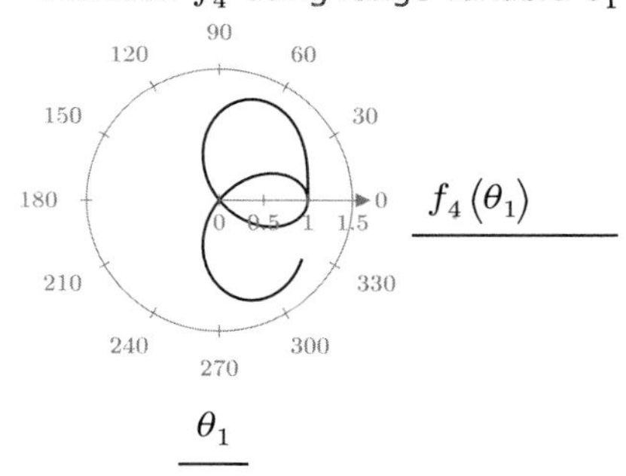

Function f_3 using range variable θ_2

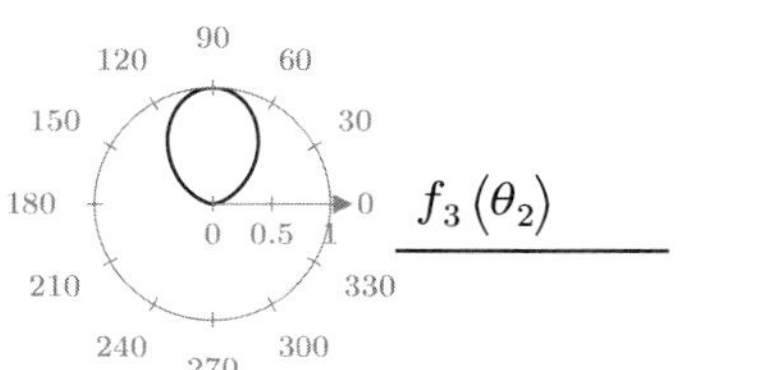

θ_2

Function f_4 using range variable θ_2

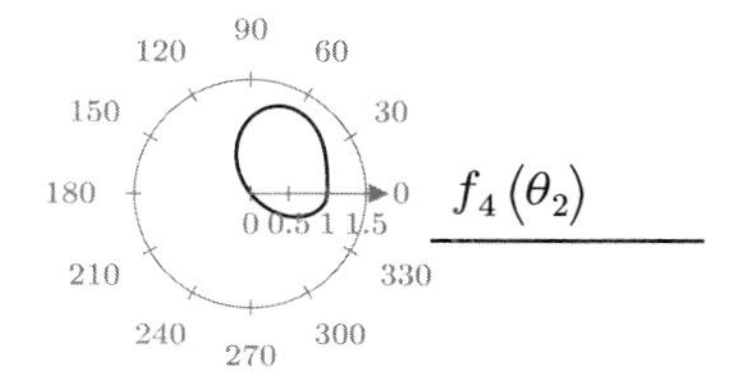

θ_2

FIGURE 8.7

Using range variables to set plot range

Polar Plot Range and Tick Marks

Polar plots have Lower Limit Tick Marks, Upper Limit Tick Marks, and Interval Tick Marks along the radial axis as shown in the below figures.

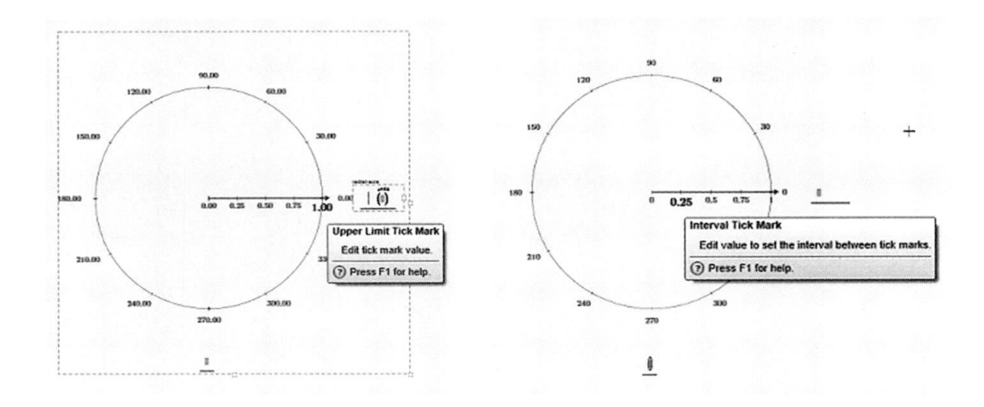

The angular axis only has an Interval Tick Mark as shown in the figure below.

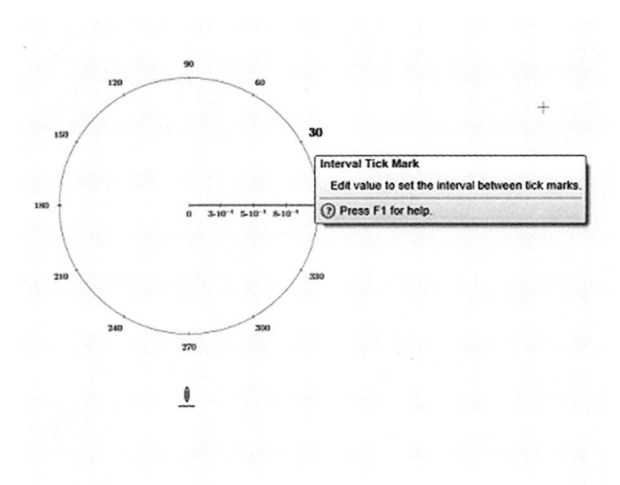

Figure 8.8 shows how changing the radial axis tick marks affects the appearance of various plots.

These plots are the same as in Figure 8.7, but in these figures the radial limits are 0 to 0.9. The range variable θ_1 is used.

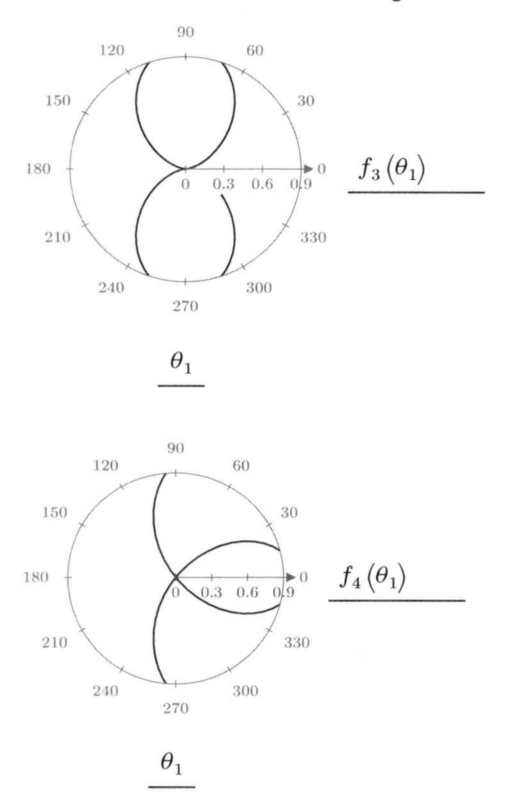

FIGURE 8.8

Setting plot range in a polar plot

Graphing with Units

Plots now have a unit placeholder where you can easily define the units for each axis.

There is a unit placeholder adjacent to the x-axis and y-axis placeholders. This unit placeholder allows you to control what unit values are plotted on each axis. The default plot values are controlled by the worksheet unit system. To change the plotted values, simply type the desired unit in the unit placeholder. Figure 8.9 shows the default plot units. Figure 8.10 shows how to change the plot units. Note that tick mark values are unitless. It is the axis expressions that have units assigned.

The weight of water is about 62.24 pound force per cubic foot.
Graph the water pressure (in psf) at various depths (in feet).

$\text{Weight} := 62.24\ pcf$
$\text{Pressure}(\text{Depth}) := \text{Weight} \cdot \text{Depth}$
$\text{Pressure}(10\ ft) = 622.4\ psf$

Set the plot limits using a range variable.

$\text{i} := 0\ ft, 1\ ft..10\ ft$

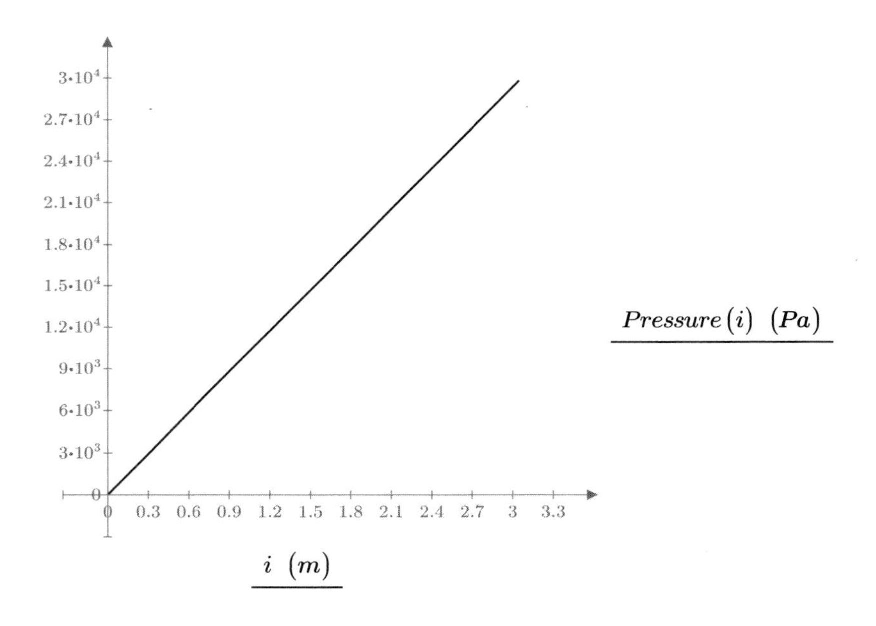

Note that the x-axis is plotting the values of meter and the y-axis is plotting the values of Pa. These default values are from the default unit system (SI).

FIGURE 8.9

Default plot units

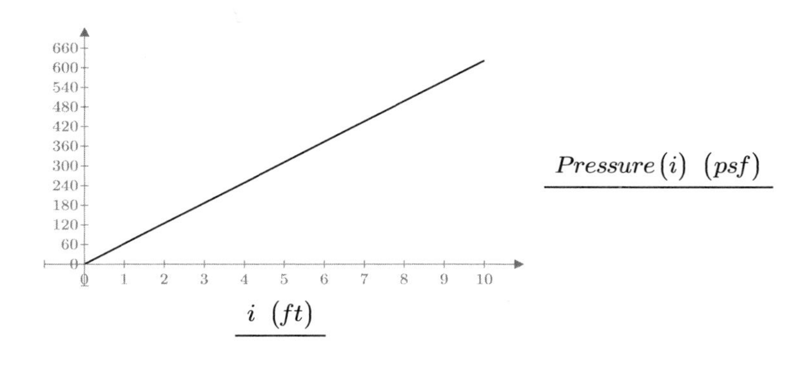

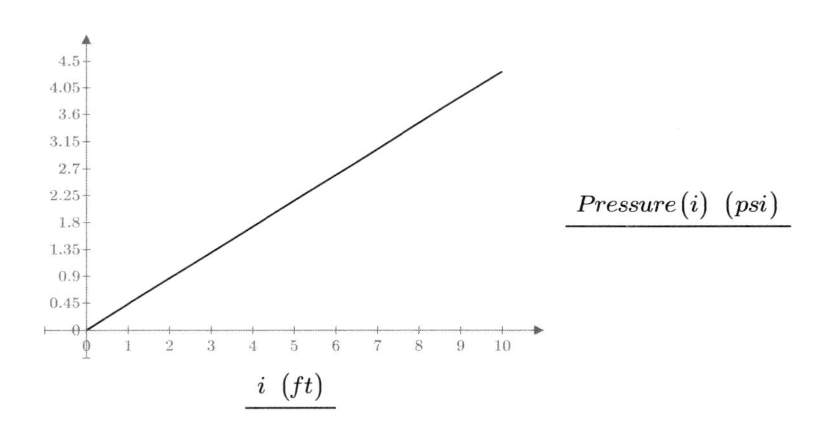

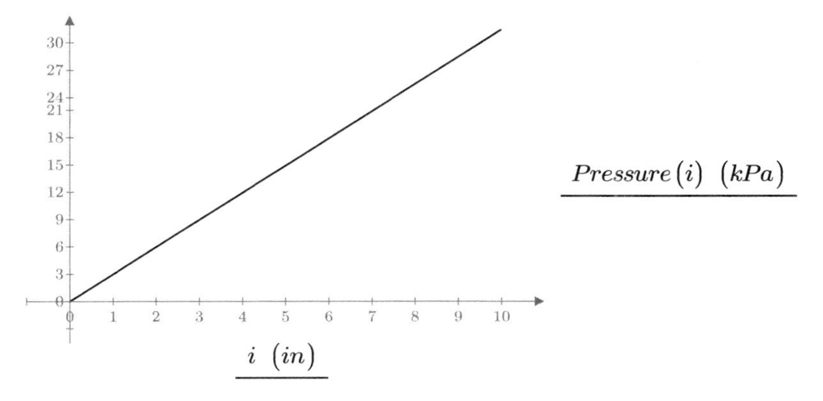

FIGURE 8.10

Changing plot units

Formatting Plots

PTC Mathcad allows you to customize many aspects of your plots. You can change the location of the placeholders on each axis, change the location of each axis, change the color, line weight, or line type of each trace, or add symbols to the trace. You can even plot in log scale.

Axis Location

> Mathcad 15 allowed only a Boxed or Crossed axis style. PTC Mathcad Prime 3.0 allows you to move either axis to any location.

For an XY plot, the axes by default cross at 0,0. To change the location of the y-axis, hover your mouse over the axis line until the cursor changes to a double-headed arrow. Click and drag the axis to the location you want and release the click. The axis snaps to the closest tick mark on the x-axis. You can change the x-axis in a similar manner. If you want to return the axes to cross at 0,0, click the **Cross Axes at 0,0** control in the **Axes** group.

Location of Axis Placeholders

The default axis placeholders are bottom center and right center. These placeholders are small regions within the plot region. To move the placeholder, click in the axis placeholder region to make it active, then move your cursor to the edge of the region until your curser changes to a move cursor. Click and drag the region. Once you begin moving the region, five shaded boxes will appear. You may then drag the region to one of these boxes. The figure below illustrates moving the y-axis region from the right side to the left side. It is also possible to move the y-axis placeholder to either side of the y-axis at the top or bottom. The x-axis placeholder can be moved in a similar manner.

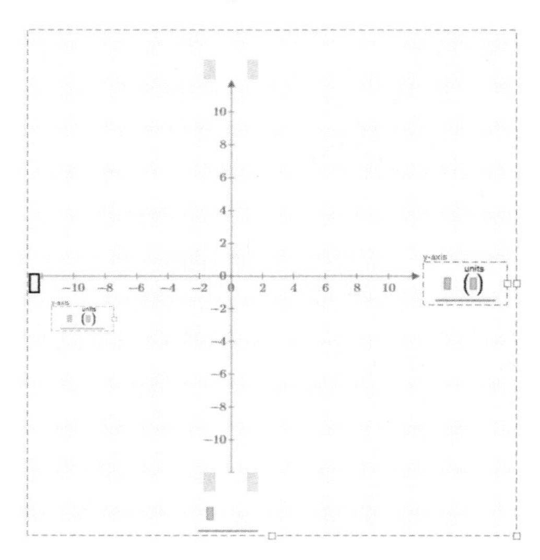

See Figure 8.11 for an example of moving axis and placeholder locations.

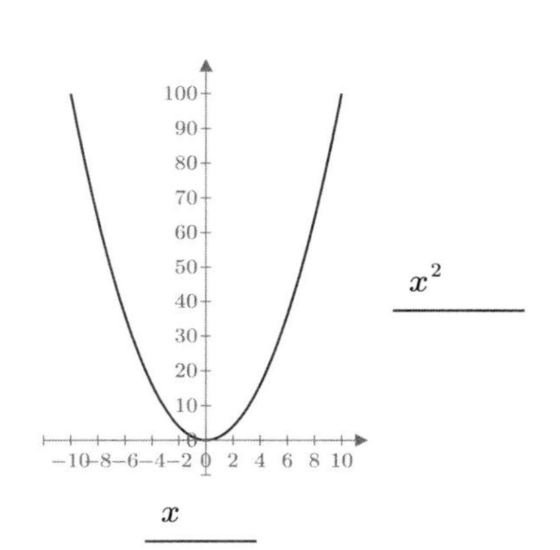

Plot after moving axes and placeholders

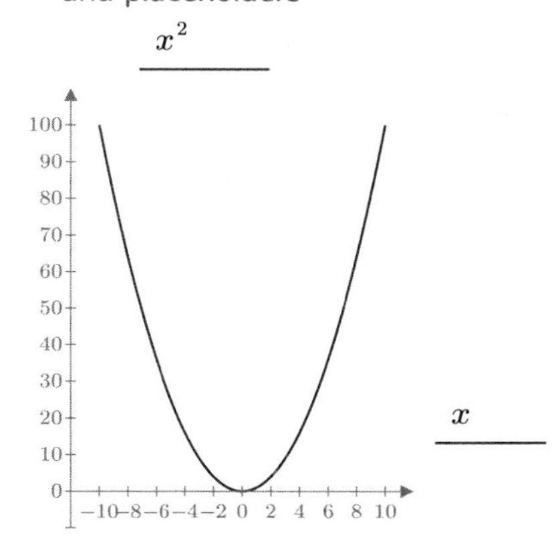

FIGURE 8.11

Location of axes and axis placeholders

Markers

Mathcad 15 only allowed two markers. PTC Mathcad Prime 3.0 allows you to add multiple markers.

Markers are vertical and horizontal lines placed on the plot at locations that you choose. See Figure 8.12. The **Markers** group on the **Plots** tab allows you to add vertical and horizontal markers. You can then drag the markers to the desired position, or you can type a value into the box to set the marker to an exact position. To delete a marker, select the value box of the marker, and then click **Delete Marker**.

When you click to add a marker, all the new markers are inserted at the same position, so it might seem like nothing is happening. They are all being stacked on top of one another.

Formatting Plot Values

The way values show on the plot axis will match the worksheet settings such as exponents, decimal places, and showing trailing zeros. For example, if the worksheet

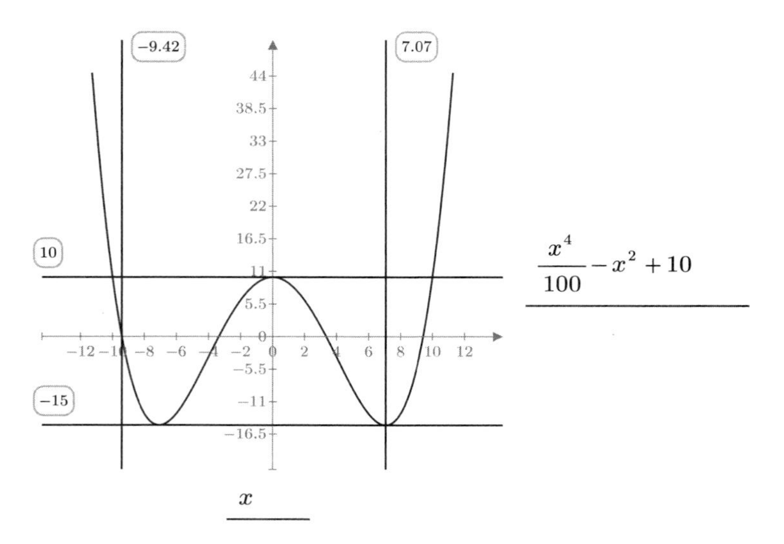

FIGURE 8.12

Markers

has a display precision of 5 with **Show Trailing Zeros** active, the plot axis values will have the same settings. You can change these settings for each axis separately. To change the formatting on the x-axis, click the x-axis placeholder or one of the three x-axis limit tick marks, and then change the desired settings in the **Format Plot Values** group. The y-axis settings are similar.

Logarithmic Scaling

Use the **Logarithmic Scaling** control in the **Axes** group to turn logarithmic scaling on or off. The axis placeholder or one of the axis limit tick marks must be selected prior to using the **Logarithmic Scaling** control. When you turn on logarithmic scaling, the number format automatically switches to Scientific (unless Engineering was previously selected). See Figure 8.13.

Trace Styles

The **Styles** group allows you to turn point symbols on or off, and to control the line style, the line color, and the line thickness of a trace. You cannot turn off a line unless a symbol is used (otherwise nothing would plot).

> You cannot display plot gridlines in PTC Mathcad Prime 3.0.

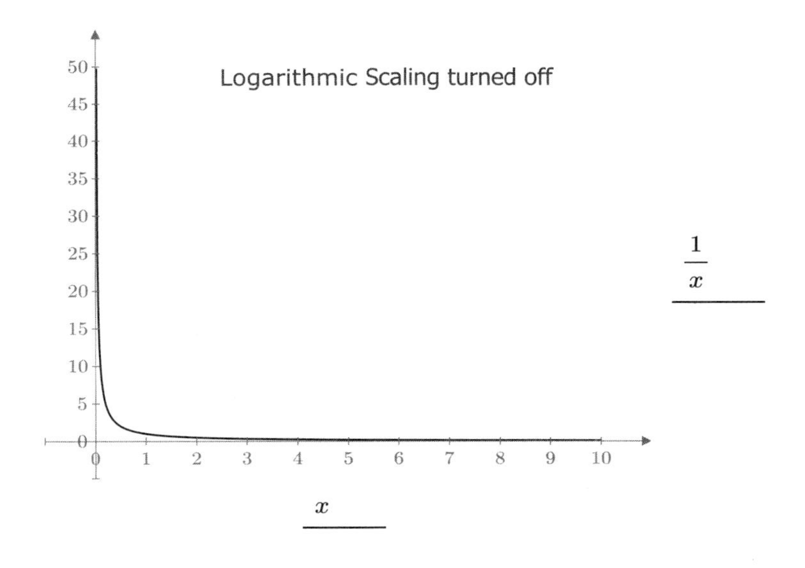

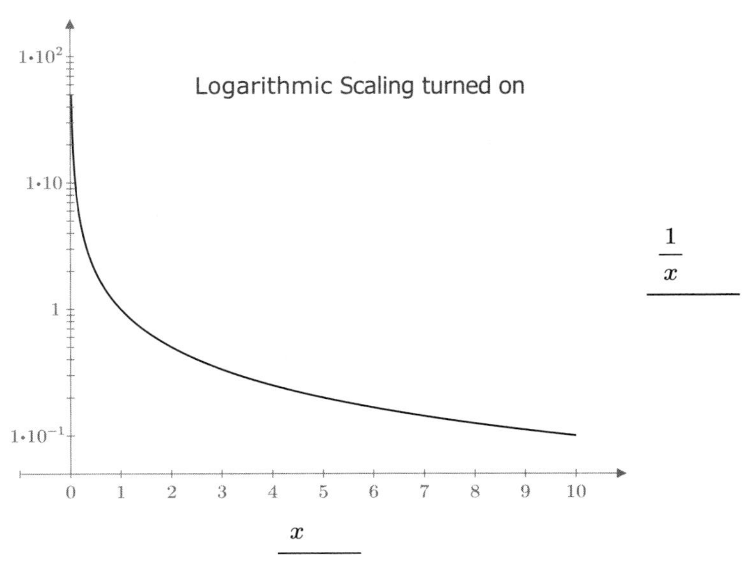

FIGURE 8.13

Logarithmic scaling

The **Plot Background** control allows you to make a plot transparent. This allows you to see the worksheet gridlines below the plot. PTC Mathcad Prime 3.0 does not have plot gridlines. The other two controls affect only the displayed plot region. Both have similar effects on the printed worksheet. The **Paper Color** control (**Plots>Styles>Plot Background>Paper Color**) allows the plot to blend in with the background of the displayed worksheet. The **White** control puts the plot on a white background that distinguishes the plot region on the displayed worksheet. See Figure 8.14.

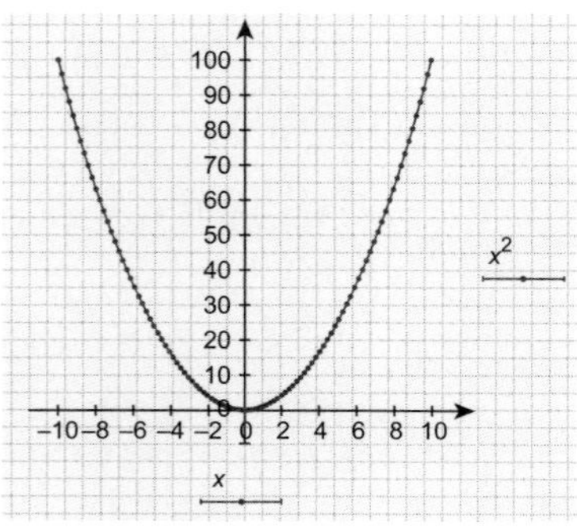

Transparent with gridlines on

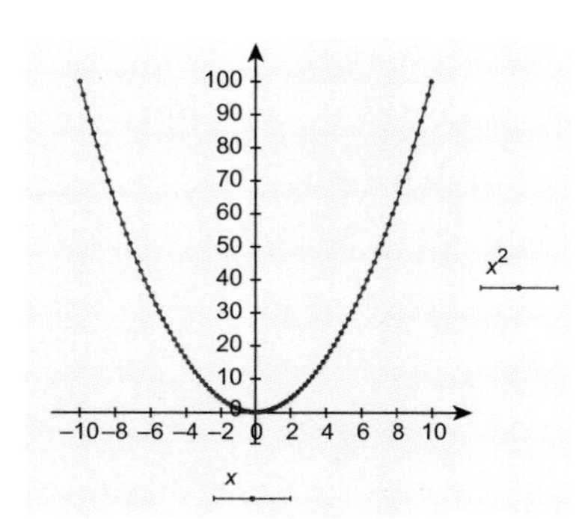

Paper Color

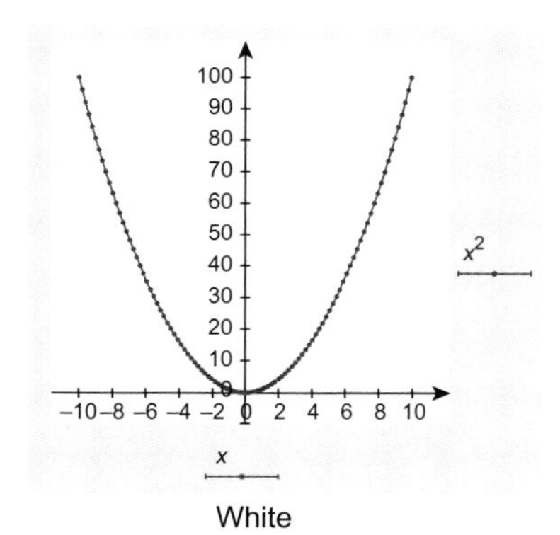

White

FIGURE 8.14

Plot background

Labels

PTC Mathcad Prime 3.0 does not have the ability to add titles or axis labels.

If you need to add titles to your plots, the only way to do this in PTC Mathcad Prime 3.0 is to use text boxes and drag the text region onto the plot region. Future versions of PTC Mathcad Prime will have the ability to add labels and titles. Figure 8.15 shows how to use text boxes to create titles and labels.

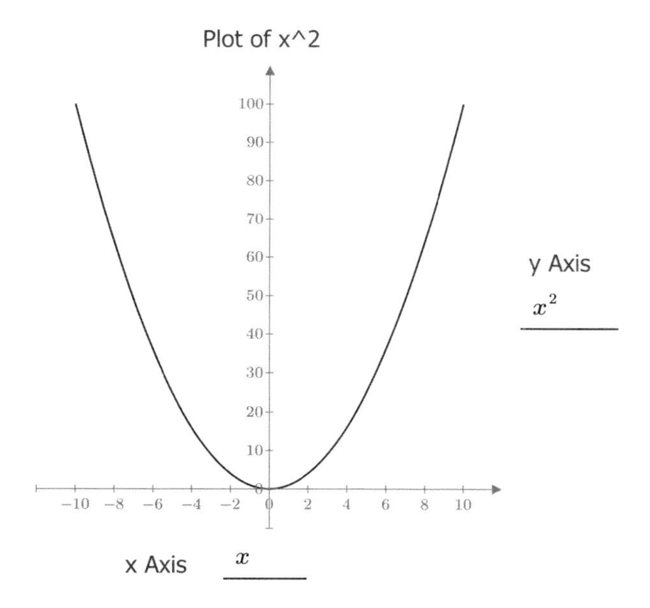

FIGURE 8.15

Titles and labels

Graphing Multiple Functions

In Mathcad 15, the keyboard shortcut to add an additional plotting expression was **COMMA**. In PTC Mathcad Prime, the keyboard shortcut is **SHIFT+ENTER**.

You can graph multiple functions or expressions on the same plot. To graph multiple expressions using the same x-axis variable, use the **Add Trace** control in the **Traces** group. The keyboard shortcut is **SHIFT+ENTER**. This places a new placeholder below the original placeholder. If you want the new placeholder to be above the current expression, then place the cursor at the leftmost insertion point prior to using the **Add Trace** control. You can now type a new function or expression in the placeholder. Each plot is called a trace. See Figure 8.16.

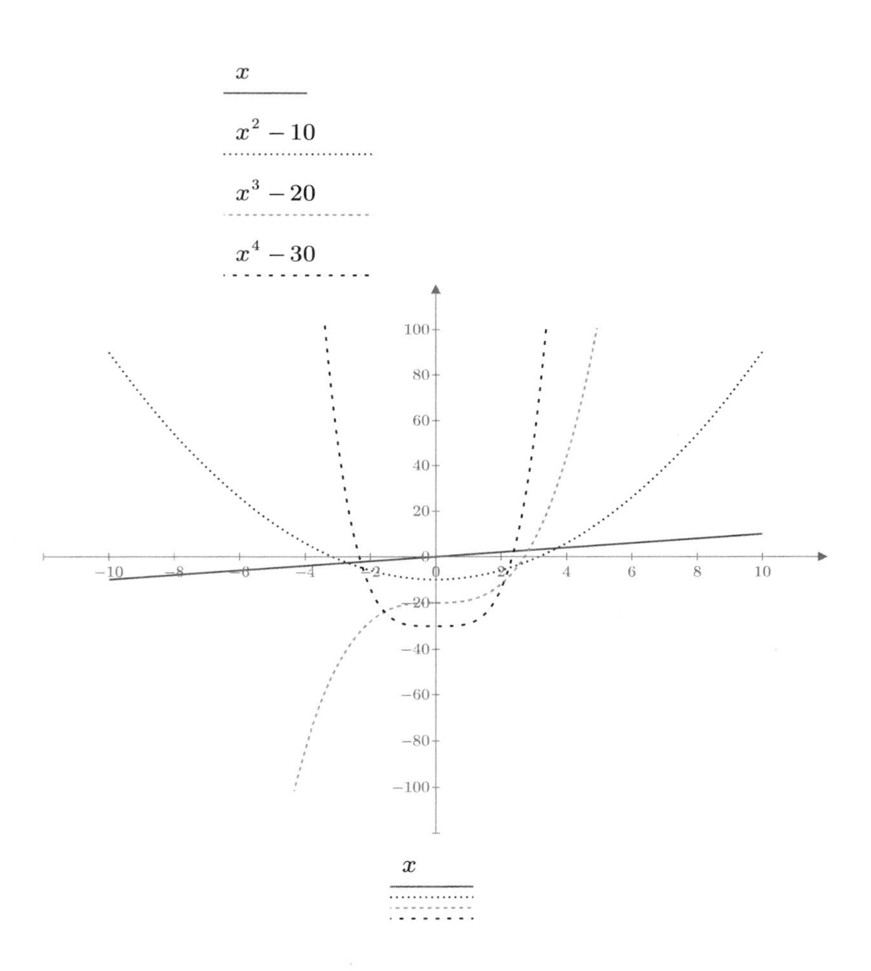

This figure is a plot of four expressions, each using the same x-axis variable. The x-axis limits are set at -10 and 10. The y-axis limits are set at -100 and 100.

FIGURE 8.16

Multiple plots

You can also use multiple variables on the x-axis and then plot corresponding expressions on the y-axis. You add new x-axis placeholders the same way you do for y-axis placeholders. To do this, use the **Add Trace** control or type **SHIFT+ENTER**. This places a new x-axis placeholder below the original placeholder. On the y-axis, create new placeholders as noted above, and use the corresponding x-axis variable name in your expression. See Figure 8.17.

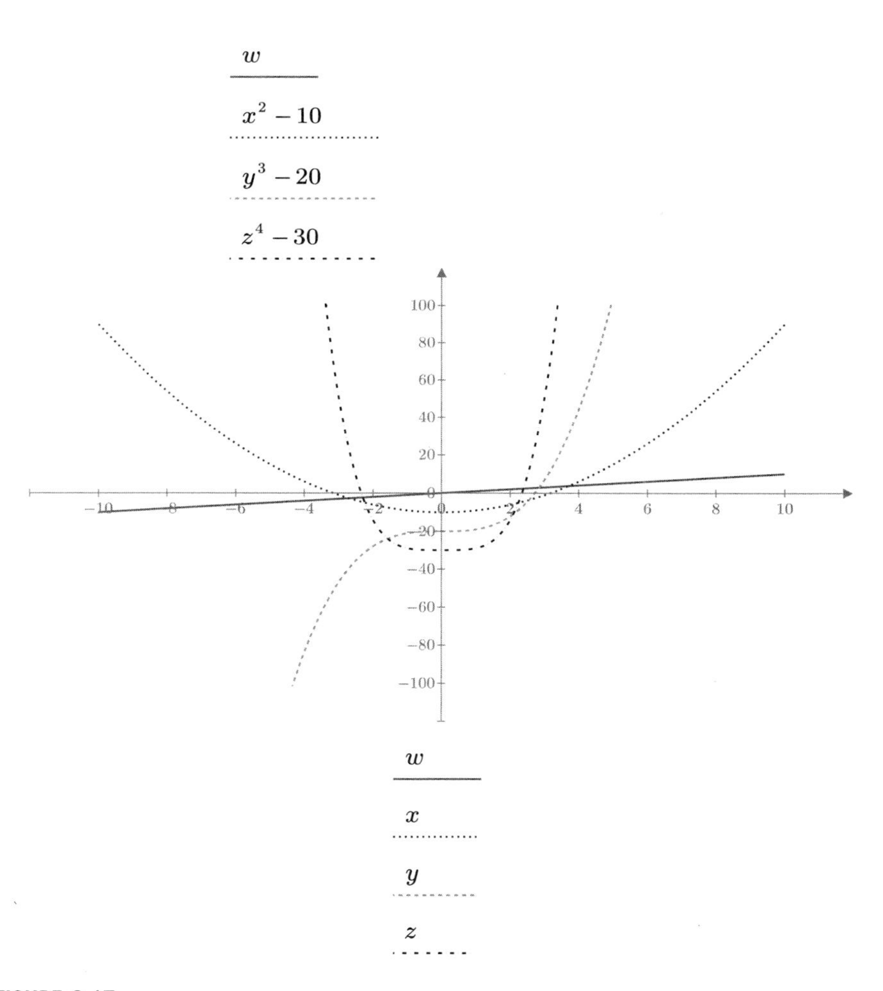

FIGURE 8.17

Multiple plots with multiple variables

It is important that the expressions on the y-axis match the corresponding x-axis placeholders. You cannot change the order.

Scale Factors

PTC Mathcad Prime 3.0 does not allow a secondary y-axis with a different scale.

Mathcad 15 allowed you to add a secondary y-axis to plot using a different scale. The current version of PTC Mathcad does not allow a secondary y-axis. This can create a challenge if the results of one expression are more than an order of magnitude different from the results of another expression. To assist with this challenge, you can use the unit placeholder as a scale factor. The scale factor value you put in the unit placeholder divides (scales) the plotted values by the scale factor. You can use a different scale factor for each trace so that the plotted values are within a reasonable distance of each other. See Figures 8.18 and 8.19.

You can use π as a scale factor to display radians scaled to π. Remember to change the label of π to **Constant** in order for it to work. See Figure 8.20.

Plotting Data Points

Up until now, we have focused on graphing functions and expressions. These plots are easily represented with lines. PTC Mathcad also allows you to plot data points. These data points can be created by using range variables, or they can be created from a vector or matrix.

Range Variables

When we discussed using range variables earlier, we were actually graphing data points. When using a range variable on the x-axis, PTC Mathcad creates a data point for each value in the range variable, and then plots the corresponding value on the y-axis. PTC Mathcad then draws a line between all the points.

Let's look at a few examples of using range variables to plot data points. See Figures 8.21 and 8.22.

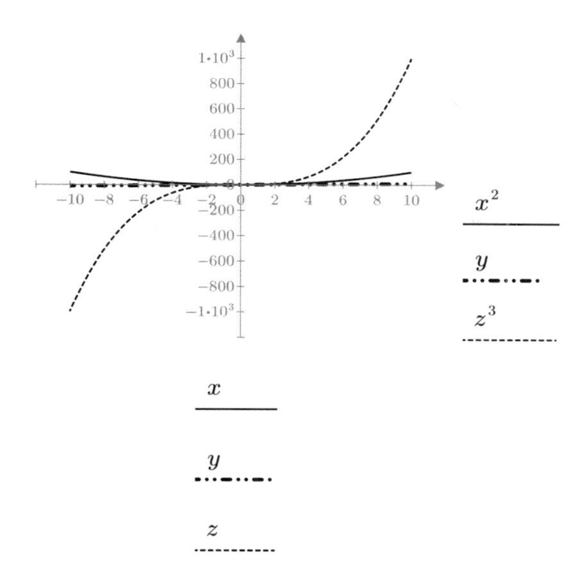

Plot after using scaling factors

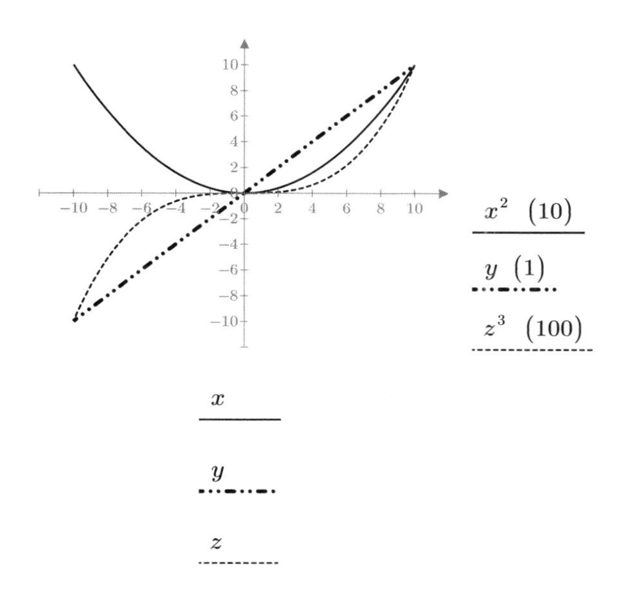

FIGURE 8.18

Scaling factors

Plot Force in lbf and pressure in psi

$$\text{Area} := 0\ ft^2, 1\ ft^2 .. 10\ ft^2 = \begin{bmatrix} 0 \\ 1 \\ 2 \\ 3 \\ 4 \\ 5 \\ 6 \\ 7 \\ 8 \\ 9 \\ 10 \end{bmatrix} ft^2$$

$$\text{Force} := 62.4\ psf \cdot \text{Area} = \begin{bmatrix} 0 \\ 62.4 \\ 124.8 \\ 187.2 \\ 249.6 \\ 312 \\ 374.4 \\ 436.8 \\ 499.2 \\ 561.6 \\ 624 \end{bmatrix} lbf$$

$$\text{Pressure} := \frac{\text{Force}}{15\ in^2} = \begin{bmatrix} 0 \\ 4.16 \\ 8.32 \\ 12.48 \\ 16.64 \\ 20.8 \\ 24.96 \\ 29.12 \\ 33.28 \\ 37.44 \\ 41.6 \end{bmatrix} psi$$

Plot values in lbf and psi

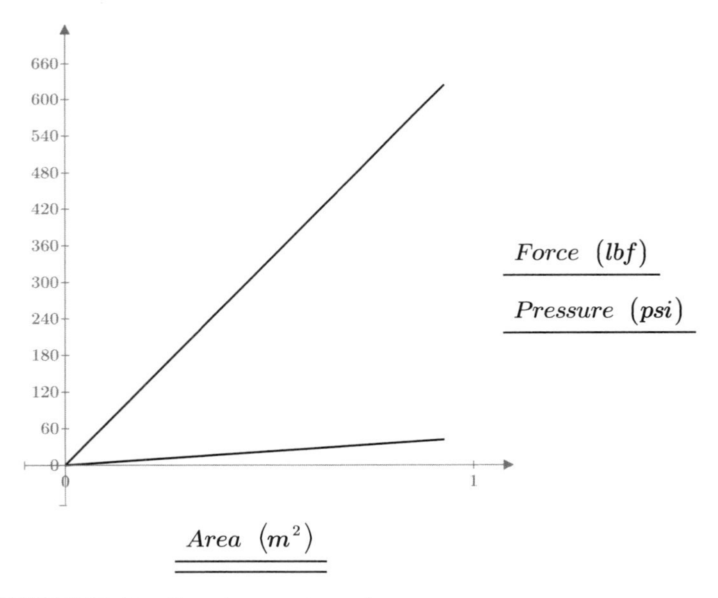

FIGURE 8.19 *(Continued on next page)*

Scaling factors with units

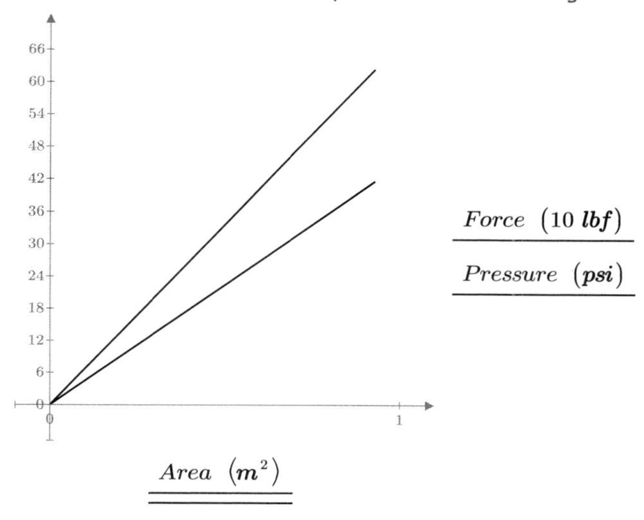

FIGURE 8.19

(*continued*)

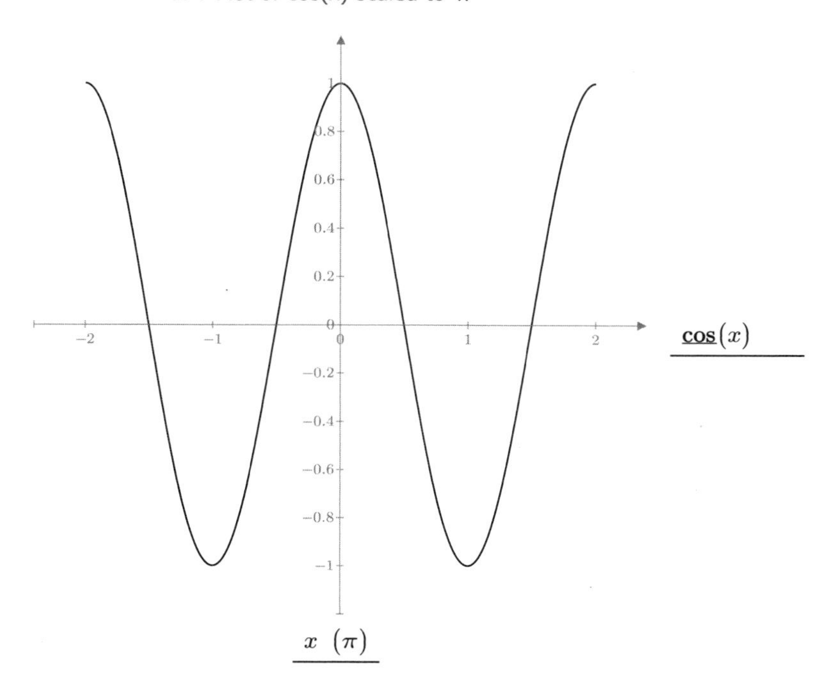

FIGURE 8.20

Using π as a scale factor

Create two range variables with a different number of values.

$$\mathrm{RV}_1 := -20, -19..20 \qquad \mathrm{RV}_2 := -20, -10..20$$

The above range variables actually create data points that are connected by straight lines. The data points are displayed in these plots (**Plots>Styles>Symbol**).

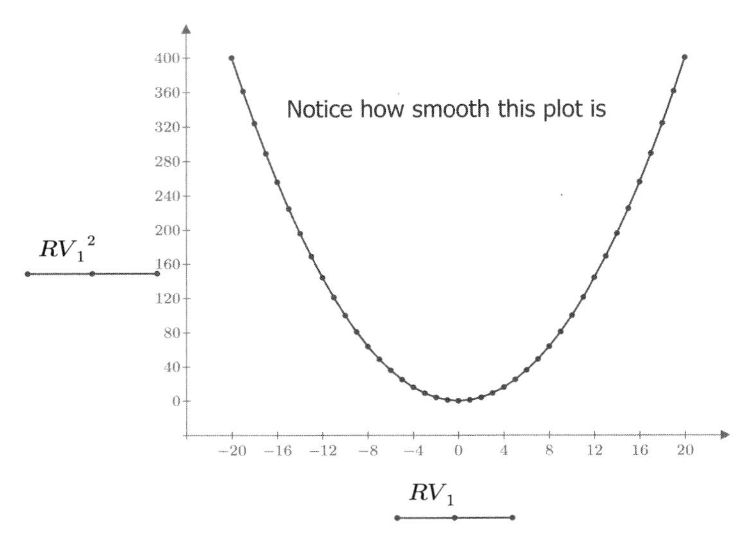

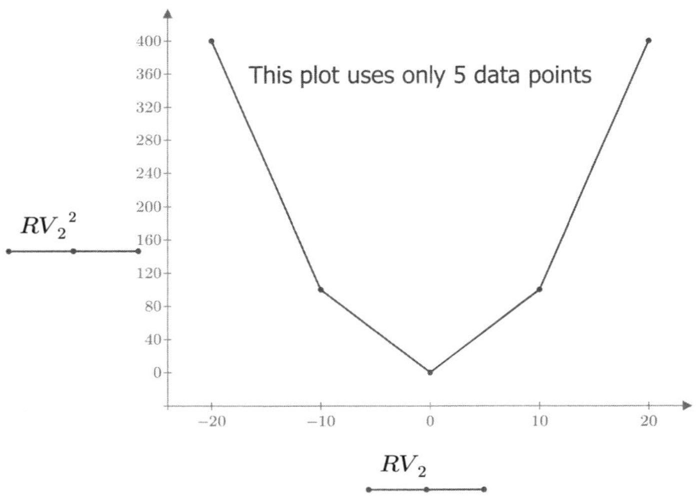

FIGURE 8.21

Plotting data points

This range variable has an increment of $\frac{\pi}{24}$.

$$\mathbf{RV_3} := \frac{-\pi}{2}, \frac{-\pi}{2} + \frac{\pi}{24} .. \frac{\pi}{2}$$

This range variable has an increment of $\frac{\pi}{8}$.

$$\mathbf{RV_4} := \frac{-\pi}{2}, \frac{-\pi}{2} + \frac{\pi}{8} .. \frac{\pi}{2}$$

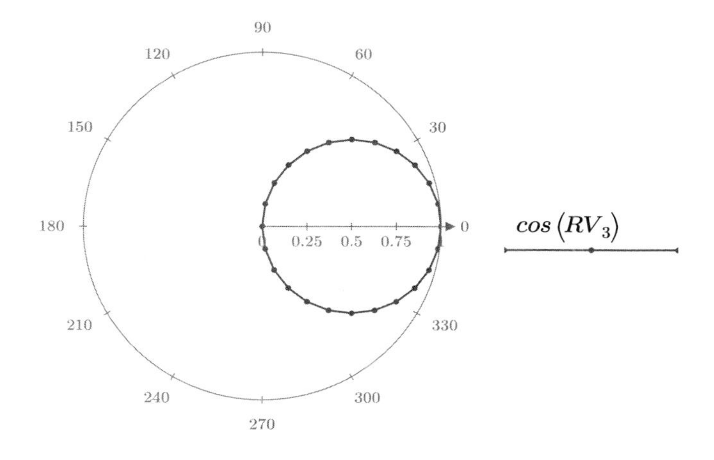

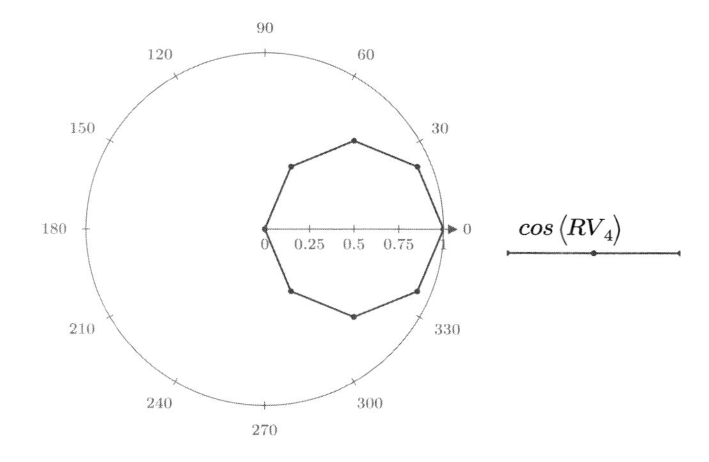

FIGURE 8.22

Plotting data points

Data Vectors

See Figure 8.23 for an example of plotting a vector of data points. See Figure 8.24 for an example of plotting matrix data points.

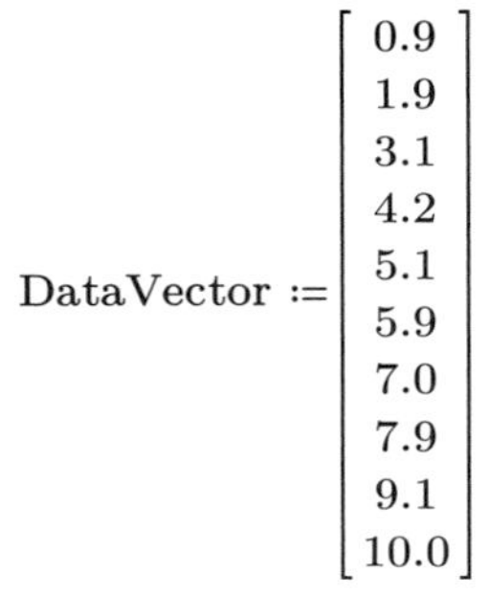

$$\text{DataVector} := \begin{bmatrix} 0.9 \\ 1.9 \\ 3.1 \\ 4.2 \\ 5.1 \\ 5.9 \\ 7.0 \\ 7.9 \\ 9.1 \\ 10.0 \end{bmatrix}$$

Note: ORIGIN is set to 1.

To plot a data vector, create a range variable to provide the x-axis value, and to tell PTC Mathcad which data points to plot. The range variable must be in integer increments.

$\underline{\text{last}}(\text{DataVector}) = 10$

$i := 1, 2 .. \underline{\text{last}}(\text{DataVector})$

Use the *last*(v) function, which returns the element number of the last element of the vector v.

With this range variable, we are telling PTC Mathcad to plot the first through the last elements of the vector "DataVector".

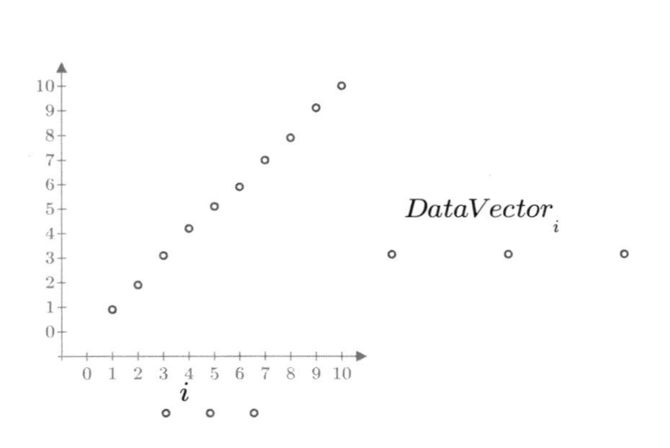

Type the name of the range variable on the x-axis. On the y-axis, type the name of the data vector and use the **vector** bolded operator (vector subscript) with the range variable as the susbscript. (Type [to get the vector operator.)

For this plot, the trace type is changed to "Points". The Symbol type is set to an open circle.

FIGURE 8.23

Plotting vector data points

$$\text{DataMatrix} := \begin{bmatrix} 0 & 0.1 \\ 1 & 0.9 \\ 2 & 1.9 \\ 3 & 3.1 \\ 4 & 4.2 \\ 5 & 5.1 \\ 6 & 4.3 \\ 7 & 4.0 \\ 10 & 2.9 \\ 11 & 2.1 \\ 12 & 0.8 \end{bmatrix}$$

To plot data from a matrix, assign each column in the matrix to a variable. To do this, use the matrix column operator "M<>". The column operator returns a vector of the matrix column numbered between the brackets. This operator is located on the Matrix toolbar, or press CTRL+6. (The ***submatrix*** function also could have been used.)

$X := \text{DataMatrix}^{\langle 1 \rangle}$

$Y := \text{DataMatrix}^{\langle 2 \rangle}$

Note: Type CTRL+6 to get the brackets.

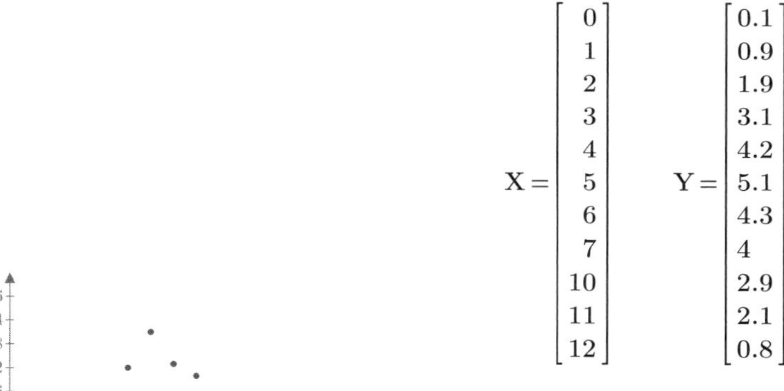

$$X = \begin{bmatrix} 0 \\ 1 \\ 2 \\ 3 \\ 4 \\ 5 \\ 6 \\ 7 \\ 10 \\ 11 \\ 12 \end{bmatrix} \quad Y = \begin{bmatrix} 0.1 \\ 0.9 \\ 1.9 \\ 3.1 \\ 4.2 \\ 5.1 \\ 4.3 \\ 4 \\ 2.9 \\ 2.1 \\ 0.8 \end{bmatrix}$$

For this plot, the trace type is changed to "Points". The data points are: (X1,Y1), (X2,Y2), etc.

FIGURE 8.24

Plotting matrix data points

Most of our discussion has focused on plotting lines and points. PTC Mathcad can also plot other types of plots. Figure 8.25 illustrates three additional trace types. These are the column trace, the bar trace, and the stem trace.

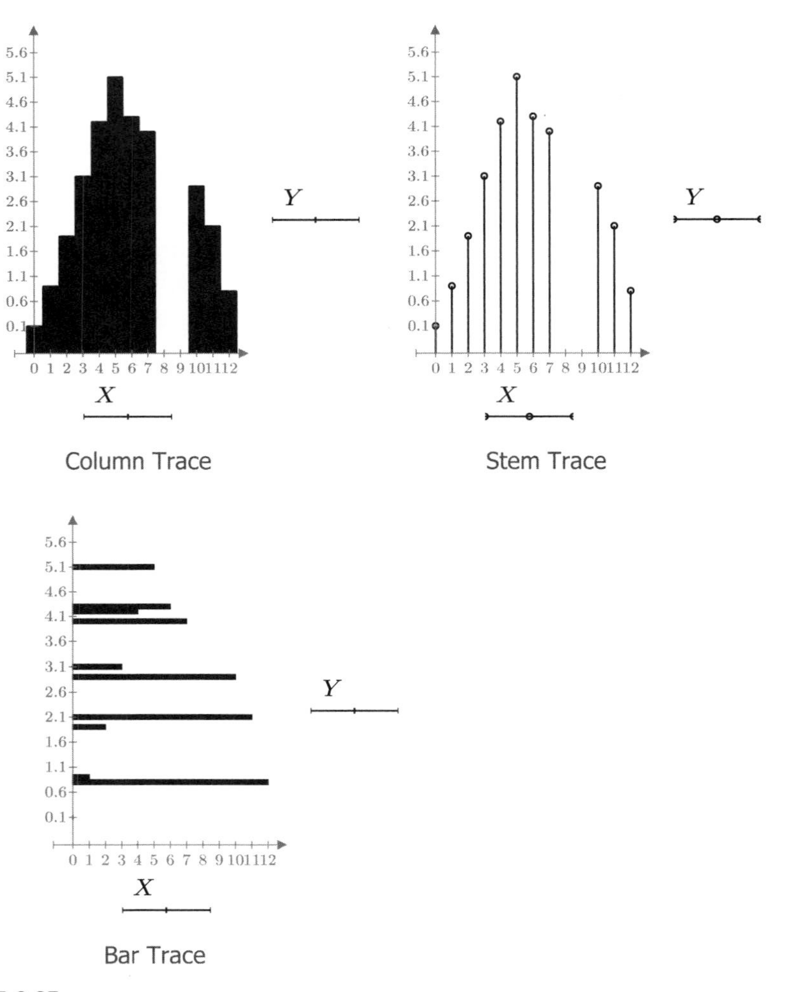

FIGURE 8.25

Additional trace types

Error Plots

An error plot requires two columns of data in the y-axis placeholder. They represent a maximum and minimum value to be plotted on the y-axis. See Figure 8.26.

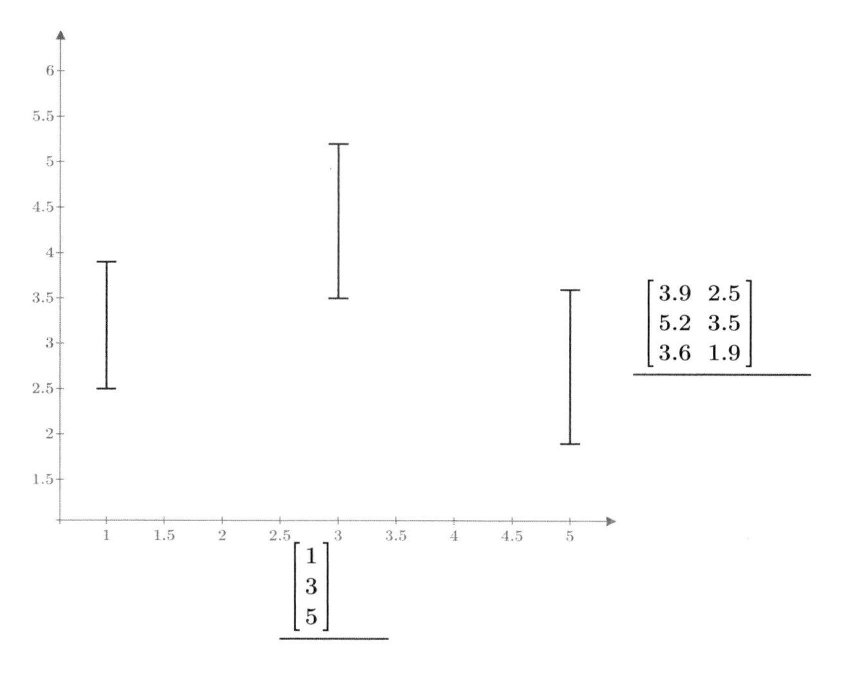

FIGURE 8.26

Error plot

Parametric Plotting

A parametric plot is one in which a function or expression is plotted against another function or expression that uses the same independent variable. See Figure 8.27 for an example.

Create the range variable

$iii := 1, 2 .. 10$

Create two functions of x

$ZZ(x) := \frac{1}{x^2}$

$YY(x) := x^2$

$$ZZ(iii) = \begin{bmatrix} 1.000 \\ 0.250 \\ 0.111 \\ 0.063 \\ 0.040 \\ 0.028 \\ 0.020 \\ 0.016 \\ 0.012 \\ 0.010 \end{bmatrix} \qquad YY(iii) = \begin{bmatrix} 1.00 \\ 4.00 \\ 9.00 \\ 16.00 \\ 25.00 \\ 36.00 \\ 49.00 \\ 64.00 \\ 81.00 \\ 100.00 \end{bmatrix}$$

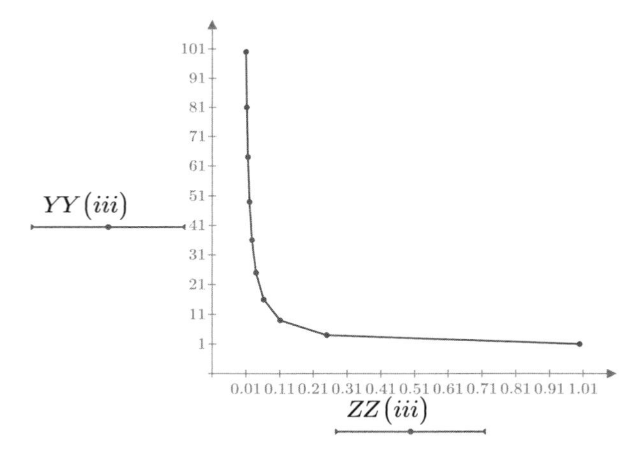

Parametric plot of ZZ(x) on the x-axis and YY(x) on the y-axis.

FIGURE 8.27

Parametric plot

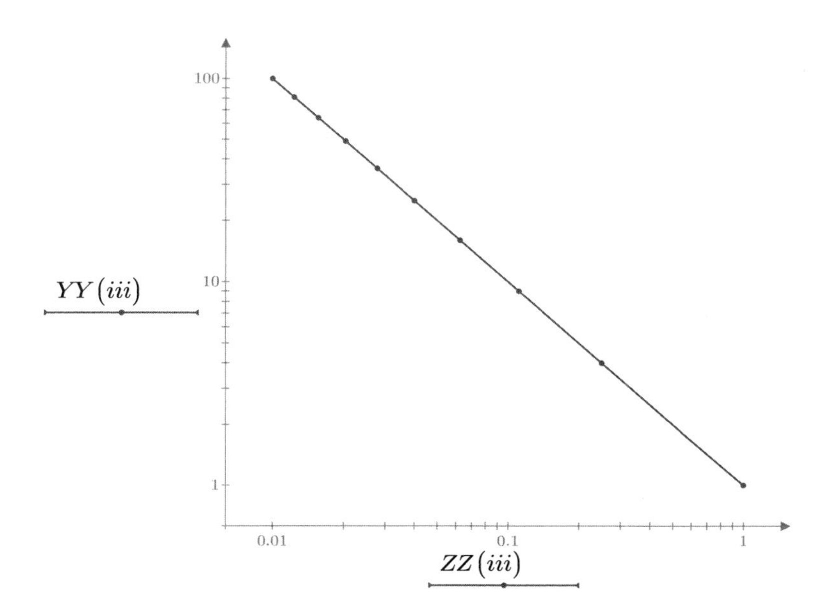

Parametric plot of ZZ(x) on the x-axis and YY(x) on the y-axis using Logarithmic scale on both axes.

FIGURE 8.27

(continued)

Trace and Zoom

In Mathcad 15 you could trace your cursor along a plot and have the program provide the x and y values. PTC Mathcad Prime 3.0 does not have this capability.

In Mathcad 15 you could temporarily zoom into a portion of a plot. PTC Mathcad Prime 3.0 does not have this capability.

Plotting over a Log Scale

If you are plotting a series of points using a log scale, you will want to have your points closer together the closer you get to zero, and further apart the further you get away from zero. For example, if you are plotting from 0.001 to 10,000, you need to have a very small increment in order to see the points near the beginning of the plot. However, if you select a small increment for a range variable, then you will be plotting millions of unnecessary points as you move toward 10,000.

The solution to this is to use a variable plotting range, using the PTC Mathcad function ***logspace***. This function creates a vector of n logarithmically spaced points. See Figures 8.28 and 8.29 for examples.

Most plot ranges occur over a uniform increment. This example uses a variable increment.

For this example, plot 21 points from a uniformly sloping line. The minimum value is 0.001. The maximum value is 10,000. Use a log scale on both axes.

$$X := \underline{\text{logspace}}(0.001, 10000, 21)$$

Create a vector of values to be plotted using the PTC Mathcad function ***logspace***. This function returns a vector of n points logarithmically spaced from the minimum to the maximum value. The function takes the form logspace(min,max,npts), where npts is the numberf of points. Notice that as the vector gets larger, so does the distance between each point.

$$X = \begin{bmatrix} 0.00100 \\ 0.00224 \\ 0.00501 \\ 0.01122 \\ 0.02512 \\ 0.05623 \\ 0.12589 \\ 0.28184 \\ 0.63096 \\ 1.41254 \\ 3.16228 \\ 7.07946 \\ 15.84893 \\ 35.48134 \\ 79.43282 \\ 177.82794 \\ 398.10717 \\ 891.25094 \\ 1995.26231 \\ 4466.83592 \\ 10000.00000 \end{bmatrix}$$

FIGURE 8.28

Plotting using a variable range

Continued from Figure 8.28

Function $(\mathbf{a}) := \mathbf{a}$

$$\mathrm{s} := 0.001, \frac{10000 - 0.001}{20} .. 10000$$

(Creates a range variable "s" with 21 uniformly spaced points and with 20 uniformly spaced intervals)

This plot uses points with the same increment

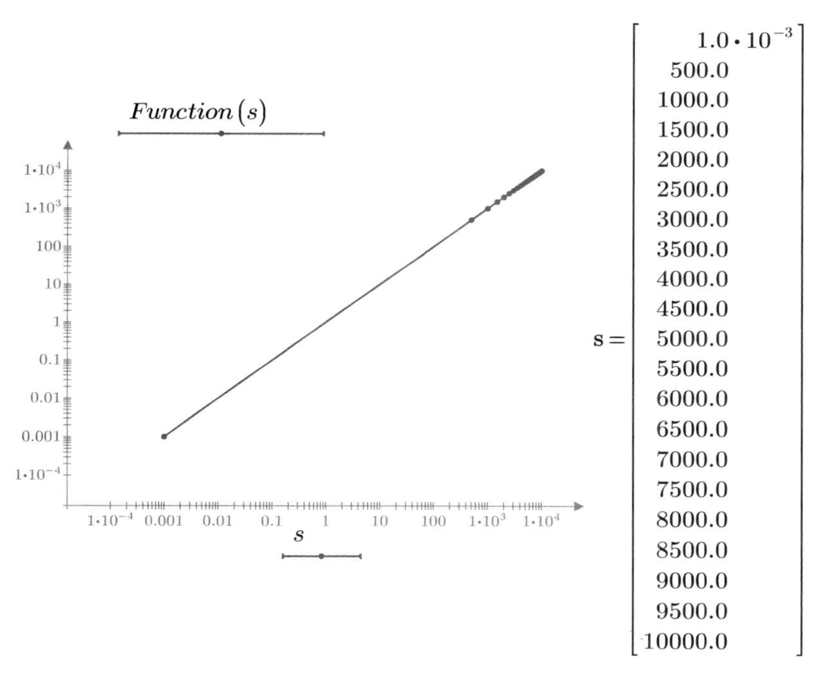

FIGURE 8.29 (*Continued on next page*)

Plotting uniform versus variable ranges on a log plot

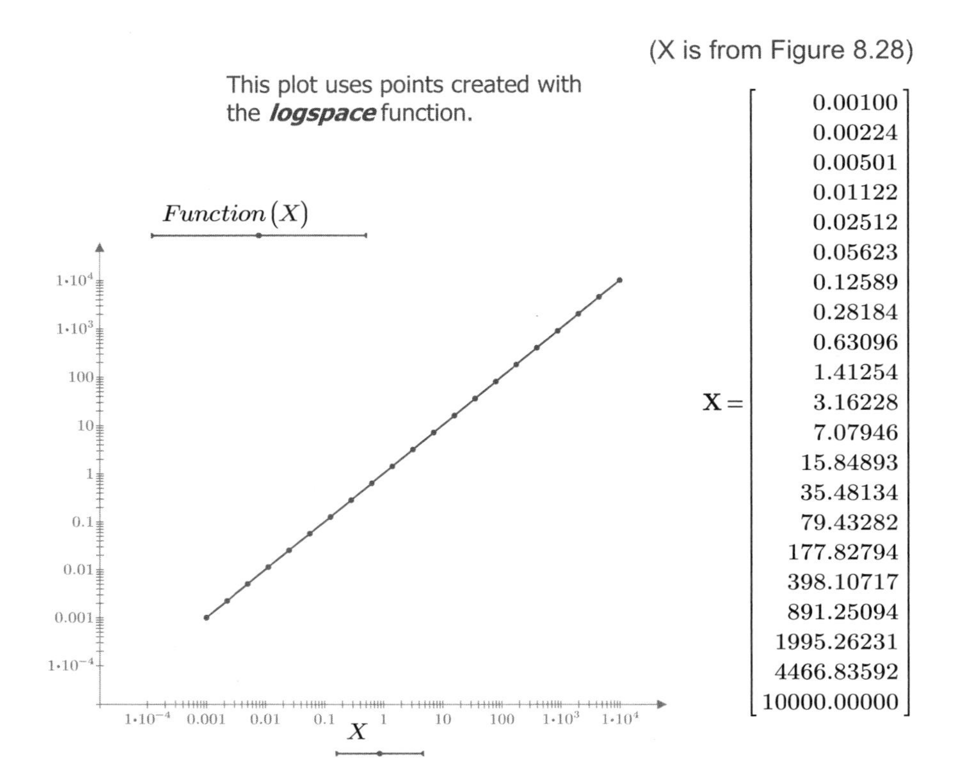

FIGURE 8.29

(*continued*)

Plotting a Family of Curves

It is possible to plot two parameters on a 2D graph. For example, you can plot F(a,b): = a^2*cos(b) over the ranges a = 0,1..4 and b = 0,0.1.. 15, with b plotted on the x-axis. To do this you need to create a data matrix, and then plot the values of b on the x-axis and the values of the data matrix on the y-axis. See Figure 8.30.

$\mathbf{Family}(a,b) := a^2 \cdot \underline{\cos}(b) \qquad a := 0,1..4 \qquad b := 0,0.2..15$

$A := a = \begin{bmatrix} 0 \\ 1 \\ 2 \\ 3 \\ 4 \end{bmatrix} \qquad B := b = \begin{bmatrix} 0 \\ 0.2 \\ 0.4 \\ 0.6 \\ \vdots \end{bmatrix}$

Note: This step converts range variables a and b to vectors and assigns them to A and B.

$i := 1..\underline{\text{last}}(B) \qquad j := 1..\underline{\text{last}}(A) \qquad F_{i,j} := \mathbf{Family}\left(A_j, B_i\right)$

$$F = \begin{bmatrix} 0 & 1 & 4 & 9 & 16 \\ 0 & 0.98 & 3.92 & 8.821 & 15.681 \\ 0 & 0.921 & 3.684 & 8.29 & 14.737 \\ 0 & 0.825 & 3.301 & 7.428 & 13.205 \\ 0 & 0.697 & 2.787 & 6.27 & 11.147 \\ 0 & 0.54 & 2.161 & 4.863 & 8.645 \\ 0 & 0.362 & 1.449 & 3.261 & 5.798 \\ 0 & 0.17 & 0.68 & 1.53 & 2.719 \\ 0 & -0.029 & -0.117 & -0.263 & -0.467 \\ 0 & -0.227 & -0.909 & -2.045 & -3.635 \\ 0 & -0.416 & -1.665 & -3.745 & -6.658 \\ 0 & -0.589 & -2.354 & -5.297 & -9.416 \\ 0 & -0.737 & -2.95 & -6.637 & -11.798 \\ & & & & \vdots \end{bmatrix}$$

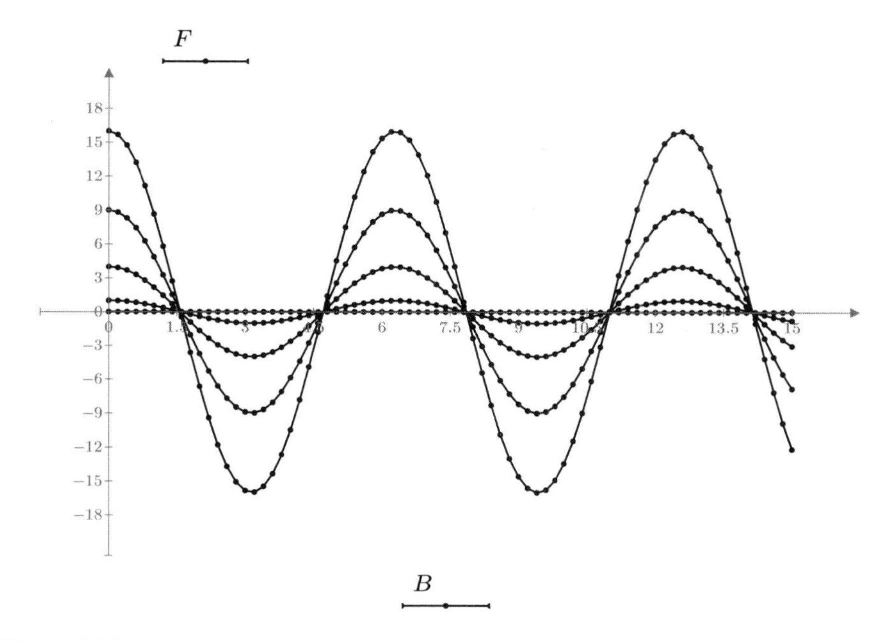

FIGURE 8.30

Plotting a family of curves

3D Plots and Contour Plots

The 3D formatting features in PTC Mathcad Prime 3.0 are very basic compared to the 3D formatting features of Mathcad 15. The additional features will hopefully make it into PTC Mathcad Prime 4.0. In Mathcad 15, contour plots were considered as a 3D plot. In PTC Mathcad Prime 3.0, contour plots are considered separately from 3D plots. You can select from a 3D plot or a contour plot. You cannot change a contour plot to a 3D plot. The 3D plot region has been improved over Mathcad 15.

PTC Mathcad treats three-dimensional plots and contour plots as two different plot types.

3D Plots

With 3D plots you can plot surfaces, curves, or data points. The topic of three dimensional plotting can become complicated. This section will only provide an introduction. The PTC Mathcad Help Center provides more thorough examples of 3D plotting.

You insert a 3D plot region by selecting **3D Plot** from the **Insert Plot** control in the **Traces** group (**Plots>Traces>Insert Plot>3D Plot**). The keyboard shortcut is **CTRL+3**. The 3D plot region is shown below.

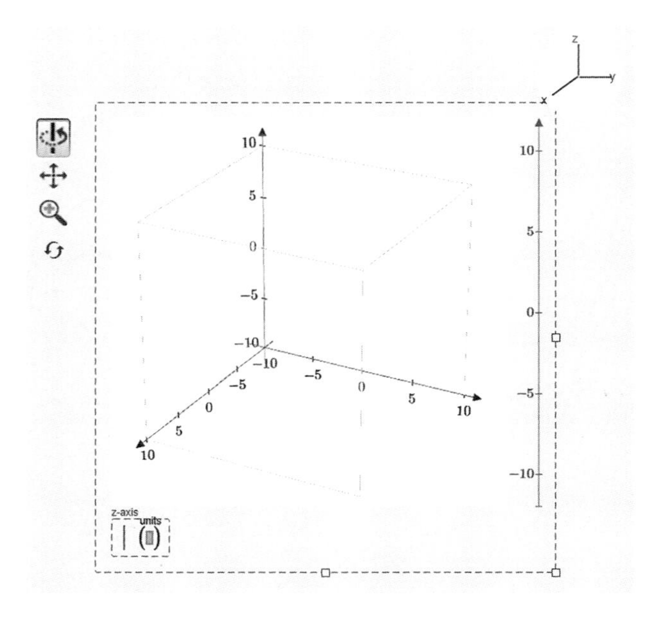

The z-axis placeholder in the lower left is where data for the plot is entered. We will discuss data shortly. The placeholder location is fixed and cannot be moved. The **Axis Selector** is visible in the upper right corner of the plot region. The vertical axis on the right side of the plot region is the Editing Axis. Use the **Axis Selector** to select one of the three axes. Once an axis has been selected, the Editing Axis can be used to modify the Lower Limit Tick Mark, the Upper Limit Tick Mark, and the Interval Tick Mark for the selected axis. The controls on the upper left allow you to rotate the plot, pan the plot, "rubber band" zoom on an area of the plot, or reset the plot to its default view. To use one of the controls, click it to make it active, then click and drag your mouse within the plot region. You can use your mouse scroll feature to zoom in or zoom out of the plot.

Three-dimensional plots can be created from a function, a matrix, or a set of vectors. Let's first discuss how to create a plot from a function. A function uses the ranges of x and y to generate values for height above the x-y plane. The order of the variables in the function or expression is important. The variable listed first in the function or expression is plotted on the x-axis. If you use a single-variable function, it will be a two-dimensional plot, plotted on the x-z plane at y=0. See Figure 8.31. The function listed in the z-axis placeholder does not need its arguments. Figures 8.31, 8.32, and 8.33 illustrate plots using functions.

To plot data points, PTC Mathcad needs three columns of data to define the x, y, and z-axis location of the plotted point: one column for the x-axis coordinates and corresponding columns of data for the y-axis and z-axis coordinates. The columns of data can be three separate column vectors, or they can be contained in a single matrix with three columns. A point is plotted at each corresponding x, y, and z coordinate. See Figures 8.34 and 8.35.

A single variable function plots as a two-dimensional plot at y=0

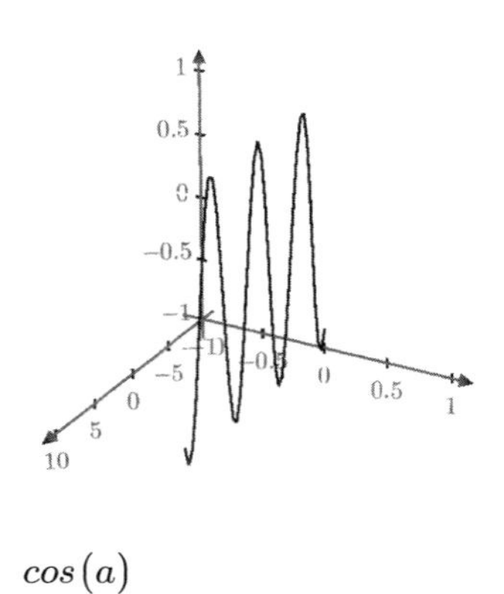

$cos(a)$

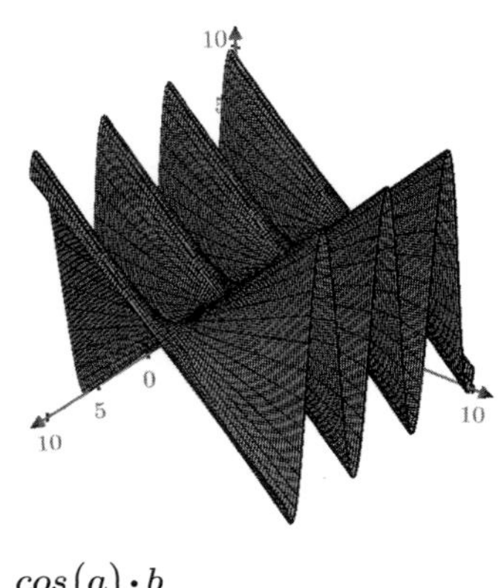

$cos(a) \cdot b$

Compare the difference of putting x first or last in the expression

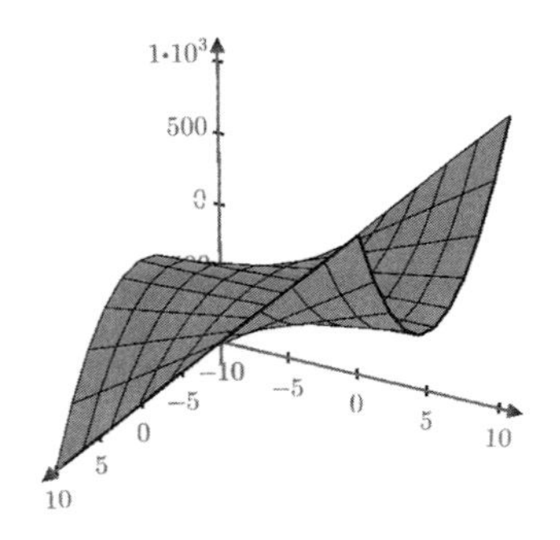

$x^2 \cdot y$

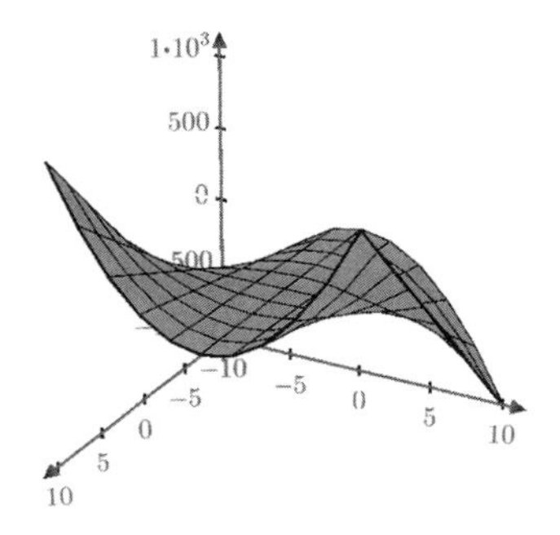

$y \cdot x^2$

FIGURE 8.31

3D plots using functions

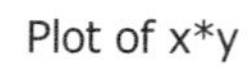

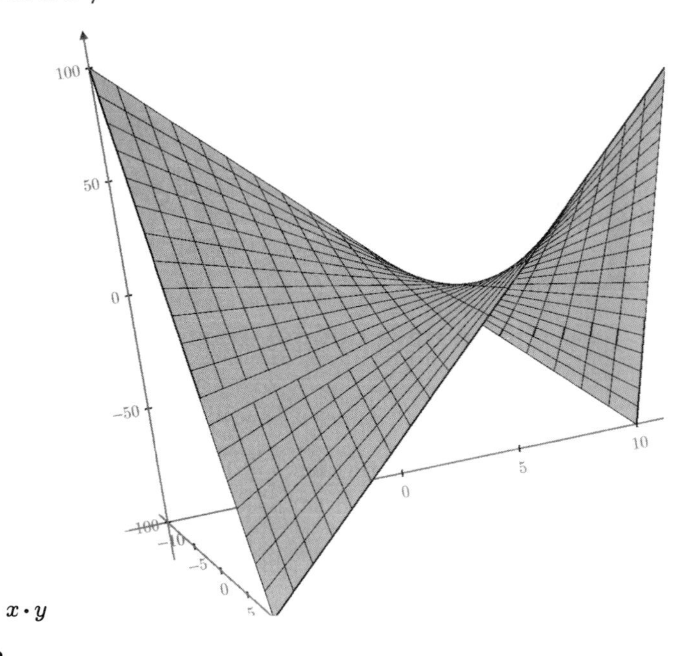

FIGURE 8.32

3D plots using functions

Plot of $x^2 + y^2$

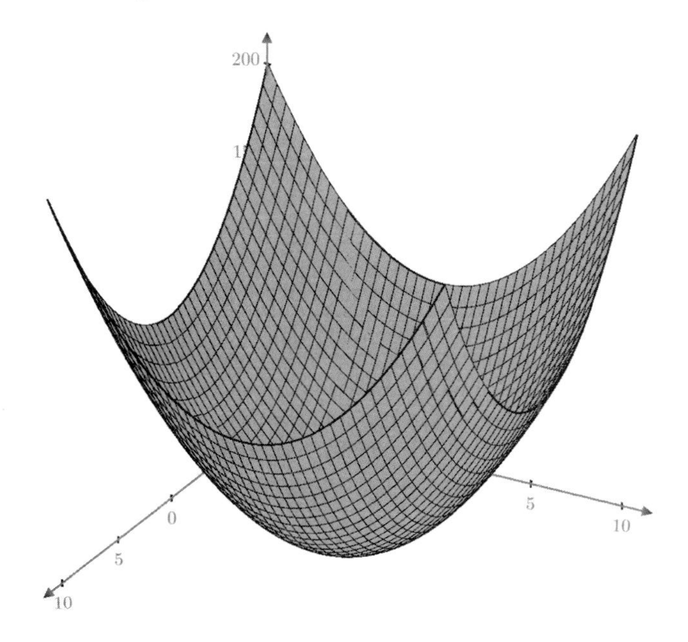

$x^2 + y^2$

FIGURE 8.33

3D plots using functions.

Create three data vectors.

$$\mathbf{X} := \begin{bmatrix} 1 \\ 1 \\ 1 \\ 2 \\ 2 \\ 2 \\ 3 \\ 3 \\ 3 \end{bmatrix} \quad \mathbf{Y} := \begin{bmatrix} 1 \\ 2 \\ 3 \\ 1 \\ 2 \\ 3 \\ 1 \\ 2 \\ 3 \end{bmatrix} \quad \mathbf{Z} := \begin{bmatrix} 2 \\ 2.5 \\ 2 \\ 3 \\ 4 \\ 3 \\ 2 \\ 1.5 \\ 2 \end{bmatrix}$$

Plot the three vectors, one plot with lines on and one plot with lines turned off.

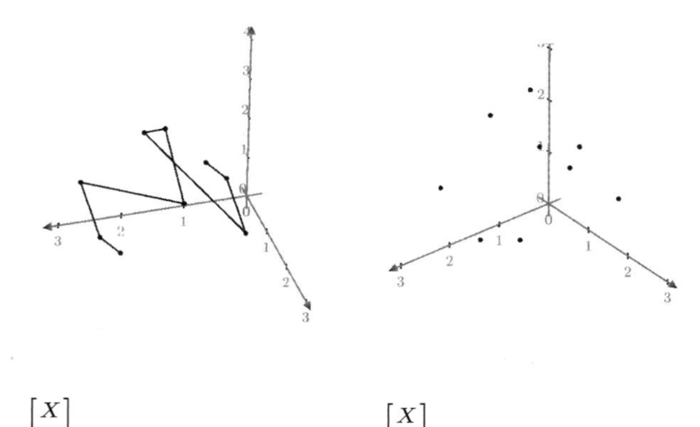

$$\begin{bmatrix} X \\ Y \\ Z \end{bmatrix} \qquad \begin{bmatrix} X \\ Y \\ Z \end{bmatrix}$$

FIGURE 8.34

3D plot using a data matrix

Place X, Y, and Z (from Figure 8.34) in a matrix.

$$\mathbf{Matrix} := \underline{\mathbf{augment}}(\mathbf{X}, \mathbf{Y}, \mathbf{Z}) = \begin{bmatrix} 1 & 1 & 2 \\ 1 & 2 & 2.5 \\ 1 & 3 & 2 \\ 2 & 1 & 3 \\ 2 & 2 & 4 \\ 2 & 3 & 3 \\ 3 & 1 & 2 \\ 3 & 2 & 1.5 \\ 3 & 3 & 2 \end{bmatrix}$$

The ***Augment*** function combines the three column vectors into a single matrix.

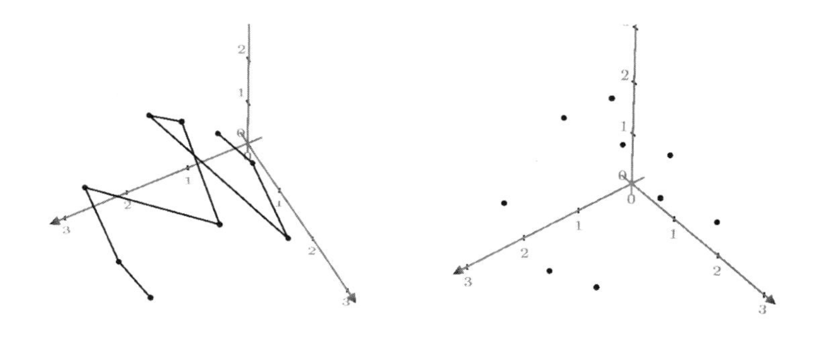

Matrix *Matrix*

FIGURE 8.35

3D plot of data points

PTC Mathcad can also plot data from an $m^x n$ matrix. The matrix row index represents the x-axis, the matrix column index represents the y-axis, and the matrix element represents the height above the x-y plane. See Figure 8.36.

$$\mathbf{Matrix} := \begin{bmatrix} 0 & 0 & 0 & 0 & 0 & 3 \\ 0 & 1 & 1 & 1 & 1 & 0 \\ 0 & 1 & 2 & 10 & 1 & 0 \\ 0 & 1 & 2 & 2 & 1 & 0 \\ 0 & 1 & 1 & 1 & 1 & 0 \\ 0 & 0 & 0 & 0 & 0 & 0 \end{bmatrix}$$

In this plot each value in the matrix represents a z value. The x and y values are the row and column indices.

ORIGIN is set to 1.

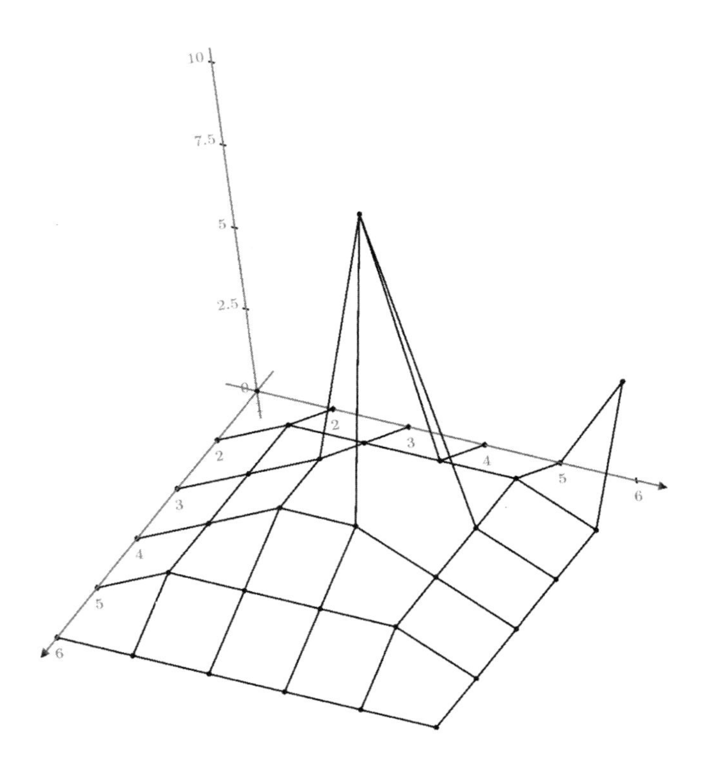

Matrix

FIGURE 8.36

3D plot using a data matrix

CreateSpace and ***CreateMesh*** are very powerful functions that create a series of data points that can be plotted by PTC Mathcad. These functions are discussed in detail in the PTC Mathcad Help Center. The figure below is from the Help Center. A simple example is illustrated in Figure 8.37 .

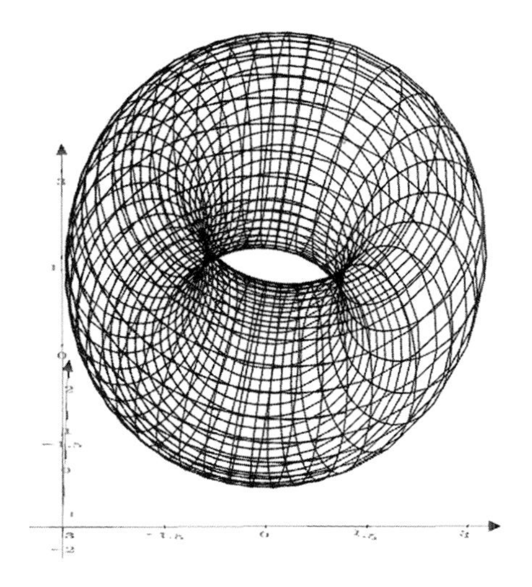

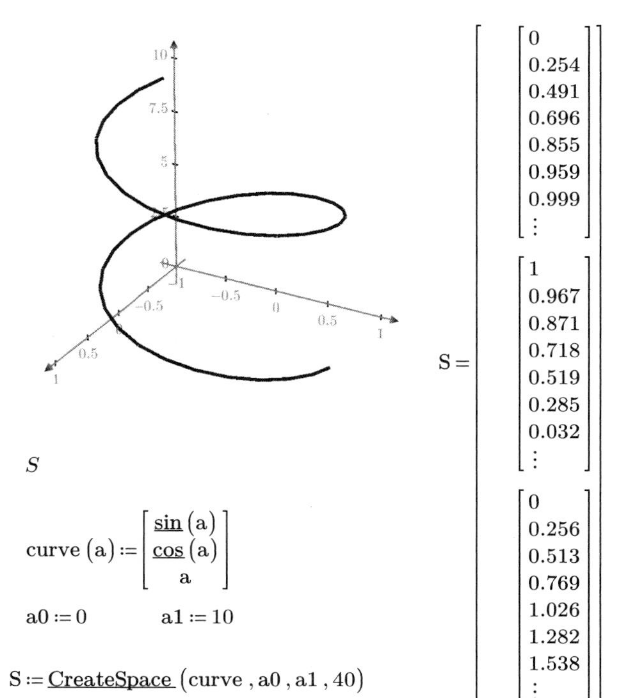

The CreateSpace function returns a nested array of three vectors representing the x, y, and z coordinates of the curve.

FIGURE 8.37

Using the CreateSpace function

You may plot multiple 3D plots by adding traces using the **Add Trace** control or by typing **SHIFT + ENTER**. See Figure 8.38.

Use the **Add Trace** control from the **Traces** group or type SHIFT+ENTER to add a second trace.

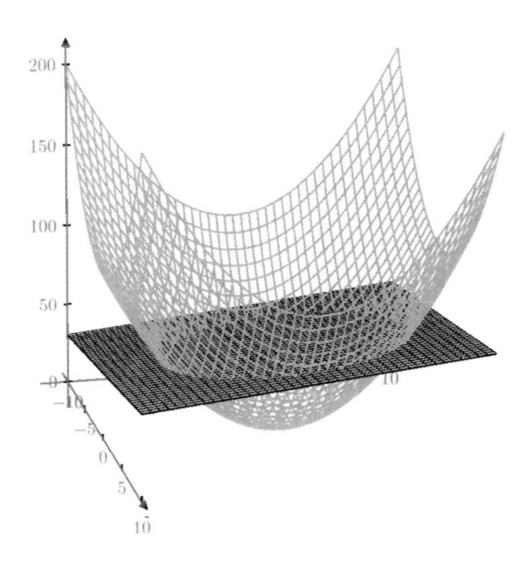

$x^2 + y^2$

$x + y + 50$

FIGURE 8.38

Multiple 3D plots

This discussion has not even scratched the surface of what is possible with 3D plotting. The PTC website has an electronic book titled *Creating Amazing Images with Mathcad 14* by Professor Byrge Birkeland that shows how to create some very amazing 3D images. The link is: http://communities.ptc.com/community/mathcad/mathcad-usage/blog/2010/05/27/creating-amazing-images-with-mathcad-14. Even though the book is written for Mathcad 14, the same functions apply to PTC Mathcad Prime 3.0. The figure below shows images from the book.

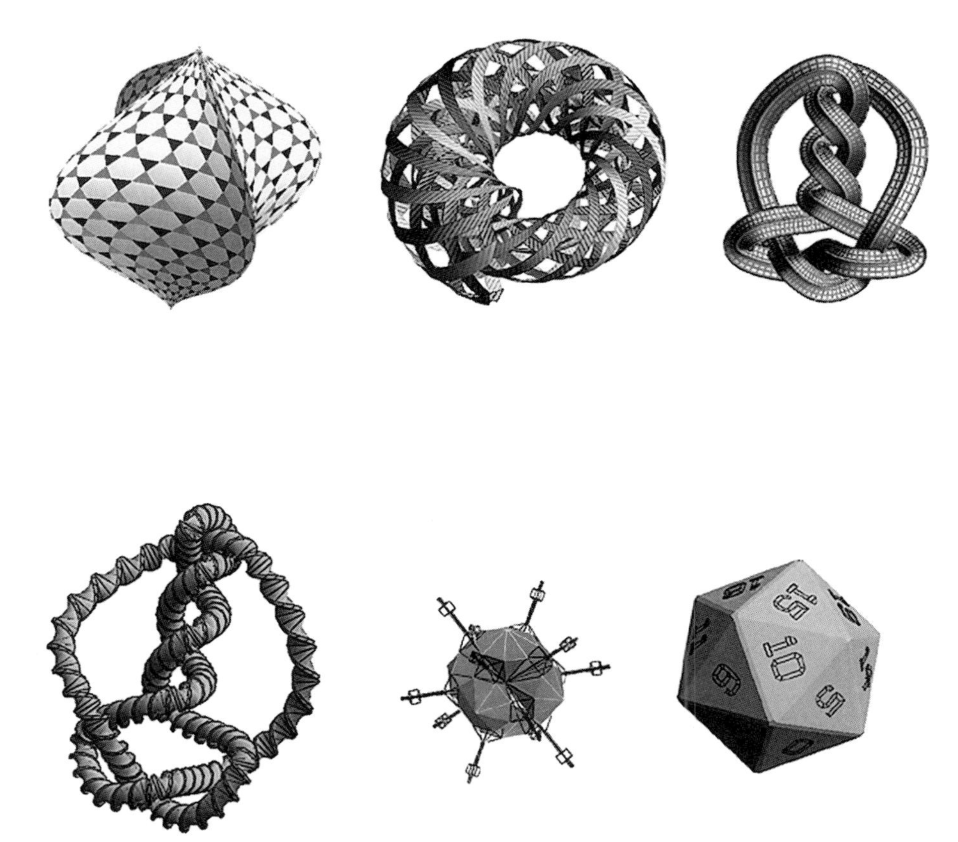

Contour Plots

A contour plot allows you to visualize three-dimensional data in a two-dimensional plot. You insert a contour plot by selecting **Contour Plot** in the **Traces** group. The keyboard shortcut is CTRL+5. You cannot switch between a contour plot and a 3D plot. The contour plot region is shown below.

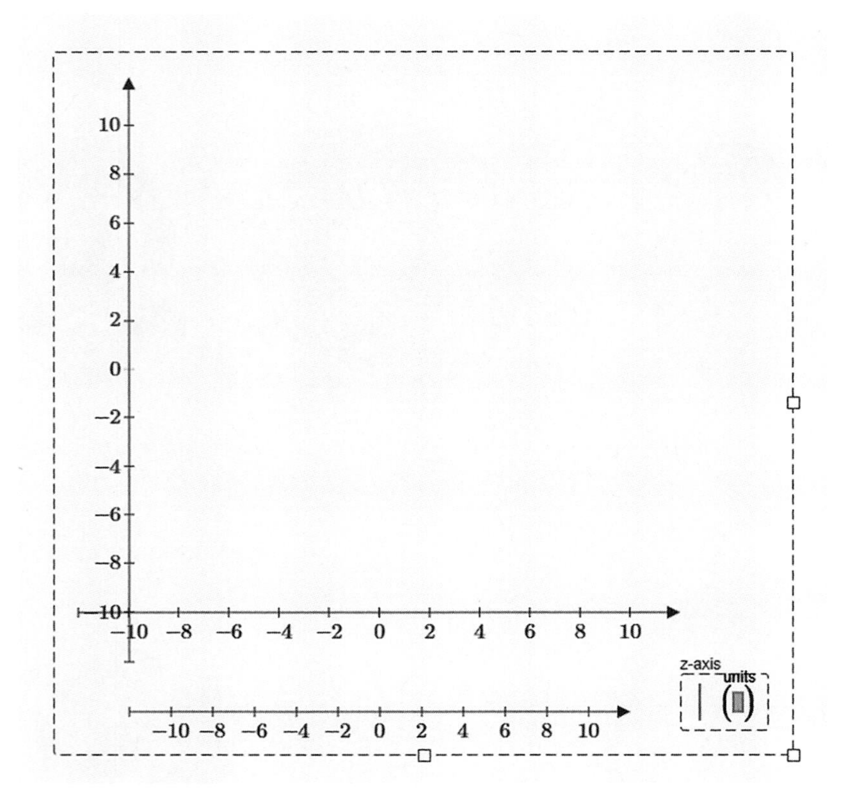

In the above figure, the second axis below the x-axis is the z-axis. When data is plotted, this will also show the color scale. The tick marks and tick mark values can be edited as in a 2D plot.

The data for a contour plot is similar to a 3D plot. You can use a function of two variables, a data matrix, or the output of the ***CreateMesh*** function. Once a contour plot is created, you can change the color scheme, the line style, trace color, and trace thickness. You can also turn the contour values on or off from the **Styles** group. See Figures 8.39 and 8.40.

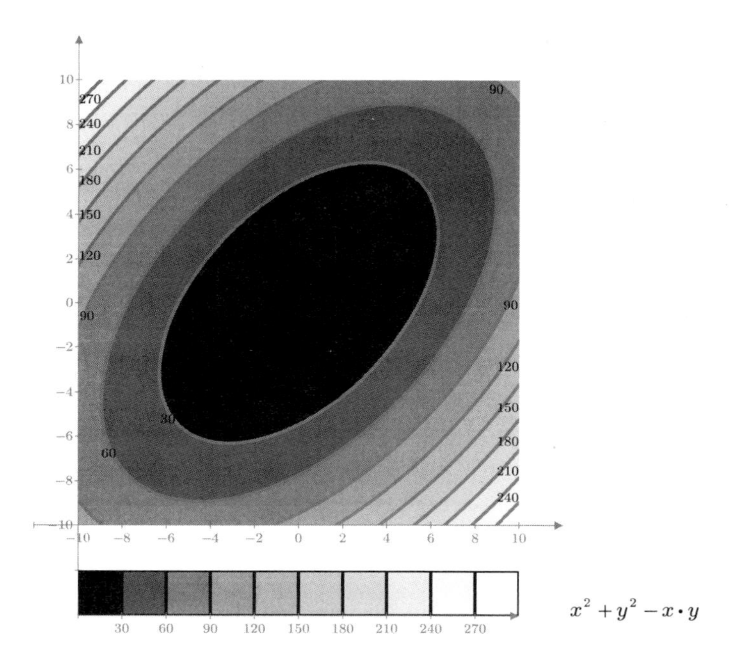

The above contour plot represents the adjacent 3D plot.

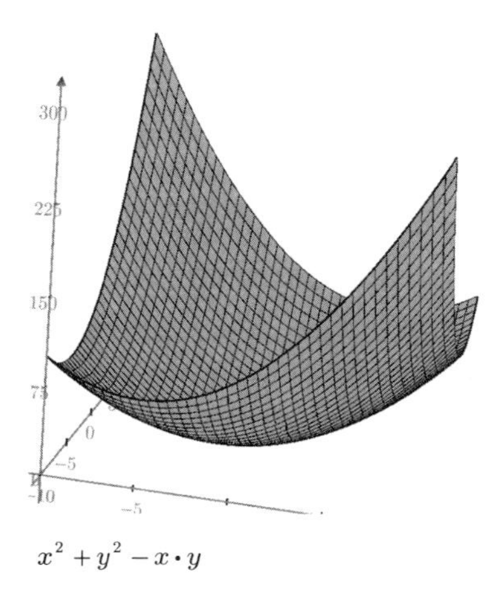

FIGURE 8.39

Contour plots

The below figure is the same as Figure 8.39. It uses a different color scheme, uses a previously defined function, and the trace thickness is different. The tick marks are also different.

$f(x, y) := x^2 + y^2 - x \cdot y$

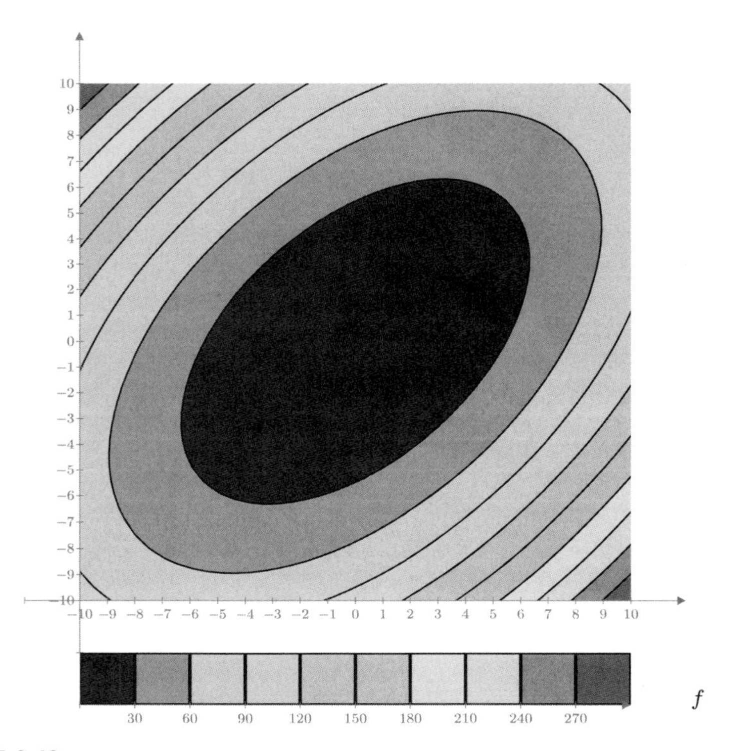

FIGURE 8.40

Contour plots

Engineering Examples

Engineering Example 8.1

For a simple span beam with uniform load (w) and length (L), plot the shear (kips), moment (Ft*kips), and deflection (inches) along the length of the beam.

The formula for shear at any point x is $V_x(w, L, x) := w \cdot \left(\frac{L}{2} - x\right)$

The formula for moment at any point x is $M_x(w, L, x) := \frac{w \cdot x}{2} \cdot (L - x)$

The formula for deflection at any point x is

$$\Delta_x(w, L, x, E, I) := \frac{w \cdot x}{24 \cdot E \cdot I} \cdot \left(L^3 - 2 \cdot L \cdot x^2 + x^3\right)$$

$$\mathbf{Length} := 20\ \boldsymbol{ft} \qquad \mathbf{Load} := 2500\ \frac{\boldsymbol{lbf}}{\boldsymbol{ft}} \qquad \mathrm{E} := 29000\ \boldsymbol{ksi} \qquad \mathrm{I} := 428\ \boldsymbol{in}^4$$

Use a range variable to set the plot points

$\mathbf{x} := 0\ \boldsymbol{ft}, 1\ \boldsymbol{ft} .. \mathbf{Length}$

Because the values of deflection are much smaller than the values for shear and moment, the results will need to be scaled.

$V_x(Load, Length, x)\ (kip)$

$M_x(Load, Length, x)\ (ft \cdot kip)$

$\Delta_x(Load, Length, x, E, I)\ \left(\frac{in}{100}\right)$

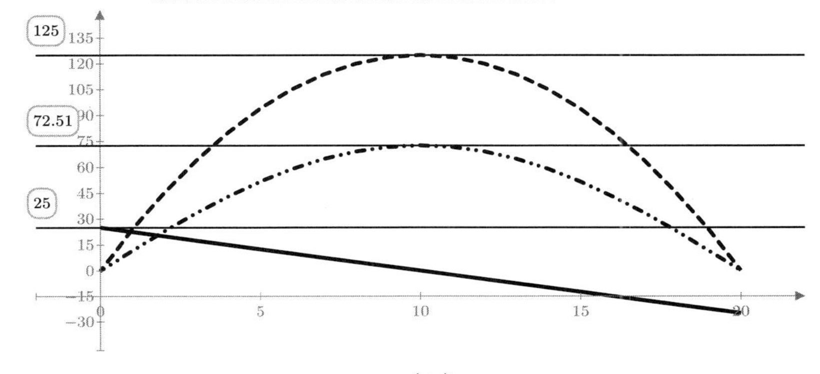

$x\ (ft)$

Engineering Example 8.2

The Manning equation $Q = \frac{1.49}{n} \cdot A_f \cdot R^{\frac{2}{3}} \cdot S^{\frac{1}{2}}$ is commonly used for computing the discharge in circular pipes flowing full or partially full. However, a difficulty in using it is computing the area of flow and wetted perimeter when the pipe is NOT either full or 1/2 full. This is because the geometric relationship between depth of flow and area of flow is not simple. There are several trigonometric formulas where the angle α can be used to compute the various characteristics required for flow problems. However, α is not generally known. We can get around this problem by computing all partial characteristics as a function of α and then plotting them against each other directly, as shown below.

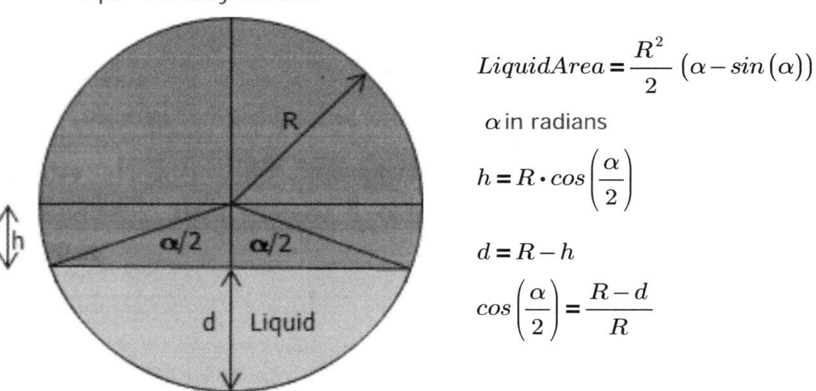

$$LiquidArea = \frac{R^2}{2}(\alpha - sin(\alpha))$$

α in radians

$$h = R \cdot cos\left(\frac{\alpha}{2}\right)$$

$$d = R - h$$

$$cos\left(\frac{\alpha}{2}\right) = \frac{R-d}{R}$$

Create a range variable of angle increments to be used in plots.

$$\alpha := 0.05\ rad, 0.1 \cdot \pi \cdot rad .. 2 \cdot \pi \cdot rad$$

The trigonometric equations given above allow calculation of the area of flow and the depth of flow for any angle α from 0 to $2 \cdot \pi$ radians. We can use these to compute the values needed for Manning's equation.

Write all independent variables in terms of the angle α.

Area of Flow

$$A_f(\alpha, D) := \frac{\left(\frac{D}{2}\right)^2}{2} \cdot (\alpha - \sin(\alpha))$$

Wetted Perimeter

$$P_w(\alpha, D) := \frac{D}{2} \cdot \alpha$$

Hydraulic Radius is the ratio of area of flow divided by the wetted perimeter.

$$R_h(\alpha, D) := \frac{A_f(\alpha, D)}{P_w(\alpha, D)}$$

$$h(\alpha, D) := \frac{D}{2} \cdot \cos\left(\frac{\alpha}{2}\right)$$

Depth of flow is radius minus h

$$Depth(\alpha, D) := \frac{D}{2} - h(\alpha, D)$$

The equations below allow calculation of the flow rate and velocity for any pipe diameter and angle α.

Flow rate (cubic feet/second) based on Manning's equation

$$Q(\alpha, D, n, S) := \frac{1.49 \cdot \frac{ft^{\frac{1}{3}}}{s}}{n} \cdot A_f(\alpha, D) \cdot R_h(\alpha, D)^{\frac{2}{3}} \cdot S^{\frac{1}{2}}$$

Velocity (ft/s) of flow based on Manning's equation

$$V(\alpha, D, n, S) := \frac{1.49 \cdot \frac{ft^{\frac{1}{3}}}{s}}{n} \cdot R_h(\alpha, D)^{\frac{2}{3}} \cdot S^{\frac{1}{2}}$$

n = Manning coefficient of roughness
S = Slope of pipe

Since all flow characteristics are in terms of the angle α, use a parametric plot using α as the independent variable. In the graph below we can get the area of flow, wetted perimeter, and velocity as a function of the depth of flow in the pipe.

Plot values for n=0.013 and S=0.007

$n := 0.013 \qquad S := 0.007 \qquad D := 1\ ft$

Relative Hydraulic Elements Chart

$$\frac{A_f(\alpha, D)}{A_f(2 \cdot \pi, D)}$$ Area of Flow

$$\frac{P_w(\alpha, D)}{P_w(2 \cdot \pi, D)}$$ Wetted Perimeter

$$\frac{V(\alpha, D, n, S)}{V(2 \cdot \pi, D, n, S)}$$ Velocity

$$\frac{Q(\alpha, D, n, S)}{Q(2 \cdot \pi, D, n, S)}$$ Flow Rate

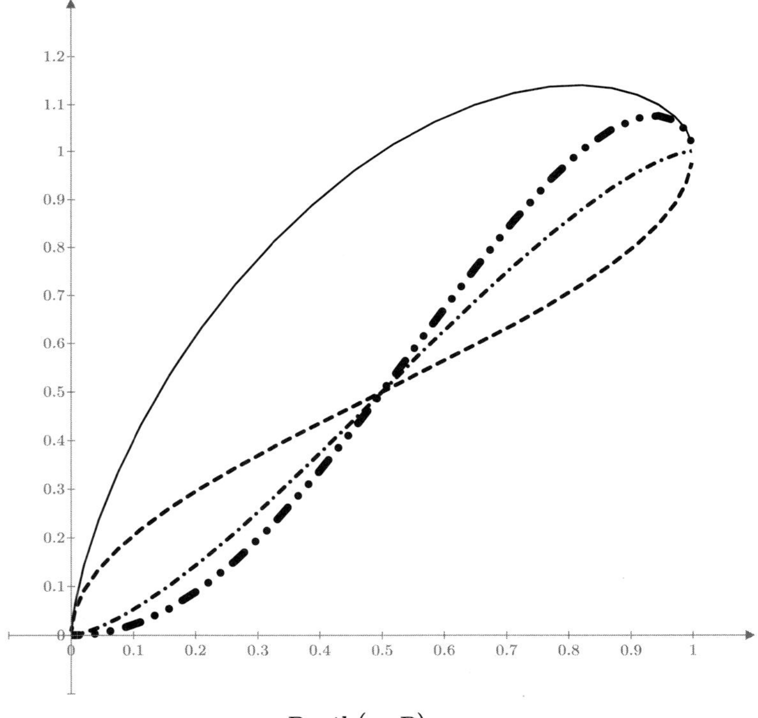

$$\frac{Depth(\alpha, D)}{D}$$

Now, let's examine the actual flow velocity as a function of depth in two pipes, one 10" and one 24" in diameter.

$D_1 := 10\ in$ $\qquad D_2 := 24\ in$ $\qquad n := 0.013$ $\qquad S := 0.007$

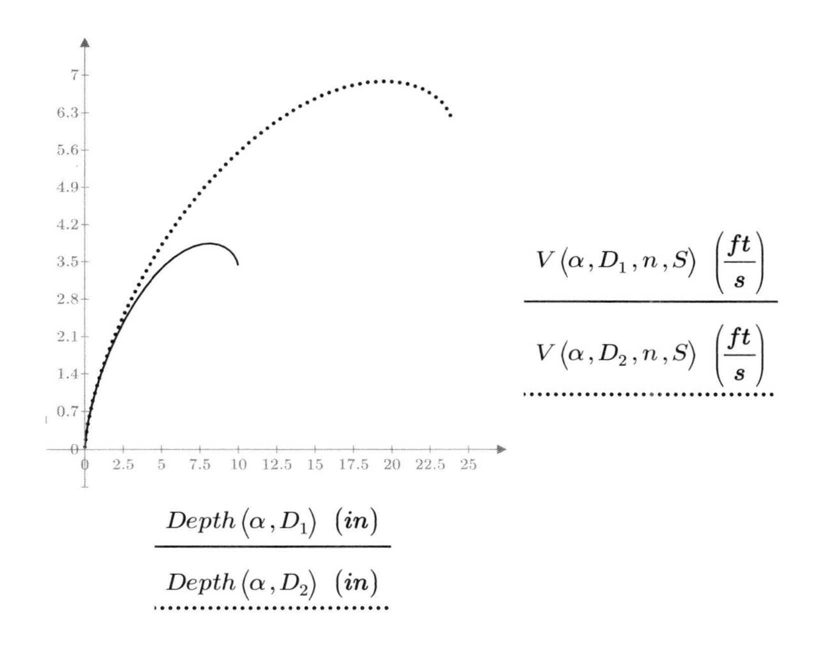

Notice that the maximum flow rate in a circular pipe does NOT occur when the pipe flows full. Note also that the velocity in the pipe actually drops as the depth of flow approaches the crown (top) of the pipe.

Using the above information, determine the diameter of pipe needed to carry 300 cfs when flowing 60% full at a slope of 0.001, with n = 0.013

Input Parameters

$S := 0.001$ $n := 0.013$ $D_3 := 9.41\ ft$ $0.6 \cdot D_3 = 5.646\ ft$

In Chapter 11, we introduce PTC Mathcad's solving functions, which can solve this solution directly. For this solution, we plot the flow and set plot markers at 300 cfs and 0.6*D. We then vary the pipe diameter until the line passes through the markers.

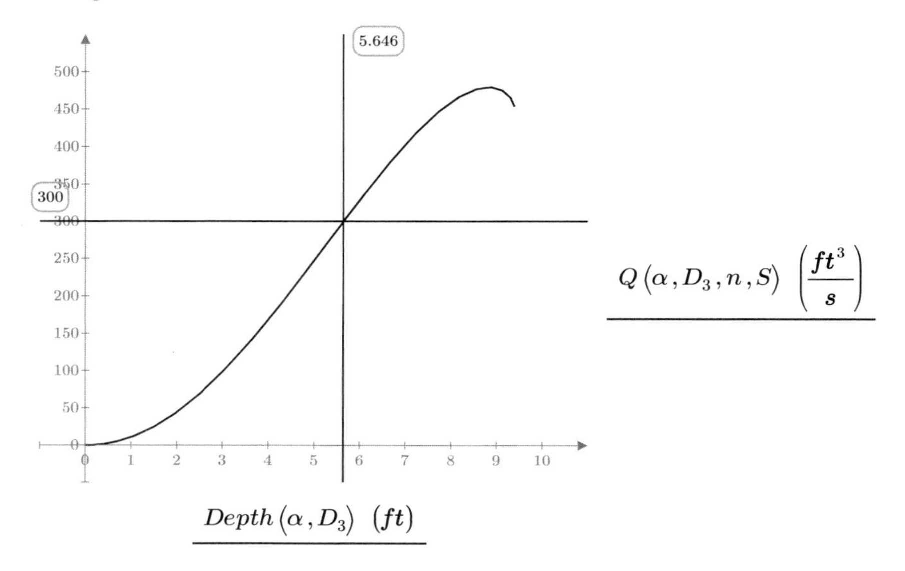

As can be seen from the plot above, a D_3 = 9.41 ft diameter pipe delivers 300 cfs when the depth of flow is 0.6 * D_3 = 5.64 ft.

Verify:
For depth of 0.6*diameter, the α is 3.544.

$$\text{Alpha} := \text{acos}\left(\frac{\frac{9.4\ ft}{2} - (.6 \cdot 9.4\ ft)}{\frac{9.4\ ft}{2}}\right) \cdot 2 = 3.544\ rad$$

$$\text{Depth}(\text{Alpha}, D_3) = 5.646\ ft \qquad \text{Diameter} := \frac{\text{Depth}(\text{Alpha}, D_3)}{.6} = 9.41\ ft$$

$$Q(\text{Alpha}, \text{Diameter}, n, S) = 299.547\ \frac{ft^3}{s}$$ **OK**

Summary

Plots are useful, easy-to-use tools. They can help you visualize equations or functions. They can also be used to help solve equations.

In Chapter 8 we:

- Showed how to create simple XY plots and polar plots.
- Showed how to plot functions and expressions.
- Discussed using range variables to set range limits.
- Showed how to plot multiple functions on the same plot.
- Showed how to plot using units, and how to change the displayed units.
- Explained how to format number value displays.
- Explained how to change the trace colors, line type, and line weight.
- Discussed plotting data information from range variables, data vectors, and data matrices.
- Explained how plots can be used to help solve systems of equations.
- Used the ***logspace*** function to create logarithmically spaced points.
- Introduced 3D plotting.

Practice

The PTC Mathcad Prime 3.0 figures and examples used in this book are available for download from the book's website. The reader is encouraged to download the files and use them to practice the concepts learned. Additional examples and problems are also provided. To access this content go to http://store.elsevier.com/9780124104105 *and click on the Resources tab, and then click on the link for the Online Companion Materials.*

1. Create separate simple XY plots of the following equations. Use expressions rather than functions on the y-axis.
   a. x^2
   b. $x^3+2x^2+3x-10$
   c. $\sin(x)$
2. Create separate polar plots of the following equations. Use expressions rather than functions on the radial axis.
   a. $x/2$
   b. $\cos(6x)$
   c. $\tan(x)$
3. Create functions for the expressions in exercises 1 and 2 and plot these functions.

4. Create two range variables for each of the above functions. One range variable should have a small increment; the other should have a large increment. Plot the above functions using each range variable. Use the controls in the **Styles** group to make each plot look different from the others.
5. The formula to calculate the bending moment at point "x" in a beam (with uniform loading) is M = (1\2*w*x)*(L − x), where w is force/length, L is total length of beam and x is distance from one end of the beam. Create a plot with distance x on the x-axis (from zero to L), and moment on the y-axis. Use w = 2 N/m and L = 10 m. Moment should be displayed as N*m. Provide a title and axis labels.
6. Plot the following data points. Use a range variable for the x-axis. Use a solid box as the symbol. Connect the data points with a dashed line.

1	19.1
2	29.5
3	40.3
4	52.4
5	59.3
6	70.5

7. Plot the following data points. Use a blue solid circle as the symbol. Do not connect the data points.

1.2	2.4
2.3	3.3
4.5	5.3
5.2	6.3
4.5	4.6
5.5	6.4

8. Plot the following equations and estimate approximate solutions where the plots intersect. $y_1(x) = 2x^2+3x-10$, $y_2(x) = -x^2+2x+20$.
9. Write a function to describe the vertical motion and a function to describe the horizontal motion of a projectile fired at 700 ft/s with a 35-degree inclination from the horizontal. Each function should be a function of time. Create a parametric plot with the following:
   a. Use a range variable to set the range of the plot. Use a range from 0 to 20 with an increment of 1.

b. Create a parametric plot with the horizontal motion on the x-axis and the vertical motion on the y-axis. Use the range variable for the argument of both functions.

c. Use units in the functions and the plots.
Hint: Remember to multiply the function argument by seconds.

10. Copy the plot from exercise 9 and plot in terms of meters instead of feet.

CHAPTER

Simple Logic Programming

Scientific and Engineering calculations must have a way to logically reach conclusions based on data calculated. PTC Mathcad programming allows you to write logic programs. These "programs" allow PTC Mathcad to choose a result based on specific parameters. Programs can be very complex, and can be used for many things. This chapter focuses solely on simple logic programming. The more complex programs are discussed in Chapter 12—Advanced Programming.

Chapter 9 will:

- Provide several simple PTC Mathcad examples to illustrate the concept of logic programming.
- List the steps necessary to create a logic program.
- Describe the logic PTC Mathcad uses to arrive at a conclusion.
- Warn the user about violating the above logic, which could cause PTC Mathcad to make an inaccurate conclusion.
- Reveal new ways of creating logic programs that are not provided in the PTC Mathcad documentation.
- Show how to use a logic program to draw and display conclusions.

Introduction to the Programming Toolbar

The purpose of this chapter is to get you comfortable with the concept of PTC Mathcad programming by using simple examples. You will soon see that you do not need to be a computer programmer to use the PTC Mathcad programming features, nor do you need to learn complex programming commands. All the operators you need to use are contained on the **Math** tab of the Ribbon Bar (**Math>Operators and Symbols>Programming**).

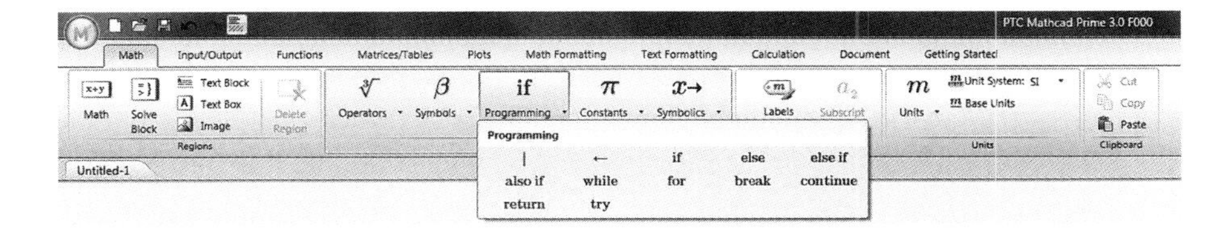

Essential PTC® Mathcad Prime® 3.0.

This chapter will focus on the following operators: ***Program***, ***if***, ***else, else if***, ***also if***, ***return***, and the ***local assignment*** operator. The remaining operators will be discussed in Chapter 12—Advanced Programming.

One important thing to remember when using PTC Mathcad programming is that all the programming operators must be inserted using the Programming controls, or by using keyboard shortcuts. You cannot type "if," "else," "else if," etc. They are operators and must be inserted. However, you can type the name of one of the programming operators and type **CTRL + J ,** and it will convert the text to a programming operator and add the appropriate placeholders.

Creating a Simple Program

> The programming features in PTC® Mathcad Prime® are considerably different than in Mathcad 15. There are new operators that make programming easier.

The easiest way to begin a PTC Mathcad program is to use the keyboard shortcut **]**. You may also click the vertical line on the **Programming** controls (upper left corner of the **Programming** controls on the **Math** tab).

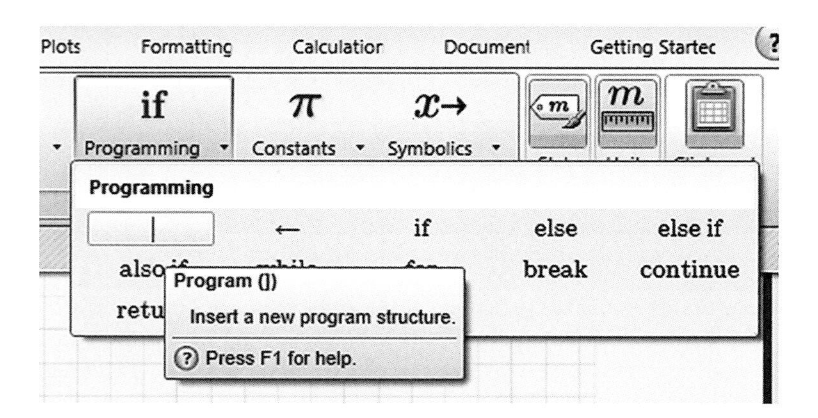

This inserts a double vertical line with a single placeholder 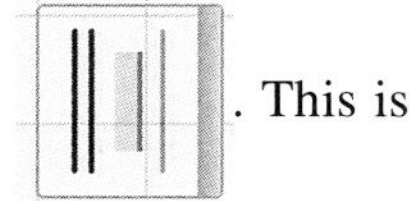. This is called the ***Program*** operator. Notice the double vertical lines on the left and the single vertical line on the right. The double vertical lines indicate various levels of the program. As a program becomes more involved a series of double lines will appear, helping you to visualize the various steps of the program. The right lines also help to indicate larger related pieces of the program.

Let's try a simple conditional program. We will place a conditional ***if*** statement in the placeholder. Do this by using the **Programming** controls and selecting "***if***" or by typing **if CTRL + J**. You may also use the keyboard shortcut **}**. This places the *if* operator followed by a placeholder with another placeholder below.

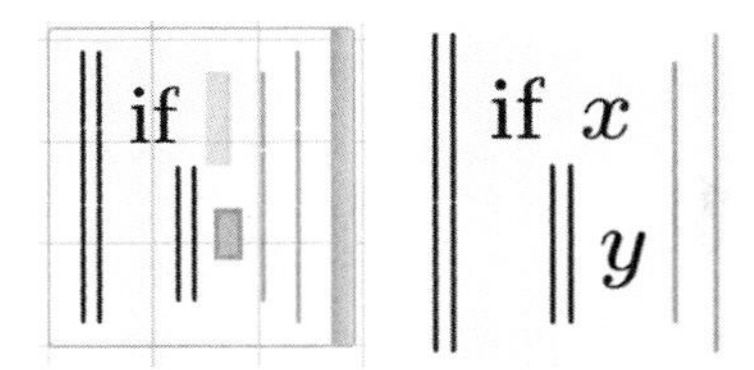

In the top placeholder we will place a conditional statement (x). In the bottom placeholder we will place a statement for PTC Mathcad to evaluate (y) if the conditional statement is true. If the conditional statement is true, PTC Mathcad will execute the statement in the bottom placeholder. If the conditional statement is false, PTC Mathcad skips the bottom placeholder and moves to the next line of the program. For this line we will use the *else* operator. The *else* operator closes a conditional statement and is evaluated only when the *if* statement is false. To insert the *else* operator, be sure your cursor is on the right hand side of the bottom placeholder statement and then use the **Programming** controls to insert the *else* operator. The keyboard shortcut is **SHIFT + CTRL + }**. PTC Mathcad only allows the *else* operator to be inserted following an *if* statement.

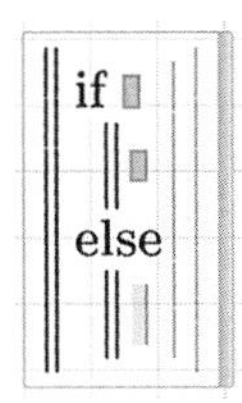

Another way to add programming operators is to create a new placeholder first and then insert the ***programming*** operator. The easiest way to insert a new placeholder is to select the entire *if* statement by using multiple presses of the **Spacebar**. Once this is done, press **ENTER**. This will create a new placeholder at the same level as the *if* operator. If the entire expression was not selected, the new placeholder will be placed at a lower programming level. You can observe the programming level by noting the location of the double vertical lines left of the placeholder.

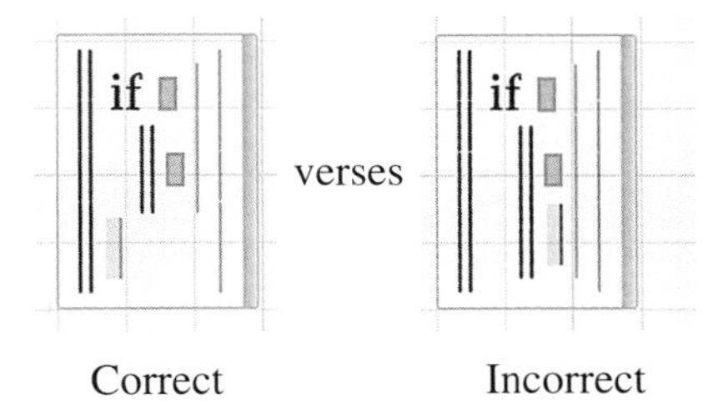

Once the placeholder is in the correct position, you can insert the ***else*** operator using the Programming controls, typing **SHIFT + CTRL + }** or by typing **else CTRL + J**.

Let's look at two simple examples. See Figures 9.1 and 9.2.

The following example is similar to Figure 7.22, which used the *if* function.

$L := 20\ ft$

$$\text{DesignLength} := \begin{array}{l} \text{if } L < 25\ ft \\ \quad 25\ ft \\ \text{else} \\ \quad L \end{array}$$

$\text{DesignLength} = 25\ ft$

To create this program follow these steps:

1. Open the **Programming** controls on the **Math** tab.
2. Type the variable name followed by the colon.
3. Click the **Program** button on the **Programming** control.
4. Select the placeholder and click the **if** button on the **Programming** control.
5. Select the bottom placeholder and click the **else** button on the **Programming** control.
6. Fill in the placeholders.

Remember. . . When creating a program, typing "if" is not the same as using the **if** operator on the **Programming** control.

Create a program to plot different functions depending on the value of x.

$f(x) := x^2 - 1$

$$h(x) := \begin{array}{l} \text{if } |x| \le 1 \\ \quad f(x) \\ \text{else} \\ \quad -f(x) \end{array}$$

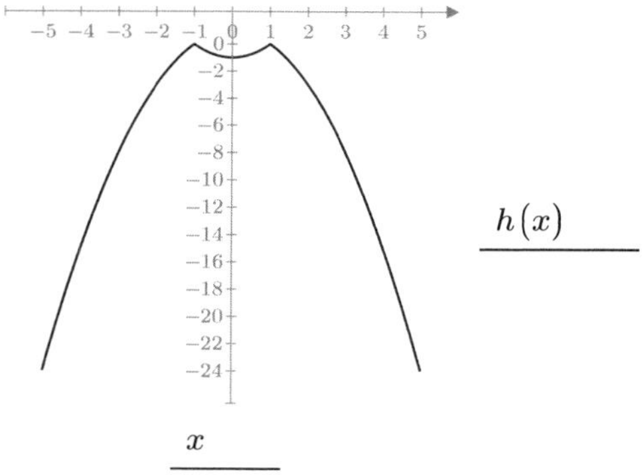

FIGURE 9.1

Simple program

Use three methods to check the program for different lengths.

Refer to Chapter 5 for a discussion of using arrays with equations and functions.

Method 1 -- **Use an expression with a range variable and vector to get multiple results.**

$i := 1, 2..3$

$$\mathrm{InputLength} := \begin{bmatrix} 5 \cdot ft \\ 25 \cdot ft \\ 50 \cdot ft \end{bmatrix}$$

Create an input vector.

$$\mathrm{DesignLength}_i := \left\| \begin{array}{l} \text{if } \mathrm{InputLength}_i < 25 \cdot ft \\ \quad \| \; 25 \cdot ft \\ \text{else} \\ \quad \| \; \mathrm{InputLength}_i \end{array} \right.$$

Define the expression. Note how both sides of the assignment operator must have the range variable i, and that i starts at ORIGIN and increments by one.

$$\mathrm{DesignLength} = \begin{bmatrix} 25 \\ 25 \\ 50 \end{bmatrix} ft$$

OK
OK
OK

Method 2 -- **Use a function. Use different arguments in the function.**

$$\mathrm{NewLength}(x) := \left\| \begin{array}{l} \text{if } x < 25 \cdot ft \\ \quad \| \; 25 \cdot ft \\ \text{else} \\ \quad \| \; x \end{array} \right.$$

$\mathrm{NewLength}(5 \cdot ft) = 25\ ft$

$\mathrm{NewLength}(25 \cdot ft) = 25\ ft$

$\mathrm{NewLength}(50 \cdot ft) = 50\ ft$

Method 3 -- **Use a function with a vector as the argument.**

$$\mathrm{ResultLength} := \overrightarrow{\mathrm{NewLength}(\mathrm{InputLength})}$$

$$\mathrm{ResultLength} = \begin{bmatrix} 25 \\ 25 \\ 50 \end{bmatrix} ft$$

Note that the ***vectorize*** operator CTRL+SHIFT+^ needs to be added to the function so that PTC Mathcad will recognize the vector with the function.

FIGURE 9.2

Check the program in Figure 9.1 for different lengths

The logic used by the program in Figure 9.1 is, "Check the conditions of the *if* statement. If they are true then evaluate the next line. If the conditions of the *if* statement are false then skip the next line. The ***else*** statement will be evaluated only if the conditions of the *if* statement are false." The *if* function could just as easily be used in place of this simple program. The benefit of using the ***program*** operator will be seen when there are multiple *if* statements, or when the programs become more complex.

Use of *else if* and *also if* Operators

PTC Mathcad Prime eliminated the ***otherwise*** operator and replaced it with the ***else*** operator. PTC Mathcad Prime also added the ***else if*** and ***also if*** operators. These new operators give programs more flexibility. PTC Mathcad Prime also added the double vertical lines on the left and the single vertical line on the right. These indicate related portions of the program.

As discussed above, the ***if*** operator opens a conditional statement and the ***else*** operator closes a conditional statement. The ***else*** operator is evaluated only when the related preceding statements are all false. The ***else if*** and ***also if*** operators are used between and related to the ***if*** and ***else*** operators.

Let's look at how the ***else if*** and ***also if*** operators work. The ***else if*** operator may only be used directly following an ***if*** statement or another ***else if*** statement. PTC Mathcad will not insert this operator if you are not following this rule. Similarly, the ***also if*** operator may only be used directly following an ***if*** statement or another ***also if*** statement. PTC Mathcad does not allow you to use both the ***else if*** and ***also if*** operators together with the same ***if*** operator.

If you try to insert the ***else***, ***else if***, or ***also if*** operators and nothing happens, then:

1. Check the location of your cursor to ensure that it is on the right hand side of the "y" statement.
2. Be sure that you are directly following an ***if***, ***else if***, or ***also if*** statement.
3. Be sure that you are not trying to follow an ***else if*** statement with an ***also if*** statement or visa versa.

The ***else if*** statement is checked only when the ***if*** statement (remember there can only be one related ***if*** statement) and all previous ***else if*** statements are false. On the other hand, the ***also if*** statement is checked regardless of whether the related ***if*** statement and all previous ***also if*** statements are true or false. It is important to remember that subsequent ***also if*** statements may overwrite the results of previous ***also if*** statements. PTC Mathcad will return the results of the last true ***if*** or ***also if*** statement. Tables 9.1 and 9.2 will help to visualize the above concepts.

Let's look at a few examples. See Figures 9.3 and 9.4.

Table 9.1 Using the ***else if*** operator (Four instances)

```
if x1
    y1
else if x2
    y2
else if x3
    y3
else
    y4
```

Statement	Is statement evaluated?	Is x true or false?	Is y evaluated?	Reasoning
if $x1$ $y1$	Yes	True	Yes	The ***if*** statement is always evaluated. Y1 is evaluated because x1 is true.
else if $x2$ $y2$	No			The ***else if*** statement is skipped because the ***if*** statement is true.
else if $x3$ $y3$	No			The ***else if*** statement is skipped because one of the previous statements is true.
else $y4$	No			The ***else*** statement is skipped because one of the previous statements is true.
if $x1$ $y1$	Yes	False	No	The ***if*** statement is always evaluated. Y1 is not evaluated because x1 is false.
else if $x2$ $y2$	Yes	True	Yes	The ***else if*** statement is evaluated because the ***if*** statement is false. Y2 is evaluated because x2 is true.
else if $x3$ $y3$	No			The ***else if*** statement is skipped because one of the previous statements is true.
else $y4$	No			The ***else*** statement is skipped because one of the previous statements is true.

(*Continued on next page*)

Table 9.1 Using the ***else if*** operator (Four instances)—Cont'd

```
if x1
  || y1
else if x2
  || y2
else if x3
  || y3
else
  || y4
```

Statement	Is statement evaluated?	Is x true or false?	Is y evaluated?	Reasoning
if $x1$ ‖ $y1$	Yes	False	No	The ***if*** statement is always evaluated. Y1 is not evaluated because x1 is false.
else if $x2$ ‖ $y2$	Yes	False	No	The ***else if*** statement is evaluated because the ***if*** statement is false. Y2 is not evaluated because x2 is false.
else if $x3$ ‖ $y3$	Yes	True	Yes	The ***else if*** statement is evaluated because all previous statements are false. y3 is evaluated because x3 is true.
else ‖ $y4$	No			The ***else*** statement is skipped because one of the previous statements is true.
if $x1$ ‖ $y1$	Yes	False	No	The ***if*** statement is always evaluated. Y1 is not evaluated because x1 is false.
else if $x2$ ‖ $y2$	Yes	False	No	The ***else if*** statement is evaluated because the ***if*** statement is false. Y2 is not evaluated because x2 is false.
else if $x3$ ‖ $y3$	Yes	False	No	The ***else if*** statement is evaluated because all previous statements are false. Y3 is not evaluated because x3 is false.
else ‖ $y4$	Yes		Yes	Y4 is evaluated because all of the previous statements are false.

Table 9.2 Using the ***also if*** operator (Five instances)

```
if x1
  || y1
also if x2
  || y2
also if x3
  || y3
else
  || y4
```

Statement	Is statement evaluated?	Is x true or false?	Is y evaluated?	Reasoning
if $x1$ ‖ $y1$	Yes	True	Yes	The ***if*** statement is always evaluated. Y1 is evaluated because x1 is true.
also if $x2$ ‖ $y2$	Yes	False	No	The ***also if*** Statement is always evaluated. Y2 is not evaluated because x2 is false.
also if $x3$ ‖ $y3$	Yes	False	No	The ***also if*** Statement is always evaluated. Y3 is not evaluated because x3 is false.
else ‖ $y4$	No			The ***else*** statement is skipped because one of the previous statements is true.
if $x1$ ‖ $y1$	Yes	True	Yes	The ***if*** statement is always evaluated. Y1 is evaluated because x1 is true.
also if $x2$ ‖ $y2$	Yes	True	Yes	The ***also if*** Statement is always evaluated. Y2 is evaluated because x2 is true.
also if $x3$ ‖ $y3$	Yes	False	No	The ***also if*** Statement is always evaluated. Y3 is not evaluated because x3 is false.
else ‖ $y4$	No			The ***else*** statement is skipped because one of the previous statements is true.
if $x1$ ‖ $y1$	Yes	False	No	The ***if*** statement is always evaluated. Y1 is not evaluated because x1 is false.
also if $x2$ ‖ $y2$	Yes	False	No	The ***also if*** Statement is always evaluated. Y2 is not evaluated because x2 is false.
also if $x3$ ‖ $y3$	Yes	True	Yes	The ***also if*** Statement is always evaluated. Y3 is evaluated because x3 is true.
else ‖ $y4$	No			The ***else*** statement is skipped because one of the previous statements is true.

(Continued on next page)

Table 9.2 Using the ***also if*** operator (Five instances)—Cont'd

```
if x1
  || y1
also if x2
  || y2
also if x3
  || y3
else
  || y4
```

Statement	Is statement evaluated?	Is x true or false?	Is y evaluated?	Reasoning
if $x1$ ‖ $y1$	Yes	True	Yes	The ***if*** statement is always evaluated. Y1 is evaluated because x1 is true.
also if $x2$ ‖ $y2$	Yes	True	Yes	The ***also if*** Statement is always evaluated. Y2 is evaluated because x2 is true.
also if $x3$ ‖ $y3$	Yes	True	Yes	The ***also if*** Statement is always evaluated. Y3 is evaluated because x3 is true.
else ‖ $y4$	No			The ***else*** statement is skipped because one of the previous statements is true.
if $x1$ ‖ $y1$	Yes	False	No	The ***if*** statement is always evaluated. Y1 is not evaluated because x1 is false.
also if $x2$ ‖ $y2$	Yes	False	No	The ***also if*** Statement is always evaluated. Y2 is not evaluated because x2 is false.
also if $x3$ ‖ $y3$	Yes	False	No	The ***also if*** Statement is always evaluated. Y3 is not evaluated because x3 is false.
else ‖ $y4$	Yes		Yes	Y4 is evaluated because all of the previous statements are false.

Assume a previous calculation returned a calculated factor. The minimum value of the factor should be 0.5. The maximum value should be 2.0. Let's write a program as a user-defined function (using the calculated factor as an argument) so that PTC Mathcad will choose a final factor between 0.5 and 2.0. 0.5<Factor<2.0.

Let's look at two examples of a program.
This first example creates inaccurate results.

```
Factor(x) := | if x > 2.0
             |   | 2.0
             | also if x > 0.5
             |   | x
             | else
             |   | 0.5
```

$\mathbf{F_1} := \mathbf{Factor}(0.25)$	$\mathbf{F_1} = 0.5$	Correct
$\mathbf{F_2} := \mathbf{Factor}(1.0)$	$\mathbf{F_2} = 1$	Correct
$\mathbf{F_3} := \mathbf{Factor}(3.0)$	$\mathbf{F_3} = 3$	Incorrect (The correct result should be 2.0)

Why did F3 give incorrect results?
It returned incorrect results because PTC Mathcad evaluates every **if** and **also if** statement, and executes it if it is true. The final true **if** statement is returned. For F3, the final true statement that is encountered is "if x is greater than 0.5 then return x." Since 3>0.5, PTC Mathcad returned 3.

Rewrite the program so that it creates correct results. Do this by ensuring that the first two statements cannot both be true.

```
RevisedFactor(x) := | if x < 0.5
                    |   | 0.5
                    | also if x > 2.0
                    |   | 2.0
                    | else
                    |   | x
```

$\mathbf{F_4} := \mathbf{RevisedFactor}(0.25) = 0.5$

$\mathbf{F_5} := \mathbf{RevisedFactor}(1.0) = 1$

$\mathbf{F_6} := \mathbf{RevisedFactor}(3.0) = 2$

The program returns correct results.

FIGURE 9.3

Using multiple *if* statements

In wood design there is a depth factor that needs to be applied to different depths of wood joists. Let's look at two examples of a conditional program to determine the proper depth factor to use.
This first example creates incorrect results.

```
Factor1(d) := | if d ≤ 3.5·in
              |    1.5
              | also if d ≤ 4.5·in
              |    1.4
              | also if d ≤ 5.5·in
              |    1.3
              | also if d ≤ 7.25·in
              |    1.2
              | also if d ≤ 9.25·in
              |    1.1
              | also if d ≤ 11.25·in
              |    1.0
              | else
              |    0.9
```

$\text{Factor}_1(2.5\cdot in) = 1$ Incorrect (The result should be 1.5)

$\text{Factor}_1(3.5\cdot in) = 1$ Incorrect (The result should be 1.5)

$\text{Factor}_1(4.5\cdot in) = 1$ Incorrect (The result should be 1.4)

$\text{Factor}_1(5.5\cdot in) = 1$ Incorrect (The result should be 1.3)

$\text{Factor}_1(7.25\cdot in) = 1$ Incorrect (The result should be 1.2)

$\text{Factor}_1(9.25\cdot in) = 1$ Incorrect (The result should be 1.1)

$\text{Factor}_1(11.25\cdot in) = 1$ Correct

$\text{Factor}_1(13.25\cdot in) = 0.9$ Correct

FIGURE 9.4 *(Continued on next page)*

Multiple *if* statements

Why did the above example give so many incorrect results?
It returned incorrect results because PTC Mathcad evaluates every **if** statement and **also if** statement and executes it if it is true. The final result is the last true statement that is encountered. The last true statement that is encountered is d <=11.25 in. Most of the depths were less than 11.25 in., so 1.0 is the value assigned to Factor1.

If we reorganize the program, so that false statements are encountered as the depth decreases, we get correct results.

```
RevisedFactor2(d) := | if d ≤ 11.25·in
                     |   | 1.0
                     | also if d ≤ 9.25·in
                     |   | 1.1
                     | also if d ≤ 7.25·in
                     |   | 1.2
                     | also if d ≤ 5.5·in
                     |   | 1.3
                     | also if d ≤ 4.5·in
                     |   | 1.4
                     | also if d ≤ 3.5·in
                     |   | 1.5
                     | else
                     |   | 0.9
```

$\mathrm{RevisedFactor}_2(2.5 \cdot in) = 1.5$ Correct

$\mathrm{RevisedFactor}_2(3.5 \cdot in) = 1.5$ Correct

$\mathrm{RevisedFactor}_2(4.5 \cdot in) = 1.4$ Correct

$\mathrm{RevisedFactor}_2(5.5 \cdot in) = 1.3$ Correct

$\mathrm{RevisedFactor}_2(7.25 \cdot in) = 1.2$ Correct

$\mathrm{RevisedFactor}_2(9.25 \cdot in) = 1.1$ Correct

$\mathrm{RevisedFactor}_2(11.25 \cdot in) = 1$ Correct

$\mathrm{RevisedFactor}_2(13.25 \cdot in) = 0.9$ Correct

FIGURE 9.4

(*Continued*)

Local Assignment

PTC Mathcad allows you to assign new variables within a program. These variables will be local to the program, meaning they will be undefined outside of the program.

Create a user-defined function that will find the two roots of a quadratic equation. Use local variables to simplify the program.

Use the left arrow button on the **Programming** control to insert the local variable operator.

The local variables are only defined within the program. They are not defined outside the program.

$$\mathbf{QuadRoots}(a,b,c) := \left\| \begin{array}{l} \mathbf{Part} \leftarrow \sqrt{b^2 - 4 \cdot a \cdot c} \\ x_1 \leftarrow \dfrac{-b + \mathbf{Part}}{2 \cdot a} \\ x_2 \leftarrow \dfrac{-b - \mathbf{Part}}{2 \cdot a} \\ \begin{bmatrix} x_1 \\ x_2 \end{bmatrix} \end{array} \right.$$

Define a local variable "Part."

Define a local variable "x1."

Define a local variable "x2."

Create an output vector comprising local variables x1 and x2. The values of x1 and x2 will be the output results of the function, but the local variables will not be recognized outside of the function.

Results of Function

$$\mathbf{QuadRoots}(1,5,6) = \begin{bmatrix} -2 \\ -3 \end{bmatrix}$$

The local variables are not recognized outside of the program.

$x_1 = ?$ $\quad\quad$ $Part = ?$

In order to make the results of the function available later on, the variable "Roots" is defined equal to the function QuadRoots.

$$\mathbf{Roots} := \mathbf{QuadRoots}(1,5,6)$$

$$\mathbf{Roots} = \begin{bmatrix} -2 \\ -3 \end{bmatrix} \quad\quad \mathbf{Roots}_1 = -2 \quad\quad \mathbf{Roots}_2 = -3$$

FIGURE 9.5

Local variables

These are called local variables, and they are assigned with the ***local assignment*** operator. The ***local assignment*** operator is represented by an arrow pointing left ←. Insert it using the **Programming** controls on the **Math** tab. The keyboard shortcut is **{**. Let's look at a simple example of a program that uses local variables. See Figure 9.5.

As you can see from Figure 9.5, local variables can help simplify a program by breaking the program down into smaller pieces. In Figure 9.6, the function is very long, but each numerator is the same. By assigning a local variable, the function definition can be simplified.

Local variables can help simplify complex equations.

The numerators in this function are all the same.

$$G_1(x,y,z) := \frac{\left(\frac{x}{y}\right)+\left(\frac{y}{2\cdot x}\right)^2+\left(\frac{x}{3\cdot z}\right)^3+\left(\frac{z}{4\cdot x}\right)^4}{x}+\frac{\left(\frac{x}{y}\right)+\left(\frac{y}{2\cdot x}\right)^2+\left(\frac{x}{3\cdot z}\right)^3+\left(\frac{z}{4\cdot x}\right)^4}{y}+\frac{\left(\frac{x}{y}\right)+\left(\frac{y}{2\cdot x}\right)^2+\left(\frac{x}{3\cdot z}\right)^3+\left(\frac{z}{4\cdot x}\right)^4}{z}$$

$$G_1(2,6,3)=2.614$$

If you define a local variable, the function is much simpler to follow.

$$G_2(x,y,z) := \left\| \begin{array}{l} a \leftarrow \left(\frac{x}{y}\right)+\left(\frac{y}{2\cdot x}\right)^2+\left(\frac{x}{3\cdot z}\right)^3+\left(\frac{z}{4\cdot x}\right)^4 \\ \frac{a}{x}+\frac{a}{y}+\frac{a}{z} \end{array} \right.$$

$$G_2(2,6,3)=2.614$$

FIGURE 9.6

Local variables

Return Operator

The ***return*** operator tells PTC Mathcad to stop the program and return the value rather than proceed to the next line of the program. Use it in conjunction with an ***if***, ***else if***, or ***also if*** statement. Figure 9.7 shows how to use the ***return*** statement to make the examples from Figure 9.4 more intuitive.

The ***return*** operator stops the program. In conjunction with an **if** statement, it can be used to stop the program if the **if** statement is true. If the **if** statement is false, then PTC Mathcad proceeds to the next line.

The first example from Figure 9.4 could be written in the following manner using the ***return*** operator:

$$\text{Factor}_2(d) := \left\| \begin{array}{l} \text{if } d \le 3.5 \cdot in \\ \quad \| \; \textbf{\textit{return}}\ 1.5 \\ \text{also if } d \le 4.5 \cdot in \\ \quad \| \; \textbf{\textit{return}}\ 1.4 \\ \text{also if } d \le 5.5 \cdot in \\ \quad \| \; \textbf{\textit{return}}\ 1.3 \\ \text{also if } d \le 7.25 \cdot in \\ \quad \| \; \textbf{\textit{return}}\ 1.2 \\ \text{also if } d \le 9.25 \cdot in \\ \quad \| \; \textbf{\textit{return}}\ 1.1 \\ \text{also if } d \le 11.25 \cdot in \\ \quad \| \; \textbf{\textit{return}}\ 1.0 \\ \text{else} \\ \quad \| \; 0.9 \end{array} \right.$$

Note: A ***return*** could have been placed in front of the last line to make it read more consistent, but it is not required.

$\text{Factor}_2(2.5 \cdot in) = 1.5$ Correct

$\text{Factor}_2(3.5 \cdot in) = 1.5$ Correct

$\text{Factor}_2(4.5 \cdot in) = 1.4$ Correct

$\text{Factor}_2(5.5 \cdot in) = 1.3$ Correct

$\text{Factor}_2(7.25 \cdot in) = 1.2$ Correct

$\text{Factor}_2(9.25 \cdot in) = 1.1$ Correct

$\text{Factor}_2(11.25 \cdot in) = 1$ Correct

$\text{Factor}_2(13.25 \cdot in) = 0.9$ Correct

FIGURE 9.7

Return operator

Boolean Operators

Boolean operators are essential when writing logical PTC Mathcad programs. They are located on the **Operators** controls of the **Math** tab (**Math>Operators and Symbols> Operators>Comparison**). Many of them may be typed from the keyboard as indicated in Table 9.3 below.

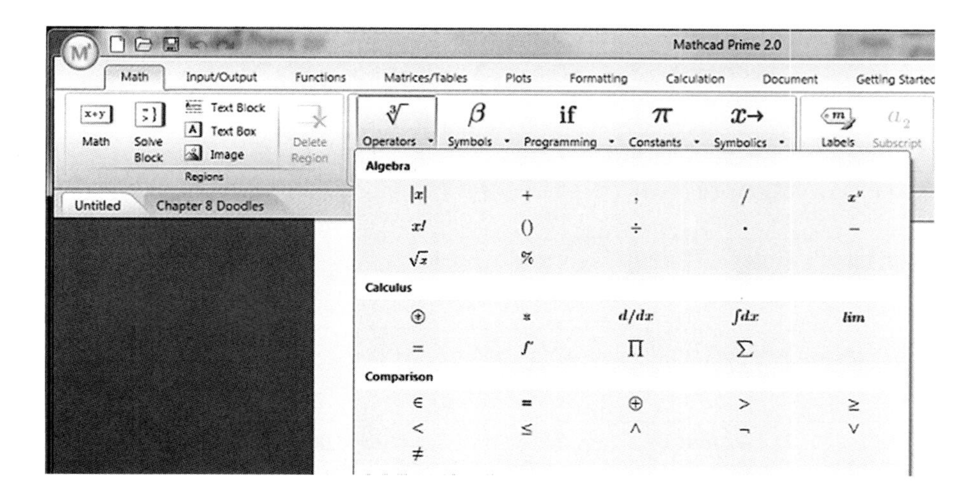

PTC Mathcad Prime added the ***Is Element Of*** operator. This allows you to determine if an element is included in the complex, real, or integer number set.

The Boolean operators can be divided into two groups: Comparison and Logical. The Comparison operators consist of $x=y$, $x \neq y$, $x<y$, $x \leq y$, $x>y$, $x \geq y$, and $x \in y$. The Logical operators consist of $x \wedge y$, $x \vee y$, $x \oplus y$, and $\neg x$. For most of these operators, PTC Mathcad returns a 1 if the expression is true and a 0 if the expression is false. The Logical operators treat any nonzero numeric value as true, and 0 as false. Table 9.3 describes the function of each operator and what values PTC Mathcad returns.

Do not confuse the ***Evaluation*** operator = with the Boolean ***Equal TO*** ==. The ***Evaluation*** operator calculates the left hand side of an expression and returns the results on the right hand side. The Boolean ***Equal TO*** is used for comparisons and does not cause an evaluation. I use the Boolean ***Equal TO*** when I want to display a function in PTC Mathcad, but do not want it to be evaluated.

When using the ***Is Element Of*** comparison operator, PTC Mathcad has special symbols to represent the set of complex numbers ($\mathbb{C}$), the set of rational numbers ($\mathbb{Q}$), the set of real numbers ($\mathbb{R}$), and the set of integers ($\mathbb{Z}$). These special symbols

Table 9.3 Boolean Operators

Operator	Operator name	Description	Keyboard shortcut
$x=y$	Boolean ***Equal TO***	Returns 1 if x is equal to y. Returns 0 otherwise.	CTRL + =
$x \neq y$	Boolean ***Inequality***	Returns 1 if x is not equal to y. Returns 0 otherwise.	< >
$x<y$	***Less Than***	Returns 1 if x is less than y. Returns 0 otherwise.	<
$x \leq y$	***Less Than or Equal To***	Returns 1 if x is less than or equal to y. Returns 0 otherwise.	< =
$x>y$	***Greater Than***	Returns 1 if x is greater than y. Returns 0 otherwise.	>
$x \geq y$	***Greater Than or Equal To***	Returns 1 if x is greater than or equal to y. Returns 0 otherwise.	> =
$x \in y$	***Is Element Of***	Returns 1 if x is an element of the number set y. Returns 0 otherwise.	
$x \wedge y$	***Logical AND***	Returns 1 if both x and y are true. Returns 0 otherwise.	
$x \vee y$	***Logical OR***	Returns 1 if either x or y are true. Returns 0 otherwise.	
$x \oplus y$	***Logical Exclusive OR (XOR)***	Returns 1 if either x or y, but not both are true (nonzero). Returns 0 otherwise.	CTRL + SHIFT + %
$\neg x$	***Logical NOT***	Returns 1 if x is false (0). Returns 0 otherwise.	

are found in the **Operators and Symbols** group on the **Math** Tab (**Math>Operators and Symbols>Symbols>Math Symbols**). See Figure 9.8.

The logical ***NOT*** operator ($\neg x$) is true if x is zero and false if x is nonzero. See Figure 9.9 for an example.

The logical ***AND*** operator ($x \wedge y$) is true if both statements are true. See Figure 9.10 for an example.

Use the **Symbols** control on the **Math** tab to insert the special symbols for use with the ***Is Element Of*** comparison.

$$\mathbf{Complex}(\mathbf{x}) := \left\| \begin{array}{l} \text{if } \mathbf{x} \in \mathbb{C} \\ \quad \| \text{"Complex"} \\ \text{else} \\ \quad \| \text{"Not element of complex numbers"} \end{array} \right.$$

$$\mathbf{Complex}\left(\sqrt{2}\right) = \text{"Complex"}$$

$$\mathbf{Complex}(7) = \text{"Complex"}$$

$$\mathbf{Complex}(7 + 3i) = \text{"Complex"}$$

$$\mathbf{Rational}(\mathbf{x}) := \left\| \begin{array}{l} \text{if } \mathbf{x} \in \mathbb{Q} \\ \quad \| \text{"Rational"} \\ \text{else} \\ \quad \| \text{"Not element of rational numbers"} \end{array} \right.$$

$Rational\left(\sqrt{3}\right) = ?$

You must evaluate this operator symbolically.

See Chapter 10 for a discussion on symbolic evaluation.

$$\mathbf{Rational}\left(\sqrt{3}\right) \rightarrow \text{"Rational"}$$

$$\mathbf{Rational}(\pi) \rightarrow \text{"Not element of rational numbers"}$$

$$\mathbf{Real}(\mathbf{x}) := \left\| \begin{array}{l} \text{if } \mathbf{x} \in \mathbb{R} \\ \quad \| \text{"Real"} \\ \text{else} \\ \quad \| \text{"Not element of real numbers"} \end{array} \right.$$

$$\mathbf{Real}\left(\sqrt{2}\right) = \text{"Real"}$$

$$\mathbf{Real}\left(\frac{-5}{6}\right) = \text{"Real"}$$

$$\mathbf{Real}(3 + 1i) = \text{"Not element of real numbers"}$$

$$\mathbf{Integer}(\mathbf{x}) := \left\| \begin{array}{l} \text{if } \mathbf{x} \in \mathbb{Z} \\ \quad \| \text{"Integer"} \\ \text{else} \\ \quad \| \text{"Not an integer"} \end{array} \right.$$

$$\mathbf{Integer}(3) = \text{"Integer"}$$

$$\mathbf{Integer}\left(\frac{1}{3}\right) = \text{"Not an integer"}$$

$$\mathbf{Integer}(3 + 2i) = \text{"Not an integer"}$$

FIGURE 9.8

Is Element Of

The logical **Not** operator returns true (1) if the value is zero. It returns false (0) if the value is non-zero.

$\neg 1 = 0 \qquad \neg 0 = 1$

Example using the **Not** operator.

$$AA(x) := \begin{array}{|l} \text{if } \neg x \\ \quad \text{“X is zero”} \\ \text{else} \\ \quad \text{“X is not zero”} \end{array}$$

$AA(2) = \text{“X is not zero”} \qquad AA(0) = \text{“X is zero”}$

This is the same result as using **if** x=0.

$$AB(x) := \begin{array}{|l} \text{if } x = 0 \\ \quad \text{“X is zero”} \\ \text{else} \\ \quad \text{“X is not zero”} \end{array}$$

$AB(2) = \text{“X is not zero”} \qquad AB(0) = \text{“X is zero”}$

FIGURE 9.9

Logical NOT

Logical **AND** operator returns true if both statements are true.

$$BB(x, y) := \begin{array}{|l} \text{if } x > 0 \land y > 0 \\ \quad \text{“This AND statement is TRUE”} \\ \text{else} \\ \quad \text{“This AND statement is false”} \end{array}$$

$BB(5, 4) = \text{“This AND statement is TRUE”}$

$BB(0, 3) = \text{“This AND statement is false”}$

$BB(5, 0) = \text{“This AND statement is false”}$

$BB(0, 0) = \text{“This AND statement is false”}$

FIGURE 9.10

Logical AND

The logical ***OR*** operator ($x \vee y$) is true if either statement is true. See Figure 9.11 for an example.

The logical ***XOR*** (exclusive OR) ($x \oplus y$) is true if either the first or the second statement is nonzero, but not both. Thus, if both statements are non-zero, the result is false. See Figure 9.12 for an example.

Logical **OR** operator returns true if either statement is true.

$$CC(x,y) := \left\| \begin{array}{l} \text{if } x>0 \vee y>0 \\ \quad \| \text{“This OR statement is TRUE”} \\ \text{else} \\ \quad \| \text{“This OR statement is false”} \end{array} \right.$$

CC(5, 4) = “This OR statement is TRUE”
CC(0, 4) = “This OR statement is TRUE”
CC(5, 0) = “This OR statement is TRUE”
CC(0, 0) = “This OR statement is false”

FIGURE 9.11

Logical OR

Logical **XOR** operator returns true if one statement is true, but not both.

$$EE(x,y) := \left\| \begin{array}{l} \text{if } x>0 \oplus y>0 \\ \quad \| \text{“The XOR statement is TRUE”} \\ \text{else} \\ \quad \| \text{“The XOR statement is false”} \end{array} \right.$$

EE(5, 6) = “The XOR statement is false” — False because both statements are true.

EE(0, 6) = “The XOR statement is TRUE” — True because one, but not both statements is true.

EE(5, 0) = “The XOR statement is TRUE” — True because one, but not both statements is true.

EE(0, 0) = “The XOR statement is false” — False because neither statement is true.

FIGURE 9.12

Logical XOR

Adding Lines to a Program

It is possible to add nested programs inside a program. Creative use of nested programs sometimes makes it easier to create conditional programs.

Remember that pressing **ENTER** will add a new line with a placeholder. Where the new line is added depends on how much of the program is selected. It is not necessary to have a new line to add the ***also if*** and ***else*** operators, but it is sometimes easier to visualize where these will be placed if they are added to a blank placeholder. The following figures illustrate these concepts and show how different forms of programming work. In Figures 9.13–9.16 we try to arrive at consistent results

$$i := 1, 2..4 \qquad \mathbf{x} := \begin{bmatrix} 3 \\ 4 \\ -7 \\ -10 \end{bmatrix} \qquad \mathbf{y} := \begin{bmatrix} 2 \\ -1 \\ 3 \\ -2 \end{bmatrix}$$

```
Test_1_i := | if x_i ≥ 0 ∧ y_i ≥ 0
            |    | 100
            | also if x_i < 0 ∧ y_i ≥ 0
            |    | 300
            | also if x_i ≥ 0 ∧ y_i < 0
            |    | 200
            | else
            |    | 400
```

This program uses the Logical **AND** operator. Both statements must be true in order for the line to be true.

The **if** and every **also if** statement are checked. The last true statement is used. If none of them are true, then the **else** statement is evaluated.

To create this program structure, type:

] if CTRL+j DOWN-ARROW Spacebar Spacebar ENTER CTRL +SHIFT+? DOWN-ARROW Spacebar Spacebar ENTER CTRL +SHIFT+? DOWN-ARROW Spacebar Spacebar ENTER else CTRL+j

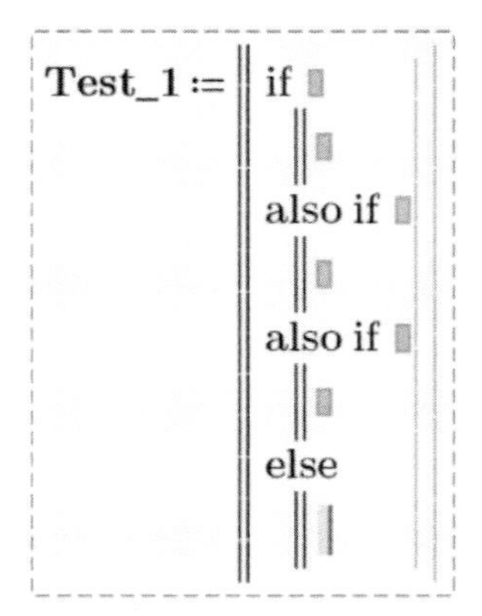

$$\text{Test\_1}_i = \begin{bmatrix} 100 \\ 200 \\ 300 \\ 400 \end{bmatrix}$$

FIGURE 9.13

Programming Form 1

$$\mathbf{Test\_2}_i := \left\| \begin{array}{l} \text{if } \mathbf{x}_i \geq 0 \\ \quad \left\| \begin{array}{l} \text{if } \mathbf{y}_i \geq 0 \\ \quad \| \; 100 \\ \text{else} \\ \quad \| \; 200 \end{array} \right. \\ \text{if } \mathbf{x}_i < 0 \\ \quad \left\| \begin{array}{l} \text{if } \mathbf{y}_i \geq 0 \\ \quad \| \; 300 \\ \text{else} \\ \quad \| \; 400 \end{array} \right. \end{array} \right.$$

This form of the program allows you to have an additional conditional statement branching off of the first conditional statement.

If the first statement is true, then the next **if** statement is evaluated.

To create this program structure, type:

] if CTRL+j DOWN-ARROW if CTRL +j DOWN-ARROW Spacebar Spacebar ENTER else CTRL+j Spacebar Spacebar Spacebar Spacebar ENTER if CTRL+j DOWN-ARROW if CTRL+j DOWN-ARROW Spacebar Spacebar ENTER else CTRL+j

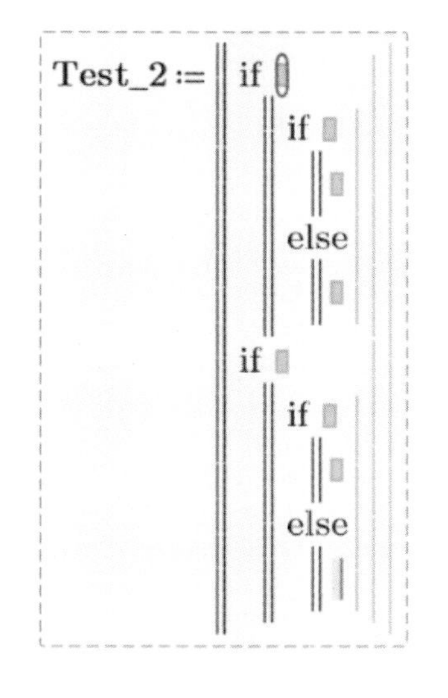

$$\mathbf{Test\_2}_i = \begin{bmatrix} 100 \\ 200 \\ 300 \\ 400 \end{bmatrix} \qquad \mathbf{Test\_1}_i = \begin{bmatrix} 100 \\ 200 \\ 300 \\ 400 \end{bmatrix}$$

FIGURE 9.14

Programming Form 2

using different forms of the programming lines. Notice how each program is a little different from the others, but still achieves the same result. In these examples, we use a range variable and vectors.

This form of the program is tricky. Essentially the **if** statement has two condition lines. Regardless of the first answer, the second condition will always be evaluated. If the second statement is true, then it doesn't matter whether or not the first statement is true. There is not an And function that will tie the two statements together.

For this reason, use a string in the first line as a comment line.

```
Test_3_i := | if  | "This line is a comment line"
            |     | x_i ≥ 0 ∧ y_i ≥ 0
            |     | 100
            | also if | "Use this line for text"
            |         | x_i ≥ 0 ∧ y_i < 0
            |     | 200
            | also if | "Add info about your program here"
            |         | x_i < 0 ∧ y_i ≥ 0
            |     | 300
            | else
            |     | 400
```

To create this program structure, type:

] if CTRL+j] ENTER DOWN-ARROW Spacebar Spacebar ENTER CTRL+SHIFT +?] ENTER DOWN-ARROW Spacebar Spacebar ENTER CTRL+SHIFT+?] ENTER DOWN-ARROW Spacebar Spacebar ENTER else CTRL+j

$$\text{Test\_3}_i = \begin{bmatrix} 100 \\ 200 \\ 300 \\ 400 \end{bmatrix} \qquad \text{Test\_1}_i = \begin{bmatrix} 100 \\ 200 \\ 300 \\ 400 \end{bmatrix}$$

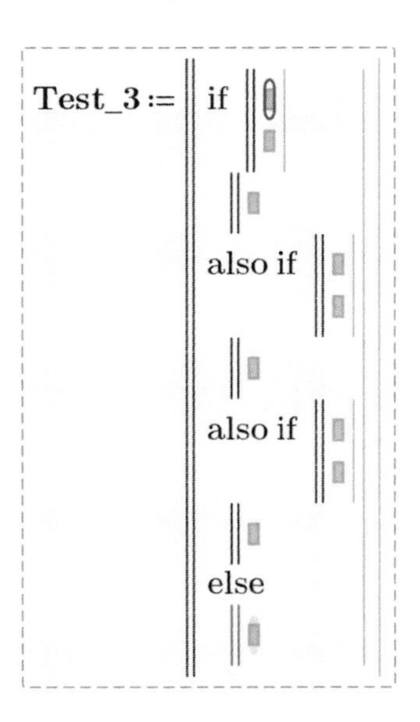

FIGURE 9.15

Programming Form 3

```
Test_4_i := | if x_i ≥ 0
            |   | if y_i ≥ 0
            |   |   | 100
            |   | else
            |   |   | 200
            | else
            |   | if y_i ≥ 0
            |   |   | 300
            |   | else
            |   |   | 400
```

To create this program structure, type:

] if CTRL+j DOWN-ARROW if CTRL+j DOWN-ARROW else CTRL+j Spacebar Spacebar Spacebar Spacebar ENTER else CTRL+j if CTRL+j DOWN-ARROW else CTRL+j

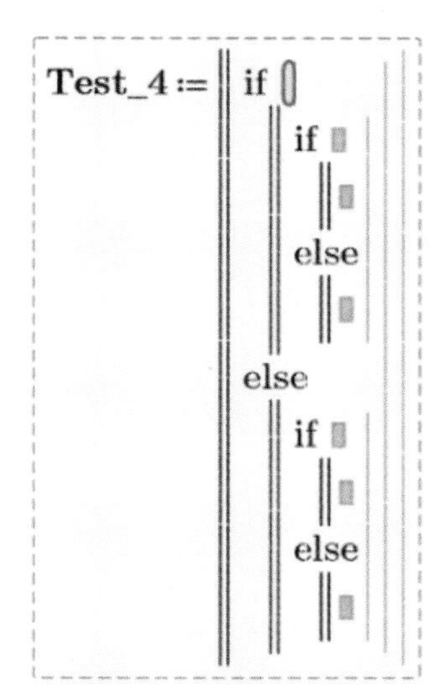

$$Test\_4_i = \begin{bmatrix} 100 \\ 200 \\ 300 \\ 400 \end{bmatrix} \qquad Test\_1_i = \begin{bmatrix} 100 \\ 200 \\ 300 \\ 400 \end{bmatrix}$$

FIGURE 9.16

Programming Form 4

Using Conditional Programs to Make and Display Conclusions

Having PTC Mathcad display a statement at the conclusion of a problem is a great feature. You can have PTC Mathcad display statements such as: "Passes" or "Fails," depending on whether or not the calculation worked. You can create other string variables that PTC Mathcad can display.

Let's look at an example. See Figure 9.17.

Create three string variables.

Passes := "Passes – Solution Works"

Fails := "Fails – Retry"

$\mathbf{i} := 1, 2 .. 3$

Zero := "Result is Zero"

$$\mathbf{Result} := \begin{bmatrix} -5 \\ 0 \\ 5 \end{bmatrix}$$

```
Test_i := | if Result_i > 0
          |   | Passes
          | also if Result_i < 0
          |   | Fails
          | else
          |   | Zero
```

$\mathbf{Test}_1$ = "Fails – Retry"

$\mathbf{Test}_2$ = "Result is Zero"

$\mathbf{Test}_3$ = "Passes – Solution Works"

FIGURE 9.17

Using conditional programming to display conclusions

Engineering Examples

Engineering Example 9.1

Engineering Example 9.1 demonstrates how to use a program to calculate the flow in a pipe or channel.

Manning's equation is an empirical relationship commonly used to predict flow in open channels. It has the form:

$$Q = \frac{1.49}{n} \cdot A \cdot R^{\frac{2}{3}} \cdot S^{\frac{1}{2}}$$ flow (Q) in $\frac{ft^3}{s}$, dimensions in feet

Create a worksheet that will calculate the flow of water based on the Manning equation and four types of channels.

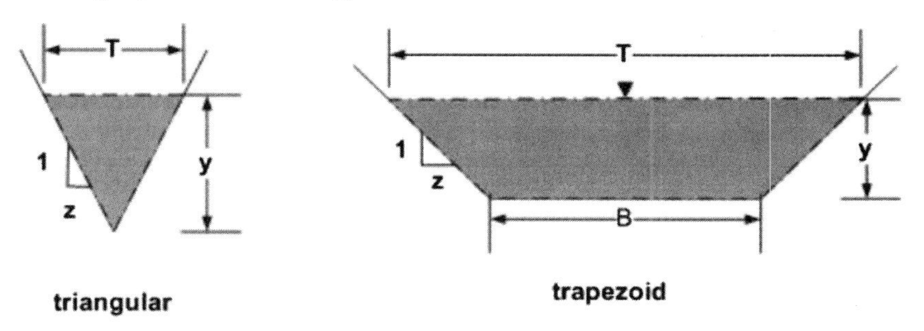

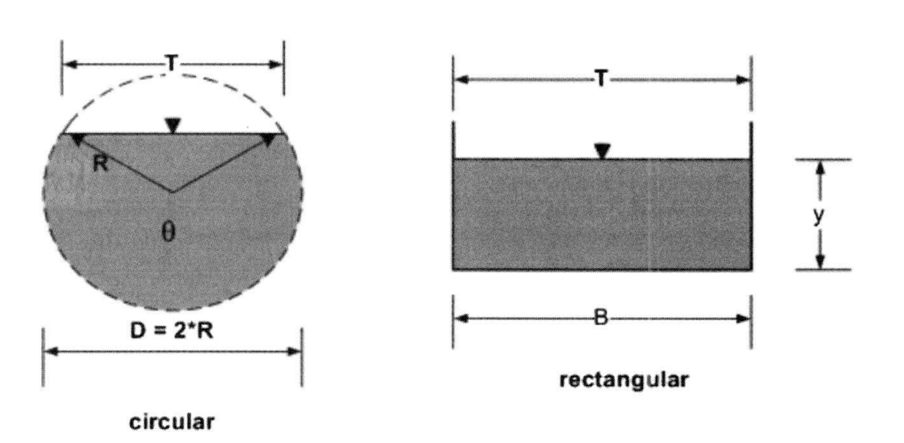

Inputs :

Bottom width of channel $\mathbf{B} := 5 \cdot ft$

Depth of flow in channel (Not needed for pipe) $\mathbf{y} := 3 \cdot ft$

Diameter of pipe $\mathbf{Diam} := 5 \cdot ft$

Side slope for trapezoidal channels (run/rise) $\mathbf{z} := 4$

Slope of channel bottom=Energy slope for uniform flow, ft/ft or m/m

Manning roughness coefficient n $\mathbf{n} := 0.015$ $\mathbf{S} := 0.0005$

θ angle of flow in circular pipe (attach units of degrees or radian)

$\theta := 150 \cdot deg$

Move one of the four definitions to the bottom for whatever shape is being used. This will then define the Shape variable.

Shape := "Triangular"

Shape := "Trapezoidal"

Shape := "Circular"

For this example
Shape:=Rectangular

Shape := "Rectangular"

Shape = "Rectangular"

The program functions below compute the flow area, wetted perimeter, and hydraulic radius.

```
Flow_Area := | "Compute flow area based on channel shape"
             | if Shape = "Triangular"
             |   | z·y^2
             | if Shape = "Trapezoidal"
             |   | (B + z·y)·y
             | if Shape = "Rectangular"
             |   | B·y
             | if Shape = "Circular"
             |   | 1/2·(Diam/2)^2·(θ − sin(θ))
```

$$\text{Flow\_Area} = 15\ ft^2$$

$$\text{Wetted\_Perimeter} := \begin{array}{|l} \text{"Compute wetted perimeter based on channel shape"} \\ \text{if Shape} = \text{"Triangular"} \\ \quad \| \; 2 \cdot y \cdot \sqrt{1+z^2} \\ \text{if Shape} = \text{"Trapezoidal"} \\ \quad \| \; B + 2 \cdot y \cdot \sqrt{1+z^2} \\ \text{if Shape} = \text{"Rectangular"} \\ \quad \| \; B + 2 \cdot y \\ \text{if Shape} = \text{"Circular"} \\ \quad \| \; \dfrac{1}{2} \cdot \dfrac{\theta}{rad} \cdot \text{Diam} \end{array}$$

$$\text{Wetted\_Perimeter} = 11\ ft$$

R in the Manning equation is the hydraulic radius, which is the cross sectional area divided by the wetted perimeter.

$$\text{Hyd\_Rad} := \frac{\text{Flow\_Area}}{\text{Wetted\_Perimeter}} \qquad \text{Hyd\_Rad} = 1.364\ ft$$

Calculate Flow Rate from the Manning equation. This is an empirical equation. Refer to Chapter 5 - Units! The equation is based on US Customary units, so attach US Customary units to the 1.49 coefficient. The value of S is unitless.

$$\text{Flow\_Rate} := \frac{1.49 \cdot \dfrac{ft^{\frac{1}{3}}}{s}}{n} \cdot \text{Flow\_Area} \cdot (\text{Hyd\_Rad})^{\frac{2}{3}} \cdot S^{\frac{1}{2}}$$

$$\text{Flow\_Rate} = 1.16\ \frac{m^3}{s}$$

Flow Rate displayed in metric units.

$$\text{Flow\_Rate} = 40.97\ \frac{ft^3}{s}$$

Flow Rate displayed in US Customary units.

Engineering Example 9.2

Engineering Example 9.2 demonstrates how to use a program to calculate a progressive income tax.

The income tax rates for a state are as follows:

0.023 for income \$0 to \$2,000
0.033 for income \$2,001 to \$4000
0.042 for income \$4001 to \$6000
0.052 for income \$6,001 to \$8000
0.060 for income \$8,001 to \$11,000
0.0698 for income over \$11,000

Write a program to calculate the taxes owed for various income levels.
Write the program so that tax rates and brackets can be easily changed.

Input the tax rates and bracket levels, and calculate maximum taxes paid in each tax bracket.

$\$:= ¤$ Be sure to assign the Unit label to \$.

$Rate_1 := 0.023$ $Max_1 := 2000 \cdot \$$ $Tax_1 := Max_1 \cdot Rate_1 = 46\ \$$

$Rate_2 := 0.033$ $Max_2 := 4000 \cdot \$$ $Tax_2 := (Max_2 - Max_1) \cdot Rate_2 = 66\ \$$

$Rate_3 := 0.042$ $Max_3 := 6000 \cdot \$$ $Tax_3 := (Max_3 - Max_2) \cdot Rate_3 = 84\ \$$

$Rate_4 := 0.052$ $Max_4 := 8000 \cdot \$$ $Tax_4 := (Max_4 - Max_3) \cdot Rate_4 = 104\ \$$

$Rate_5 := 0.06$ $Max_5 := 11000 \cdot \$$ $Tax_5 := (Max_5 - Max_4) \cdot Rate_5 = 180\ \$$

$Rate_6 := 0.0698$

$$\text{Taxes(Income)} := \begin{array}{l} \text{if Income} > \text{Max}_5 \\ \quad \text{return } ((\text{Tax}_1 + \text{Tax}_2 + \text{Tax}_3 + \text{Tax}_4 + \text{Tax}_5) + (\text{Income} - \text{Max}_5) \cdot \text{Rate}_6) \\ \text{also if Income} > \text{Max}_4 \\ \quad \text{return } ((\text{Tax}_1 + \text{Tax}_2 + \text{Tax}_3 + \text{Tax}_4) + (\text{Income} - \text{Max}_4) \cdot \text{Rate}_5) \\ \text{also if Income} > \text{Max}_3 \\ \quad \text{return } ((\text{Tax}_1 + \text{Tax}_2 + \text{Tax}_3) + (\text{Income} - \text{Max}_3) \cdot \text{Rate}_4) \\ \text{also if Income} > \text{Max}_2 \\ \quad \text{return } ((\text{Tax}_1 + \text{Tax}_2) + (\text{Income} - \text{Max}_2) \cdot \text{Rate}_3) \\ \text{also if Income} > \text{Max}_1 \\ \quad \text{return } ((\text{Tax}_1) + (\text{Income} - \text{Max}_1) \cdot \text{Rate}_2) \\ \text{else} \\ \quad \text{Income} \cdot \text{Rate}_1 \end{array}$$

$$\text{IncomeInput} := \begin{bmatrix} 1000 \\ 2000 \\ 3000 \\ 4000 \\ 5000 \\ 6000 \\ 7000 \\ 8000 \\ 9000 \\ 10000 \\ 11000 \\ 20000 \end{bmatrix} \cdot \$$$

Use vectorize. See Chapter 6.

$$\text{Tax} := \overrightarrow{\text{Taxes(IncomeInput)}}$$

$$\text{IncomeInput} = \begin{bmatrix} 1000 \\ 2000 \\ 3000 \\ 4000 \\ 5000 \\ 6000 \\ 7000 \\ 8000 \\ 9000 \\ 10000 \\ 11000 \\ 20000 \end{bmatrix} ¤ \qquad \text{Tax} = \begin{bmatrix} 23 \\ 46 \\ 79 \\ 112 \\ 154 \\ 196 \\ 248 \\ 300 \\ 360 \\ 420 \\ 480 \\ 1108.2 \end{bmatrix} ¤$$

Summary

Logical programs are used in place of the ***if*** function. They are easier to visualize than the ***if*** function, and they are easier to write and check. Chapter 12 will discuss programming in much more detail.

In Chapter 9 we:

- Explained how to create and add lines to a program.
- Emphasized the need to use the Programming Tool Bar to insert the programming operators. Typing the operators will not work unless you type the operator and use CTRL+J.
- Warned about checking the logic in the program so that a subsequent true statement does not change the result.
- Introduced the ***local assignment*** operator.
- Discussed the use of the ***return*** operator.
- Learned about Boolean logical operators.
- Showed how to insert the ***if*** operator into different locations within the program.
- Demonstrated how to use logic programs to draw and display conclusions.

Practice

The PTC Mathcad Prime 3.0 figures and examples used in this book are available for download from the book's website. The reader is encouraged to download the files and use them to practice the concepts learned. Additional examples and problems are also provided. To access this content go to http://store.elsevier.com/9780124104105 *and click on the Resources tab, and then click on the link for the Online Companion Materials.*

1. Use the ***if*** function for the following logic:
   a. Create variable x = 3.
   b. If x = 3 the variable "Result" = 40.
   c. If x is not equal to 3 then "Result" = 100.
2. Create a PTC Mathcad program to achieve the same result as above.
3. Rewrite the program to place the ***if*** operator in a different location.
4. Write three PTC Mathcad programs that use the logical operator ***And***.
5. Write three PTC Mathcad programs that use the logical operator ***Or***.
6. Create and solve five problems from your field of study. At the conclusion of each problem, write a logical program to have PTC Mathcad display a concluding statement depending on the result of the problem. Change input variables to ensure that the display is accurate for all conditions.

PART

Power Tools for Your PTC Mathcad Toolbox

You have just filled your PTC Mathcad toolbox with many simple, yet useful tools. Part III will now add some very powerful PTC Mathcad tools.

The purpose of Part II was to introduce features and functions that were not too confusing. It also gave simple examples. The features and functions introduced in Part III are more complex and require more advanced examples. We postponed the discussion of these topics until after you had a thorough understanding of PTC Mathcad basics.

Part III discusses the powerful topics of symbolic calculations, root finding, solve blocks, advanced PTC Mathcad programming, calculus, and differential equations.

CHAPTER

Introduction to Symbolic Calculations

10

You may be one of the many people who do not understand the power of using symbolic calculations. If so, by the end of this chapter you will have a new appreciation and understanding of the benefits of having this power tool in your PTC Mathcad toolbox. Symbolic calculations return algebraic results rather than numeric results. PTC Mathcad has a sophisticated symbolic processor that can solve very complex problems algebraically. The symbolic processor is different from the numeric processor.

The intent of this chapter is to whet your appetite for symbolic calculations. We will only cover a few of the topics in symbolic calculations. The topics we will discuss are useful to help solve some basic engineering problems. We will not get into solving complex mathematical equations.

As always, the PTC Mathcad Help Center and Tutorials provide excellent examples of topics not covered in this chapter.

Chapter 10 will:

- Introduce the ***symbolic evaluation*** operator.
- Tell how to get numeric rather than algebraic results.
- Show how to use keywords and modifiers.
- Discuss how the "solve" keyword can be used to solve equations.
- Tell how to create user-defined functions using symbolic calculations.
- Show how to use the "explicit" keyword to display the values of variables used in your expressions.
- Demonstrate how to use more than one symbolic keyword.
- Discuss the use of units in symbolic calculations.

Getting Started with Symbolic Calculations

Up until this point we have been using the equal sign to evaluate expressions. The equal sign is called the ***evaluation*** operator. It is used to evaluate numerical expressions or functions and provide numerical results. In order to use it, all variables must be previously assigned. In this chapter we introduce the ***symbolic evaluation*** operator. Unlike numerical evaluation, with symbolic evaluation you can evaluate expressions without having values assigned to variables. The results are displayed showing the relationships between variables, rather than as numeric results. The ***symbolic evaluation*** operator also reduces fractions and radicals to their simplest form. As we shall soon see, this opens up many possibilities.

Essential PTC® Mathcad Prime® 3.0.

Numeric	Symbolic	Symbolic with Numeric
$\frac{1}{3}+\frac{1}{5}=0.53333$	$\frac{1}{3}+\frac{1}{5}\rightarrow\frac{8}{15}$	$\frac{1}{3}+\frac{1}{5}\rightarrow\frac{8}{15}=0.533333333333333$
$\sqrt{3}+\sqrt{27}=6.9282$	$\sqrt{3}+\sqrt{27}\rightarrow 4\cdot\sqrt{3}$	$\sqrt{3}+\sqrt{27}\rightarrow 4\cdot\sqrt{3}=6.92820323027551$
$2\,x+3\,x=?$	$2\,x+3\,x\rightarrow 5\cdot x$	$2\,x+3\,x\rightarrow 5\cdot x=?$
$\cos\left(\frac{\pi}{4}\right)=0.70711$	$\cos\left(\frac{\pi}{4}\right)\rightarrow\frac{\sqrt{2}}{2}$	$\cos\left(\frac{\pi}{4}\right)\rightarrow\frac{\sqrt{2}}{2}=0.707106781186548$
$\pi=3.14159$	$\pi\rightarrow\pi$	$\pi\rightarrow\pi=3.14159265358979$

FIGURE 10.1

Symbolic evaluation operator

The ***symbolic evaluation*** operator is an arrow pointing to the right →. This operator is located on the **Symbolics** control on the **Math** tab (**Math>Operators and Symbols>Symbolics>Operators**). The keyboard shortcut is **CTRL + PERIOD**. Let's look at a few simple expressions and compare the difference between the ***evaluation*** operator and the ***symbolic evaluation*** operator. See Figure 10.1.

When using the ***symbolic evaluation*** operator, the results depend on whether or not a variable has been assigned prior to using the operator. See Figure 10.2.

You can also use the ***symbolic evaluation*** operator to evaluate functions. See Figure 10.3. By creating a vector and using the ***vectorize*** operator you can compute the result for multiple input values. See Figure 10.4.

If an expression contains a decimal number then PTC Mathcad returns a decimal result. This is referred to as "floating-point" format. In the next section we discuss this in more detail and show how to control the precision of floating-point results. See Figure 10.5.

If the ***evaluation*** operator (=) is used in the definition of a variable, then the numeric result is retained in memory. As a result, the ***symbolic evaluation*** operator will return a numeric result. See Figures 10.6 and 10.7.

Keywords and Modifiers

> The ***symbolic evaluation*** operator in PTC® Mathcad Prime® now has a placeholder above the arrow where the keywords and modifies are placed. These may be stacked or listed on the same line.

Comparing results if variables are defined prior to using the symbolic evaluation operator.

$x^2 + 5 \rightarrow x^2 + 5$ Symbolic evaluation prior to definition of "x".

$\mathrm{x} := 2$ Assign a value to x.

$\mathrm{x}^2 + 5 = 9$ Numeric evaluation.

$\mathrm{x}^2 + 5 \rightarrow 9$ Symbolic evaluation after definition of "x".

FIGURE 10.2

Evaluating previously defined variables

$\mathrm{f(y)} := \mathrm{y}^2 + 5\ \mathrm{y} + 6$ Define function

$\mathrm{f(2)} = 20$ Numeric evaluation

$\mathrm{f}(z) \rightarrow z^2 + 5 \cdot z + 6$ $\mathrm{f}(y) \rightarrow y^2 + 5 \cdot y + 6$ Symbolic evaluation

$\mathrm{v} := \sqrt{2}$ Once a variable is assigned, PTC Mathcad substitutes the value of the variable in place of the variable name.

$\mathrm{f(v)} \rightarrow 5 \cdot \sqrt{2} + 8 = 15.0710678118655$

$\mathrm{f}\left(\sqrt{2}\right) \rightarrow 5 \cdot \sqrt{2} + 8 = 15.0710678118655$

$\mathrm{f}(\underline{\sin}(z)) \rightarrow \underline{\sin}(z)^2 + 5 \cdot \underline{\sin}(z) + 6$

$\mathrm{f}(\underline{\sin}(a)) \rightarrow \underline{\sin}(a)^2 + 5 \cdot \underline{\sin}(a) + 6 = ?$ The variable "a" is not defined, so it cannot be evaluated numerically.

FIGURE 10.3

Symbolic evaluation using functions

$$\text{vector} := \begin{bmatrix} 0 \\ \frac{\pi}{4} \\ \frac{\pi}{2} \\ \frac{3}{4}\cdot\pi \\ \pi \end{bmatrix} \qquad \underline{\cos}(\text{vector}) \rightarrow \begin{bmatrix} 1 \\ \frac{\sqrt{2}}{2} \\ 0 \\ -\frac{\sqrt{2}}{2} \\ -1 \end{bmatrix}$$

Define function

$$f(x) := x^2$$

$$f(\text{vector}) \rightarrow \begin{bmatrix} 0 \\ \frac{\pi^2}{16} \\ \frac{\pi^2}{4} \\ \frac{9\cdot\pi^2}{16} \\ \pi^2 \end{bmatrix}$$

Use the ***vectorize*** operator (CTRL+SHIFT+^)

$$\overrightarrow{f(\underline{\cos}(\text{vector}))} \rightarrow \begin{bmatrix} 1 \\ \frac{1}{2} \\ 0 \\ \frac{1}{2} \\ 1 \end{bmatrix}$$

FIGURE 10.4

Symbolic evaluation using vector input

$$\frac{1}{2}+\frac{1}{3} \rightarrow \frac{5}{6}$$

Using a decimal forces a decimal solution.

$$0.5+\frac{1}{3} \rightarrow 0.83333333333333333333$$

FIGURE 10.5

Returning decimal results

If the ***evaluation*** operator (=) is used in the definition of a variable, then the ***symbolic evaluation*** operator will return a numeric result.

This example numerically evaluates the expression in the definition of variable "g." The ***symbolic evaluation*** returns a numeric result.

$$g := \frac{1}{2} + \frac{1}{3} \rightarrow \frac{5}{6} = 0.833333333333333$$

$g = 0.833333333333333$ $\qquad$ $g \rightarrow 0.83333333333333337$

This example does not numerically evaluate the variable "h," so the ***symbolic evaluation*** still provides a fraction result.

$$h := \frac{1}{2} + \frac{1}{3} \rightarrow \frac{5}{6}$$

$h = 0.833333333333333$ $\qquad$ $h \rightarrow \frac{5}{6}$

FIGURE 10.6

Numeric results

$$\underline{var}\left(\begin{bmatrix} 1.2 \\ 3.7 \\ 2.5 \\ 1.9 \end{bmatrix}\right) = 0.841875000000000$$

The **var** function computes the variance of a set of numbers.

$$\underline{var}\left(\begin{bmatrix} 1.2 \\ 3.7 \\ 2.5 \\ 1.9 \end{bmatrix}\right) \rightarrow \underline{var}\left(\begin{bmatrix} 1.2 \\ 3.7 \\ 2.5 \\ 1.9 \end{bmatrix}\right)$$

The ***symbolic*** operator does not evaluate the **var** function.

Assign the **var** function to the variable V

$$V := \underline{var}\left(\begin{bmatrix} 1.2 \\ 3.7 \\ 2.5 \\ 1.9 \end{bmatrix}\right) \qquad V \rightarrow \underline{var}\left(\begin{bmatrix} 1.2 \\ 3.7 \\ 2.5 \\ 1.9 \end{bmatrix}\right)$$

Still does not evaluate symbolically.

If the ***evaluation*** operator (=) is used in the definition of the variable, then the ***symbolic evaluation*** operator will return a numeric result.

$$V := \underline{var}\left(\begin{bmatrix} 1.2 \\ 3.7 \\ 2.5 \\ 1.9 \end{bmatrix}\right) = 0.841875 \qquad V \rightarrow 0.84187500000000015$$

FIGURE 10.7

Forcing a numeric result

The ***symbolic evaluation*** operator has a placeholder located above the arrow.

$$x^2 - 4 \xrightarrow{\;\blacksquare\;} x^2 - 4$$

The placeholder is used to insert keywords and modifiers. For example, if we type **solve** in the placeholder in the above example, PTC Mathcad solves the equation for x and returns the two solutions to the equation.

$$x^2 - 4 \xrightarrow{solve} \begin{bmatrix} 2 \\ -2 \end{bmatrix}$$

PTC Mathcad has over 20 keywords, and over 30 modifiers. These keywords and modifies allow you to control what solutions the ***symbolic evaluation*** operator returns. They may be used by themselves or combined. When combining keywords, separate each keyword with a comma. The list of keywords and modifiers is located on the **Symbolics** controls on the **Math** tab (**Math>Operators and Symbols>Symbolics**).

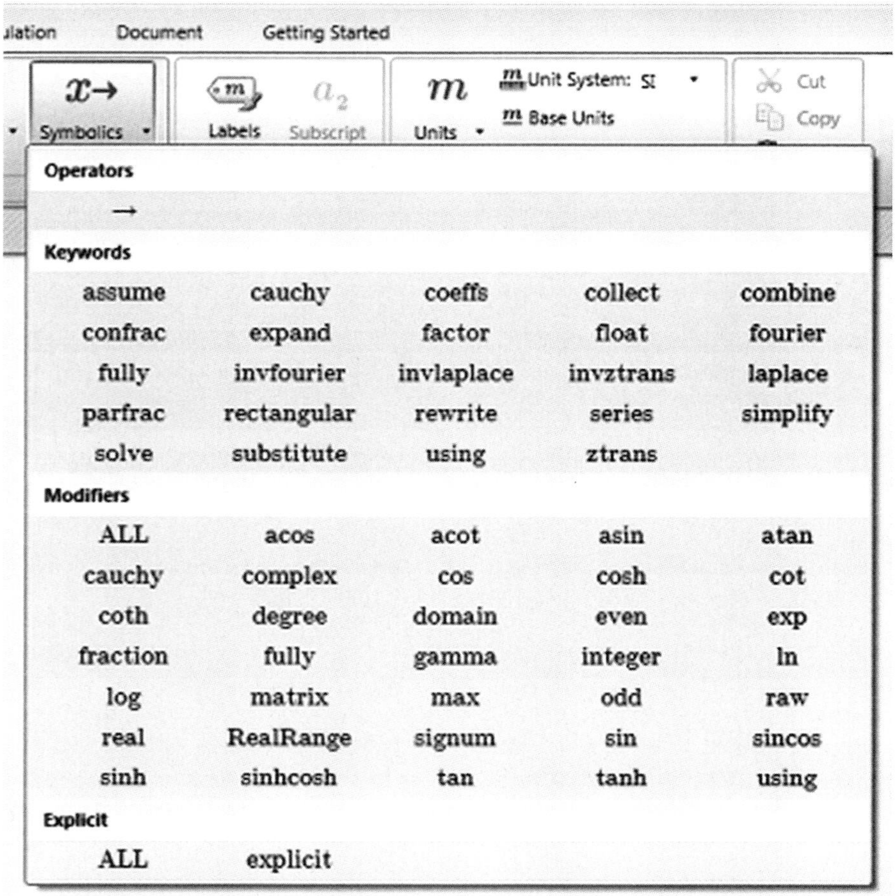

ulation Document Getting Started

Symbolics | Labels | Subscript | Units | Unit System: SI | Base Units | Cut | Copy

Operators

→

Keywords

assume	cauchy	coeffs	collect	combine
confrac	expand	factor	float	fourier
fully	invfourier	invlaplace	invztrans	laplace
parfrac	rectangular	rewrite	series	simplify
solve	substitute	using	ztrans	

Modifiers

ALL	acos	acot	asin	atan
cauchy	complex	cos	cosh	cot
coth	degree	domain	even	exp
fraction	fully	gamma	integer	ln
log	matrix	max	odd	raw
real	RealRange	signum	sin	sincos
sinh	sinhcosh	tan	tanh	using

Explicit

ALL	explicit

Float

In previous examples, we have shown how PTC Mathcad provides a floating-point result by including a decimal in an expression, or by including the ***evaluation*** operator in the definition of a variable, prior to using the ***symbolic evaluation*** operator. There is another way to have the symbolic processor return numeric results. This is accomplished by using the keyword "float." Using the "float" keyword skips the display of symbolic results. When using floating-point calculations, PTC Mathcad uses a default level of precision equal to 20. You can control the level of precision by typing a comma following the "float" keyword and typing an integer, which defines the desired precision. It is important to understand that the level of precision is very different than specifying the number of decimal places you want displayed in numeric calculations. The level of precision specified is the number that PTC Mathcad uses for all future calculations. It does not retain a number in memory with a higher level of precision. The precision can be between 1 and 250. See Figure 10.8.

$\pi = 3.14159265358979$ $\qquad e = 2.71828182845905$

$\pi \rightarrow \pi$ $\qquad e \rightarrow e$

$\pi \xrightarrow{float} 3.1415926535897932385$ $\qquad e \xrightarrow{float} 2.7182818284590452354$

$\pi \xrightarrow{float,4} 3.142 = 3.142$ $\qquad e \xrightarrow{float,4} 2.718$

$\pi \xrightarrow{float,8} 3.1415927$ $\qquad e \xrightarrow{float,8} 2.7182818$

$$\pi \xrightarrow{float,50} 3.1415926535897932384626433832795028841971693993751$$

$$e \xrightarrow{float,50} 2.7182818284590452353602874713526624977572470937$$

$\sqrt{3} + \sqrt{27} = 6.92820323027551$ Numeric evaluation with display precision set to 15.

$$\sqrt{3} + \sqrt{27} \xrightarrow{float,20} 6.9282032302755091741$$

$$\cos\left(\frac{3}{8} \cdot \pi\right) \rightarrow \frac{\sqrt{2-\sqrt{2}}}{2} \xrightarrow{float,10} 0.3826834324$$

FIGURE 10.8

Using the keyword "float"

I suggest keeping the precision at 20. The accuracy of the symbolic processor goes down when you choose a lower precision. If you want a precision less than 20, never use less than 5.

In PTC Mathcad, the constant π is a numeric approximation of π and not its exact value. Thus, the ***sin***$(\pi) =$ returns a result very close to zero, but does not return a result of 0. Figure 10.9 illustrates this and shows how results vary when using different levels of precision in floating-point calculations.

The following numeric evaluations use the "General" result format.
If "Decimal" is used the result shows 0.

$\underline{\sin}(\pi) \rightarrow 0$ Symbolic evaluation gives exact result.

Numeric evaluation gives approximate result.

$\underline{\sin}(\pi) = 1.22460635382238 \cdot 10^{-16}$

Test results based on different levels of precision.

$\text{Float}_1 := \pi$

$\text{Float}_2 := \pi \xrightarrow{float} 3.14159265358979323850$

$\text{Float}_3 := \pi \xrightarrow{float,4} 3.142$

$\text{Float}_4 := \pi \xrightarrow{float,8} 3.1415927$

$\underline{\sin}(\text{Float}_1) = 1.22460635382238 \cdot 10^{-16}$

$\underline{\sin}(\text{Float}_1) \rightarrow 0$ Float1 was never numerically evaluated. The symbolic evaluation uses the exact value of π and not a numerical approximation.

Note how the accuracy is affected by using float less than 20.

$\underline{\sin}(\text{Float}_2) = 1.22460635382238 \cdot 10^{-16}$

$\underline{\sin}(\text{Float}_3) = -0.000407346398941$

$\underline{\sin}(\text{Float}_4) = -0.000000046410207$

FIGURE 10.9

Understanding level of precision

Solve

The keyword "solve" is very powerful. It allows you to solve some very complex expressions. Let's first look at a few simple examples. See Figure 10.10.

Most examples in Figure 10.10 assume that the right side of the expression is 0. In these cases you do not need to type = 0. If the right side of the expression is not equal to zero, then you can set the restraint by using the Boolean ***Equal To*** operator. Remember that this is inserted by typing **CTRL + =**. It can also be found on the Ribbon Bar (**Math>Operators and Symbols>Operators>Comparison>=**).

$$e^x = -1 \xrightarrow{solve} \pi \cdot 1i$$

$$3\,x^2 - 9\,x - 12 \xrightarrow{solve} \begin{bmatrix} -1 \\ 4 \end{bmatrix} \qquad 6 \cdot x - 2 \xrightarrow{solve} \frac{1}{3}$$

$$x^2 + x + 3 \xrightarrow{solve} \begin{bmatrix} -\frac{1}{2} + \frac{\sqrt{11} \cdot 1i}{2} \\ -\frac{1}{2} - \frac{\sqrt{11} \cdot 1i}{2} \end{bmatrix} \qquad x^3 - 1 \xrightarrow{solve} \begin{bmatrix} 1 \\ -\frac{1}{2} + \frac{\sqrt{3} \cdot 1i}{2} \\ -\frac{1}{2} - \frac{\sqrt{3} \cdot 1i}{2} \end{bmatrix}$$

$x = 2$

$\text{clear}(x)$ $\quad x = ?$

This clears the x assignment so the below expression will work.

The above expressions assumed that the right hand side of the expression was set to zero.

The below example sets the expression equal to -1.

$$e^x = -1 \xrightarrow{solve} \pi \cdot 1i$$

FIGURE 10.10

Using keyword "solve"

The above examples also used a single variable. If the expression has multiple variables, you must tell PTC Mathcad which variable to solve for. You do this by adding a comma after the keyword "solve" and then typing the name of the variable you want to solve for. See Figure 10.11.

Refer to Figure 10.12. If the values of "a", "b", and "c" are defined prior to using the "solve" expression, then there is now only a single unknown, and PTC Mathcad solves the solution to the expression. If the expression is rewritten as a function, then "a", "b", and "c" become arguments to the function and it does not matter if they were previously defined. Notice the power of defining the symbolic expression as a function. You can see the symbolic solution, and you can reuse the function unlimited times to solve for many combinations of "a", "b", and "c."

Figure 10.13 illustrates solving a fourth-order polynomial.

$$a \cdot x^2 + b \cdot x + c \xrightarrow{solve\,,x} \begin{bmatrix} -\dfrac{\dfrac{b}{2} + \dfrac{\sqrt{b^2 - 4 \cdot a \cdot c}}{2}}{a} \\ -\dfrac{\dfrac{b}{2} - \dfrac{\sqrt{b^2 - 4 \cdot a \cdot c}}{2}}{a} \end{bmatrix}$$

$$a \cdot x^2 + b \cdot x + c \xrightarrow{solve\,,a} -\frac{c + b \cdot x}{x^2}$$

$$a \cdot x^2 + b \cdot x + c \xrightarrow{solve\,,b} -\frac{a \cdot x^2 + c}{x}$$

$$a \cdot x^2 + b \cdot x + c \xrightarrow{solve\,,c} -(a \cdot x^2) - b \cdot x$$

FIGURE 10.11

Solving an equation with multiple variables

$\mathrm{a} := 1 \qquad \mathrm{b} := 5 \qquad \mathrm{c} := 6$

Because a, b, and c are now defined, PTC Mathcad can provide a numeric solution.

$$\mathrm{a} \cdot x^2 + \mathrm{b} \cdot x + \mathrm{c} \xrightarrow{solve\,,\,x} \begin{bmatrix} -2 \\ -3 \end{bmatrix}$$

If a, b, and c are added to the function, then PTC Mathcad can provide a symbolic solution again. Create a function to solve quadradic equations.

$$\mathrm{f}(\mathrm{a},\mathrm{b},\mathrm{c}) := a \cdot x^2 + b \cdot x + c \xrightarrow{solve\,,\,x} \begin{bmatrix} -\dfrac{\dfrac{b}{2} + \dfrac{\sqrt{b^2 - 4 \cdot a \cdot c}}{2}}{a} \\ -\dfrac{\dfrac{b}{2} - \dfrac{\sqrt{b^2 - 4 \cdot a \cdot c}}{2}}{a} \end{bmatrix}$$

The function *f* can now be used to solve for x with any values of a, b, and c.

$$\mathrm{f}(1,5,6) = \begin{bmatrix} -3 \\ -2 \end{bmatrix} \qquad \mathrm{f}(1,0,-9) = \begin{bmatrix} -3 \\ 3 \end{bmatrix}$$

$$\mathrm{f}(2,3,5) = \begin{bmatrix} -0.75 - 1.391941090707751\mathrm{i} \\ -0.75 + 1.391941090707751\mathrm{i} \end{bmatrix}$$

FIGURE 10.12

Assigning the symbolic expression to a function

clear (x) $x = ?$

$$x^4 + 9\,x^3 + 29\,x^2 + 41\,x + 20 \xrightarrow{solve} \begin{bmatrix} -1 \\ -4 \\ -2+1i \\ -2-1i \end{bmatrix}$$

$$x^4 + 9\,x^3 + 29\,x^2 + 41\,x + 20 \xrightarrow{solve\,,float} \begin{bmatrix} -1.0 \\ -4.0 \\ -2.0+1.0i \\ -2.0-1i \end{bmatrix}$$

$$x^4 + 6\,x^3 + 7\,x^2 - 6\,x - 8 \xrightarrow{solve} \begin{bmatrix} 1 \\ -1 \\ -2 \\ -4 \end{bmatrix} = \begin{bmatrix} 1.00000 \\ -1.00000 \\ -2.00000 \\ -4.00000 \end{bmatrix}$$

$$x^4 - x^2 + 1 \xrightarrow{solve} \begin{bmatrix} \frac{\sqrt{3}}{2} + \frac{1}{2}\cdot 1i \\ \frac{\sqrt{3}}{2} - \frac{1i}{2} \\ -\frac{\sqrt{3}}{2} + \frac{1}{2}\cdot 1i \\ -\frac{\sqrt{3}}{2} - \frac{1i}{2} \end{bmatrix}$$

$$x^4 - x^2 + 1 \xrightarrow{solve\,,float} \begin{bmatrix} 0.86602540378443864676+0.5i \\ 0.86602540378443864676-0.5i \\ -0.86602540378443864676+0.5i \\ -0.86602540378443864676-0.5i \end{bmatrix}$$

$$x^4 - x^2 + 1 \xrightarrow{solve} \begin{bmatrix} \frac{\sqrt{3}}{2} + \frac{1}{2}\cdot 1i \\ \frac{\sqrt{3}}{2} - \frac{1i}{2} \\ -\frac{\sqrt{3}}{2} + \frac{1}{2}\cdot 1i \\ -\frac{\sqrt{3}}{2} - \frac{1i}{2} \end{bmatrix} = \begin{bmatrix} 0.86603+0.5i \\ 0.86603-0.5i \\ -0.86603+0.5i \\ -0.86603-0.5i \end{bmatrix}$$

FIGURE 10.13

Solving fourth-order polynomials

Explicit

If the name of the variable has been previously defined in the worksheet, PTC Mathcad gives an error when using the "solve" keyword. For example, in Figure 10.14, the variable "x" has been previously assigned the value of 2. When using the keyword "solve" PTC Mathcad first substitutes the value of x in the expression and then performs the keyword operation. Since the value of 2 has been substituted for x, PTC Mathcad can no longer solve for x and thus gives an error. There are two way of resolving this issue. The first is to use the PTC Mathcad function ***clear***(x). This function clears the value of "x" for the remainder of the worksheet. See Figure 10.14. Using the ***clear*** function may have consequences further down in the worksheet. A better way to resolve a previous definition is to use the keyword "explicit." This keyword temporarily suppresses the value of x for a single symbolic evaluation. See Figure 10.15.

$x := 2$ $x = 2$ $x \rightarrow 2$

$x^2 - d^2 \xrightarrow{solve,\, x} ?$ Does not provide a solution because x is previously defined.

The PTC Mathcad function **clear**(x) clears the variable "x" both numerically and symbolically.

$\text{clear}(x)$ $x = ?$ $x \rightarrow x$

$$x^2 - d \xrightarrow{solve,\, x} \begin{bmatrix} \sqrt{d} \\ -\sqrt{d} \end{bmatrix}$$

FIGURE 10.14

Symbolically solving if variable has been previously assigned

$\mathbf{x} := 2$

$x^2 - d \xrightarrow{\boxed{solve}, x} ?$ Does not provide a solution because x is previously defined.

Use the keyword "explicit" to suppress the previous definition of "x".

$$x^2 - d \xrightarrow{solve, \mathbf{x}, explicit} \begin{bmatrix} \sqrt{d} \\ -\sqrt{d} \end{bmatrix}$$

$$a \cdot x^2 + b \cdot x + c \xrightarrow{\boxed{solve}, x} ?$$

$$\mathbf{a} \cdot x^2 + \mathbf{b} \cdot x + \mathbf{c} \xrightarrow{solve, \mathbf{x}, explicit} \begin{bmatrix} -\dfrac{\dfrac{b}{2} + \dfrac{\sqrt{b^2 - 4 \cdot a \cdot c}}{2}}{a} \\ -\dfrac{\dfrac{b}{2} - \dfrac{\sqrt{b^2 - 4 \cdot a \cdot c}}{2}}{a} \end{bmatrix}$$

$$\mathbf{g}(\mathbf{a}, \mathbf{b}, \mathbf{c}) := a \cdot x^2 + b \cdot x + c \xrightarrow{\boxed{solve}, x} ?$$

$$\mathbf{g}(\mathbf{a}, \mathbf{b}, \mathbf{c}) := a \cdot x^2 + b \cdot x + c \xrightarrow{solve, \mathbf{x}, explicit} \begin{bmatrix} -\dfrac{\dfrac{b}{2} + \dfrac{\sqrt{b^2 - 4 \cdot a \cdot c}}{2}}{a} \\ -\dfrac{\dfrac{b}{2} - \dfrac{\sqrt{b^2 - 4 \cdot a \cdot c}}{2}}{a} \end{bmatrix}$$

$$\mathbf{g}(1, 5, 6) = \begin{bmatrix} -3 \\ -2 \end{bmatrix}$$

FIGURE 10.15

Using the keyword "explicit" to suppress a previous definition

Assume

Using the keyword "assume" with the keyword "solve" allows you to constrain the region over which PTC Mathcad finds a result. The keyword "assume" must be followed by a modifier which tells PTC Mathcad the constraints. The following table lists the types of modifier constraints that may be used.

Modifying expression	Assumption
x = real	x is a real number.
x = integer	x is an integer.
$x > a$	x is a real number greater than a.
$x \geq a$	x is a real number greater than or equal to a.
$x < b$	x is a real number less than b.
$x \leq b$	x is a real number less than or equal to b.
x = RealRange(a, b)	x is a real number in the range $a < x < b$, where $a < b$.
n = even	n is an even integer.
n = odd	n is an odd integer.

Figure 10.16 illustrates the use of these modifiers with the "solve" and "assume" keywords. PTC Mathcad 3.0 appears to have a bug using these modifiers. The modifiers do not constrain the solutions.

Fully

The keyword "fully" returns a detailed solution to an equation. See Figures 10.17 and 10.18 for a description of the "fully" keyword.

In Figure 10.18 the _n is a generated variable that represents an arbitrary integer. The underscore is used in the variable name to avoid name conflicts with other variables in the worksheet. The $\_n \in \mathbb{Z}$ means that _n is a subset of the set of integers.

$$\left(x^3-1\right)\left(x^2-2\right) \xrightarrow{solve} \begin{bmatrix} 1 \\ \sqrt{2} \\ -\sqrt{2} \\ -\frac{1}{2}+\frac{\sqrt{3}\cdot 1i}{2} \\ -\frac{1}{2}-\frac{\sqrt{3}\cdot 1i}{2} \end{bmatrix}$$

PTC Mathcad Prime 3.0 appears to have a bug because the modifiers do not seem to be working.

$$\left(x^3-1\right)\left(x^2-2\right) \xrightarrow{solve\,,\,assume\,,\,\mathbf{x}=real} \begin{bmatrix} 1 \\ \sqrt{2} \\ -\sqrt{2} \\ -\frac{1}{2}+\frac{\sqrt{3}\cdot 1i}{2} \\ -\frac{1}{2}-\frac{\sqrt{3}\cdot 1i}{2} \end{bmatrix}$$

$$\left(x^3-1\right)\left(x^2-2\right) \xrightarrow{solve\,,\,assume\,,\,\mathbf{x}=integer} \begin{bmatrix} 1 \\ \sqrt{2} \\ -\sqrt{2} \\ -\frac{1}{2}+\frac{\sqrt{3}\cdot 1i}{2} \\ -\frac{1}{2}-\frac{\sqrt{3}\cdot 1i}{2} \end{bmatrix}$$

$$\left(x^3-1\right)\left(x^2-2\right) \xrightarrow{solve\,,\,assume\,,\,\mathbf{x}>1} \begin{bmatrix} 1 \\ \sqrt{2} \\ -\sqrt{2} \\ -\frac{1}{2}+\frac{\sqrt{3}\cdot 1i}{2} \\ -\frac{1}{2}-\frac{\sqrt{3}\cdot 1i}{2} \end{bmatrix}$$

FIGURE 10.16

Using “assume”

$\text{clear}(a) \qquad \text{clear}(x)$

$$(3\ a-7)\cdot x=1 \xrightarrow{solve\,,\,x} \frac{1}{3\cdot a-7}$$

The above solution is undefined when the denominater equals zero.

The keyword "fully" provides a detailed solution that identifies this.

$$(3\ a-7)\cdot x=1 \xrightarrow{solve\,,\,x\,,\,fully} \left\| \begin{array}{l} \text{if } a \neq \frac{7}{3} \\ \quad \left\| \frac{1}{3\cdot a-7} \right. \\ \text{else if } a = \frac{7}{3} \\ \quad \left\| undefined \right. \end{array} \right.$$

Assign the above expression to a function.

$$\text{f}(\text{a}) := (3\ a-7)\cdot x=1 \xrightarrow{solve\,,\,x\,,\,fully} \left\| \begin{array}{l} \text{if } a \neq \frac{7}{3} \\ \quad \left\| \frac{1}{3\cdot a-7} \right. \\ \text{else if } a = \frac{7}{3} \\ \quad \left\| undefined \right. \end{array} \right.$$

$\text{f}(3) \rightarrow \frac{1}{2} \qquad \text{f}\left(\frac{7}{3}\right) \rightarrow undefined$

$\text{f}(3) = 0.5$

$\text{f}\left(\frac{7}{3}\right) = \boldsymbol{NaN}$

FIGURE 10.17

Using "fully" keyword

$$\underline{\sin}(x) \xrightarrow{solve} 0$$

$$\underline{\cos}(x) \xrightarrow{solve} \frac{\pi}{2}$$

$$\underline{\cos}(x) \xrightarrow{solve, x, fully} \begin{Vmatrix} \text{if } \_n \in \mathbb{Z} \\ \quad \begin{Vmatrix} \frac{\pi}{2} + \pi \cdot \_n \end{Vmatrix} \\ \text{else} \\ \quad \begin{Vmatrix} undefined \end{Vmatrix} \end{Vmatrix}$$

$$\underline{\tan}(x) \xrightarrow{solve, fully} \begin{Vmatrix} \text{if } \_n \in \mathbb{Z} \\ \quad \begin{Vmatrix} \pi \cdot \_n \end{Vmatrix} \\ \text{else} \\ \quad \begin{Vmatrix} undefined \end{Vmatrix} \end{Vmatrix}$$

$$\underline{\sin}(x) = \underline{\cos}(x) \xrightarrow{solve} \frac{\pi}{4}$$

$$\underline{\sin}(x) = \underline{\cos}(x) \xrightarrow{solve, fully} \begin{Vmatrix} \text{if } \_n \in \mathbb{Z} \\ \quad \begin{Vmatrix} \frac{\pi}{4} + \pi \cdot \_n \end{Vmatrix} \\ \text{else} \\ \quad \begin{Vmatrix} undefined \end{Vmatrix} \end{Vmatrix}$$

$$\underline{\tan}(x) = \underline{\sin}(x) \xrightarrow{solve, fully} \begin{Vmatrix} \text{if } \_n \in \mathbb{Z} \\ \quad \begin{Vmatrix} \pi \cdot \_n \end{Vmatrix} \\ \text{else} \\ \quad \begin{Vmatrix} undefined \end{Vmatrix} \end{Vmatrix}$$

FIGURE 10.18

Using “fully” to solve equations with periodic solutions

Using

The modifier "using" can be used following the keyword "fully" to replace the generated variable _n with a designated variable. For example, to replace the _n with the variable "j" type the following: **fully, using, _n CTRL + = j**. If the value of "j" is previously defined, PTC Mathcad substitutes the numeric value of "j". If this happens you can use the keyword "explicit" to temporarily suppress the value of "j." See Figure 10.19.

The modifier "using" following the keyword "fully" can replace the generated variable _n with a designated variable name.

$$\cos(x) \xrightarrow{solve, x, fully, using, \_n = j} \begin{cases} \text{if } j \in \mathbb{Z} \\ \quad \frac{\pi}{2} + \pi \cdot j \\ \text{else} \\ \quad undefined \end{cases}$$

Define the variable "j" $j := 2$

$$\tan(x) \xrightarrow{solve, fully, using, \_n = \mathrm{j}} 2 \cdot \pi$$

Use keyword "explicit" to suppress the definition of j.

$$\tan(x) \xrightarrow{solve, explicit, fully, using, \_n = \mathrm{j}} \begin{cases} \text{if } j \in \mathbb{Z} \\ \quad \pi \cdot j \\ \text{else} \\ \quad undefined \end{cases}$$

FIGURE 10.19

Keyword "using"

Solving a System of Equations

PTC Mathcad will solve a system of equations using the ***symbolic evaluation*** operator. To do this, place the equations and constraints in a column vector. Use the Boolean Equal to set the constraints. Following the "solve" keyword, list the variables to solve for separated by commas. The solution to the equations is returned in a row vector that corresponds to the variables in the list. The list of variables to solve for may also be included in a column vector. See Figures 10.20 and 10.21.

Place equations and constraints in a column vector.
List the variables to solve for.

$$\begin{bmatrix} 2\,x+y=9 \\ 3\,x-y=16 \end{bmatrix} \xrightarrow{solve\,,x\,,y} \begin{bmatrix} 5 & -1 \end{bmatrix}$$

$$\mathbf{Vector_1} := \begin{bmatrix} 2\,x+y=9 \\ 3\,x-y=16 \end{bmatrix}$$

Ignore the warning that x is undefined.

$$\mathbf{Vector_1} \xrightarrow{solve\,,x\,,y} \begin{bmatrix} 5 & -1 \end{bmatrix}$$

$$\mathbf{Vector_1} \xrightarrow{solve\,,\begin{bmatrix} x \\ y \end{bmatrix}} \begin{bmatrix} 5 & -1 \end{bmatrix}$$

Assign the solution to a variable so that the results can be used later in the worksheet.

$$\mathbf{Solution_1} := \mathbf{Vector_1} \xrightarrow{solve\,,\begin{bmatrix} x \\ y \end{bmatrix}} \begin{bmatrix} 5 & -1 \end{bmatrix}$$

$$\mathbf{Solution_1} = \begin{bmatrix} 5 & -1 \end{bmatrix} \qquad \mathbf{Solution_1}_{1,1} = 5 \qquad \mathbf{Solution_1}_{1,2} = -1$$

FIGURE 10.20

Solving a system of equations

Create a column vector with equations and constraints.

$$\mathbf{Vector_2} := \begin{bmatrix} w+x+y+z=0 \\ -w+x+y-z=1 \\ w+x-y+z=3 \\ w+x-y-z=-1 \end{bmatrix} \qquad \mathbf{Vector_3} := \begin{bmatrix} 2\,w-x+5\,y+z=-3 \\ 3\,w+2\,w+2\,y-6\,z=-32 \\ w+3\,x+3\,y-z=-47 \\ 5\,w-2\,x-3\,y+3\,z=49 \end{bmatrix}$$

The list of variables to solve can also be in a column vector rather than a list.

$$\mathbf{Vector_2} \xrightarrow{solve,\begin{bmatrix} w \\ x \\ y \\ z \end{bmatrix}} \begin{bmatrix} -\frac{5}{2} & 2 & -\frac{3}{2} & 2 \end{bmatrix}$$

$$\mathbf{Solution_2} := \mathbf{Vector_2} \xrightarrow{solve,\begin{bmatrix} w \\ x \\ y \\ z \end{bmatrix}} \begin{bmatrix} -\frac{5}{2} & 2 & -\frac{3}{2} & 2 \end{bmatrix}$$

$\mathbf{Solution_2}_{1,1} = -2.5$ $\qquad$ $\mathbf{Solution_2}_{1,2} = 2$

$\mathbf{Solution_2}_{1,3} = -1.5$ $\qquad$ $\mathbf{Solution_2}_{1,4} = 2$

$$\mathbf{Vector_3} \xrightarrow{solve,\begin{bmatrix} w \\ x \\ y \\ z \end{bmatrix}} \begin{bmatrix} \frac{31}{54} & -\frac{1135}{108} & -\frac{829}{216} & \frac{979}{216} \end{bmatrix}$$

FIGURE 10.21

Solving a system of equations with four unknowns

Expand, Simplify, and Factor

You can also use the ***symbolic evaluation*** operator to expand, simplify, and factor equations.

When PTC Mathcad expands an equation, it multiplies the various elements and expands the exponents. Let's look at how PTC Mathcad can expand algebraic equations. See Figures 10.22 and 10.23.

$\text{clear}(x)$

Use the keyword "expand" to expand the below equations.

$$(\mathrm{x}+1)^{3}+(\mathrm{x}+3)^{2}+\mathrm{x}-3 \xrightarrow{expand} x^{3}+4\cdot x^{2}+10\cdot x+7$$

$$\ln\left(\frac{8\,x^{4}}{5}\right) \xrightarrow{expand} \ln\left(x^{4}\right)-\ln(5)+\ln(8)$$

$$(2\cdot a\cdot \mathrm{c}\cdot x+2\cdot \mathrm{c}-d)^{2} \xrightarrow{expand} 144\cdot a^{2}\cdot x^{2}-24\cdot a\cdot d\cdot x+288\cdot a\cdot x+d^{2}-24\cdot d+144$$

FIGURE 10.22

Expanding expressions

Use the keyword "expand" to expand the below equations.

$$(\cot(x)+1)^{2}+(\tan(x)+2)^{2} \xrightarrow{expand} 4\cdot\tan(x)+\frac{2}{\tan(x)}+\frac{1}{\tan(x)^{2}}+\tan(x)^{2}+5$$

$$(a\cdot x-\mathrm{b}\cdot y)\cdot(\mathrm{c}\cdot z+d\cdot w) \xrightarrow{expand} 6\cdot a\cdot x\cdot z-30\cdot y\cdot z-5\cdot d\cdot w\cdot y+a\cdot d\cdot w\cdot x$$

$$\frac{1}{2}+\frac{1}{3}+\frac{1}{4} \xrightarrow{expand} \frac{13}{12}=1.08333333333333$$

FIGURE 10.23

Expanding expressions

When PTC Mathcad simplifies an equation, it tries to reduce the equation to the simplest form. Figure 10.24 shows the results of some simplifications.

When you factor an expression, PTC Mathcad breaks the equation into all of its parts. Figure 10.25 gives examples of what happens when you use the keyword "factor."

Use the keyword "simplify" to simplify the below equations.

$$(\cot(x)+1)^2+(\tan(x)+2)^2 \xrightarrow{simplify} \cot(x)^2+2\cdot\cot(x)+\tan(x)^2+4\cdot\tan(x)+5$$

$$\frac{x^2+5\cdot x+6}{x+3} \xrightarrow{simplify} x+2$$

$$1-\sin(x)^2 \xrightarrow{simplify} \cos(x)^2$$

$$\sin(x)^2+\cos(x)^2 \xrightarrow{simplify} 1$$

FIGURE 10.24

Simplifying equations

Use the keyword "factor" to break the following expressions down into the smallest parts.

$$21 \xrightarrow{factor} 3\cdot 7$$

$$462 \xrightarrow{factor} 2\cdot 3\cdot 7\cdot 11$$

$$x^2+5\cdot x+6 \xrightarrow{factor} (x+3)\cdot(x+2)$$

$$x^3-x^2-17\cdot x-15 \xrightarrow{factor} (x-5)\cdot(x+3)\cdot(x+1)$$

$$x^4-4\cdot x^3-7\cdot x^2+22\cdot x+24 \xrightarrow{factor} (x-3)\cdot(x-4)\cdot(x+2)\cdot(x+1)$$

FIGURE 10.25

Factoring equations

Coeffs and Collect

The keyword "coeffs" places the coefficients of a polynomial in a column vector with the top row representing the constant term. If a term is missing, PTC Mathcad

$$4\,x^4+3\,x^3+x^2+2 \xrightarrow{coeffs} \begin{bmatrix}2\\0\\1\\3\\4\end{bmatrix}$$

$$4\,x^4+3\,x^3+x^2+2 \xrightarrow{coeffs\,,\,degree} \begin{bmatrix}2&0\\0&1\\1&2\\3&3\\4&4\end{bmatrix}$$

$$4\,x^4+3\,x^3+x^2+6\,x^{-4}+x^{-2}+2 \xrightarrow{coeffs\,,\,degree} \begin{bmatrix}6&-4\\0&-3\\1&-2\\0&-1\\2&0\\0&1\\1&2\\3&3\\4&4\end{bmatrix}$$

$$4\,x^4\cdot y+3\,x^3\cdot y^2+x^2\cdot y^3+y^4+2 \xrightarrow{coeffs\,,\,x\,,\,degree} \begin{bmatrix}y^4+2&0\\0&1\\y^3&2\\3\cdot y^2&3\\4\cdot y&4\end{bmatrix}$$

$$4\,x^4\cdot y+3\,x^3\cdot y^2+x^2\cdot y^3+y^4+2 \xrightarrow{coeffs\,,\,y\,,\,degree} \begin{bmatrix}2&0\\4\cdot x^4&1\\3\cdot x^3&2\\x^2&3\\1&4\end{bmatrix}$$

FIGURE 10.26

Finding the coefficients of a polynomial

places a 0 in the row. If the modifier "degree" is used following the keyword "coeffs" then PTC Mathcad returns an array with two columns. The first column contains the coefficients and the second row contains the corresponding exponent. If the expression has negative exponents they are placed prior to the constant term and the smallest exponent is placed in the top row with zeros added for any skipped exponents in the expression. When an expression contains several variables, you must tell PTC Mathcad which variable you want the coefficients for. See Figure 10.26.

The keyword "collect" collects coefficients of variables with matching exponents. If the expression has more than one variable, you must specify which variable you want to collect the coefficients for. The result is a new expression. It returns similar results to the keyword "coeffs" but expressed in a different form. See Figure 10.27.

$$2\,x^2 + 3\,x + 5 + 6\,x^2 - 5\,x - 2 \xrightarrow{collect} 8 \cdot x^2 - 2 \cdot x + 3$$

$$2\,\mathrm{x}^2 \cdot y + 3\,\mathrm{x} \cdot y^2 + 5 + 6\,\mathrm{x}^2 \cdot y - 5\,\mathrm{x} - 2 \xrightarrow{collect\,,\mathrm{x}} 8 \cdot y \cdot x^2 + \left(3 \cdot y^2 - 5\right) \cdot x + 3$$

$$2\,\mathrm{x}^2 \cdot y + 3\,\mathrm{x} \cdot y^2 + 5 + 6\,\mathrm{x}^2 \cdot y - 5\,\mathrm{x} - 2 \xrightarrow{collect\,,y} 3 \cdot x \cdot y^2 + 8 \cdot x^2 \cdot y + 3 - 5 \cdot x$$

The keyword "collect" returns similar results to the keyword "coeffs".

$$2\,x^2 \cdot y + 3\,x \cdot y^2 + 5 + 6\,x^2 \cdot y - 5\,x - 2 \xrightarrow{coeffs\,,x\,,degree} \begin{bmatrix} 3 & 0 \\ 3 \cdot y^2 - 5 & 1 \\ 8 \cdot y & 2 \end{bmatrix}$$

$$2\,\mathrm{x}^2 \cdot y + 3\,\mathrm{x} \cdot y^2 + 5 + 6\,\mathrm{x}^2 \cdot y - 5\,\mathrm{x} - 2 \xrightarrow{coeffs\,,y\,,degree} \begin{bmatrix} 3 - 5 \cdot x & 0 \\ 8 \cdot x^2 & 1 \\ 3 \cdot x & 2 \end{bmatrix}$$

FIGURE 10.27

Collecting terms of a polynomial

Substitute

The keyword "substitute" allows you to substitute new values or variables in place of the variables in the expression. You must tell PTC Mathcad which variable you want to substitute and what value or variable you want to replace it with. Use the Boolean ***equal to*** operator **CTRL + =** to define the value to substitute. See Figure 10.28. If you use the modifier "raw," PTC Mathcad will not simplify the expression allowing you to see the substitution.

$$4\,x^2+3\cdot x+5 \xrightarrow{substitute\,,\,x=y} 4\cdot y^2+3\cdot y+5$$

$$4\,x^2+3\cdot x+5 \xrightarrow{substitute\,,\,x=2} 27$$

The modifier "raw" returns results that show the substitution.

$$4\,x^2+3\cdot x+5 \xrightarrow{substitute\,,\,x=2\,,\,raw} 2\cdot 3+2^2\cdot 4+5=27$$

FIGURE 10.28

Using the keyword "Substitute"

Combine and Rewrite

In its simplest use, the keyword "combine" combines the exponents of an expression.

$$x^m\cdot x^n \xrightarrow{combine} x^{m+n}$$

The real power of this keyword comes from using the available modifiers (atan, exp, fully, ln, log, sincos, sinhcosh, and using). See Figure 10.29.

$$x^m \cdot x^n \xrightarrow{combine} x^{m+n}$$

$$3 \cdot \underline{\sin}(x) \cdot \underline{\cos}(x) \xrightarrow{combine} 3 \cdot \underline{\cos}(x) \cdot \underline{\sin}(x)$$

$$3 \cdot \underline{\sin}(x) \cdot \underline{\cos}(x) \xrightarrow{combine, sincos} \frac{3 \cdot \underline{\sin}(2 \cdot x)}{2}$$

$$2 \cdot \underline{\ln}(3) - 4 \cdot \underline{\ln}(5) \xrightarrow{combine} 2 \cdot \underline{\ln}(3) - 4 \cdot \underline{\ln}(5)$$

$$2 \cdot \underline{\ln}(3) - 4 \cdot \underline{\ln}(5) \xrightarrow{combine, \underline{\ln}} \underline{\ln}\left(\frac{9}{625}\right)$$

Compare with the "simplify" keyword.

$$2 \cdot \underline{\ln}(3) - 4 \cdot \underline{\ln}(5) \xrightarrow{simplify} \underline{\ln}(9) - \underline{\ln}(625)$$

FIGURE 10.29

Using the keyword “combine”

The keyword "rewrite" is very powerful because it allows you to rewrite an expression using specific types of identities. This keyword must be followed by a modifier. The allowable modifiers are listed in the following table.

Modifiers for the keyword "rewrite"	
Modifier	**Rewrites expression in terms of these elementary functions**
acos	Inverse cosine
acot	Inverse cotangent
asin	Inverse sine
atan	Inverse tangent
cos	Cosine
cosh	Hyperbolic cosine
cot	Cotangent
coth	Hyperbolic cotangent
exp	Exponential
gamma	Gamma
ln	Natural logarithm
log	Logarithm base 10
signum	Signum
sin	Sine
sincos	Sine or cosine
sinh	Hyperbolic sine
sinhcosh	Hyperbolic sine or hyperbolic cosine
tan	Tangent
tanh	Hyperbolic tangent

See Figure 10.30 for examples of using the keyword "rewrite."

$$\underline{\sin}(x) \xrightarrow{rewrite,\, \underline{\cos}} \underline{\sin}(x)$$

$$\underline{\sin}(x)^2 \xrightarrow{rewrite,\, \underline{\cos}} 1-\underline{\cos}(x)^2$$

$$\underline{\sin}(x) \xrightarrow{rewrite,\, \underline{\exp}} -\frac{e^{x \cdot 1i} \cdot 1i}{2}+\frac{e^{-(x \cdot 1i)} \cdot 1i}{2}$$

$$\underline{\sin}(x) \xrightarrow{rewrite,\, \underline{\tan}} \frac{2 \cdot \underline{\tan}\left(\frac{x}{2}\right)}{\underline{\tan}\left(\frac{x}{2}\right)^2+1}$$

$$\underline{\tan}(x) \xrightarrow{rewrite,\, sincos} \frac{\underline{\sin}(x)}{\underline{\cos}(x)}$$

$$\underline{\sinh}(x) \xrightarrow{rewrite,\, \underline{\sin}} -(\underline{\sin}(x \cdot 1i) \cdot 1i)$$

$$\underline{\cosh}(x) \xrightarrow{rewrite,\, \underline{\cos}} \underline{\cos}(x \cdot 1i)$$

FIGURE 10.30

Using the keyword "rewrite"

Series

The keyword "series" will return the Taylor series of a function about a point. PTC Mathcad by default uses the point 0. To expand the series around a different point, specify a value using the Boolean ***equal to*** operator. Refer to the PTC Mathcad Help for additional information regarding the use of the keyword "series." See Figure 10.31.

$$e^{x} \xrightarrow{series} 1+x+\frac{x^{2}}{2}+\frac{x^{3}}{6}+\frac{x^{4}}{24}+\frac{x^{5}}{120}$$

$$\sin(x) \xrightarrow{series} x-\frac{x^{3}}{6}+\frac{x^{5}}{120}$$

$$\frac{1}{1-x} \xrightarrow{series} 1+x+x^{2}+x^{3}+x^{4}+x^{5}$$

$$e^{x} \xrightarrow{series,\,x=3} e^{3}+e^{3}\cdot(x-3)+\frac{e^{3}\cdot(x-3)^{2}}{2}+\frac{e^{3}\cdot(x-3)^{3}}{6}+\frac{e^{3}\cdot(x-3)^{4}}{24}+\frac{e^{3}\cdot(x-3)^{5}}{120}$$

$$\sin(x) \xrightarrow{series,\,x=1} \sin(1)+\cos(1)\cdot(x-1)-\frac{\sin(1)\cdot(x-1)^{2}}{2}-\frac{\cos(1)\cdot(x-1)^{3}}{6}+\frac{\sin(1)\cdot(x-1)^{4}}{24}+\frac{\cos(1)\cdot(x-1)^{5}}{120}$$

$$\frac{1}{1-x} \xrightarrow{series,\,x=2} -3+x-(x-2)^{2}+(x-2)^{3}-(x-2)^{4}+(x-2)^{5}$$

FIGURE 10.31

Taylor series expansion

Explicit

We use the keyword "explicit" following the keyword "solve" to temporarily suppress the assignment of a variable. The keyword "explicit" also allows you to display

The "explicit" keyword allows you to show intermediate results in your calculations.

Given the following input variable definitions:

$C_1 := 5.1 \cdot N$ $\quad C_2 := 4.5 \cdot N$ $\quad D_1 := 3.2 \cdot m$ $\quad D_2 := 4.8 \cdot m$

A normal equation would look like this:

$M_1 := C_1 \cdot D_1 + C_2 \cdot D_2$ $\quad M_1 = 37.92\ N \cdot m$

Using explicit calculations, you are able to show the values that PTC Mathcad uses to calculate the results:

$$M_2 := C_1 \cdot D_1 + C_2 \cdot D_2 \xrightarrow{explicit, C_1, D_1, C_2, D_2} 5.1 \cdot N \cdot 3.2 \cdot m + 4.5 \cdot N \cdot 4.8 \cdot m \rightarrow 16.32 \cdot N \cdot m + 21.6 \cdot N \cdot m$$

$M_2 = 37.92\ N \cdot m$

$$LineLoad_1 := \frac{C_1}{D_1} + \frac{C_2}{D_2} \qquad LineLoad_1 = 2.53125\ \frac{N}{m}$$

$$LineLoad_2 := \frac{C_1}{D_1} + \frac{C_2}{D_2} \xrightarrow{explicit, C_1, C_2, D_1, D_2} \frac{5.1 \cdot N}{3.2 \cdot m} + \frac{4.5 \cdot N}{4.8 \cdot m} \rightarrow \frac{0.9375 \cdot N}{m} + \frac{1.59375 \cdot N}{m}$$

$$LineLoad_2 = 2.53125\ \frac{N}{m}$$

FIGURE 10.32

Using the keyword "explicit"

the values of the variables used in your expressions. To use this feature, type "explicit" as a keyword followed by a comma and a list of the variables you want to display. See Figures 10.32—10.35 and Figure 4.24.

Time

$$T := 3 \cdot s$$

Velocity

$$V := 5 \cdot \frac{m}{s}$$

Initial Distance

$$ID := 5.2\ m$$

A normal equation would look like this:

$$Distance_1 := ID + T \cdot V \qquad Distance_1 = 20.2\ m$$

In the above example, if the definitions for InitialDist, Time, and Velocity were on a different page, you would not know what values PTC Mathcad is using for the three variables. With the keyword "explicit," you can have PTC Mathcad display the input variables.

$$Distance_2 := ID + T \cdot V \xrightarrow{explicit, ID, T, V} 5.2\ m + 3 \cdot s \cdot 5 \cdot \frac{m}{s} \rightarrow 15 \cdot m + 5.2 \cdot m$$

$$Distance_2 = 20.2\ m$$

FIGURE 10.33

Using the keyword "explicit"

You can also include previous results in the list following "explicit." The results vary depending on what variables are included in the modify list. See the examples below.

$E_1 := 3.0 \cdot m$ $\quad E_2 := 4.2 \cdot m$ $\quad E_3 := 5.5 \cdot m$ $\quad E_4 := 6.2 \cdot m$

$F_1 := \frac{E_1}{E_2}$ $\quad F_1 = 0.714285714285714$ $\quad F_2 := \frac{E_3}{E_4}$ $\quad F_2 = 0.887096774193548$

$$G_1 := \frac{F_1}{F_2} \xrightarrow{explicit, F_1, F_2} \frac{\frac{E_1}{E_2}}{\frac{E_3}{E_4}}$$

By listing F1 and F2, PTC Mathcad substitutes the definitions of F1 and F2.

$G_1 = 0.8052$

$$G_2 := \frac{F_1}{F_2} \xrightarrow{explicit, E_1, E_2, E_3, E_4} \frac{F_1}{F_2}$$

Listing only E1, E2, E3, and E4 does not have any effect, because they do not appear in the current definition.

$G_2 = 0.8052$

By listing F1 and F2 in addition to E1, E2, E3, and E4, PTC Mathcad substitutes all values used.

$$G_3 := \frac{F_1}{F_2} \xrightarrow{explicit, E_1, E_2, E_3, E_4, F_1, F_2} \frac{\frac{3.0 \cdot m}{4.2 \cdot m}}{\frac{5.5 \cdot m}{6.2 \cdot m}} \rightarrow 0.80519480519480519481$$

$G_3 = 0.8052$

FIGURE 10.34

Using the keyword “explicit”

This Figure is very similar to Figure 10.34. In this example, the values of F3 and F4 are evaluated in addition to being defined.

Define and evaluate F3.

$$\mathbf{F_3} := \frac{\mathbf{E_1}}{\mathbf{E_2}} \xrightarrow{explicit, E_1, E_2} \frac{3.0 \cdot m}{4.2 \cdot m} \rightarrow 0.71428571428571428571$$

Define and evaluate F4.

$$\mathbf{F_4} := \frac{\mathbf{E_3}}{\mathbf{E_4}} \xrightarrow{explicit, E_3, E_4} \frac{5.5 \cdot m}{6.2 \cdot m} \rightarrow 0.88709677419354838871$$

Define and evaluate G5. Because F3 and F4 have already been evaluated, the numeric results of F3 and F4 are displayed.

$$\mathbf{G_5} := \frac{\mathbf{F_3}}{\mathbf{F_4}} \xrightarrow{explicit, F_3, F_4} \frac{0.71428571428571428571}{0.88709677419354838871} \rightarrow 0.805194805194805194797$$

FIGURE 10.35

Using the keyword "explicit"

Using More than One Keyword

In previous examples we have used multiple keywords separated by commas. Multiple keywords can be listed above a single ***symbolic evaluation*** operator, or they may be separated into multiple ***symbolic evaluation*** operators so that intermediate results become visible. This second method is referred to as chaining. See Figure 10.36.

One benefit of chaining is that you can provide intermediate results.

$$(4\,\mathrm{x} + 5\,y - 3\,\mathrm{x} - 3\,y)^3 \xrightarrow{expand} x^3 + 6 \cdot x^2 \cdot y + 12 \cdot x \cdot y^2 + 8 \cdot y^3$$

$$(4\,\mathrm{x} + 5\,y - 3\,\mathrm{x} - 3\,y)^3 \xrightarrow{simplify} (x + 2 \cdot y)^3 \xrightarrow{expand} x^3 + 6 \cdot x^2 \cdot y + 12 \cdot x \cdot y^2 + 8 \cdot y^3$$

FIGURE 10.36

Chaining

When keywords are placed above a single ***symbolic evaluation*** operator, the second keyword may control the behavior of the first keyword. If this is the case, then the second keyword must be kept above a single ***symbolic evaluation*** operator. If these keywords are split between multiple ***symbolic evaluation*** operators, the results will be different. Figure 10.37 shows the effect of splitting keywords to a second ***symbolic evaluation*** operator.

$$e^{x} \xrightarrow{series\,,substitute\,,x=2} \frac{109}{15}$$

The keyword "substitute" does not affect the behavior of "series," thus it may be split to another operator.

$$e^{x} \xrightarrow{series} 1+x+\frac{x^2}{2}+\frac{x^3}{6}+\frac{x^4}{24}+\frac{x^5}{120} \xrightarrow{substitute\,,x=2} \frac{109}{15}$$

The keyword "float" does not affect the behavior of "solve," thus it may be split to another operator.

$$x^4-x^2+1 \xrightarrow{solve\,,float} \begin{bmatrix} 0.86602540378443864676+0.5\mathrm{i} \\ 0.86602540378443864676-0.5\mathrm{i} \\ -0.86602540378443864676+0.5\mathrm{i} \\ -0.86602540378443864676-0.5\mathrm{i} \end{bmatrix}$$

$$x^4-x^2+1 \xrightarrow{solve} \begin{bmatrix} \frac{\sqrt{3}}{2}+\frac{1}{2}\cdot 1\mathrm{i} \\ \frac{\sqrt{3}}{2}-\frac{1\mathrm{i}}{2} \\ -\frac{\sqrt{3}}{2}+\frac{1}{2}\cdot 1\mathrm{i} \\ -\frac{\sqrt{3}}{2}-\frac{1\mathrm{i}}{2} \end{bmatrix} \xrightarrow{float} \begin{bmatrix} 0.86602540378443864676+0.5\mathrm{i} \\ 0.86602540378443864676-0.5\mathrm{i} \\ -0.86602540378443864676+0.5\mathrm{i} \\ -0.86602540378443864676-0.5\mathrm{i} \end{bmatrix}$$

The keyword "fully" does affect the behavior of "solve," thus if it is split to another operator, the results are different.

$$\mathrm{clear}(j)$$

$$\underline{\cos}(x) \xrightarrow{solve\,,x\,,fully\,,using\,,\_n=j} \left\| \begin{array}{l} \text{if } j\in\mathbb{Z} \\ \quad \left\| \frac{\pi}{2}+\pi\cdot j \right. \\ \text{else} \\ \quad \left\| undefined \right. \end{array} \right.$$

$$\underline{\cos}(x) \xrightarrow{solve\,,x} \frac{\pi}{2} \xrightarrow{fully} \frac{\pi}{2}$$

FIGURE 10.37

Effect of splitting multiple keywords

It is also possible to stack multiple keywords. Stacking has no effect on the results. You can use stacking to make the string of keywords more clear or if you want your expression to take less horizontal space on the worksheet. The top keyword matches the first keyword in a list of keywords. To stack keywords, type **SHIFT + ENTER** after typing the first keyword. This creates a new placeholder below the keyword where the second keyword or modifier may be added. See Figure 10.38.

$$\underline{\cos}(x)+y \xrightarrow{solve,\, x,\, fully,\, using,\, \_n=j} \left\| \begin{array}{l} \text{if } j \in \mathbb{Z} \\ \quad \left\| \begin{bmatrix} \pi+\underline{\text{acos}}(y)+2\cdot\pi\cdot j \\ \pi-\underline{\text{acos}}(y)+2\cdot\pi\cdot j \end{bmatrix} \right. \\ \text{else} \\ \quad \left\| \, undefined \right. \end{array} \right.$$

Stacking keywords. Use SHIFT+ENTER to stack.

$$\underline{\cos}(x)+y \xrightarrow{\begin{array}{l} solve \\ x \\ fully \\ using \\ \_n=j \end{array}} \left\| \begin{array}{l} \text{if } j \in \mathbb{Z} \\ \quad \left\| \begin{bmatrix} \pi+\underline{\text{acos}}(y)+2\cdot\pi\cdot j \\ \pi-\underline{\text{acos}}(y)+2\cdot\pi\cdot j \end{bmatrix} \right. \\ \text{else} \\ \quad \left\| \, undefined \right. \end{array} \right.$$

FIGURE 10.38

Stacking keywords and modifiers

Units with Symbolic Calculations

The ***symbolic evaluation*** operator does not recognize units. It passes units through as undefined variables. Thus, you can use units with your expressions; however, they will not be truly understood. For instance, 10 ft/s * 5 sec is returned 50 ft, but you cannot convert the 50 ft to meters. Another example is adding 10 ft + 5 sec. The ***symbolic evaluation*** operator returns 10 ft + 5 sec, not recognizing the error of adding units from different unit dimensions. In order to process the units, you need to do a numeric evaluation of the result. Do this by assigning a variable to the symbolic calculation and then numerically evaluating the variable. See Figure 10.39 for examples.

You may attach units to symbolic calculations.

$\mathbf{SampleUnits_1} := 1 \cdot ft + 5 \cdot in + 2 \cdot m$

The symbolic processor just passes units through. It does not evaluate them.

$\mathbf{SampleUnits_1} \rightarrow 2 \cdot m + 5 \cdot in + ft$

$\mathbf{SampleUnits_1} = 2.4318\ m$

You must use the numeric processor to evaluate the units.

$10 \cdot ft + 5 \cdot s \rightarrow 10 \cdot ft + 5 \cdot s$

The symbolic processor did not recognize the error of adding units in different unit dimensions. Once a variable is assigned to the expression, the numeric processor recognizes the error.

$\mathbf{SampleUnits_2} := 10 \cdot ft + 5 \cdot s \rightarrow 5 \cdot s + 10 \cdot ft$

$\mathbf{SampleUnits_3} := 10 \cdot ft + 5 \cdot s \rightarrow 5 \cdot s + 10 \cdot ft = ?$

$\mathbf{SampleUnits_3} := 10 \cdot ft + 5 \cdot s \rightarrow 5 \cdot s + 10 \cdot ft = ?$

These units are not compatible.

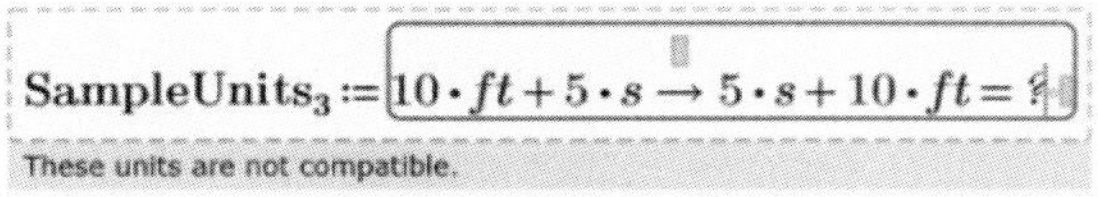

FIGURE 10.39 *(Continued on next page)*

Using units with symbolic calculations

The symbolic processor cancels common units ((ft/s)*(s)=ft), but it does not convert ft to meters.

$$10 \cdot \frac{ft}{s} \cdot 20 \cdot s + \frac{1}{2} \cdot 50 \cdot \frac{m}{s^2} \cdot 20^2 \cdot s^2 \rightarrow 10000 \cdot m + 200 \cdot ft$$

Assign a user-defined function.

$$\text{Distance}\left(v_o, a, t\right) := v_o \cdot t + \frac{1}{2} \cdot a \cdot t^2$$

The symbolic processor does not convert units.

$$\text{Distance}\left(10 \cdot \frac{ft}{s}, 50 \cdot \frac{m}{s^2}, 20 \cdot s\right) \rightarrow 10000 \cdot m + 200 \cdot ft$$

The numeric processor does convert units.

$$\text{Distance}\left(10 \cdot \frac{ft}{s}, 50 \cdot \frac{m}{s^2}, 20 \cdot s\right) = \left(1.006096 \cdot 10^4\right) m$$

Because all the units were consistent, the symbolic processor returned meters.

$$\text{Distance}\left(10 \cdot \frac{m}{s}, 50 \cdot \frac{m}{s^2}, 20 \cdot s\right) \rightarrow 10200 \cdot m$$

The symbolic processor does not have a units placeholder to convert results to other units.

$$\text{Distance}\left(10 \cdot \frac{m}{s}, 50 \cdot \frac{m}{s^2}, 20 \cdot s\right) = \left(3.34645669291339 \cdot 10^4\right) ft$$

FIGURE 10.39

(*continued*)

If you attach units using the **Units** control on the Ribbon Bar, then PTC Mathcad will attach the **Unit** label style to your units. See Chapter 16 for a discussion about math label styles. When using symbolics, if you attach units by typing the name of your unit, then by default PTC Mathcad will not change the display of the typed unit. When using symbolics, there is a setting on the toolbar that will allow PTC Mathcad to recognize built-in units and constants and automatically modify the display of these units to show the **Unit** label style. This control is (**Calculation>Worksheet Settings>Calculation Options>Units/Constants in Symbolics**). Figure 10.40 shows how the display of units looks with this control off and with it on.

Input using units from ribbon controls.

$1\,m+2\,m \rightarrow 3\cdot m$

$1\,m+2\,ft \rightarrow m+2\cdot ft$

Input using text.

$1\,m+2\,m \rightarrow 3\cdot m$

$1\,m+2\,ft \rightarrow m+2\cdot ft$

Input using a combination of text and ribbon.

$1\,m+2\,m \rightarrow m+2\cdot m$

PTC Mathcad views these as two different variables.

With **Units/Constants in Symbolics** control active **Calculation>Worksheet Settings>Calculation Options>Units/Constants in Symbolics**.

$1\,m+2\,m \rightarrow m+2\cdot m$

Prior to activating the control.

$1\,m+2\,m \rightarrow 3\cdot m$

After activating the control, PTC Mathcad added the Unit label style to the first m, and recognizes them both as the same type of variable.

FIGURE 10.40

Units/constants in symbolics

Additional Topics to Study

The topic of symbolic calculations could fill an entire book. In this chapter, we have introduced you to a few symbolic calculation concepts. Even though the discussion of these concepts has been brief, these concepts will be a valuable addition to your PTC Mathcad toolbox. More discussion of symbolic calculations will occur in the following chapters.

Some of the topics that we did not discuss in this chapter that may be of interest to you for further study include:

- Partial fractions
- Calculus (see Chapter 13)
- Symbolic transformations
- Complex input
- Symbolic functions
- Symbolic optimization

Summary

In Chapter 10 we:

- Introduced the ***symbolic evaluation*** operator.
- Discussed the differences between numeric and symbolic calculations.
- Learned how to force the ***symbolic evaluation*** operator to return numeric results.
- Illustrated how to use keywords with the ***symbolic evaluation*** operator.
- Used the keyword "solve" with various modifiers to solve algebraic equations.
- Introduced several keywords and modifiers.
- Discussed using the "explicit" keyword to display values of variables used in your calculations.
- Recommended studying the PTC Mathcad Help to learn more advanced features of symbolic calculations.

Practice

The PTC Mathcad Prime 3.0 figures and examples used in this book are available for download from the book's website. The reader is encouraged to download the files and use them to practice the concepts learned. Additional examples and problems are also provided. To access this content go to http://store.elsevier.com/9780124104105 *and click on the Resources tab, and then click on the link for the Online Companion Materials.*

1. Write symbolic equations to solve for x with the following functions. Assign each to a user-defined function.

a. $x^3 + 6x^2 - x - 30$
b. $y = x^2(x - 1)2$
c. $3x^2 + 4y^2 = 12$

2. Write symbolic equations to evaluate the following functions. Assign each to a user-defined function.
   a. $\cos(\pi/3) + \tan(2\pi/3)$
   b. $[(21/11)+ (5/3)]/3$
   c. $\pi + \pi/4 + \ln(2)$
3. Write symbolic equations to expand the following functions. Assign each to a user-defined function.
   a. $(x + 2)(x-3)$
   b. $(x - 1)(x + 2)(x - 5)$
   c. $(x + 3 * i)(x - 3 * i)$
4. Write symbolic equations to simplify the following functions. Assign each to a user-defined function.
   a. $\sin(\pi/2 - x)$
   b. $\sec(x)$
   c. $(x^4 - 8x^3 - 7x^2 + 122x - 168)/(x - 2)$
5. Write symbolic equations to factor the following functions. Assign each to a user-defined function.
   a. $x^3 + 5x^2 + 6x$
   b. $x * y + y * x^2$
   c. $2x^2 + llx-21$
6. Use the "float" keyword to evaluate the following functions. Use various levels of precision.
   a. $\cos(\pi/6)$
   b. $\tan(3\pi/4) + COS(\pi\ /7) - \sin(\pi/5)$
   c. $(\pi + e3)(\tan(3\pi/8))$
7. Use the equations from exercise 6 and show intermediate results before using the "float" keyword.
8. From your field of study, create three expressions using assigned variables. Use the "explicit" keyword to show intermediate results.
9. Open the PTC Mathcad Help and review additional topics on symbolics under the topic Symbolics and from the Symbolic Operators section under the topic Operators.

CHAPTER

11 Solving Engineering Equations

Once you begin using the solving features of PTC Mathcad, you will wonder how you ever got along without them. Solving engineering equations is one of PTC Mathcad's most useful power tools. It ranks up there with the use of units as one of PTC Mathcad's best features. In Chapter 10, we introduced the keyword "solve" with the ***symbolic evaluation*** operator. In this chapter, we will add more solving tools to your PTC Mathcad toolbox. The intent of this chapter is to illustrate how engineering problems can be solved using the PTC Mathcad functions ***root***, ***polyroots***, and ***Find***. Because these are some of the most useful functions for engineers, this chapter will use many examples to illustrate their use. As discussed in Chapter 8, when trying to solve for an equation, it is very useful to plot the equations. This helps to visualize the solution before using PTC Mathcad to solve the equations. Some of the functions discussed require initial guess values, thus a plot is useful to help select the initial guess.

Chapter 11 will:

- Introduce the ***root*** function.
- Discuss the two different forms that it takes: unbracketed and bracketed.
- Give examples of each form.
- Show how a plot is useful to determine the initial guess.
- Discuss the ***polyroots*** function.
- Give examples of the ***polyroots*** function.
- Show how to use a solve block using ***find, maximize,*** and ***minimize*** to solve multiple equations with multiple unknowns.
- Note when to use ***minerr*** instead of ***find***.
- Illustrate how to use units in solving equations.
- Provide several engineering examples which illustrate the use of each method of solving equations.

Essential PTC® Mathcad Prime® 3.0.

Root Function

The ***root*** function is used to find a single solution to a single function with a single unknown. In later sections, we will discuss finding all the solutions to a polynomial function. We will also discuss solving multiple equations with multiple unknowns. For now, we will focus on using the ***root*** function.

If a function has several solutions, then the solution that PTC Mathcad finds is based on the initial guess you give PTC Mathcad. Because of this, it is helpful to plot the function prior to giving PTC Mathcad the initial guess.

The ***root*** function takes the form ***root***(f(var), var, [a, b]). It returns the value of var to make the function f equal to zero. The real numbers a and b are optional. If they are specified (bracketed), ***root*** finds var on this interval. The values of a and b must meet these requirements: $a < b$ and f(a) and f(b) must be of opposite signs. This is because the function must cross the x-axis in this interval. If you do not specify the numbers a and b (unbracketed), then var must be defined with an initial guess prior to using the root function.

When plotting the function, use a different variable name on the x-axis than the variable you define for your initial guess. If you do not use a different variable name, the plot will not work because PTC Mathcad will only plot the value var on the x-axis. The variable used on the x-axis needs to be a previously undefined variable.

Let's look at some simple examples of using ***root***. See Figures 11.1 and 11.2.

If you do not use a plot to determine your initial guesses, PTC Mathcad may not arrive at the solution you expect. If your initial guesses are close to a maximum or minimum point or have multiple solutions between the bracketed initial guesses, then PTC Mathcad may not arrive at a solution, or may arrive at a different solution than you wanted. If PTC Mathcad is not arriving at a solution, you can refer to PTC Mathcad Help for ideas on how to help resolve the issue.

Complex numbers may be used as an initial guess to arrive at a complex solution.

$\mathbf{SampleFunction}(\mathbf{x}) := \mathbf{x}^2 + 5 \cdot \mathbf{x} - 6$ Define the function.

Find the root closest to x=0

The first form of *root* (bracketed) needs two initial guesses such that a<b, and SampleFunction(a)<0 and SampleFunction(b)>0.

$\mathbf{Root_{1a}} := \mathbf{root}(\mathbf{SampleFunction}(\mathbf{x}), x, 0, 2)$

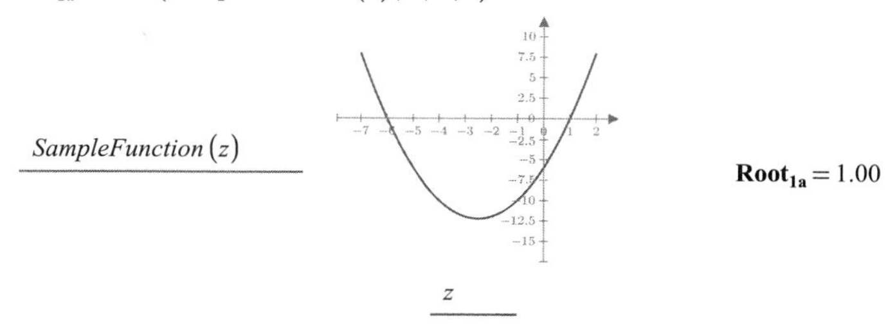

$\mathbf{Root_{1a}} = 1.00$

The second form of *root* (unbracketed) does not need the two values a and b, but you must define a guess value for x prior to using the *root* function.

$\mathbf{x} := 0$ Guess value for x

$\mathbf{Root_{1b}} := \mathbf{root}(\mathbf{SampleFunction}(\mathbf{x}), \mathbf{x})$ $\mathbf{Root_{1b}} = 1.00$

Find the root closest to x= - 5 Bracketed

$\mathbf{Root_{1c}} := \mathbf{root}(\mathbf{SampleFunction}(\mathbf{x}), x, -7, -5)$ $\mathbf{Root_{1c}} = -6.00$

Unbracketed

$\mathbf{x} := -5$ New guess value for x

$\mathbf{Root_{1d}} := \mathbf{root}(\mathbf{SampleFunction}(\mathbf{x}), \mathbf{x})$ $\mathbf{Root_{1d}} = -6.00$

Check to see if results are zero at calculated roots.

$\mathbf{SampleFunction}(\mathbf{Root_{1a}}) = 0.00$ $\mathbf{SampleFunction}(\mathbf{Root_{1c}}) = 0.00$ **OK**

Here is a way to input your guess value as the argument of a function.

$\mathbf{Test}(\mathbf{x}) := \mathbf{root}(\mathbf{SampleFunction}(\mathbf{x}), \mathbf{x})$ This function provides an initial guess to "SampleFunction."

$\mathbf{Root_{1e}} := \mathbf{Test}(0)$ $\mathbf{Root_{1e}} = 1.00$ Root of "SampleFunction" with x= 0 as the guess value.

$\mathbf{Root_{1f}} := \mathbf{Test}(-5)$ $\mathbf{Root_{1f}} = -6.00$ Root of "SampleFunction" with x= -5 as the guess value.

FIGURE 11.1

Using the function *root*

$F_2(x) := 2^x - 9 \cdot x^2 + 120$

Define the function.

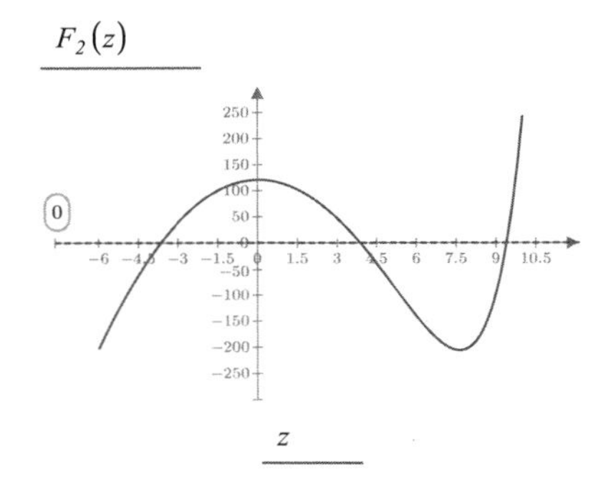

From the plot, use the values 7 and 10 for a and b.

$\mathbf{Root_{2a}} := \mathbf{root}(\mathbf{F_2(x)}, x, 7, 10)$

$\mathbf{Root_{2a}} = 9.40$

From the plot, use the values 3 and 4 for a and b.

$\mathbf{Root_{2b}} := \mathbf{root}(\mathbf{F_2(x)}, x, 3, 4)$

$\mathbf{Root_{2b}} = 3.87$

From the plot, use the values -6 and -2 for a and b.

$\mathbf{Root_{2c}} := \mathbf{root}(\mathbf{F_2(x)}, x, -6, -2)$

$\mathbf{Root_{2c}} = -3.65$

Check to see if results are zero at calculated roots.

$F_2(\mathbf{Root_{2a}}) = 0.00 \quad F_2(\mathbf{Root_{2b}}) = 0.00 \quad F_2(\mathbf{Root_{2c}}) = -2.842 \cdot 10^{-14}$ **OK**

Input the guess value as part of a function.

This function provides an initial guess to "F2".

$\mathbf{TT(x)} := \mathbf{root}(\mathbf{F_2(x)}, \mathbf{x})$

Root of "F2" with x=-6 as the guess value.

$\mathbf{Root_{2d}} := \mathbf{TT}(-6) \qquad \mathbf{Root_{2d}} = -3.65$

Root of "F2" with x=3 as the guess value.

$\mathbf{Root_{2e}} := \mathbf{TT}(3) \qquad \mathbf{Root_{2e}} = 3.87$

Root of "F2" with x=10 as the guess value.

$\mathbf{Root_{2f}} := \mathbf{TT}(10) \qquad \mathbf{Root_{2f}} = 9.40$

FIGURE 11.2

Another example of using the function *root*

Polyroots Function

The ***polyroots*** function is used to solve for all solutions to a polynomial equation at the same time. The solution is returned in a vector containing the roots of the polynomial.

In order to use this function, you need to create a vector of coefficients of the polynomial. Include all coefficients in the vector even if they are zero. The coefficients of the polynomial $f(x) = 4x^3 - 2x^2+3x, - 5$, are $(4, -2, 3, -5)$. For the PTC Mathcad vector however, you begin with the constant term. The PTC Mathcad vector looks like this:

$$Vector := \begin{bmatrix} -5 \\ 3 \\ -2 \\ 4 \end{bmatrix}$$

The ***polyroots*** function takes the following form: ***polyroots***(v), where v is the vector of polynomial coefficients. The result is a vector containing the roots of the polynomial. Figure 11.3 shows the solution of the same simple quadratic equation as was used in Figure 11.1.

To use the ***polyroots*** function, create a vector of each of the coefficients of the polynomial. Begin with the constant term as the first element. Use zero for all zero coefficients.

Use ***polyroots*** to find the roots of the same equation as in Figure 11.1.

$$\mathbf{SampleFunction_{3a}}(\mathbf{x}) := \mathbf{x}^2 + 5 \cdot \mathbf{x} - 6$$

$$\mathbf{V_{3a}} := \begin{bmatrix} -6 \\ 5 \\ 1 \end{bmatrix}$$

Create a vector of coefficients beginning with the constant term.

$$\mathbf{Solution\_3a} := \underline{\mathbf{polyroots}}(\mathbf{V_{3a}}) \qquad \mathbf{Solution\_3a} = \begin{bmatrix} -6.00 \\ 1.00 \end{bmatrix}$$

Check to see if results are zero at calculated roots.

$$\mathbf{SampleFunction_{3a}}\left(\mathbf{Solution\_3a}_1\right) = 0.00 \quad \mathbf{SampleFunction_{3a}}\left(\mathbf{Solution\_3a}_2\right) = 0.00$$

OK

FIGURE 11.3

Using the function *polyroots*

Figure 11.4 illustrates a larger polynomial. Notice that this polynomial has two imaginary roots. The check results are displayed twice. The first result uses the General result format. The second result uses the Decimal result format with the number of decimal places set to 15.

Figure 11.5 uses the ***symbolic evaluation*** operator to have PTC Mathcad create the polynomial vector automatically.

$$\mathbf{SampleFunction_{4a}}(\mathbf{x}) := -\mathbf{x}^5 - 2\cdot\mathbf{x}^4 + \mathbf{x}^3 - \mathbf{x}^2 + 5$$

Create a vector of coefficients.
Remember that the constant goes first.

$$\mathbf{V_{4a}} := \begin{bmatrix} 5 \\ 0 \\ -1 \\ 1 \\ -2 \\ -1 \end{bmatrix}$$

$$\mathbf{Solution\_4a} := \underline{\mathbf{polyroots}}(\mathbf{V_{4a}})$$

$$\mathbf{Solution\_4a} = \begin{bmatrix} -2.436964498961 \\ -1.216885522117 \\ 0.260206634327 + 1.191574629143i \\ 0.260206643463 - 1.191574625107i \\ 1.133436743287 \end{bmatrix}$$

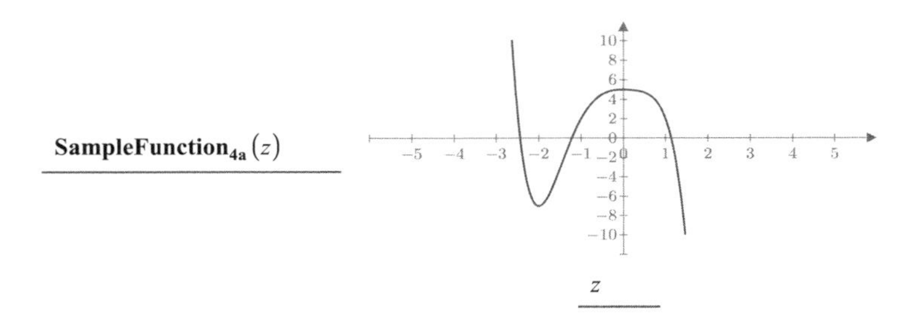

Check $\quad \mathbf{Check_{4a}} := \mathbf{SampleFunction_{4a}}(\mathbf{Solution\_4a})$

$$\mathbf{Check_{4a}} = \begin{bmatrix} 2.59\cdot10^{-8} \\ -7.68\cdot10^{-9} \\ -7.68\cdot10^{-9} - 1.22i\cdot10^{-15} \\ 4.28\cdot10^{-8} - 1.90i\cdot10^{-7} \\ 1.43\cdot10^{-7} \end{bmatrix}$$

Display results with decimals.

$$\mathbf{Check_{4a}} = \begin{bmatrix} 0.000000025932199 \\ -0.000000007677749 \\ -0.000000007677752 - 0.000000000000001i \\ 0.000000042822379 - 0.000000190174987i \\ 0.000000143379047 \end{bmatrix}$$

FIGURE 11.4

Using the function *polyroots* for a larger polynomial

PTC Mathcad can automate the generation of the coefficient vector.
Use the following steps:

1. Type the name of the variable to be used for the vector, followed by a colon.
2. Type the function name using a previously undefined variable name as the argument.
3. Click the *coeffs* control (**Math>Operators and Symbols>Symbolics>coeffs**)
4. Type a comma and then type the name of the argument.
5. Click outside the region.

For this example, use the same equation as in Figure 11.4.

$$\mathbf{SampleFunction_{4a}}(z) \rightarrow z^3 - 2 \cdot z^4 - z^5 - z^2 + 5$$

$$\mathbf{V_{5a}} := \mathbf{SampleFunction_{4a}}(z) \xrightarrow{coeffs\,,\,z} \begin{bmatrix} 5 \\ 0 \\ -1 \\ 1 \\ -2 \\ -1 \end{bmatrix} \qquad \mathbf{V_{5a}} = \begin{bmatrix} 5.00 \\ 0.00 \\ -1.00 \\ 1.00 \\ -2.00 \\ -1.00 \end{bmatrix}$$

PTC Mathcad generates the coefficient vector. If SampleFunction4a changes, the vector will update automatically.

$$\mathbf{Solution\_5a} := \underline{\mathbf{polyroots}}(\mathbf{V_{5a}}) \qquad \mathbf{Solution\_5a} = \begin{bmatrix} -2.436964499 \\ -1.216885522 \\ 0.260206634 + 1.191574629i \\ 0.260206643 - 1.191574625i \\ 1.133436743 \end{bmatrix}$$

FIGURE 11.5

Using the ***symbolic evaluation*** operator to have PTC Mathcad create the polynomial vector

Solve Blocks

> Solve blocks no longer use the Given and Find keywords. The solve block in PTC® Mathcad Prime® is a container that contains the guess values, the constraints, and the solutions. You are able to move the solve block with all of its contained regions as a single entity.

Until now, we have been solving single equations and single variables. PTC Mathcad is able to solve for multiple equations using multiple unknowns.

Several methods can be used to solve for multiple equations. The first method that we will use is called a solve block. A solve block is a large region that contains three areas: 1) Guess Values, 2) Constraints, and 3) Solver (for the solutions). The solve block is inserted from the **Regions** group on the **Math** tab. The keyboard shortcut is **CTRL + 1**.

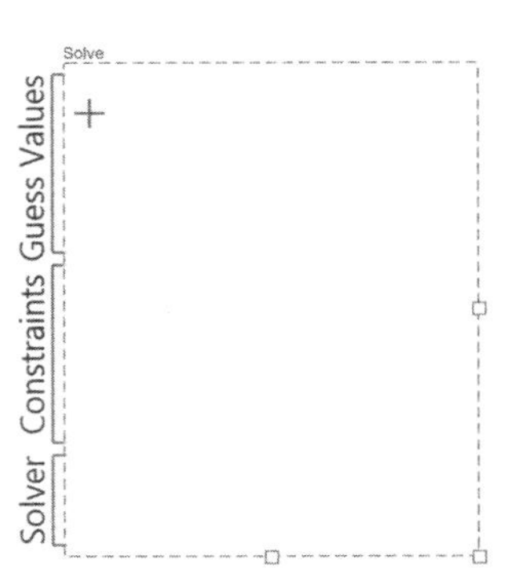

A solve block is illustrated in Figure 11.6.

Follow these steps to create a solve block:

1. In the Guess Values region, provide an initial guess for each variable you are solving for.
2. In the Constraints region, type a list of constraint equations using Boolean operators.
3. In the Solver region, type the function ***find***() and list the desired solution variables as arguments.

If you are solving for n unknowns, you must have at least n equations in the constraints area. Equality equations must be defined using the Boolean equals. These must be inserted from the **Operator** controls on the **Math** tab or use **CTRL + EQUAL**. If there is more than one solution, the solution PTC Mathcad calculates is based on your initial guesses. If you modify your guesses, then PTC Mathcad may arrive at a different solution. If one of the solutions is negative and the other is positive, you can add an additional constraint telling PTC Mathcad that you want the solution to be greater than zero. See Figure 11.7.

You can also assign the equations as functions and use the functions as the constraints. See Figure 11.8.

Solve blocks should not contain the following:

- Range variables.
- Constraints involving the "not equal to" operator.
- Other solve blocks.
- Assignment statements using the ***definition*** operator (:=).

Find the intersection points of a parabola and a line using a solve block.
Plot both equations to see approximate solutions.

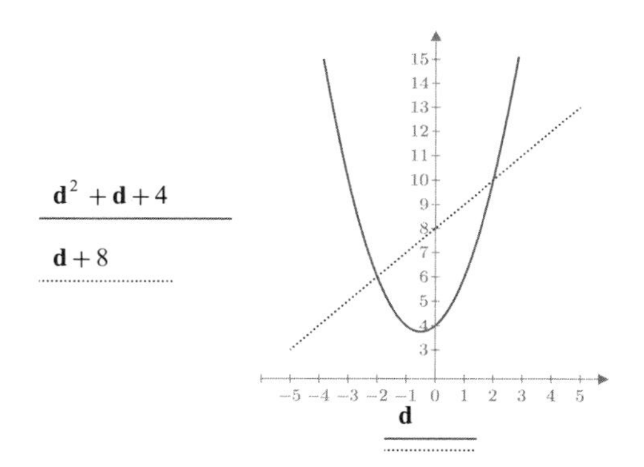

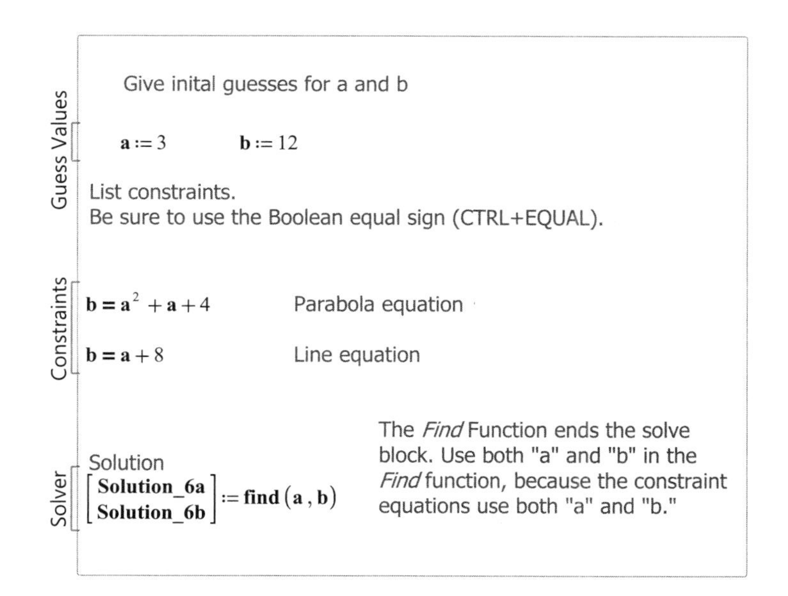

$Solution\_6a = 2.00$ $Solution\_6b = 10.00$

Check Result

$Solution\_6a^2 + Solution\_6a + 4 = 10.00$ **OK**

$Solution\_6a + 8 = 10.00$ **OK**

FIGURE 11.6 *(Continued on next page)*

Using a solve block to solve two equations

Have PTC Mathcad solve for the second solution. Change the initial guess and try again.

Guess Values

$\mathbf{a} := -3 \qquad \mathbf{b} := 6$

Constraints

$\mathbf{b} = \mathbf{a}^2 + \mathbf{a} + 4$

$\mathbf{b} = \mathbf{a} + 8$

Solver

$\begin{bmatrix} \mathbf{Solution\_6c} \\ \mathbf{Solution\_6d} \end{bmatrix} := \mathbf{find}(\mathbf{a}, \mathbf{b})$

$\mathbf{Solution\_6c} = -2.00 \qquad \mathbf{Solution\_6d} = 6.00$ PTC Mathcad selects the other solution.

Check Result

$\mathbf{Solution\_6c}^2 + \mathbf{Solution\_6c} + 4 = 6.00$ **OK**

$\mathbf{Solution\_6c} + 8 = 6.00$ **OK**

FIGURE 11.6

(continued)

Find the intersection point of a parabola and a line.

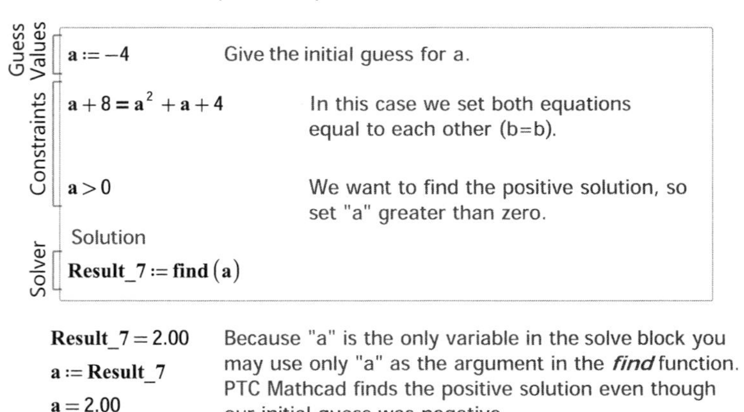

Guess Values
a := −4
Give the initial guess for a.
Constraints
a + 8 = a² + a + 4
In this case we set both equations equal to each other (b=b).
a > 0
We want to find the positive solution, so set "a" greater than zero.
Solution
Solver
Result_7 := find (a)
Result_7 = 2.00
a := Result_7
a = 2.00
Because "a" is the only variable in the solve block you may use only "a" as the argument in the find function. PTC Mathcad finds the positive solution even though our initial guess was negative.
Check Result

$\mathbf{a}^2 + \mathbf{a} + 4 = 10.00$ **OK**

$\mathbf{a} + 8 = 10.00$ **OK**

FIGURE 11.7

Forcing PTC Mathcad to solve for a positive root

$\mathbf{SampleFunction_{8a}(d) := d^2 + d + 4}$ Define two functions

$\mathbf{SampleFunction_{8b}(d) := d + 8}$ Remember that the argument "d" is only used to define the function. We can use "d" or any other argument when solving the functions. In this case we use the variable "f".

Solver Constraints Guess Values

$\mathbf{f := -100}$ Initial guess of solution

$\mathbf{SampleFunction_{8a}(\mathit{f}) = SampleFunction_{8b}(\mathit{f})}$

$\mathbf{Result\_8a := find(f)}$

$\mathbf{Result\_8a} = -2.00$

Check

$\mathbf{SampleFunction_{8a}(Result\_8a) - SampleFunction_{8b}(Result\_8a)} = 0.00$ **OK**

Now add an additional constraint so that "f" is greater than zero

Solver Constraints Guess Values

$\mathbf{f := -100}$

$\mathbf{SampleFunction_{8a}(\mathit{f}) = SampleFunction_{8b}(\mathit{f})}$

$f > 0$

$\mathbf{Result\_8b := find(f)}$

Check $\mathbf{Result\_8b} = 2.00$ PTC Mathcad finds the positive solution

$\mathbf{SampleFunction_{8a}(Result\_8b) - SampleFunction_{8b}(Result\_8b)} = 0.00$ **OK**

FIGURE 11.8

Solve block using functions instead of equations

lsolve Function

The ***lsolve*** function is used similarly to a solve block because you can solve for x equations and x unknowns. It is also similar to the ***polyroots*** function because the input arguments are vectors representing the coefficients of the equations. The use of this function is illustrated in Figure 11.9.

We can use a solve block to solve for four equations and four unknowns.

$w := 1 \quad x := 1 \quad y := 1 \quad z := 1$

Guess Values

$w := 1 \quad x := 1 \quad y := 1 \quad z := 1$

Constraints

$2.4 \cdot w + 6.1 \cdot x + 6.6 \cdot y + 2.1 \cdot z = 2.5$

$-7.5 \cdot w - 1.5 \cdot x - 2.3 \cdot y + 4.5 \cdot z = 14$

$4.3 \cdot w + 4.2 \cdot x + 1.8 \cdot y + 2.1 \cdot z = 10.2$

$5.8 \cdot w + 3.4 \cdot x + 7.8 \cdot y + 1.5 \cdot z = 0.1$

Solver

$\text{Solution}_A := \text{find}(w, x, y, z)$

$$\text{Solution}_A = \begin{bmatrix} 0.7686 \\ 0.3739 \\ -1.4480 \\ 3.7766 \end{bmatrix}$$

There is another way to solve for this condition. It is a function called ***lsolve***.

This function is similar to the function ***polyroots*** in that you use an array as input. To use this function, create a Matrix (M) of the coefficients of the four unknowns. Then create a vector of the constraints (V). The function takes the form ***lsolve***(M,V)

$$\text{Matrix} := \begin{bmatrix} 2.4 & 6.1 & 6.6 & 2.1 \\ -7.5 & -1.5 & -2.3 & 4.5 \\ 4.3 & 4.2 & 1.8 & 2.1 \\ 5.8 & 3.4 & 7.8 & 1.5 \end{bmatrix} \quad \text{Vector} := \begin{bmatrix} 2.5 \\ 14 \\ 10.2 \\ 0.1 \end{bmatrix}$$

With a solve block, you must give initial guess values. You do not need to give initial guesses with the ***lsolve*** function.

$\text{Solution}_B := \underline{\text{lsolve}}(\text{Matrix}, \text{Vector})$

$$\text{Solution}_B = \begin{bmatrix} 0.7686 \\ 0.3739 \\ -1.4480 \\ 3.7766 \end{bmatrix}$$

Check

$$\text{Matrix} \cdot \text{Solution}_B = \begin{bmatrix} 2.50 \\ 14.00 \\ 10.20 \\ 0.10 \end{bmatrix} \quad \text{Vector} = \begin{bmatrix} 2.50 \\ 14.00 \\ 10.20 \\ 0.10 \end{bmatrix}$$

OK

FIGURE 11.9

Using function *lsolve*

Solve Blocks using *Maximize* and *Minimize*

The ***maximize*** and ***minimize*** functions allow you to find the point at which the input function is at its maximum or minimum. These functions do not need to be used within a solve block, unless you want to provide constraints with the functions. When used within a solve block these functions solve for the maximum and minimum values that satisfy the constraints of the solve block. These functions take the form of ***maximize***(f, var1, var2, ...). The argument f is a previously defined function that uses the arguments var1, var2, etc. The solution returns a vector with the values of var1, var2, etc, which satisfy the constraints in the solve block. Note that the solution does not return the actual maximum or minimum values of the function; it returns the values of the variables that when used in the function will return the maximum or minimum solution.

To use the ***maximize*** function in a solve block, follow these steps:

1. Define a function using the variables you are solving for.
2. Give initial guess values for the variables you are solving for.
3. Type a list of constraint equations using Boolean operators.
4. Type `maximize(`, then type the name of the function (do not include a parenthesis after the name of the function), then type the variables you are solving for separated by commas.

The ***minimize*** function is used exactly the same way. See Figure 11.10. Note that these functions are very sensitive to boundary conditions and the initial guess value. In order to see this, try the equation shown in Figure 11.10 and vary the initial guess value and the boundary conditions. Additional examples of the ***maximize*** and ***minimize*** functions are found in the Engineering Examples.

If ***maximize*** or ***minimize*** do not return a result, and the error box states, "Could not find a maximum or minimum," try adjusting the initial guess values. A plot may be helpful to select the initial guess values.

TOL, CTOL, and Minerr

TOL and CTOL are built-in PTC Mathcad variable names. They can be set in the **Worksheet Settings** group on the **Calculation** tab (**Calculation>Worksheet Settings>TOL**). The values can also be redefined within the worksheet.

Define Function

$\textbf{Function}(\mathbf{x}) := 0.5 \cdot \mathbf{x}^3 - 8 \cdot \mathbf{x}^2 + 15 \cdot \mathbf{x} + 100$

Define Boundary Conditions

$\textbf{Bound}_1 := -5 \qquad \textbf{Bound}_2 := 15$

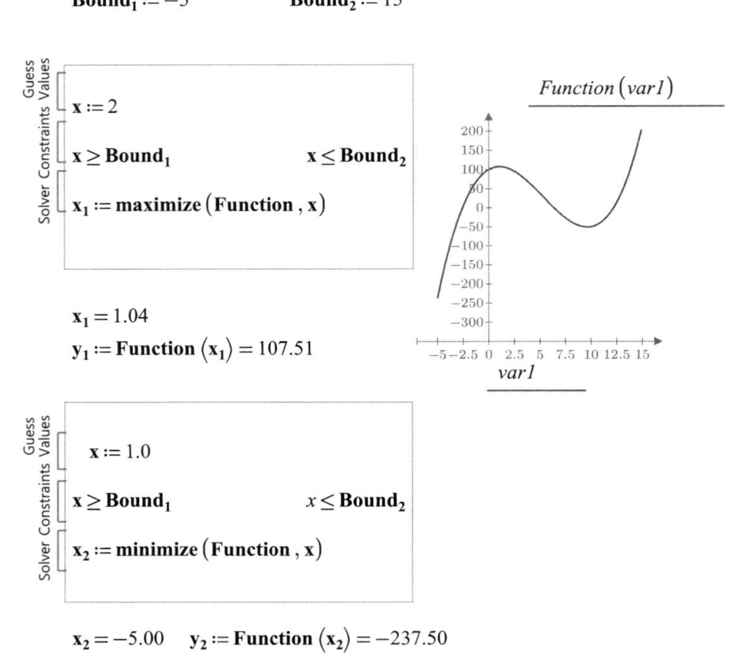

Note that the ***maximize*** and ***minimize*** functions are sensitive to the initial guess and to the boundary conditions. When the above initial guess value is moved to the right of the maximum, the minimum value of x is found at 9.63.

The functions may also be used without boundary conditions and outside of a solve block to find the local max and min. The solution is very dependant on the inital guess value. If the initial guess value is left of the local maximum, then ***minimize*** will not find a solution for the minimum value. Similarly, if the initial guess is to the right of the local minimum, then maximize will not find a solution for the maximum values. You must have boundaries to find solutions that are not at local maximum and minimum values.

$\textbf{maximize}(\textbf{Function}, \mathbf{x}) = 1.04 \qquad \textbf{minimize}(\textbf{Function}, \mathbf{x}) = ?$

$\textbf{minimize}(\textbf{Function}, \mathbf{x}) = ?$

Could not find a maximum or minimum.

FIGURE 11.10

Using *maximize* and *minimize*

When using solve blocks and the ***root*** function, PTC Mathcad iterates a solution. When the difference between the two most recent iterations is less than TOL (convergence tolerance), PTC Mathcad arrives at a solution. The default value of TOL is 0.001. If you are not satisfied with the solutions arrived at, you can redefine TOL to be a smaller number. This will increase the precision of the result, but it will also increase the calculation time, or may make it impossible for PTC Mathcad to arrive at a solution.

The CTOL (constraint tolerance) built-in variable tells PTC Mathcad how closely a constraint in a solve block must be met for a solution to be acceptable. The default value of CTOL is 0.001. Thus if a solve block constraint is $x > 3$, this constraint is satisfied if $x > 2.9990$.

The ***minerr*** function is used in place of the ***find*** function in a solve block. You may want to use ***minerr*** if PTC Mathcad cannot iterate a solution. The ***find*** function iterates a solution until the difference in the two most recent iterations is less than TOL. The ***minerr*** function uses the last iteration even if it falls outside of TOL. This function is useful when ***find*** fails to find a solution. If you use ***minerr***, it is important to check the solution to see if the solution falls within your acceptable limits.

For more information on these topics, refer to the PTC Mathcad Help. It has a good discussion of these topics in much greater depth.

Using Units

You may use units with ***root***, ***polyroots***, and solve blocks. These will be illustrated in the following Engineering Examples.

Engineering Examples

Engineering Example 11.1 summarizes five different ways of solving for the time it takes to travel a specific distance. Engineering Example 11.2 calculates the current flow in a multi-loop circuit. Engineering Example 11.3 calculates the flow of water in a pipe network. Engineering Example 11.4 demonstrates how PTC Mathcad can be used to solve chemistry equations. Engineering Example 11.5 uses a solve block to solve the same problem presented in Engineering Example 8.2. Engineering Example 11.6 uses the ***maximize*** function to maximize the volume of a box with a fixed surface area. Engineering Example 11.7 uses given cost and revenue functions to calculate how many products to produce in order to obtain the maximum and minimum profits.

Engineering Example 11.1: Object in Motion

The general equation for the distance traveled by an object is given by the function $\mathbf{Dist} = v_0 \cdot t + \frac{1}{2} \cdot a \cdot t^2$. Let's look at the different ways of solving for this equation for the time needed to travel a specific distance.

Given: $\mathbf{v_0} := 10 \frac{m}{s}$, $\mathbf{a} := 6 \frac{m}{s^2}$, and $\mathbf{Dist} := 5000\ m$

First, use symbolic evaluation

$$\mathbf{Time\_2a} := \mathbf{v_0} \cdot t + \frac{1}{2} \cdot \mathbf{a} \cdot t^2 - \mathbf{Dist} \xrightarrow{solve\ ,\ t} \begin{bmatrix} \dfrac{s^2 \cdot \left(\dfrac{5 \cdot \sqrt{601}}{s} + \dfrac{5}{s} \right)}{3} \\ -\dfrac{s^2 \cdot \left(\dfrac{5}{s} - \dfrac{5 \cdot \sqrt{601}}{s} \right)}{3} \end{bmatrix} \xrightarrow{float} \begin{bmatrix} -42.525502240437542846 \cdot s \\ 39.192168907104209513 \cdot s \end{bmatrix}$$

$$\mathbf{Time\_2a} = \begin{bmatrix} -42.53 \\ 39.19 \end{bmatrix} s$$

Same as above, but without the symbolic solution.

$$\mathbf{Time\_2b} := \mathbf{v_0} \cdot t + \frac{1}{2} \cdot \mathbf{a} \cdot t^2 - \mathbf{Dist} \xrightarrow[float]{solve\ ,\ t} \begin{bmatrix} -42.525502240437542846 \cdot s \\ 39.192168907104209513 \cdot s \end{bmatrix}$$

$$\mathbf{Time\_2b} = \begin{bmatrix} -42.53 \\ 39.19 \end{bmatrix} s$$

$\mathbf{Time\_2b}_1 = -42.53\ s$ $\mathbf{Time\_2b}_2 = 39.19\ s$

Second, create a function using symbolic evaluation

Create the function.

$$\mathbf{TimeFunction\_3}(\mathbf{v_0}, \mathbf{a}, \mathbf{Dist}) := v_0 \cdot t + \frac{1}{2} \cdot a \cdot t^2 - Dist \xrightarrow{solve,\, t} \begin{bmatrix} -\dfrac{2 \cdot \left(\dfrac{v_0}{2} + \dfrac{\sqrt{v_0^2 + 2 \cdot a \cdot Dist}}{2}\right)}{a} \\ -\dfrac{2 \cdot \left(\dfrac{v_0}{2} - \dfrac{\sqrt{v_0^2 + 2 \cdot a \cdot Dist}}{2}\right)}{a} \end{bmatrix}$$

$\mathbf{Time\_3a} := \mathbf{TimeFunction\_3}(\mathbf{v_0}, \mathbf{a}, \mathbf{Dist})$ Apply the arguments to the function.

$$\mathbf{Time\_3a} = \begin{bmatrix} -42.53 \\ 39.19 \end{bmatrix} s$$

This method allows you to use other input variables very easily.

$$\mathbf{Time\_3b} := \mathbf{TimeFunction\_3}\left(20 \cdot \frac{m}{s}, 20 \cdot \frac{m}{s^2}, 100000 \cdot m\right)$$

$$\mathbf{Time\_3b} = \begin{bmatrix} -101.00 \\ 99.00 \end{bmatrix} s$$

Third, use the function *root*

Solving for the positive root.

$$\mathbf{Time\_4}_1 := \mathbf{root}\left(\mathbf{v_0} \cdot \mathbf{t} + \frac{1}{2} \cdot \mathbf{a} \cdot \mathbf{t}^2 - \mathbf{Dist}, t, 0 \cdot s, 100 \cdot s\right) \qquad \mathbf{Time\_4}_1 = 39.19\ s$$

Solving for the negative root.

$$\mathbf{Time\_4}_2 := \mathbf{root}\left(\mathbf{v_0} \cdot \mathbf{t} + \frac{1}{2} \cdot \mathbf{a} \cdot \mathbf{t}^2 - \mathbf{Dist}, t, -100 \cdot sec, 0 \cdot sec\right) \quad \mathbf{Time\_4}_2 = -42.53\ s$$

$$\mathbf{Time\_4} = \begin{bmatrix} 39.19 \\ -42.53 \end{bmatrix} s$$

Fourth, use the function *polyroots*

$v_0 = 10.00\ \frac{m}{s}$ $\quad$ $Dist = 5000.00\ m$ $\quad$ $a = 6.00\ \frac{m}{s^2}$

$$Vector := \begin{bmatrix} -Dist \\ v_0 \\ \frac{1}{2} \cdot a \end{bmatrix} = \begin{bmatrix} -5000.00\ m \\ 10.00\ \frac{m}{s} \\ 3.00\ \frac{m}{s^2} \end{bmatrix}$$

$$Time\_5 := \underline{polyroots}(Vector) = \begin{bmatrix} -42.53 \\ 39.19 \end{bmatrix} s$$

$Time\_5_1 = -42.53\ s$ $\quad$ $Time\_5_2 = 39.19\ s$

Fifth, use a solve block

Guess Values

$t := 100\ s$ Set guess value to find a positive solution

Constraints

$v_0 \cdot t + \frac{1}{2} \cdot a \cdot t^2 - Dist = 0$

Solver

$Time\_6_1 := find(t) = 39.19\ s$

$Time\_6_1 = 39.19\ s$

Guess Values

$t := -30\ s$ Initial guess value to find a negative solution

Constraints

$v_0 \cdot t + \frac{1}{2} \cdot a \cdot t^2 - Dist = 0$

Solver

$Time\_6_2 := find(t) = -42.53\ s$

$Time\_6_2 = -42.53\ s$

Guess Values

$t := -1\ s$ An initial negative guess for time would normally find a negative solution, but adding an additional constraint finds the positive solution.

Constraints

$v_0 \cdot t + \frac{1}{2} \cdot a \cdot t^2 - Dist = 0$ Set constraint to find the positive solution

$t > 0$

Solver

$Time\_6_3 := find(t)$

$Time\_6_3 = 39.19\ s$

Engineering Example 11.2: Electrical Network

Kirchoff's current law states that the algebraic sum of voltage drops around any closed path within a circuit equals the sum of the voltage sources. SV=S(I*R). Using Kirchoff's current law, calculate the current flowing in this network.

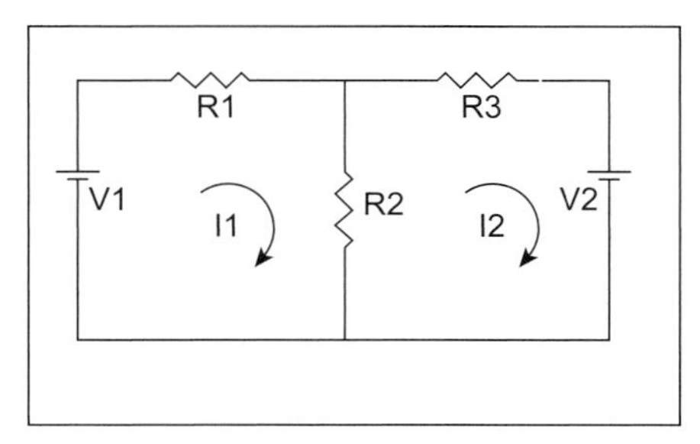

Resistance	Voltage	Initial Guess for Currents
$\mathbf{R_1} := 4 \cdot \Omega$	$\mathbf{V_1} := 75 \cdot \boldsymbol{volt}$	$\mathbf{I_1} := 15 \cdot \boldsymbol{A}$
$\mathbf{R_2} := 15 \cdot \Omega$	$\mathbf{V_2} := 40 \cdot \boldsymbol{volt}$	$\mathbf{I_2} := 15 \cdot \boldsymbol{A}$
$\mathbf{R_3} := 2 \cdot \Omega$		

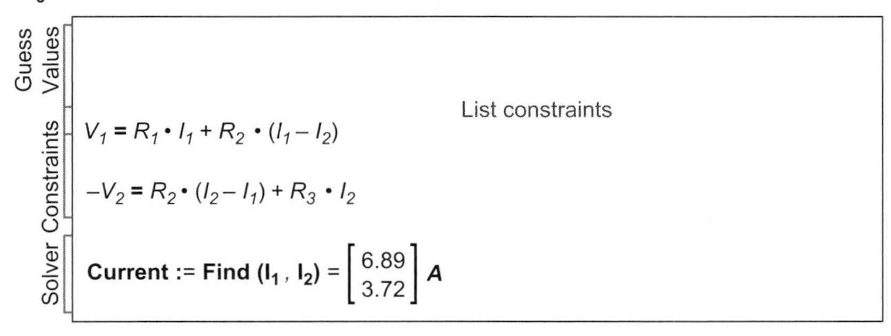

$$V_1 = R_1 \cdot I_1 + R_2 \cdot (I_1 - I_2)$$

$$-V_2 = R_2 \cdot (I_2 - I_1) + R_3 \cdot I_2$$

$$\mathbf{Current} := \mathbf{Find}\,(\mathbf{I_1}, \mathbf{I_2}) = \begin{bmatrix} 6.89 \\ 3.72 \end{bmatrix} \boldsymbol{A}$$

$\mathbf{I_1} := \mathbf{Current_1}$ $\quad$ $\mathbf{I_2} := \mathbf{Current_2}$ $\quad$ $\mathbf{I_1} = 6.89\ \boldsymbol{A}$ $\quad$ $\mathbf{I_2} = 3.72\ \boldsymbol{A}$

Check

$\mathbf{R_1} \cdot \mathbf{I_1} + \mathbf{R_2} \cdot (I_1 - \mathbf{I_2}) - \mathbf{V_1} = 0.0000000\ \boldsymbol{V}$ $\quad$ **OK**

$\mathbf{R_2} \cdot (\mathbf{I_2} - \mathbf{I_1}) + \mathbf{R_3} \cdot I_2 + \mathbf{V_2} = 0.0000000\ \boldsymbol{V}$ $\quad$ **OK**

Engineering Example 11.3: Pipe Network

In a pipe network, two conditions must be satisfied: 1) The algebraic sum of the pressure drops around any closed loop must be zero (the pressure at any point is the same no matter how you get there), and 2) The flow entering a junction must equal the flow leaving it.

Use the Darcy-Weisbach equation for head loss $\mathbf{h_L} = f \cdot \frac{L}{D} \cdot \frac{V^2}{2 \cdot g}$.

The factor f is a function of the relative roughness of the pipe. For this example assume f=0.02.

Define new unit and assign Unit label.

$cfs := \frac{ft^3}{s}$ $\quad$ $\mathbf{f} := 0.02$ $\quad$ $\mathbf{TOL} = 0.001000000000000$

For this example, TOL=0.001 is adequate.

Calculate the flow in each pipe given that:

Flow into A is 1.2 cfs
Flow out of C is 0.8 cfs
Flow out of E is 0.4 cfs

Pipe lengths and diameters

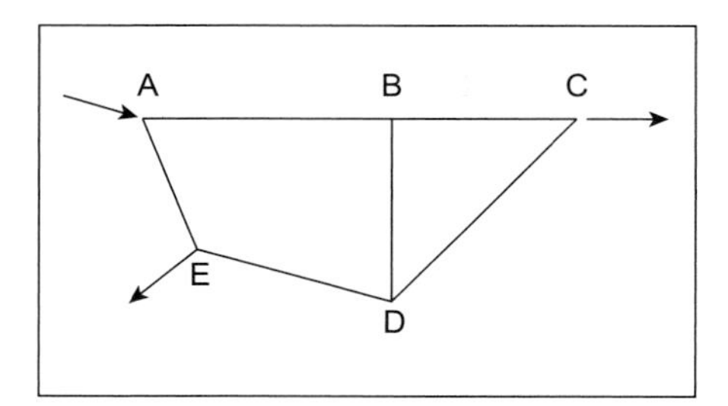

		Initial Guess
$\mathbf{L_{AB}} := 3000 \cdot ft$	$\mathbf{D_{AB}} := 4 \cdot in$	$\mathbf{Q_{AB}} := .6 \cdot cfs$
$\mathbf{L_{BC}} := 4000 \cdot ft$	$\mathbf{D_{BC}} := 5 \cdot in$	$\mathbf{Q_{BC}} := .6 \cdot cfs$
$\mathbf{L_{AE}} := 2000 \cdot ft$	$\mathbf{D_{AE}} := 3 \cdot in$	$\mathbf{Q_{AE}} := .6 \cdot cfs$
$\mathbf{L_{ED}} := 3000 \cdot ft$	$\mathbf{D_{ED}} := 8 \cdot in$	$\mathbf{Q_{ED}} := .6 \cdot cfs$
$\mathbf{L_{DC}} := 5000 \cdot ft$	$\mathbf{D_{DC}} := 6 \cdot in$	$\mathbf{Q_{DC}} := .6 \cdot cfs$
$\mathbf{L_{BD}} := 3000 \cdot ft$	$\mathbf{D_{BD}} := 7 \cdot in$	$\mathbf{Q_{BD}} := .6 \cdot cfs$

Define a user-defined function for head loss based on the variables L, D, and Q. Velocity is a function of Flow and Area where V=Q/A.

$$\mathbf{h_L}(\mathbf{L},\mathbf{D},\mathbf{Q}) := \mathbf{f}\cdot\frac{\mathbf{L}}{\mathbf{D}}\cdot\frac{1}{2\cdot g}\cdot\left(\frac{\mathbf{Q}}{\frac{\pi\cdot\mathbf{D}^2}{4}}\right)^2$$

Guess Values

Write equation for head loss at Point C.

$$\left(h_L\left(L_{AB},D_{AB},Q_{AB}\right)+h_L\left(L_{BC},D_{BC},Q_{BC}\right)\right)-\left(h_L\left(L_{AE},D_{AE},Q_{AE}\right)+h_L\left(L_{ED},D_{ED},Q_{ED}\right)+h_L\left(L_{DC},D_{DC},Q_{DC}\right)\right)=0$$

Write equation for head loss at Point D.

$$\left(h_L\left(L_{AB},D_{AB},Q_{AB}\right)+h_L\left(L_{BD},D_{BD},Q_{BD}\right)\right)-\left(h_L\left(L_{AE},D_{AE},Q_{AE}\right)+h_L\left(L_{ED},D_{ED},Q_{ED}\right)\right)=0$$

Constraints

Write equation for head loss at Point B.

$$\left(h_L\left(L_{BC},D_{BC},Q_{BC}\right)\right)-\left(h_L\left(L_{BD},D_{BD},Q_{BD}\right)+h_L\left(L_{DC},D_{DC},Q_{DC}\right)\right)=0$$

Write relationship between various flows into each of the points.

$$1.2\cdot cfs = Q_{AB}+Q_{AE} \qquad Q_{AB}=Q_{BD}+Q_{BC}$$

$$Q_{AE}=0.4\cdot cfs+Q_{ED} \qquad Q_{ED}+Q_{BD}=Q_{DC}$$

$$Q_{BC}+Q_{DC}=0.8\cdot cfs$$

Solver

$$\begin{bmatrix}\mathbf{Q_{AB}}\\\mathbf{Q_{BC}}\\\mathbf{Q_{AE}}\\\mathbf{Q_{ED}}\\\mathbf{Q_{DC}}\\\mathbf{Q_{BD}}\end{bmatrix} := \mathbf{Find}\left(\mathbf{Q_{AB}},\mathbf{Q_{BC}},\mathbf{Q_{AE}},\mathbf{Q_{ED}},\mathbf{Q_{DC}},\mathbf{Q_{BD}}\right)$$

$\mathbf{Q_{AB}} = 0.749210\ cfs$

$\mathbf{Q_{BC}} = 0.351147\ cfs$

$\mathbf{Q_{AE}} = 0.450790\ cfs$

$\mathbf{Q_{ED}} = 0.050790\ cfs$

$\mathbf{Q_{DC}} = 0.448853\ cfs$

$\mathbf{Q_{BD}} = 0.398063\ cfs$

Check results

$$(\mathbf{h_L}(\mathbf{L_{AB}}, \mathbf{D_{AB}}, \mathbf{Q_{AB}}) + \mathbf{h_L}(\mathbf{L_{BC}}, \mathbf{D_{BC}}, \mathbf{Q_{BC}})) - (\mathbf{h_L}(\mathbf{L_{AE}}, \mathbf{D_{AE}}, \mathbf{Q_{AE}}) + \mathbf{h_L}(\mathbf{L_{ED}}, \mathbf{D_{ED}}, \mathbf{Q_{ED}}) + \mathbf{h_L}(\mathbf{L_{DC}}, \mathbf{D_{DC}}, \mathbf{Q_{DC}})) = 0.00\ ft$$

$$(\mathbf{h_L}(\mathbf{L_{AB}}, \mathbf{D_{AB}}, \mathbf{Q_{AB}}) + \mathbf{h_L}(\mathbf{L_{BD}}, \mathbf{D_{BD}}, \mathbf{Q_{BD}})) - (\mathbf{h_L}(\mathbf{L_{AE}}, \mathbf{D_{AE}}, \mathbf{Q_{AE}}) + \mathbf{h_L}(\mathbf{L_{ED}}, \mathbf{D_{ED}}, \mathbf{Q_{ED}})) = 0.00\ ft$$

$$(\mathbf{h_L}(\mathbf{L_{BC}}, \mathbf{D_{BC}}, \mathbf{Q_{BC}})) - (\mathbf{h_L}(\mathbf{L_{BD}}, \mathbf{D_{BD}}, \mathbf{Q_{BD}}) + \mathbf{h_L}(\mathbf{L_{DC}}, \mathbf{D_{DC}}, \mathbf{Q_{DC}})) = (1.46 \cdot 10^{-14})\ ft$$

$\mathbf{Q_{AB}} + \mathbf{Q_{AE}} = 1.200000\ cfs$ **Equals 1.2 cfs OK**

$\mathbf{Q_{AE}} - \mathbf{Q_{ED}} = 0.400000\ cfs$ **Equals 0.4 cfs OK**

$\mathbf{Q_{BC}} + \mathbf{Q_{DC}} = 0.800000\ cfs$ **Equals 0.8 cfs OK**

Engineering Example 11.4: Chemistry

You can get solutions to chemistry problems by writing out the needed equations and solving them using a solve block. See the example below.

The problem below involves a "closed system", that is, no exchange between the system and the outside, for example a closed beaker.

Find the concentration of carbonate CO_3 in equilibrium with a solution (closed to the atmosphere) containing 10^{-3} M soluble calcium $[Ca^{+2}]$ and solid calcite $CaCO_3$.

Problem interpretation: The calcite has dissolved up to the limit imposed by the solubility product. This is the point at which the $[Ca^{+2}]$ is measured and found to be 10^{-3} M. The only other cations in the water are H^+ from the dissociation of water. The other anions in the water are hydroxide $[OH^-]$ from water and the carbonate species created by the dissolution of $CaCO_3$. The dissolved carbonate molecules have reacted to form carbonate, bicarbonate and some carbonic acid according to their respective equilibrium relationships.

First, define values for the constants to be used:

Soluble calcium in water $\mathbf{Ca} := 10^{-3} \cdot \frac{mole}{liter}$

First dissociation constant for carbonic acid: $\mathbf{K_{a1}} := 10^{-6.3} \cdot \frac{mole}{liter}$

Second dissociation constant for carbonic acid: $\mathbf{K_{a2}} := 10^{-10.35} \cdot \frac{mole}{liter}$

Solubility product for calcite: $\mathbf{K_{so}} := 10^{-8.34} \cdot \left(\frac{mole}{liter}\right)^2$

Ion product of water: $\mathbf{K_w} := 10^{-14} \cdot \left(\frac{mole}{liter}\right)^2$

Unknowns: H^+, OH^-, H_2CO_3, HCO_3^-, $CO_3^=$. Therefore we need five equations.

Provide initial guesses for each unknown in the system of equations:

$H := 10^{-7.0} \cdot \frac{mole}{liter}$ $H2CO_3 := 10^{-9} \cdot \frac{mole}{liter}$ $CO_3 := 10^{-2.9} \cdot \frac{mole}{liter}$

$OH := \frac{K_w}{H}$ $HCO_3 := 10^{-3.01} \cdot \frac{mole}{liter}$

Note: It sometimes helps to select initial guesses for H^+ and OH^- so that they satisfy the ion product of water:

Guess Values

Constraints

$H \cdot OH = K_w$ Ionization of water

$H \cdot HCO_3 = K_{a1} \cdot H2CO_3$ First dissociation for carbonic acid

$H \cdot CO_3 = K_{a2} \cdot HCO_3$ Second dissociation for carbonic acid

$Ca + H = HCO_3 + 2 \cdot CO_3 + OH$ Charge balance on the system

$Ca = H2CO_3 + HCO_3 + CO_3$ Mass balance on carbon, all carbon species originated from calcium carbonate dissolution.

NOTE : Either the charge balance or mass balance on carbonate ions can be used as one of the equations to be solved. Make the other region disabled.

$Ca \cdot CO_3 = K_{so}$ solubility product

Solver

$$\begin{bmatrix} H2CO_{3equil} \\ HCO_{3equil} \\ CO_{3equil} \\ H_{equil} \\ OH_{equil} \end{bmatrix} := Find(H2CO_3, HCO_3, CO_3, H, OH)$$

$$\mathbf{H2CO_{3equil}} = \left(1.91 \cdot 10^{-5}\right) \frac{mole}{liter}$$

$$\mathbf{HCO_{3equil}} = \left(9.90 \cdot 10^{-4}\right) \frac{mole}{liter}$$

$$\mathbf{CO_{3equil}} = \left(4.57 \cdot 10^{-6}\right) \frac{mole}{liter}$$

$$\mathbf{H_{equil}} = \left(9.67 \cdot 10^{-9}\right) \frac{mole}{liter}$$

$$\mathbf{OH_{equil}} = \left(1.03 \cdot 10^{-6}\right) \frac{mole}{liter}$$

$$\mathbf{pH} := -\log\left(\frac{\mathbf{H_{equil}}}{\frac{mole}{liter}}\right) = 8.01$$

$$\mathbf{pOH} := -\log\left(\frac{\mathbf{OH_{equil}}}{\frac{mole}{liter}}\right) = 5.99$$

There are several ways to check the validity of our solution, for example the pH + pOH must equal 14.

$$\mathbf{pH} + \mathbf{pOH} = 14.00$$

Now check to see if the solution is correct by substituting the values back into the original equations.

$$\frac{\mathbf{H_{equil}} \cdot \mathbf{OH_{equil}}}{\mathbf{K_w}} = 1.00 \qquad \frac{\mathbf{Ca}}{\mathbf{H2CO_{3equil}} + \mathbf{HCO_{3equil}} + \mathbf{CO_{3equil}}} = 0.99$$

$$\frac{\mathbf{H_{equil}} \cdot \mathbf{HCO_{3equil}}}{\mathbf{K_{a1}} \cdot \mathbf{H2CO_{3equil}}} = 1 \qquad \frac{\mathbf{Ca} \cdot \mathbf{CO_{3equil}}}{\mathbf{K_{so}}} = 1.00 \qquad \frac{\mathbf{H_{equil}} \cdot \mathbf{CO_{3equil}}}{\mathbf{K_{a2}} \cdot \mathbf{HCO_{3equil}}} = 1.00$$

This example was provided by Dr. Dixie Griffin, a retired professor of Civil Engineering at Louisian Tech University. He adds the following note: "The use of PTC Mathcad to solve chemistry problems has been one of my most enlightening discoveries, as well as a time savings. An interesting aside here is that such problems often give answers varying over several to many orders of magnitude. Doing such problems by hand would be practically impossible. The graphical solutions give approximate solutions but require a knowledge of chemistry not possessed by the average civil engineering undergraduate."

Engineering Example 11.5: Determining the Flow Properties of a Circular Pipe Flowing Partially Full

In Engineering Example 8.2, we showed how to use plotting as a means to solve for the pipe diameter needed to carry 300 cfs flowing at a depth of 60% of the diameter.

In this example, we use a solve block to solve the same problem. Refer to Engineering Example 8.2 for a description of the problem and associated equations.

Known values

Pipe slope $S := 0.001 \cdot \frac{ft}{ft}$

Manning's coefficient $n := 0.013$

Flow $Q := 300 \cdot \frac{ft^3}{s}$

Guess Values

$$D := 15\,ft \quad A_f := 42\,ft^2 \quad \alpha := 1 \cdot \pi \quad d := 0.6\,D \quad R := \frac{D}{2}$$

Note: It sometimes works better to put the governing equations as a ratio equal to one.

Constraints

$$\frac{Q}{\frac{1.49\,\frac{ft^{\frac{1}{3}}}{s}}{n} \cdot A_f \cdot \left(\frac{A_f}{R \cdot \alpha}\right)^{\frac{2}{3}} \cdot S^{\frac{1}{2}}} = 1$$

$$\frac{A_f}{\left(\frac{R^2}{2} \cdot (\alpha - sin(\alpha))\right)} = 1 \qquad \frac{d}{D} = 0.6$$

$$R = \frac{D}{2} \qquad \frac{cos\left(\frac{\alpha}{2}\right)}{\frac{R-d}{R}} = 1$$

Solver

$$\begin{bmatrix} A_{Solve} \\ d_{Solve} \\ \alpha_{Solve} \\ D_{Solve} \\ R_{Solve} \end{bmatrix} := \text{Find}(A_f, d, \alpha, D, R) = \begin{bmatrix} 43.62 \\ 5.65 \\ 203.07 \\ 9.42 \\ 4.71 \end{bmatrix} \begin{bmatrix} ft^2 \\ ft \\ deg \\ ft \\ ft \end{bmatrix}$$

Pipe diameter $D_{Solve} = 9.42\,ft$ Depth of flow $d_{Solve} = 5.65\,ft$

Area of flow $A_{Solve} = 43.62\,ft^2$ Relative depth $\frac{d_{Solve}}{D_{Solve}} = 0.60$

Angle $\alpha_{Solve} = 203.07\,deg$ Radius $R_{Solve} = 4.71\,ft$

Check the governing equations

$$\frac{Q}{\frac{1.49\cdot\frac{ft^{\frac{1}{3}}}{s}}{n}\cdot A_{Solve}\cdot\left(\frac{A_{Solve}}{\frac{D_{Solve}}{2}\cdot\alpha_{Solve}}\right)^{\frac{2}{3}}\cdot S^{\frac{1}{2}}}=1.00 \qquad \frac{\cos\left(\frac{\alpha_{Solve}}{2}\right)}{\frac{R_{Solve}-d_{Solve}}{R_{Solve}}}=1.00$$

$$\frac{A_{Solve}}{\left(\frac{R_{Solve}^{2}}{2}\cdot\left(\alpha_{Solve}-\sin\left(\alpha_{Solve}\right)\right)\right)}=1.00 \qquad \frac{R_{Solve}}{\frac{D_{Solve}}{2}}=1.00 \qquad \frac{d_{Solve}}{D_{Solve}}=0.60$$

The solution is the same as determined in Engineering Example 8.2, D=9.42 ft.

Engineering Example 11.6: Box Volume

Maximize the volume of a box with a total surface area of 200 cm^2 . The length of the top and bottom is to be twice the width.

Define Function

$$\mathbf{Volume}(w,l,h):=w\cdot l\cdot h \qquad \mathbf{Area}:=200\cdot cm^{2}$$

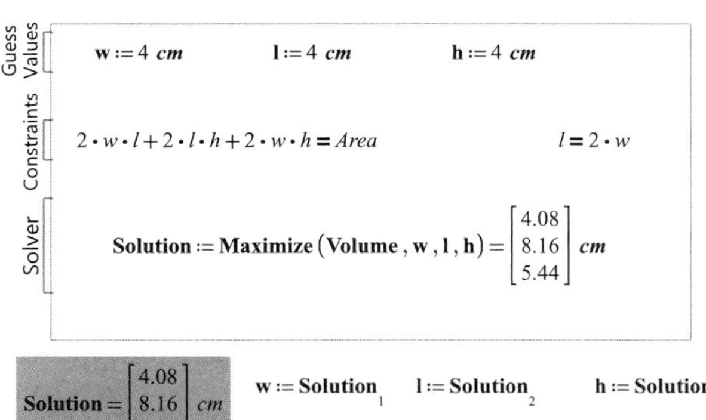

Guess Values
w := 4 cm
l := 4 cm
h := 4 cm
Constraints
2 · w · l + 2 · l · h + 2 · w · h = Area
l = 2 · w
Solver
Solution := Maximize (Volume , w , l , h) = [4.08 8.16 5.44] cm

$$\mathbf{Solution}=\begin{bmatrix}4.08\\8.16\\5.44\end{bmatrix}cm \qquad w:=\mathbf{Solution}_1 \qquad l:=\mathbf{Solution}_2 \qquad h:=\mathbf{Solution}_3$$

$$w=4.08\ cm \qquad l=8.16\ cm \qquad h=5.44\ cm$$

$$\mathbf{Volume}(w,l,h)=181.44\ cm^{3}$$

Verify Constraints

$$2\cdot w\cdot l+2\cdot l\cdot h+2\cdot w\cdot h=200.00\ cm^{2}$$

$$2\cdot w=8.16\ cm \qquad l=8.16\ cm$$

Engineering Example 11.7: Maximize Profit

The profit of a certain manufactured product is determined by subtracting the cost of the product from the revenue of the product. The revenue function for a certain product was determined to be 10*n where n is the number of products produced in thousands. The cost function for the same product was determined to be $n^3 - 7 \cdot n^2 + 18 \cdot n$.

Determine the number of products that should be produced to maximize the profit. Plot the revenue, cost, and profit to verify the solutions.

Define Functions

$\mathbf{Rev}(\mathbf{n}) := 10 \cdot \mathbf{n}$ $\mathbf{Cost}(\mathbf{n}) := \mathbf{n}^3 - 7 \cdot \mathbf{n}^2 + 18 \cdot \mathbf{n}$ $\mathbf{Profit}(\mathbf{n}) := \mathbf{Rev}(\mathbf{n}) - \mathbf{Cost}(\mathbf{n})$

Initial Guess $\mathbf{n} := 3$

In this example there are no constraints, so a solve block is not required.

$\mathbf{n_{max}} := \mathbf{maximize}(\mathbf{Profit}, \mathbf{n})$

$\mathbf{n_{max}} = 4.00$

$\mathbf{Profit}(\mathbf{n_{max}}) = 16.00$

$\mathbf{n_{min}} := \mathbf{minimize}(\mathbf{Profit}, \mathbf{n})$

$\mathbf{n_{min}} = 0.67$

$\mathbf{Profit}(\mathbf{n_{min}}) = -2.52$

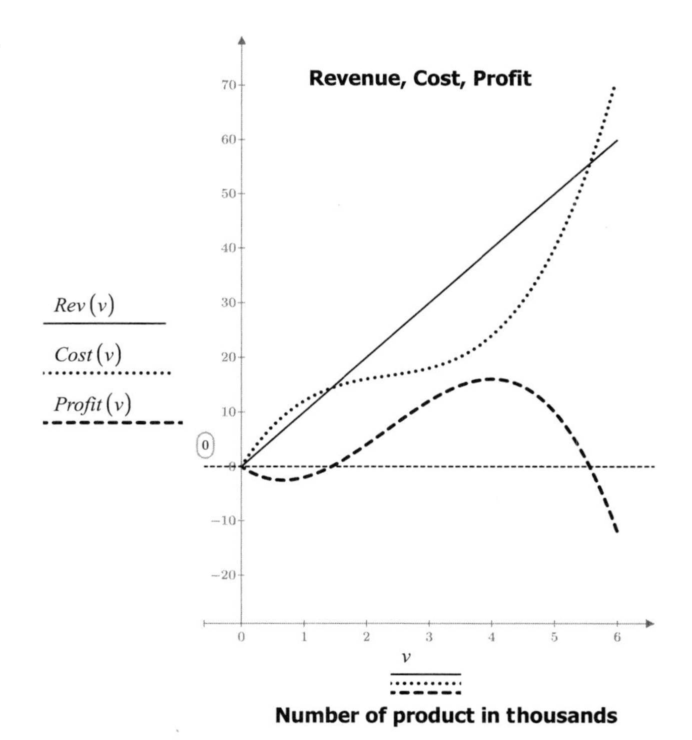

If we set restraints for n between 0 and 6, then the minimum will move to 6.

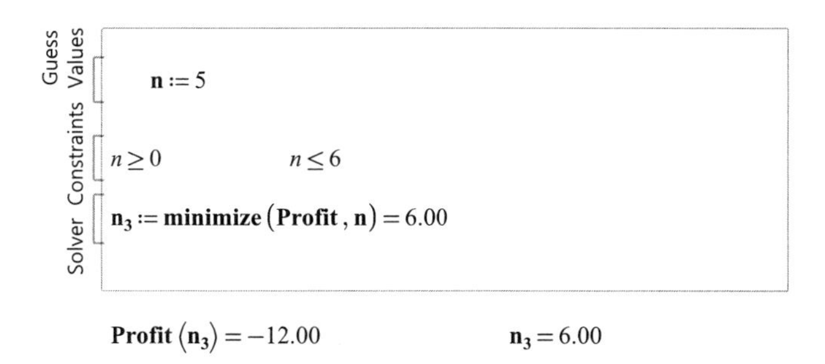

$\mathbf{Profit}(\mathbf{n_3}) = -12.00$ $\qquad$ $\mathbf{n_3} = 6.00$

Summary

The PTC Mathcad solve features are essential power tools to have in your toolbox. They should be kept sharp and used often.

In Chapter 11 we:

- Introduced the ***root*** function and showed how it can be used to solve for a single variable with a single function.
- Showed how the ***root*** function can be used with an initial guess prior to using ***root***, or how two guesses can be included as arguments in the ***root*** function.
- Encouraged the use of a plot to help define the initial guess values for the ***root*** function.
- Introduced the ***polyroots*** function and showed how it can be used to solve for all the roots of a polynomial equation.
- Introduced solve blocks and showed how they can be used to find solutions to multiple unknowns using multiple equations.
- Emphasized the rules of using ***find, maximize,*** and ***minimize***.
- Discussed the built-in variables TOL and CTOL.
- Told how the ***minerr*** function can be used in place of the ***find*** function if PTC Mathcad is not finding a solution.
- Provided several engineering examples to illustrate the use of the solve features.

Practice

The PTC Mathcad Prime 3.0 figures and examples used in this book are available for download from the book's website. The reader is encouraged to download the files and use them to practice the concepts learned. Additional examples and problems are also provided. To access this content go to http://store.elsevier.com/9780124104105 *and click on the Resources tab, and then click on the link for the Online Companion Materials.*

1. Create ten polynomial equations. Use at least 2nd- and 3rd-order equations. Plot each equation. Use both the bracketed ***root*** function and unbracketed ***root*** function to solve for each equation.
2. Use the ***polyroots*** function to solve for each equation used in exercise 1.
3. Create three 5th-order equations, and use the ***polyroots*** function to solve for all roots. Use the symbolics keyword "coeffs" to create the polynomial vector.
4. From your field of study, create four problems where you can use a solve block to solve for at least two unknowns with at least two constraints.

CHAPTER 12

Advanced Programming

PTC Mathcad programs can be much more powerful than the logic programs discussed in Chapter 9. This chapter expands the topic of programming.

Chapter 12 will:

- Use local variables as counters.
- Introduce the topic of looping.
- Discuss the use of the programming commands: ***for***, ***while***, ***break***, ***continue***, ***return***, and ***try-on-error***. Several examples will be given for the use of these features.

Local Definition

We discussed local variables in Chapter 9 as a way to create a variable that was local to the program in order to simplify it. We will now discuss the use of a local variable as a counter to keep track of how many times certain things are done within a program. Local variables can also be used to count how many true statements are in a program. Figure 12.1 illustrates this concept.

Essential PTC® Mathcad Prime® 3.0.

Local variables can be used as counters.

$$\text{Counter}(x) := \left\| \begin{array}{l} m \leftarrow 0 \\ \text{if } x > 4 \\ \quad \| \; m \leftarrow m + 1 \\ \text{if } x > 3 \\ \quad \| \; m \leftarrow m + 1 \\ \text{if } x > 2 \\ \quad \| \; m \leftarrow m + 1 \\ \text{if } x > 1 \\ \quad \| \; m \leftarrow m + 1 \\ \text{if } x > 0 \\ \quad \| \; m \leftarrow m + 1 \\ m \end{array} \right.$$

Define a local variable "m" equal to zero.

This program adds 1 to the local variable "m" for every true statement. It does not loop; it goes down only one time.

$\text{Counter}(-1) = 0$

$\text{Counter}(1) = 1$

$\text{Counter}(3) = 3$

$\text{Counter}(5) = 5$

Check the program for different arguments.

In order to check a program for several different arguments, you can use a range variable or a vector. The example below uses both a range variable and a vector to provide different arguments to the function "Counter."

Using a Range Variable

$i := -1, 0..7$

$$i = \begin{bmatrix} -1 \\ 0 \\ 1 \\ 2 \\ 3 \\ 4 \\ 5 \\ 6 \\ 7 \end{bmatrix} \qquad \text{Counter}(i) = \begin{bmatrix} 0 \\ 0 \\ 1 \\ 2 \\ 3 \\ 4 \\ 5 \\ 5 \\ 5 \end{bmatrix}$$

Using a Vector

$$b := \begin{bmatrix} -1 \\ 0 \\ 1 \\ 2 \\ 3 \\ 4 \\ 5 \\ 6 \\ 7 \end{bmatrix} \qquad \overrightarrow{\text{Counter}(b)} = \begin{bmatrix} 0 \\ 0 \\ 1 \\ 2 \\ 3 \\ 4 \\ 5 \\ 5 \\ 5 \end{bmatrix}$$

The ***vectorize*** operator (CTRL+SHIFT+^) must be used in order to use the vector with this program.

FIGURE 12.1

Local variables as counters

Looping

Looping allows program steps to be repeated a certain number of times, or until a certain criteria is met. This book will introduce the concept of looping, but an in-depth study of the topic is beyond its scope. The PTC Mathcad Help and Tutorial provide excellent coverage on this subject.

There are two types of loop structures. The first allows a program to execute a specific number of times. This is called a ***for*** loop. The second type of loop structure will execute until a specific condition is met. This is called a ***while*** loop.

For Loops

You insert the ***for*** loop into a program by clicking on the **for** button in the **Programming** controls on the **Math** tab. Do not type the word "for." The keyboard shortcut is **CTRL + SHIFT + "**. A ***for*** loop takes the following form:

```
for x ∈ y
    z
```

PTC Mathcad evaluates z for each value of x over the range y. The placeholders have these meanings:

- x is a local variable within the program defined by the ***for*** symbol. It initially takes the first value of y. The value of x changes for each step in the value of y.
- y is referred to as an iteration variable. It is usually a range variable, but it can be a vector, or a series of scalars or arrays separated by commas. Each time the program loops, the next value of y is used and this value is assigned to the variable x.
- z is any valid PTC Mathcad expression or sequence of expressions.

The number of times a ***for*** loop executes is controlled by y.

Let's first look at two examples using range variables. In the examples, think of the $\in$ symbol as a local variable definition. The local variable "i" is a range variable, and the value of variable "i" increments for each loop. There are two examples in Figure 12.2.

Figure 12.3 shows a ***for*** loop using a list as the iteration variable. Notice that the list of numbers is not consecutive, and the numbers do not increment by the same value every time. When PTC Mathcad loops, it moves to the next value in the list and assigns that value to the local variable "i." Figure 12.4 is the same as Figure 12.3 except that the iteration variable is a previously defined vector containing the same numbers as the list in Figure 12.3.

Figure 12.5 uses two local variables. It uses the second variable "b" as a means to sum all the different values of "a."

Chapter 6 showed a method of converting a range variable to a vector by using the ***evaluation*** operator on the same line as the ***definition*** operator. Figure 12.6 shows how to use a ***for*** loop to convert a range variable to a vector. Figure 12.7 shows how to create a vector in a similar way to creating a range variable.

$$j := 1, 2..5$$

$$\text{ForLoop}_1(x) := \left\| \begin{array}{l} a \leftarrow 0 \\ \text{for } i \in 1, 2..x \\ \quad \| \; a \leftarrow a + 1 \end{array} \right.$$

Define a local variable "a" equal to zero.

"i" is a local range variable that takes on the values 1,2,3... until x. For this example, i is not used anywhere. It only controls when the loop stops.

$$j = \begin{bmatrix} 1 \\ 2 \\ 3 \\ 4 \\ 5 \end{bmatrix} \qquad \text{ForLoop}_1(j) = \begin{bmatrix} 1 \\ 2 \\ 3 \\ 4 \\ 5 \end{bmatrix}$$

For the first loop "a" takes the value 0+1=1.
For the second loop "i" increments to 2, and "a" becomes 1+1=2.
For the third loop "i" increments to 3, and "a" becomes 2+1=3.
The loop continues until "i" reaches the value of x.

For this next example, increment "a" by the value of i.

$$\text{ForLoop}_2(x) := \left\| \begin{array}{l} a \leftarrow 0 \\ \text{for } i \in 1, 2..x \\ \quad \| \; a \leftarrow a + i \end{array} \right.$$

Define "a" as a local variable.

"i" is a local range variable that takes on the values 1,2,3... until x.

$$j = \begin{bmatrix} 1 \\ 2 \\ 3 \\ 4 \\ 5 \end{bmatrix} \qquad \text{ForLoop}_2(j) = \begin{bmatrix} 1 \\ 3 \\ 6 \\ 10 \\ 15 \end{bmatrix}$$

For the first loop, "i" is 1, and "a" takes the value 0+1=1.
For the second loop, "i" is 2 and "a" becomes 1+2=3.
For the third loop, "i" is 3, and "a" becomes 3+3=6.
For the fourth loop, "i" is 4, and "a" becomes 6+4=10.
For the fifth loop, "i" is 5, and "a" becomes 10+5=15.

FIGURE 12.2

For loop using range variables

$$\text{ForLoop}_3(x) := \left\| \begin{array}{l} a \leftarrow 0 \\ \text{for } i \in [1\ \ 2\ \ 5\ \ 3] \\ \quad \| \ a \leftarrow x \cdot (a + i) \\ a \end{array} \right.$$

In this example the iteration variable is a list that controls the number of loops. The list also sets the assigned values of "i" .

The local variable "a" is incremented by the value of "i" and then multiplied by the argument "x".

$$\text{ForLoop}_3(2) = 58$$

For x=2, the flow is 2*(0+1), 2*(2+2), 2*(8+5), 2*(26+3)
The final result is 2*(26+3)=58

$$j = \begin{bmatrix} 1 \\ 2 \\ 3 \\ 4 \\ 5 \end{bmatrix} \qquad \text{ForLoop}_3(j) = \begin{bmatrix} 11 \\ 58 \\ 189 \\ 476 \\ 1015 \end{bmatrix}$$

FIGURE 12.3

For loop using a list

$$\text{Vector} := \begin{bmatrix} 1 \\ 2 \\ 5 \\ 3 \end{bmatrix}$$

$$\text{ForLoop}_4(x) := \left\| \begin{array}{l} a \leftarrow 0 \\ \text{for } i \in \text{Vector} \\ \quad \| \ a \leftarrow x \cdot (a + i) \end{array} \right.$$

This is exactly the same as in the previous figure except that the iteration is controlled by a previously defined vector.

$$\text{ForLoop}_4(2) = 58$$

$$j = \begin{bmatrix} 1 \\ 2 \\ 3 \\ 4 \\ 5 \end{bmatrix} \qquad \text{ForLoop}_4(j) = \begin{bmatrix} 11 \\ 58 \\ 189 \\ 476 \\ 1015 \end{bmatrix}$$

FIGURE 12.4

For loop using a vector

This example sums each value of "a".

$$\text{ForLoop}_5(x) := \left\| \begin{array}{l} a \leftarrow 0 \\ b \leftarrow 0 \\ \text{for } i \in 1,2..x \\ \quad \left\| \begin{array}{l} a \leftarrow a + i \\ b \leftarrow b + a \end{array} \right. \\ b \end{array} \right.$$

$$j = \begin{bmatrix} 1 \\ 2 \\ 3 \\ 4 \\ 5 \end{bmatrix} \qquad \text{ForLoop}_5(j) = \begin{bmatrix} 1 \\ 4 \\ 10 \\ 20 \\ 35 \end{bmatrix}$$

Define "a" and "b" as local variables.

"i" is a local variable that takes on the values 1,2,3 until "x."

For the first loop "i" is 1, and "a" takes the value 0+1=1, thus b=0+1=1.
For the next loop "i" is 2, and "a" becomes 1+2=3 and b=1+3=4.
For the third loop "i" is 3, and "a" becomes 3+3=6 and b=4+6=10.
For the fourth loop "i" is 4, and "a" becomes 6+4=10 and b=10+10=20.
For the fifth loop "i" is 5, and "a" becomes 10+5=15 and b=20+15=35.

FIGURE 12.5

For loop using two local variables

Use a ***For*** loop to convert a range variable to a range vector.

Because this is a user-defined function that will be used many times, we will assign the Function label to it. This is done from the **Labels** control on the **Math** tab.

$$\underline{\text{Range2Vec}}(\text{Range}) := \left\| \begin{array}{l} \text{Count} \leftarrow \mathit{ORIGIN} \\ \text{for } i \in \text{Range} \\ \quad \left\| \begin{array}{l} \text{Vec}_{\mathit{Count}} \leftarrow i \\ \text{Count} \leftarrow \text{Count} + 1 \end{array} \right. \\ \text{Vec} \end{array} \right.$$

This function converts the argument range variable to a vector by using the ***for*** loop. The counter is used as the vector index and begins at ORIGIN. This first value is assigned to the first value of the range variable. The process repeats until the end of the range variable. The vector "Vec" is a local variable.

$$\text{RV1} := 1, 1.2..2.0 \qquad \text{NewVector\_1} := \underline{\text{Range2Vec}}(\text{RV1})$$

$$\text{NewVector\_1} = \begin{bmatrix} 1.00 \\ 1.20 \\ 1.40 \\ 1.60 \\ 1.80 \\ 2.00 \end{bmatrix} \qquad \text{NewVector\_1}_5 = 1.80$$

FIGURE 12.6

For loop to convert a range variable to a vector

This function creates a vector with a range of values. This is based on a function in the February 2002 Mathcad Advisor Newsletter.
The ***for*** loop creates a range variable for iterating from the three input arguments.
The vector "Vec" is a local variable.

Because this is a user-defined function that will be used many times, we will assign the Function label to it. This is done from the **Labels** control on the **Math** tab.

$$\underline{\text{RangeVec}}(s_1, s_2, e) := \left\| \begin{array}{l} \text{Count} \leftarrow ORIGIN \\ \text{for } i \in s_1, s_2 .. e \\ \quad \left\| \begin{array}{l} \text{Vec}_{Count} \leftarrow i \\ \text{Count} \leftarrow \text{Count} + 1 \end{array} \right. \\ \text{Vec} \end{array} \right.$$

Create a variable with a range from 0 to 1 with 0.1 increment.

$\text{RangeVariable\_1} := 0, 0.1 .. 1.0$

$\text{NewVector\_2} := \underline{\text{RangeVec}}(0, 0.1, 1.0)$

Refer to the discussion of range variables and vectors in Chapters 1 and 6.
These two variables appear to be identical, but they behave very differently.

$$\text{RangeVariable\_1} = \begin{bmatrix} 0 \\ 0.1 \\ 0.2 \\ 0.3 \\ 0.4 \\ 0.5 \\ 0.6 \\ 0.7 \\ 0.8 \\ 0.9 \\ 1 \end{bmatrix} \qquad \text{NewVector\_2} = \begin{bmatrix} 0 \\ 0.1 \\ 0.2 \\ 0.3 \\ 0.4 \\ 0.5 \\ 0.6 \\ 0.7 \\ 0.8 \\ 0.9 \\ 1 \end{bmatrix}$$

$RangeVariable\_1_3 = ?$ $\qquad$ $\text{NewVector\_2}_3 = 0.2$

$RangeVariable\_1_3 = ?$

This value must be a vector.

FIGURE 12.7

For loop to create a range vector

While Loops

The ***while*** loop is different than the ***for*** loop. A ***for*** loop executes a fixed number of times, based on the size of the iteration variable. A ***while*** loop continues to execute while a specified condition is true. Once the condition is not true, the execution stops. If the condition is always true, PTC Mathcad will go into an infinite loop. For this reason, you must be careful when using a ***while*** loop. You insert the ***while*** loop into a program by clicking on the **while** button on the **Programming** controls on the **Math** tab. Do not type the word "while."

$$\text{WhileLoop}_1(x) := \begin{vmatrix} a \leftarrow 0 \\ \text{while } a < x \\ \quad \| \; a \leftarrow a + 1 \\ a \end{vmatrix}$$

Defines local variable.

Checks value of local variable.

Increments local variable by one.

Displays the value of "a" after the program stops.

$$j = \begin{bmatrix} 1 \\ 2 \\ 3 \\ 4 \\ 5 \end{bmatrix} \qquad \text{WhileLoop}_1(j) = \begin{bmatrix} 1 \\ 2 \\ 3 \\ 4 \\ 5 \end{bmatrix}$$

For the first loop "a" is zero, so the "while" statement is true and the program continues to the next line where "a" is incremented to the value 1.
On the second loop "a" is now 1 and the "while" statement is checked again. If it is true, then the program continues to the next line, and "a" is incremented to two. If the statement is false, then the program execution stops and the value of "a" is returned (1).
The loop continues until a >= x.

FIGURE 12.8

While loop

Figure 12.8 shows a simple ***while*** loop. The program continues to loop until "a" is equal to or greater than the argument "x." When this happens, the loop stops and PTC Mathcad returns the value of "a."

Figure 12.9 uses local variable "a" to control the number of loops. Local variable "b" is multiplied by 2 every time the program loops. Once "a" becomes equal to or greater than the argument "x," the loops stops and the program returns the value of "b."

Figure 12.10 shows three ways to create a user-defined function for factorial. PTC Mathcad already has the ***factorial*** operator in the **Operators** control on the **Math** tab, but it is interesting to use factorial to demonstrate the ***while*** loop.

$$\text{WhileLoop}_2(x) := \begin{array}{|l} a \leftarrow 0 \\ b \leftarrow 1 \\ \text{while } a < x \\ \quad \begin{array}{|l} a \leftarrow a + 1 \\ b \leftarrow b \cdot 2 \end{array} \\ b \end{array}$$

Defines local variables.

Checks value of local variable "a".

Increments "a" by one.

Every time the program loops it multiplies "b" by 2.

Displays the value of "b" after the program stops.

$$j = \begin{bmatrix} 1 \\ 2 \\ 3 \\ 4 \\ 5 \end{bmatrix} \qquad \text{WhileLoop}_2(j) = \begin{bmatrix} 2 \\ 4 \\ 8 \\ 16 \\ 32 \end{bmatrix}$$

For the first loop "a" is zero, so the "while" statement is true and the program continues to the next line where "a" is incremented to the value 1 and b is multiplied by 2.
On the second loop "a" is now 1 and the "while" statement is checked again. If a is less than x, then the program continues to the next line, and "a" is incremented to two and "b" becomes 2*2=4. If a is not less than x, then the program execution stops and the value of "b" is returned (4).

FIGURE 12.9

While loop

PTC Mathcad already has a built-in operator to calculate factorial, but let's look at a way to create a user-defined function for factorial.

$$\text{FactorialNew}(x) := \begin{array}{|l} f \leftarrow x - 1 \\ \text{while } f > 0 \\ \quad \begin{array}{|l} x \leftarrow x \cdot f \\ f \leftarrow f - 1 \end{array} \\ x \end{array}$$

Assigns local variable "f" as one less than the argument "x".

Checks if "f" is greater than zero.

If "f" is greater than zero, reassigns "x" the value x*f.

Reduces the value of "f" by one.

Returns x if "f" is not greater than zero.

$$j = \begin{bmatrix} 1 \\ 2 \\ 3 \\ 4 \\ 5 \end{bmatrix} \qquad \text{FactorialNew}(j) = \begin{bmatrix} 1 \\ 2 \\ 6 \\ 24 \\ 120 \end{bmatrix}$$

Assuming x=4:
On the first loop, "f" is assigned f=4-1=3.
Since "f" is greater than zero, x=4*3=12, "f" is reduced by one to f=3-1=2, and the program returns to the while statement.
On the next loop, f=2 is greater than zero, so x=12*2=24, and f=2-1=1.
On the next loop, f=1 is greater than zero, so x=24*1=24, and f=1-1=0.
On the next loop, "f" is not greater than zero (the statement is false), so the execution stops and the value of x=24 is returned.

The above function is not perfect. If a non-integer is used a non-integer is returned. Some **if** statements are needed to resolve this issue.

FIGURE 12.10 *(Continued on next page)*

Factorial function

Note:
There are additional ways to create the factorial function.
The following are listed in PTC Mathcad Help.

$$\mathbf{Fac_1(n)} := \left\| \begin{array}{l} f \leftarrow 1 \\ \text{while } n \leftarrow n-1 \\ \quad \| f \leftarrow f \cdot (n+1) \\ f \end{array} \right.$$

This *while* statement is not a conditional statement, but the loop stops when the value of n=0 (a false condition).

Assuming n=4:
On the first loop the argument "n" is reduced by 1 and becomes 3. The value of f becomes f=1*(3+1)=4.
On the second loop, n=3-1=2, and f=4*(2+1)=12.
On the third loop, n=2-1=1, and f=12*(2+1)=24.
On the fourth loop, n=1-1=0. This means that this is a false condition for the *while* loop and the loop stops and returns the value of f=24.

$\mathbf{Fac_1(4)} = 24$

The following is a recursive function, which means that the function refers to itself. Refer to "Recursive Functions" in PTC Mathcad Help.

$$\mathbf{Fac_2(n)} := \left\| \begin{array}{l} \text{if } n > 1 \\ \quad \| n \cdot Fac_2(n-1) \\ \text{else} \\ \quad \| 1 \end{array} \right.$$

$\mathbf{Fac_2(5)} = 120$

$5! = 120$

FIGURE 12.10

(Continued)

The ***while*** loop in Figure 12.11 will cause an infinite loop because the condition will always be true. If this condition happens, you can interrupt the loop by clicking the **Stop All Calculations** control in the **Controls** group of the **Calculation** tab (**Calculation>Controls>Stop All Calculations**).

The following causes an infinite loop because "a" can never be anything but zero.

$$\mathbf{WhileLoop_3(x)} := \left\| \begin{array}{l} a \leftarrow 0 \\ \text{while } a \le x \\ \quad \| a \leftarrow a \cdot 2 \end{array} \right.$$

If this condition happens, press the **Stop All Calculations** control on the **Calculation** tab.

FIGURE 12.11

Avoid infinite loops

Break and *Continue* Operators

The ***break*** and ***continue*** operators give more control to programs. They tell PTC Mathcad what to do if specific conditions are met and usually precede an ***if*** statement. These operators are used to check for specific conditions during the time that PTC Mathcad is looping with either the ***for*** loop or the ***while*** loop.

The ***break*** operator stops the loop if the condition is true, while the ***continue*** operator allows PTC Mathcad to skip the current loop iteration if a true statement is encountered. The ***continue*** operator is useful if you want to continue looping even if a condition is true. For example, you may want to extract only even numbers. By using the ***continue*** operator, you can skip over odd numbers. Without the ***continue*** operator, PTC Mathcad would stop if it encountered an odd number.

See Figure 12.12 for an example of the ***break*** operator. See Figure 12.13 for an example of the ***continue*** operator. These examples are relatively simple, and are intended to introduce the concept of using these two operators. More advanced examples are given in PTC Mathcad Help.

This example is similar to Figures 12.8 and 12.9, but it allows you to give starting and ending numbers. In order to prevent an infinite loop, if the starting number is less than or equal to zero, the program is stopped by the ***break*** statement.

```
WhileLoop4(w, x) := | i ← 1
                    | while w ≤ x
                    |   | if w ≤ 0
                    |   |   | break
                    |   | a_i ← w
                    |   | w ← w·2
                    |   | i ← i + 1
                    | a
```

Define local variable.

Continue looping as long as w <= x.

Note: To add the ***break*** operator, add the ***if*** statement first, then click on the first placeholder and add the ***break*** operator. The ***break*** operator stops the loop if x<=0.

This creates a vector using the value of "i" as the vector index and assigns it the value of "w".

Changes the values of "w" and "i".

Returns the vector "a".

$\text{WhileLoop}_4(0, 32) = 0$

$\text{WhileLoop}_4(-3, 32) = 0$

$\text{WhileLoop}_4(1, 32) = \begin{bmatrix} 1 \\ 2 \\ 4 \\ 8 \\ 16 \\ 32 \end{bmatrix}$ $\quad \text{WhileLoop}_4(5, 100) = \begin{bmatrix} 5 \\ 10 \\ 20 \\ 40 \\ 80 \end{bmatrix}$

FIGURE 12.12

Using the ***break*** operator to prevent infinite loops

This example illustrates how the ***continue*** operator can skip over certain conditions within a loop. In this example, the program skips over elements that are not integers. The function uses argument x to divide into integers from 1 to argument y, and selects only the integers.

```
WhileLoop5(x, y) := | i ← 0
                    | while i ≤ y
                    |   | j ← (i+1)/x
                    |   | i ← i+1
                    |   | if floor(j) ≠ j
                    |   |   | continue
                    |   | a_i ← j
                    | a
```

Local variable "i" begins with 0, in order to increment before the ***continue*** statement.

Local variable "j" begins with j=(0+1)/x.

Increment "i" by 1 to i=0+1=1.

When you see the ***continue*** operator it means, "Continue with the loop if this statement is false. If the statement is true, then stop this loop and go back to the top of the loop." This statement is checking to see if "j" is an integer.

The local variable "a" is a vector with all the values of j.

Assume x=2 and y=5:

On the first loop, i=0 and is less than 5 so the ***while*** loop is executed. The local variable j=(0+1)/2=0.5. Increment "i" by 1 to i=0+1=1. Check the ***continue*** statement. Floor(j)=0 is not equal to 0.5 so the statement is true. Since the statement is true, the ***continue*** operator tells PTC Mathcad to stop the current loop and begin at the top of the loop.

On the second loop, i=1 and is less than 5 so the ***while*** loop is executed again. j=(1+1)/2=1.0. i=1+1=2. (Note: We had to increment i above the ***continue*** statement, because if the ***continue*** statement is true, the rest of the loop is skipped and "i" would not have been incremented. This would have created an infinite loop.) Floor(1)=1 is equal to 1.0 so the statement is false, and the loop continues. The vector "a" has element a2 assigned the value 1.0. Because we did not assign element a1, it is left at 0.0.

FIGURE 12.13 *(Continued on next page)*

Using the ***continue*** operator

Return Operator

We introduced the ***return*** operator in Chapter 9. This operator is used in conjunction with an ***if*** statement. It stops program execution when the ***if*** statement is true, and returns the value listed. See Figure 9.7 for an example.

On the third loop, i=2 and is less than 5 so the ***while*** loop is executed again. j=(2+1)/2=1.5. i=2+1=3. Floor(1.5)=1 is not equal to 1.5. so the ***continue*** statement is true and the loop begins again at the top.
On the fourth loop, i=3 and is less than 5 so the ***while*** loop is executed again. j=(3+1)/2=2.0. i=3+1=4. Floor(2.0)=2 is equal to 2.0, so the ***continue*** statement is false, and the loop continues. The vector "a" has element a4 assigned the value 2.0. Element a3 is left at zero.

On the fourth loop, i=3 and is less than 5 so the ***while*** loop is executed again. j=(3+1)/2=2.0. i=3+1=4. Floor(2.0)=2 is equal to 2.0, so the ***continue*** statement is false, and the loop continues. The vector "a" has element a4 assigned the value 2.0. Element a3 is left at zero.
On the fifth loop, i=4 and is less than 5 so the ***while*** loop is executed again. j=(4+1)/2=2.5. i=4+1=5. Floor(2.5)=2 is not equal to 2.5. so the ***continue*** statement is true and the loop begins again at the top.

On the fifth loop, i=4 and is less than 5 so the ***while*** loop is executed again. j=(4+1)/2=2.5. i=4+1=5. Floor(2.5)=2 is not equal to 2.5. so the ***continue*** statement is true and the loop begins again at the top.

On the sixth loop, i=5 is less than or equal to 5 so the ***while*** loop is executed again. j=(5+1)/2=3.0. i=5+1=6. Floor (3.0)=3 is equal to 3.0, so the ***continue*** statement is false, and the loop continues. The vector "a" has element a6 assigned the value 3.0. Element a5 is left at zero.

On the next loop i=6 is not less than or equal to 5, so the ***while*** loop is stopped. The vector "a" is returned.

$$\text{WhileLoop}_5(2,5)=\begin{bmatrix}0.00\\1.00\\0.00\\2.00\\0.00\\3.00\end{bmatrix}\qquad \text{WhileLoop}_5(1,9)=\begin{bmatrix}1.00\\2.00\\3.00\\4.00\\5.00\\6.00\\7.00\\8.00\\9.00\\10.00\end{bmatrix}$$

$$\text{WhileLoop}_5(3,9)=\begin{bmatrix}0.00\\0.00\\1.00\\0.00\\0.00\\2.00\\0.00\\0.00\\3.00\end{bmatrix}$$

FIGURE 12.13

(continued)

Try-On-Error Operator

The ***try-on-error*** operator is very useful to tell PTC Mathcad what to do if it encounters an error. A very common error occurs when PTC Mathcad tries to divide by zero. There will be many times when an input value to a function causes this to occur. The ***try-on-error*** operator gives you the ability to tell PTC Mathcad what to do when this happens. See Figure 12.14 for several examples of using the ***try-on-error*** operator.

The ***try-on-error*** operator has the form x *on error* y. If y causes an error then x is evaluated.

try
‖
on error
‖

This figure gives three methods of dealing with the function 24/x.

Example 1

$\mathrm{Example}_1(x) := \frac{24}{x}$

Example 2

$\mathrm{Example}_2(x) :=$ try
‖ $\frac{24}{x}$
on error
‖ "Division by zero"

Example 3

$\mathrm{Example}_3(x) :=$ try
‖ $\frac{24}{x}$
on error
‖ ∞

Note: To insert the built-in constant infinity type CRTL +SHIFT+z.

$\mathrm{Result} := 0$

$\mathrm{Example}_1(\mathrm{Result}) = ?$

$\mathrm{Example}_1(\mathrm{Result}) = ?$

This expression is divided by zero. It cannot be computed.

$\mathrm{Example}_2(\mathrm{Result}) =$ "Division by zero"

$\mathrm{Example}_3(\mathrm{Result}) = 1 \cdot 10^{307}$

$j := -1, 0..3$

$\mathrm{Example}_1(j) = ?$

$\mathrm{Example}_1(j) = ?$

This expression is divided by zero. It cannot be computed.

$\mathrm{Example}_2(j) = \begin{bmatrix} -24.00 \\ \text{"Division by zero"} \\ 24.00 \\ 12.00 \\ 8.00 \end{bmatrix}$

$\mathrm{Example}_3(j) = \begin{bmatrix} -24 \\ 1 \cdot 10^{307} \\ 24 \\ 12 \\ 8 \end{bmatrix}$

Example 1 produces errors.
Example 2 returns a string.
Example 3 returns PTC Mathcad's numeric equivalent of infinity.

FIGURE 12.14

The ***try-on-error*** operator

Engineering Example 12.1

Engineering Example 12.1 demonstrates how to create Intensity Duration Frequency (IDF) curves. It is a great example because it summarizes everything that we have discussed in the book. It uses advanced programming, functions, reading from files, plotting functions, and plotting data points.

Based on Problem 5.5 from CHIN, DAVID A., WATER RESOURCES ENGINEERING, 2nd, ©2007. Reproduced by permission of Pearson Education, Inc. Upper Saddle River, New Jersey.

Development of Intensity Duration Frequency (IDF) curves

Rainfall data (cm of rain) was compiled over several years by determining the maximum amount of rainfall that occurred in a given time period during each given year.

Use the rainfall data to develop IDF curves for 5 year, 10 year, and 25 year return periods.

This will occur in several steps:

1. Calculate the rainfall intensity in cm/hr for each time period.
2. Rank the rainfall intensity data by sorting heaviest to lightest for each time period. The lightest intensity will receive the highest rank and the highest intensity will receive the lowest rank. These numbers will be used to calculate the return period. The heaviest rains have a lower probability of occuring.
3. Use the rankings to calculate return periods using the Wiebull formula.
4. Plot the rainfall intensity vs. return period for each rainfall time period.
5. Create a data matrix giving rainfall intensities for 5, 10 and 25 year return periods and 1, 2, 4, 6, 10, 12, and 24 hour time periods.
6. Plot a 5 year, 10 year, and 25 year IDF curve (using data points)
7. Using the IDF curves, model the curves to fit a function of the form:
   $$i = \frac{a}{t+b}\ .$$
8. Plot the IDF curves using the mathematical formula and compare it to the data points.

The rainfall data for the years 1963 to 2007 is contained in an Excel spreadsheet. Use the PTC Mathcad function ***READFILE*** to get the data and assign it to the variable "Rain."

$$\text{Rain} := \underline{\text{READFILE}}\left(\text{“RainfallData.xls”}, \text{“Excel”}\right)$$

Rain =

NaN	"1 –hr"	"2–hr"	"4–hr"	"6–hr"	"10–hr"	"12 hr"	"24 hr"
1963.0	1.7	2.9	4.6	4.8	4.8	4.8	5.1
1964.0	1.9	2.6	2.6	3.0	3.7	3.7	4.8
1965.0	1.9	2.3	2.7	2.9	3.1	3.1	3.2
1966.0	1.9	2.4	2.8	4.0	4.7	4.9	5.7
1967.0	1.9	1.9	2.2	2.5	2.5	2.5	3.0
1968.0	2.1	2.4	3.0	3.5	4.0	4.2	5.1
1969.0	1.7	2.5	2.7	2.8	3.0	3.0	4.0
1970.0	1.7	3.6	3.6	3.6	4.3	4.3	4.7
1971.0	1.9	2.1	2.8	3.0	3.8	3.8	3.8
1972.0	2.4	3.7	4.5	4.5	4.6	4.6	5.4
1973.0	1.0	1.3	1.6	1.7	1.7	1.7	1.8
1974.0	2.7	2.8	3.2	3.7	3.7	3.8	4.4
1975.0	1.3	1.5	2.0	2.6	3.5	3.8	4.1
1976.0	2.8	3.2	3.4	4.1	5.2	5.7	6.0
1977.0	2.0	2.1	2.8	4.1	5.6	6.0	7.9
1978.0	1.8	2.7	2.8	2.8	2.8	2.8	2.8
1979.0	1.5	2.6	2.8	3.1	3.8	4.5	6.4
1980.0	1.4	2.5	3.1	3.7	4.2	4.3	4.5
1981.0	1.6	1.8	2.2	2.6	3.6	3.6	4.1
							⋮

Use the ***submatrix*** function to extract information from the matrix.

$\underline{\text{rows}}(\text{Rain}) = 46$

$\text{OneHr} := \underline{\text{submatrix}}(\text{Rain}, 2, \underline{\text{rows}}(\text{Rain}), 2, 2) \cdot cm$

$\text{TwoHr} := \underline{\text{submatrix}}(\text{Rain}, 2, \underline{\text{rows}}(\text{Rain}), 3, 3) \cdot cm$

$\text{FourHr} := \underline{\text{submatrix}}(\text{Rain}, 2, \underline{\text{rows}}(\text{Rain}), 4, 4) \cdot cm$

$\text{SixHr} := \underline{\text{submatrix}}(\text{Rain}, 2, \underline{\text{rows}}(\text{Rain}), 5, 5) \cdot cm$

$\text{TenHr} := \underline{\text{submatrix}}(\text{Rain}, 2, \underline{\text{rows}}(\text{Rain}), 6, 6) \cdot cm$

$\text{TwelveHr} := \underline{\text{submatrix}}(\text{Rain}, 2, \underline{\text{rows}}(\text{Rain}), 7, 7) \cdot cm$

$\text{TwentyFourHr} := \underline{\text{submatrix}}(\text{Rain}, 2, \underline{\text{rows}}(\text{Rain}), 8, 8) \cdot cm$

$$\text{OneHr} = \begin{bmatrix} 1.7 \\ 1.9 \\ 1.9 \\ 1.9 \\ 1.9 \\ 2.1 \\ 1.7 \\ 1.7 \\ 1.9 \\ 2.4 \\ 1 \\ 2.7 \\ \vdots \end{bmatrix} cm$$

Step 1: Compute the rainfall intensities (cm/hr)

Divide the rainfall amounts by the corresponding time.

$$\text{OneHrRain} := \frac{\text{OneHr}}{1 \cdot hr}$$

$$\text{TwoHrRain} := \frac{\text{TwoHr}}{2 \cdot hr}$$

$$\text{FourHrRain} := \frac{\text{FourHr}}{4 \cdot hr}$$

$$\text{SixHrRain} := \frac{\text{SixHr}}{6 \cdot hr}$$

$$\text{TenHrRain} := \frac{\text{TenHr}}{10 \cdot hr}$$

$$\text{TwelveHrRain} := \frac{\text{TwelveHr}}{12 \cdot hr}$$

$$\text{TwentyFourHrRain} := \frac{\text{TwentyFourHr}}{24 \cdot hr}$$

$$\text{OneHrRain} = \begin{bmatrix} 1.7 \\ 1.9 \\ 1.9 \\ 1.9 \\ 1.9 \\ 2.1 \\ 1.7 \\ 1.7 \\ 1.9 \\ \vdots \end{bmatrix} \frac{cm}{hr}$$

$$\text{TenHrRain} = \begin{bmatrix} 0.48 \\ 0.37 \\ 0.31 \\ 0.47 \\ 0.25 \\ 0.4 \\ 0.3 \\ 0.43 \\ 0.38 \\ \vdots \end{bmatrix} \frac{cm}{hr}$$

Step 2: Sort data and assign a rank

We need to compute the corresponding rank of each intensity with the largest one getting the lowest rank. First sort the data. The sort function sorts from smallest to largest. Then reverse the sort so the data is sorted from largest to smallest. Finally, compute the ranks of the data, sorted in this way. The rank is the mean of the indices corresponding to the given rainfall intensity. For example the intensity of 1.7 cm/hr is found in the 25th, 26th, 27th, 28th, and 29th element. The mean of these indices is 27. Thus the intensity of 1.7 cm/hr is given the rank of 27. The intensity of 3.7 cm/hr is found in the 1st element only. Thus the intensity of 3.7 cm/hr is given the rank of 1.

Create a function to rank the data.

```
RankData(x) := | Sorted ← reverse(sort(x))
               | for i ∈ 1..rows(x)
               |   | m ← match(x_i, Sorted)
               |   | Rank_i ← mean(m)
               | Rank
```

The ***match(z, A)*** function looks in the sorted vector, "Sorted", for a given value, x_i, from the original vector, x, and returns the index (indices) of its position in the sorted vector.

Note: PTC Mathcad compares the values in its default value of m/s. In order for the match function to work, the Convergence Tolerance (CTOL) for this worksheet needed to be changed to $1 \cdot 10^{-7}$. Go to **Worksheet Settings** on the **Calculation** tab.

Test the ***match*** function

$\text{Sorted} := \text{reverse}(\text{sort}(\text{OneHrRain}))$

$\text{OneHrRain}_1 = 1.7\ \frac{cm}{hr}$

$\text{Check\_1} := \text{match}(\text{OneHrRain}_1, \text{Sorted}) = \begin{bmatrix} 25 \\ 26 \\ 27 \\ 28 \\ 29 \end{bmatrix}$ **OK**

$\text{mean}(\text{Check\_1}) = 27$

$\text{Check\_2} := \text{match}\left(3.7 \cdot \frac{cm}{hr}, \text{Sorted}\right) = [1.00]$ **OK**

Sorted = ($\frac{cm}{hr}$)

1	3.7
2	3.1
3	2.9
4	2.8
5	2.7
6	2.7
7	2.5
8	2.4
9	2.2
10	2.1
11	2.1
12	2
13	2
14	1.9
15	1.9
16	1.9
17	1.9
18	1.9
19	1.9
20	1.9
21	1.9
22	1.8
23	1.8
24	1.8
25	1.7
26	1.7
27	1.7
28	1.7
29	1.7
30	1.6
⋮	⋮

$$RankOneHr := RankData(OneHrRain)$$

$$RankTwoHr := RankData(TwoHrRain)$$

$$RankFourHr := RankData(FourHrRain)$$

$$RankSixHr := RankData(SixHrRain)$$

$$RankTenHr := RankData(TenHrRain)$$

$$RankTwelveHr := RankData(TwelveHrRain)$$

$$RankTwentyFourHr := RankData(TwentyFourHrRain)$$

$$RankOneHr = \begin{bmatrix} 27.0 \\ 17.5 \\ 17.5 \\ 17.5 \\ 17.5 \\ 10.5 \\ 27.0 \\ 27.0 \\ 17.5 \\ 8.0 \\ 45.0 \\ 5.5 \\ \vdots \end{bmatrix} \quad RankTenHr = \begin{bmatrix} 8.5 \\ 21.5 \\ 32.5 \\ 9.5 \\ 38.5 \\ 17.5 \\ 33.0 \\ 13.0 \\ 21.5 \\ 10.5 \\ 44.5 \\ 21.5 \\ \vdots \end{bmatrix}$$

Step 3: Calculate return periods using the Wiebull formula

To calculate the return period using the Wiebull formula, use the number of data years plus 1.

$$\text{Return}(x) := \frac{\underline{\text{rows}}(x)+1}{x} \cdot yr \qquad \underline{\text{rows}}(\text{RankOneHr}) = 45$$

$$\text{ReturnOneHr} := \text{Return}(\text{RankOneHr})$$

$$\text{ReturnTwoHr} := \text{Return}(\text{RankTwoHr})$$

$$\text{ReturnFourHr} := \text{Return}(\text{RankFourHr})$$

$$\text{ReturnSixHr} := \text{Return}(\text{RankSixHr})$$

$$\text{ReturnTenHr} := \text{Return}(\text{RankTenHr})$$

$$\text{ReturnTwelveHr} := \text{Return}(\text{RankTwelveHr})$$

$$\text{ReturnTwentyFourHr} := \text{Return}(\text{RankTwentyFourHr})$$

$$\text{ReturnOneHr} = \begin{bmatrix} 1.704 \\ 2.629 \\ 2.629 \\ 2.629 \\ 2.629 \\ 4.381 \\ 1.704 \\ 1.704 \\ 2.629 \\ 5.750 \\ 1.022 \\ 8.364 \\ \vdots \end{bmatrix} yr$$

$$\text{ReturnTenHr} = \begin{bmatrix} 5.412 \\ 2.140 \\ 1.415 \\ 4.842 \\ 1.195 \\ 2.629 \\ 1.394 \\ 3.538 \\ 2.140 \\ 4.381 \\ 1.034 \\ 2.140 \\ \vdots \end{bmatrix} yr$$

Step 4: Plot the rainfall intensity vs. return period

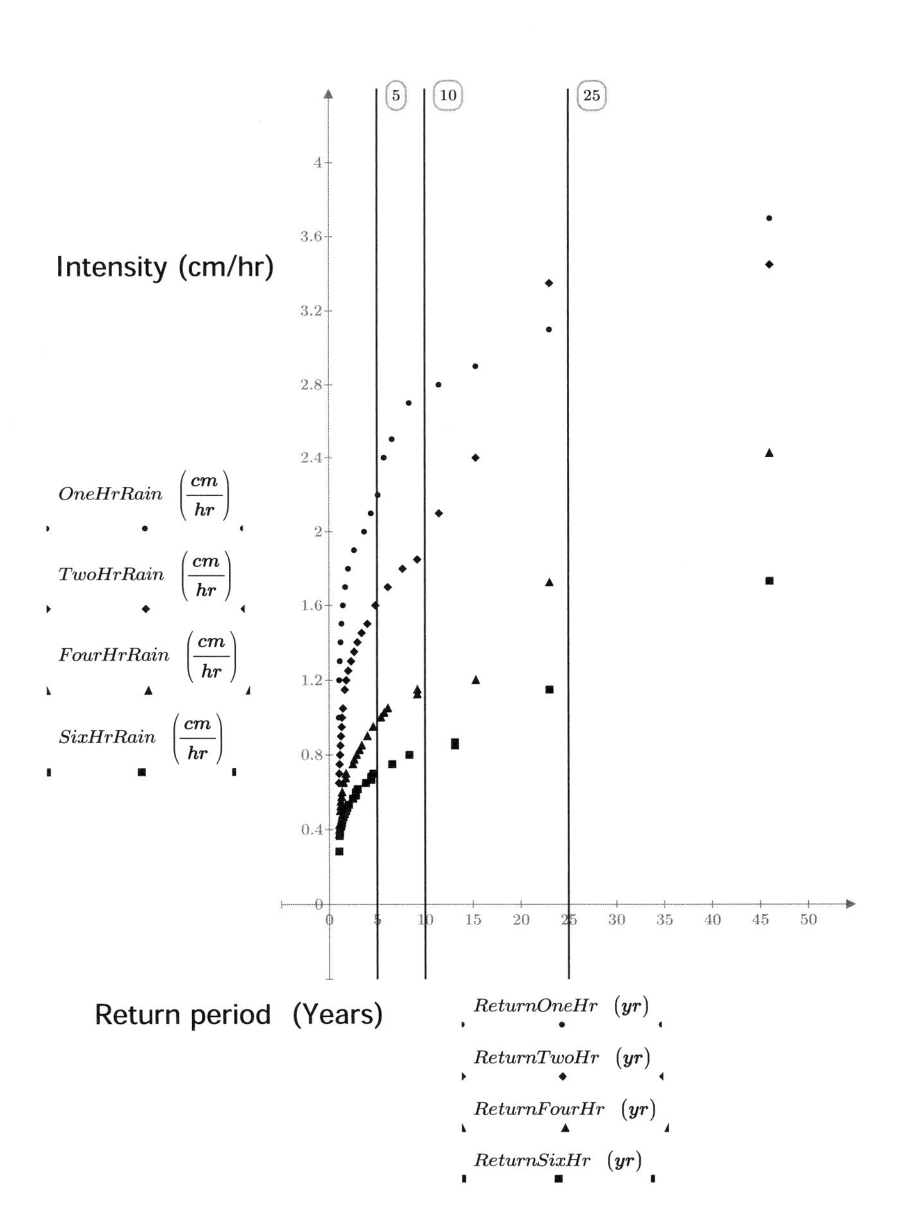

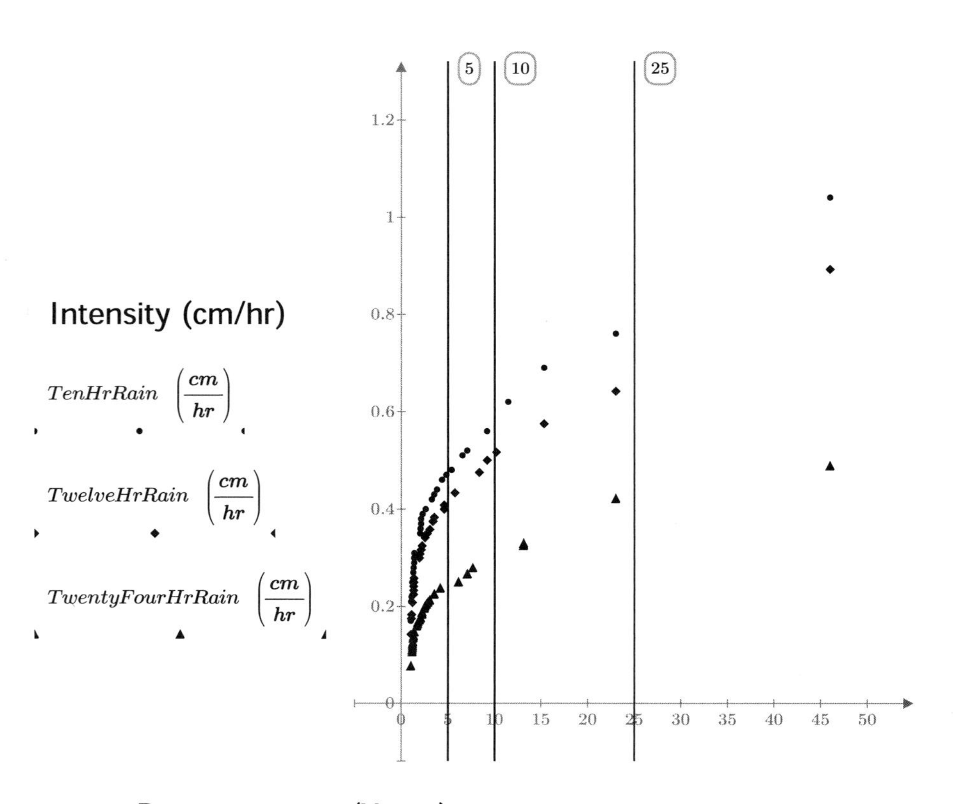
Intensity (cm/hr)
TenHrRain (cm/hr)
TwelveHrRain (cm/hr)
TwentyFourHrRain (cm/hr)
5
10
25
1.2
1
0.8
0.6
0.4
0.2
0
0
5
10
15
20
25
30
35
40
45
50
Return period (Years)
ReturnTenHr (yr)
ReturnTwelveHr (yr)
ReturnTwentyFourHr (yr)

Step 5: Create a data matrix

For a return period of 5 years, read the intensity of 1hr, 2hr, 4hr, 6hr, 10hr, 12hr, and 24hr.
Repeat for a return period of 10 years, and repeat for a return period of 25 years.
Put the data into a table. This is done manually.

Column 1 is duration in hours.
Column 2 is 5 year return period.
Column 3 is 10 year return period.
Column 4 is 25 year return period.

Duration (hr)	FiveYear $\left(\frac{cm}{hr}\right)$	TenYear $\left(\frac{cm}{hr}\right)$	TwentyFiveYear $\left(\frac{cm}{hr}\right)$
1	2.2	2.73	3.2
2	1.6	1.98	3.35
4	1.01	1.12	1.85
6	0.696	0.83	1.24
10	0.478	0.59	0.81
12	0.41	0.499	0.689
24	0.23	0.297	0.431

$$\text{Duration} = \begin{bmatrix} 1 \\ 2 \\ 4 \\ 6 \\ 10 \\ 12 \\ 24 \end{bmatrix} hr \qquad \text{FiveYear} = \begin{bmatrix} 2.2 \\ 1.6 \\ 1.01 \\ 0.696 \\ 0.478 \\ 0.41 \\ 0.23 \end{bmatrix} \frac{cm}{hr}$$

$$\text{TenYear} = \begin{bmatrix} 2.73 \\ 1.98 \\ 1.12 \\ 0.83 \\ 0.59 \\ 0.499 \\ 0.297 \end{bmatrix} \frac{cm}{hr} \qquad \text{TwentyFiveYear} = \begin{bmatrix} 3.2 \\ 3.35 \\ 1.85 \\ 1.24 \\ 0.81 \\ 0.689 \\ 0.431 \end{bmatrix} \frac{cm}{hr}$$

Step 6: Plot a 5, 10, and 25 year IDF curve (using data points)

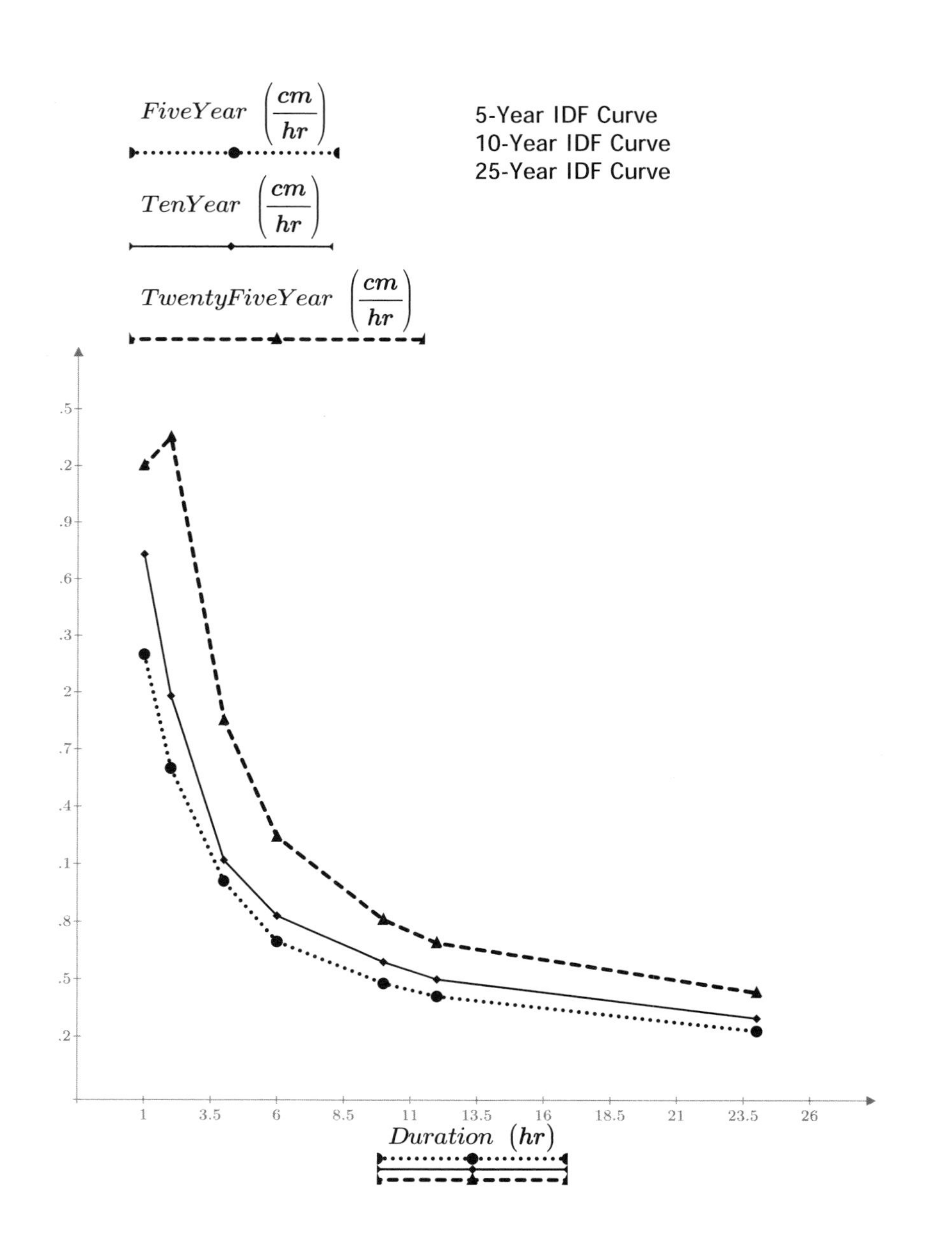

Step 7: Model IDF curves to fit a function of the form: $i = \frac{a}{t+b}$

Write the equation for the IDF curve in a linear form (y(x)=mx+b), where $\frac{1}{i}$ is the dependent variable, $\frac{1}{a}$ is the slope, and $\frac{b}{a}$ is the intercept.

$$\frac{1}{i} = \frac{t+b}{a} = \frac{t}{a} + \frac{b}{a}$$

$$\text{SlopeIDF}_5 := \underline{\text{slope}}\left(\text{Duration}, \frac{1}{\text{FiveYear}}\right) = 16.968\ \frac{1}{m}$$

$$a_5 := \frac{1}{\text{SlopeIDF}_5} = 5.893\ cm$$

$$b/a_5 := \underline{\text{intercept}}\left(\text{Duration}, \frac{1}{\text{FiveYear}}\right) = 122088.311\ \frac{s}{m}$$

$$b_5 := b/a_5 \cdot a_5 = 1.999\ hr$$

$$\text{SlopeIDF}_{10} := \underline{\text{slope}}\left(\text{Duration}, \frac{1}{\text{TenYear}}\right) = 13.01\ \frac{1}{m}$$

$$a_{10} := \frac{1}{\text{SlopeIDF}_{10}} = 7.686\ cm$$

$$b/a_{10} := \underline{\text{intercept}}\left(\text{Duration}, \frac{1}{\text{TenYear}}\right) = 121315.417\ \frac{s}{m}$$

$$b_{10} := b/a_{10} \cdot a_{10} = 2.59\ hr$$

$$\text{SlopeIDF}_{25} := \underline{\text{slope}}\left(\text{Duration}, \frac{1}{\text{TwentyFiveYear}}\right) = 9.128\ \frac{1}{m}$$

$$a_{25} := \frac{1}{\text{SlopeIDF}_{25}} = 10.956\ cm$$

$$b/a_{25} := \underline{\text{intercept}}\left(\text{Duration}, \frac{1}{\text{TwentyFiveYear}}\right) = 81193.247\ \frac{s}{m}$$

$$b_{25} := b/a_{25} \cdot a_{25} = 2.471\ hr$$

Create a Plot Range Variable. $t := 0 \cdot hr, 1 \cdot hr .. 25 \cdot hr$

Create a user-defined function.

$$\text{Prediction}(\text{SlopeIDF}, x, a, b) := \text{SlopeIDF} \cdot x + \frac{b}{a}$$

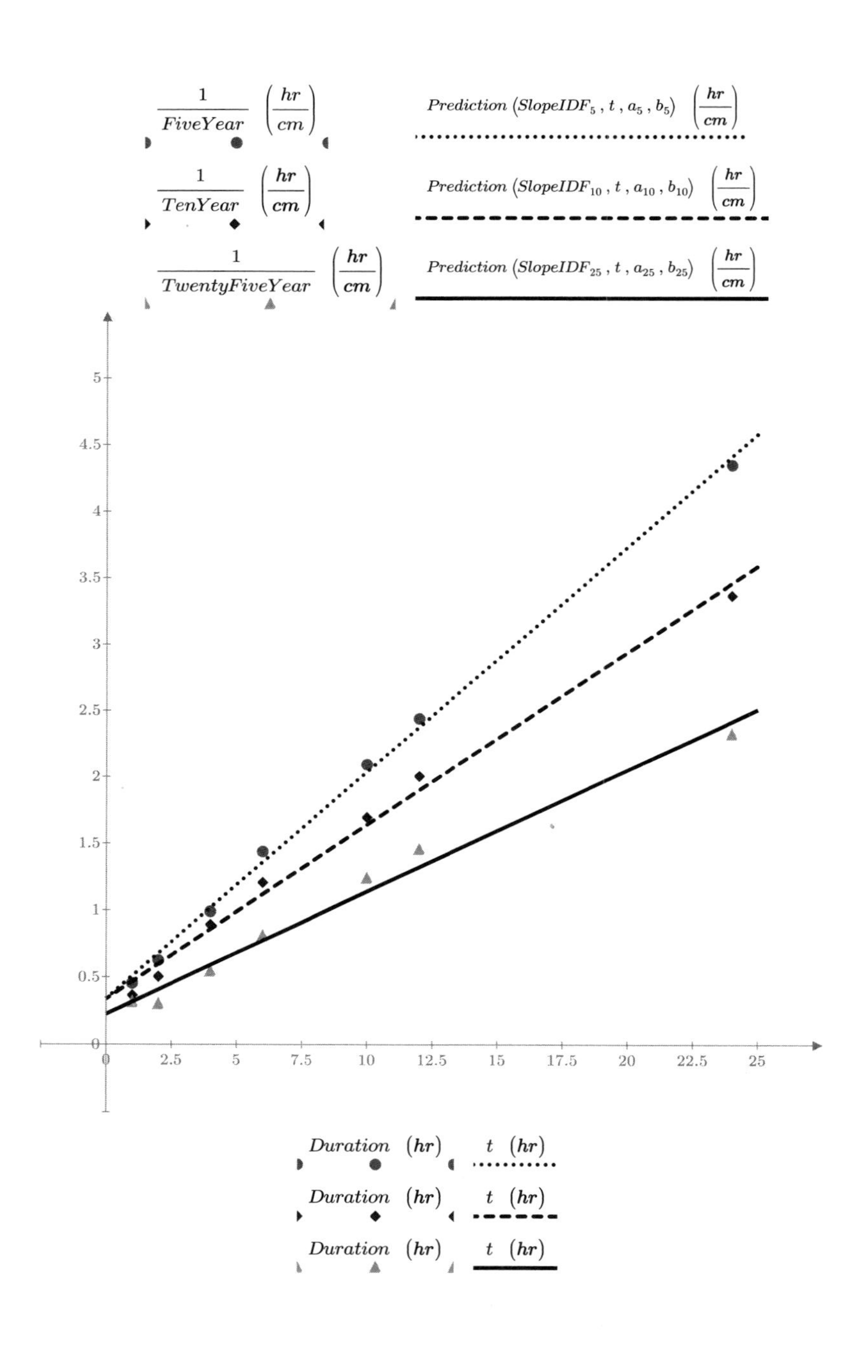

1/FiveYear (hr/cm)
1/TenYear (hr/cm)
1/TwentyFiveYear (hr/cm)
Prediction (SlopeIDF5, t, a5, b5) (hr/cm)
Prediction (SlopeIDF10, t, a10, b10) (hr/cm)
Prediction (SlopeIDF25, t, a25, b25) (hr/cm)
0
0.5
1
1.5
2
2.5
3
3.5
4
4.5
5
2.5
5
7.5
10
12.5
15
17.5
20
22.5
25
Duration (hr)
t (hr)
Duration (hr)
t (hr)
Duration (hr)
t (hr)

Step 8: Plot the IDF curves using the mathematical formula

Using the values of a and b plot the 5yr, 10yr, and 25yr IDF curves using $i = \frac{a}{t+b}$

$a_5 = 5.893\ cm \qquad b_5 = 1.999\ hr$

$a_{10} = 7.686\ cm \qquad b_{10} = 2.59\ hr$

$a_{25} = 10.956\ cm \qquad b_{25} = 2.471\ hr$

$$\text{Intensity}(t, a, b) := \frac{a}{t+b}$$

$\text{Intensity}(1 \cdot hr, a_5, b_5) = 1.965\ \frac{cm}{hr}$ $\quad \text{Intensity}(10 \cdot hr, a_5, b_5) = 0.491\ \frac{cm}{hr}$

$\text{Intensity}(1 \cdot hr, a_{10}, b_{10}) = 2.141\ \frac{cm}{hr}$ $\quad \text{Intensity}(10 \cdot hr, a_{10}, b_{10}) = 0.61\ \frac{cm}{hr}$

$\text{Intensity}(1 \cdot hr, a_{25}, b_{25}) = 3.156\ \frac{cm}{hr}$ $\quad \text{Intensity}(10 \cdot hr, a_{25}, b_{25}) = 0.878\ \frac{cm}{hr}$

Check Plot

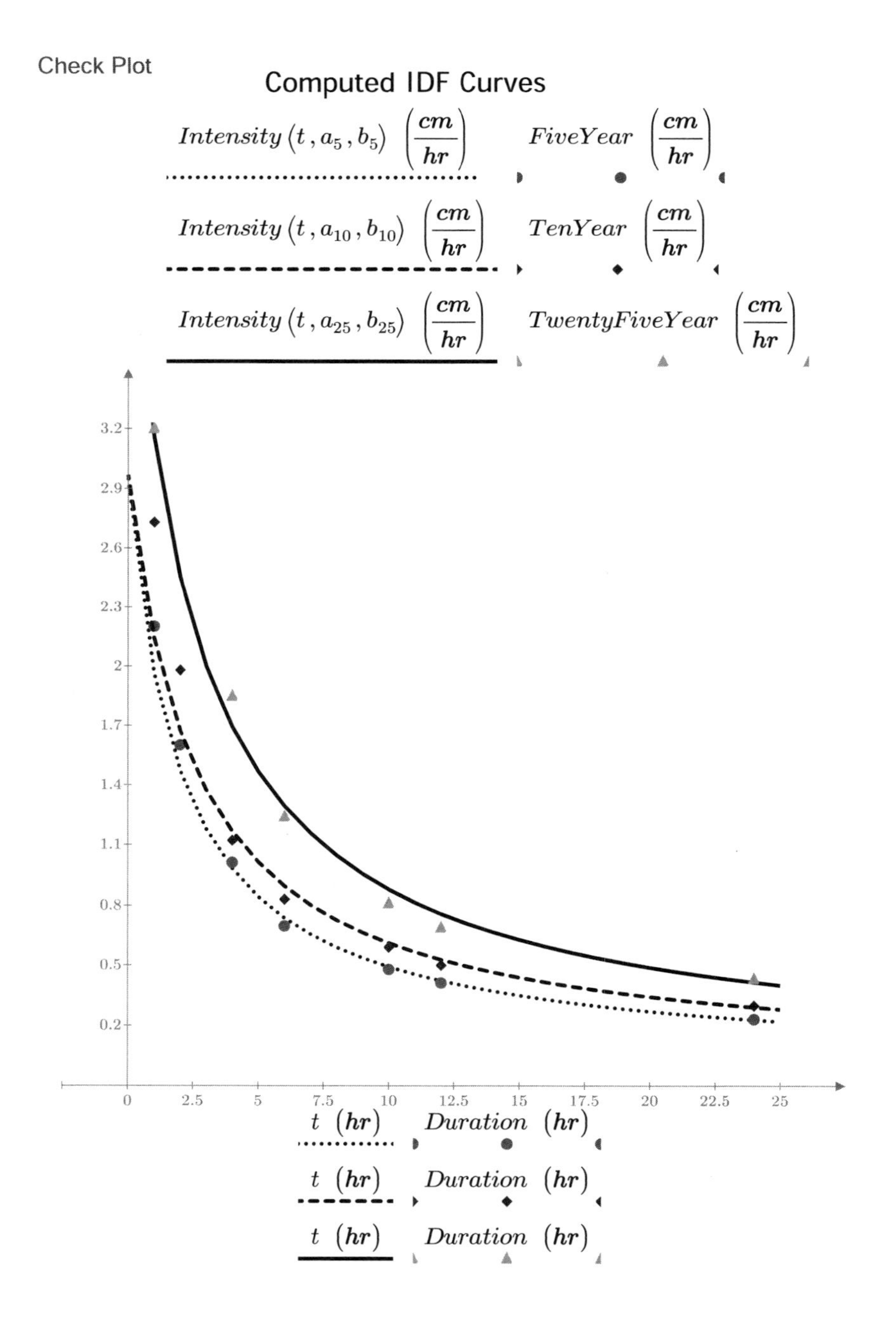
Computed IDF Curves
Intensity (t, a5, b5) (cm/hr)
FiveYear (cm/hr)
Intensity (t, a10, b10) (cm/hr)
TenYear (cm/hr)
Intensity (t, a25, b25) (cm/hr)
TwentyFiveYear (cm/hr)
3.2
2.9
2.6
2.3
2
1.7
1.4
1.1
0.8
0.5
0.2
0
2.5
5
7.5
10
12.5
15
17.5
20
22.5
25
t (hr)
Duration (hr)
t (hr)
Duration (hr)
t (hr)
Duration (hr)

Summary

Looping was the main concept covered in this chapter. This concept, combined with the logical programming discussed in Chapter 9 can make some very powerful programs.

There are two types of loops: ***for*** loops and ***while*** loops. The ***for*** loop loops a specific number of times and is controlled by an iteration variable. The ***while*** loop will loop until a specific condition is met. If the condition is never met, the loop will go indefinitely.

The ***break*** and ***continue*** operators are used to control when and how a program loops. The ***break*** operator stops the execution when a specific condition is met. The ***continue*** operator allows PTC Mathcad to skip over a portion of a loop and move back to the beginning of the loop.

The ***return*** operator is used in conjunction with an ***if*** statement. It stops the execution of the program and returns the value following the ***return*** operator.

The ***try-on-error*** operator is very useful when a particular operation could cause an error, because you can tell PTC Mathcad what to do when the error occurs.

Practice

The PTC Mathcad Prime 3.0 figures and examples used in this book are available for download from the book's website. The reader is encouraged to download the files and use them to practice the concepts learned. Additional examples and problems are also provided. To access this content go to http://store.elsevier.com/9780124104105 *and click on the Resources tab, and then click on the link for the Online Companion Materials.*

1. From your field of study, create 10 programs that use the features discussed in this chapter.
2. Create five functions or expressions that use the ***try-on-error*** operator.

CHAPTER

Calculus and Differential Equations

13

PTC Mathcad will easily perform differentiation and integration operations. PTC Mathcad can also perform a number of differential equation solutions for both ordinary differential equations (ODEs) and partial differential equations (PDEs). The use of symbolics is essential for working with calculus. Please review Chapter 10 before proceeding with this chapter. Most of the solutions for calculus problems involve solving with symbolics first.

Differentiation

> Mathcad 15 used two differentiation operators: one for the 1st derivative and another for the nth derivative. PTC® Mathcad Prime® uses a single derivative operator. If you want a first order derivative, the placeholder for higher order derivatives can be left blank. PTC Mathcad Prime also allows for a prime notation for derivatives.

The ***derivative*** operator is found in the **Operators** list on the **Math** tab (**Math>Operators and Symbols>Operators>Calculus>d/dx**). The keyboard shortcut is **CTRL + SHIFT + D**. The ***derivative*** operator has three placeholders.

$$\frac{d}{d\blacksquare^{\blacksquare}}\blacksquare$$

If you are evaluating a first order derivative the exponent placeholder may be left blank.

$$\frac{d}{dx^{\blacksquare}}x^3 \quad \frac{d}{dx}x^3$$

Essential PTC® Mathcad Prime® 3.0.

Use the exponent placeholder for higher order derivatives.

$$\frac{d^{2}}{dx^{2}}x^{3} \quad \frac{d^{2}}{dx^{2}}x^{3}$$

The following examples are intentionally very simple. They are intended to help you easily understand the use of differentiation in PTC Mathcad. Each example will use the same functions and arguments.

Figure 13.1 illustrates a simple use of the ***derivative*** operator to calculate the numeric derivative of a function at a single point. The point is defined prior to the use of the ***derivative*** operator.

You can create a user-defined function that includes the ***derivative*** operator. Figure 13.2 creates a function using the ***derivative*** operator and the same functions as used in Figure 13.1. Each time the function is used PTC Mathcad calculates the derivative at the value of the function argument. Figure 13.2 also illustrates the difference between using range variables and vectors as the arguments of the function. The range variable allows the display of multiple results, but the results are not allowed to be assigned. The use of a vector creates an issue. The function is expecting to calculate the derivative at a single point, and a vector provides multiple points. You get an error stating the argument should be a scalar. Use the ***vectorize*** operator (**CTRL + SHIFT + ^**) to force PTC Mathcad to do an element-by-element calculation. This provides multiple results, and the result is a column vector with each result available.

A better way to use the ***derivative*** operator is to use symbolics. One advantage of using the ***symbolic evaluation*** operator (**CTRL + PERIOD**) is that you see the

The keyboard shortcut for the ***derivative*** operator is **CTRL+SHIFT+D**.

Calculate the derivative at a single point.

$x := 2$

$$\frac{d}{dx}x^{2} = 4.000 \qquad \frac{d}{dx}\left(x^{3} + 3 \cdot x^{2} + 4 \cdot x - 2\right) = 28.000$$

$x := 4$

$$\frac{d}{dx}x^{2} = 8.000 \qquad \frac{d}{dx}\left(x^{3} + 3 \cdot x^{2} + 4 \cdot x - 2\right) = 76.000$$

FIGURE 13.1

Numeric differentiation at a single point

Assign the ***derivative*** operator to a function.

$f(x) := \frac{d}{dx} x^2$ $\quad$ $g(x) := \frac{d}{dx}(x^3 + 3 \cdot x^2 + 4 \cdot x - 2)$

$f(2) = 4.000$ $\quad$ $g(2) = 28.000$

$f(4) = 8.000$ $\quad$ $g(4) = 76.000$

Use a range variable for input. The results are displayed but they cannot be assigned to a variable.

$RV := 0, 2 .. 10$

$$f(RV) = \begin{bmatrix} 0.000 \\ 4.000 \\ 8.000 \\ 12.000 \\ 16.000 \\ 20.000 \end{bmatrix} \quad g(RV) = \begin{bmatrix} 4.000 \\ 28.000 \\ 76.000 \\ 148.000 \\ 244.000 \\ 364.000 \end{bmatrix}$$

$R\_1 := f(RV)$

$Result := f(RV)$

This value must be a scalar.

Use a vector for input.

$$Vector := 0, 2 .. 10 = \begin{bmatrix} 0.000 \\ 2.000 \\ 4.000 \\ 6.000 \\ 8.000 \\ 10.000 \end{bmatrix}$$

The vector is created by placing the ***evaluation*** operator on the same line as the ***definition*** operator.

$f(Vector) = ?$

This value must be a scalar.

Error message: "This value must be a scalar."

$f(Vector) = ?$ $\quad$ $g(Vector) = ?$

Use the ***vectorize*** operator (**CTRL+SHIFT+^**) to do an element-by-element differentiation. Because the input is now a vector rather than a range variable, the results can be assigned to a variable, and individual results may be obtained.

$$R\_2 := \overrightarrow{f(Vector)} = \begin{bmatrix} 0.000 \\ 4.000 \\ 8.000 \\ 12.000 \\ 16.000 \\ 20.000 \end{bmatrix} \quad R\_3 := \overrightarrow{g(Vector)} = \begin{bmatrix} 4.000 \\ 28.000 \\ 76.000 \\ 148.000 \\ 244.000 \\ 364.000 \end{bmatrix}$$

$R\_2_2 = 4.000$ $\quad$ $R\_3_2 = 28.000$

$R\_2_3 = 8.000$ $\quad$ $R\_3_3 = 76.000$

FIGURE 13.2

Assign the ***derivative*** operator to a function

algebraic result displayed. Another advantage is that the user-defined function is now a function of the algebraic result. This means that every time you use the function, PTC Mathcad does not need to recalculate the derivative of the function. This makes it so that you do not need to use the ***vectorize*** operator. See Figure 13.3.

Use **CTRL+PERIOD** for the ***symbolic evaluation*** operator.

$$F(x) := \frac{d}{dx} x^2 \rightarrow 2 \cdot x \quad G(x) := \frac{d}{dx}\left(x^3 + 3 \cdot x^2 + 4 \cdot x - 2\right) \rightarrow 3 \cdot x^2 + 6 \cdot x + 4$$

The above functions calculate the derivative of the function algebraically and assign the resulting function to the user-defined function. In other words, F(x) is now assigned as $2 \cdot x$ and G(x) is now assigned as $3 \cdot x^2 + 6 \cdot x + 4$.

This allows the use of a vector input without the need to use the ***vectorize*** operator.

$$\text{Vector} = \begin{bmatrix} 0.000 \\ 2.000 \\ 4.000 \\ 6.000 \\ 8.000 \\ 10.000 \end{bmatrix}$$ From Figure 13.2.

$$R\_4 := F(\text{Vector}) = \begin{bmatrix} 0.000 \\ 4.000 \\ 8.000 \\ 12.000 \\ 16.000 \\ 20.000 \end{bmatrix} \quad R\_5 := G(\text{Vector}) = \begin{bmatrix} 4.000 \\ 28.000 \\ 76.000 \\ 148.000 \\ 244.000 \\ 364.000 \end{bmatrix}$$

$R\_4_2 = 4.000$ $\quad R\_5_2 = 28.000$

$R\_4_3 = 8.000$ $\quad R\_5_3 = 76.000$

FIGURE 13.3

Using symbolic differentiation

You can also create a user-defined function that provides the derivative of a previously defined function.

$a(x) := x^2 \quad b(x) := x^3 + 3\cdot x^2 + 4\cdot x - 2$ Define two functions a and b.

$F1(x) := \frac{d}{dx} a(x) \rightarrow 2\cdot x$ This user-defined function takes the derivative of a previously defined function.

$G1(x) := \frac{d}{dx} b(x) \rightarrow 3\cdot x^2 + 6\cdot x + 4$

$$R\_6 := F1(Vector) = \begin{bmatrix} 0.000 \\ 4.000 \\ 8.000 \\ 12.000 \\ 16.000 \\ 20.000 \end{bmatrix} \qquad R\_7 := G1(Vector) = \begin{bmatrix} 4.000 \\ 28.000 \\ 76.000 \\ 148.000 \\ 244.000 \\ 364.000 \end{bmatrix}$$

An even more powerful function allows you to assign the function to be differentiated as part of its definition. Assign the Function label.

$\underline{F2}(f, x) := \frac{d}{dx} f(x)$ The argument f is a function.

$\underline{F2}(a, 2) = 4.000 \qquad \underline{F2}(b, 2) = 28.000$

$\underline{F2}(a, z) \rightarrow 2\cdot z \qquad \underline{F2}(b, z) \rightarrow 3\cdot z^2 + 6\cdot z + 4$

$$R\_8 := \underline{F2}(a, Vector) \rightarrow \begin{bmatrix} 0 \\ 4 \\ 8 \\ 12 \\ 16 \\ 20 \end{bmatrix} \qquad R\_9 := \overrightarrow{\underline{F2}(b, Vector)} = \begin{bmatrix} 4.000 \\ 28.000 \\ 76.000 \\ 148.000 \\ 244.000 \\ 364.000 \end{bmatrix}$$

$R\_8_2 = 4.000 \qquad R\_9_2 = 28.000$

FIGURE 13.4

Using previously defined functions

Figure 13.4 shows how to get the derivative of a previously defined function.

To get higher order derivatives, simply use the exponent placeholder on the ***derivative*** operator.

$$\frac{d}{dx^{\blacksquare}} x^2$$

$$F3(x) := \frac{d^2}{dx^2} x^2 \rightarrow 2 \qquad G3(x) := \frac{d^2}{dx^2}\left(x^3 + 3 \cdot x^2 + 4 \cdot x - 2\right) \rightarrow 6 \cdot x + 6$$

$$F4(x) := \frac{d^3}{dx^3} x^2 \rightarrow 0 \qquad G4(x) := \frac{d^3}{dx^3}\left(x^3 + 3 \cdot x^2 + 4 \cdot x - 2\right) \rightarrow 6$$

$$G5(x) := \frac{d^4}{dx^4}\left(x^3 + 3 \cdot x^2 + 4 \cdot x - 2\right) \rightarrow 0$$

FIGURE 13.5

Higher order derivatives

To get higher order derivatives, use the exponent placeholder on the ***derivative*** operator. See Figure 13.5.

PTC Mathcad also allows for the use of prime notation for derivatives. This is referred to as the ***prime*** operator. It is found in the **Operators** list on the **Math** tab. The keyboard shortcut is **CTRL + '**.

Add additional ***prime*** operators for calculating higher order derivatives.

When using the ***prime*** operator for a function, remember that the ***prime*** operator is used on the function before listing the arguments. You may also use the ***prime*** operator without listing the function arguments. The concepts are illustrated in Figure 13.6.

Figures 13.7 and 13.8 provide examples of plotting a function and its first and second derivatives.

a and b are functions defined in Figure 13.4.

Define a new function using the ***prime*** operator. The ***prime*** operator is used on the function name, not the arguments. The keyboard shortcut for the ***prime*** operator is **CTRL** + '.

$$\mathbf{H1a}(\mathbf{x}) := \mathbf{a}'(\mathbf{x}) \rightarrow 2 \cdot x \qquad \mathbf{H1b} := \mathbf{a}' \rightarrow \mathit{function} \qquad \mathbf{Vector} = \begin{bmatrix} 0.000 \\ 2.000 \\ 4.000 \\ 6.000 \\ 8.000 \\ 10.000 \end{bmatrix}$$

$$\mathbf{H1a}(2) = 4.000 \qquad \mathbf{H1b}(2) = 4.000$$

$$\mathbf{R\_10} := \mathbf{H1a}(\mathbf{Vector}) = \begin{bmatrix} 0.000 \\ 4.000 \\ 8.000 \\ 12.000 \\ 16.000 \\ 20.000 \end{bmatrix} \qquad \mathbf{R\_11} := \mathbf{H1b}(\mathbf{Vector}) = \begin{bmatrix} 0.000 \\ 4.000 \\ 8.000 \\ 12.000 \\ 16.000 \\ 20.000 \end{bmatrix}$$

$$\mathbf{H2a}(\mathbf{x}) := \mathbf{b}'(x) \rightarrow 3 \cdot x^2 + 6 \cdot x + 4 \qquad \mathbf{H2b} := \mathbf{b}' \rightarrow \mathit{function}$$

$$\mathbf{H2a}(2) = 28.000 \qquad \mathbf{H2b}(2) = 28.000$$

$$\mathbf{R\_12} := \mathbf{H2a}(\mathbf{Vector}) = \begin{bmatrix} 4.000 \\ 28.000 \\ 76.000 \\ 148.000 \\ 244.000 \\ 364.000 \end{bmatrix} \qquad \mathbf{R\_13} := \mathbf{H2b}(\mathbf{Vector}) = \begin{bmatrix} 4.000 \\ 28.000 \\ 76.000 \\ 148.000 \\ 244.000 \\ 364.000 \end{bmatrix}$$

FIGURE 13.6

Using prime notation

$\text{clear}(x)$

$\mathbf{a}(\mathbf{x}) := \mathbf{x}^2$

$\mathbf{a}'(\mathbf{x}) := \mathbf{a}'(\mathbf{x}) \rightarrow 2 \cdot x$

$\mathbf{a}''(\mathbf{x}) := \mathbf{a}''(x) \rightarrow 2$

Note: The prime symbol on the left of the ***definition*** operator is the single quote symbol and is part of the variable name. The prime symbol on the right of the ***definition*** operator is the ***prime*** operator (**CTRL+'**).

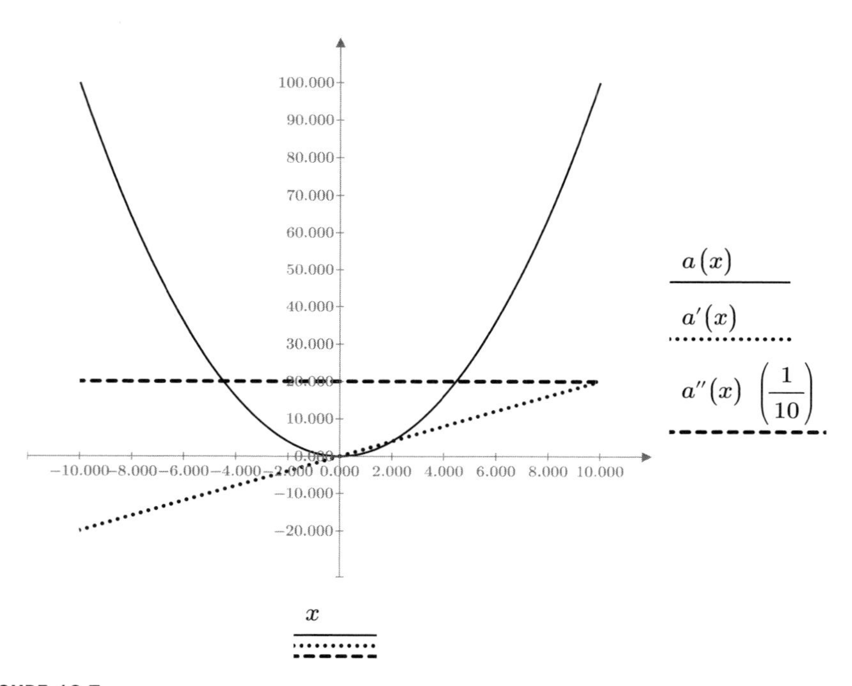

FIGURE 13.7

Plotting derivatives

$$\mathrm{H3}(x) := \frac{x^3+1}{x^3-8}$$

$$\mathrm{FirstDeriv}(x) := \frac{d}{dx}\mathrm{H3}(x) \rightarrow \frac{3\cdot x^2}{x^3-8} - \frac{3\cdot x^2\cdot(x^3+1)}{(x^3-8)^2} \xrightarrow{simplify} -\frac{27\cdot x^2}{(x^3-8)^2}$$

$$\mathrm{SecondDeriv}(x) := \frac{d^2}{dx^2}\mathrm{H3}(x) \xrightarrow{simplify} \frac{108\cdot x\cdot(x^3+4)}{(x^3-8)^3}$$

$a := -4, -3.9..4$ Set plot range

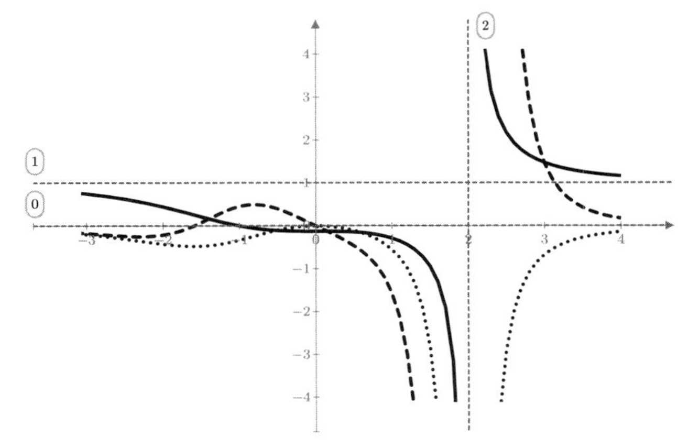

H3(a)

FirstDeriv(a)

SecondDeriv(a)

a

FIGURE 13.8

Plotting derivatives

Integration

Mathcad 15 used both an indefinite integral operator and a definite integral operator. PTC Mathcad Prime uses a single integral operator. To use the indefinite integral operator, just leave the placeholders for the definite integral operator blank.

The ***integral*** operator is found in the **Operators** list on the **Math** tab (**Math>Operators and Symbols>Operators>Calculus>** $\int dx$). The keyboard shortcut is **CTRL + SHIFT + I**. The ***integral*** operator has four placeholders.

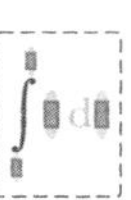

The definite integral looks like the following:

$$\int_a^b f(x)\,dx$$

For an indefinite integral, leave the placeholders on the integral blank:

$$\int f(x)\,dx$$

Indefinite integrals may only be solved symbolically. Figure 13.9 illustrates how to use symbolics to solve indefinite integrals.

Indefinite integrals may only be solved symbolically.

$$\mathbf{Integrate_1(x)} := \int x^2\,dx \rightarrow \frac{x^3}{3}$$

$I_{1a} := \mathbf{Integrate_1}(5) = 41.667 \qquad I_{1b} := \mathbf{Integrate_1}(1) = 0.333$

$I_{1c} := I_{1a} - I_{1b} = 41.333$

$$\mathbf{Integrate_2(x)} := \int x^3 + 2\cdot x^2 + x - 3\,dx \rightarrow \frac{x^4}{4} + \frac{2\cdot x^3}{3} + \frac{x^2}{2} - 3\cdot x$$

$I_{2a} := \mathbf{Integrate_2}(5) = 237.083 \qquad I_{2b} := \mathbf{Integrate_2}(1) = -1.583$

$I_{2c} := I_{2a} - I_{2b} = 238.667$

$$\mathbf{Integrate_3(x)} := \int \underline{\sin}(x)\,dx \rightarrow -\underline{\cos}(x)$$

$I_{3a} := \mathbf{Integrate_3}(5) = -0.284 \qquad I_{3b} := \mathbf{Integrate_3}(1) = -0.540$

$I_{3c} := I_{3a} - I_{3b} = 0.257$

$$\mathbf{Integrate_4(x)} := \int \left(30 - x^2\right)^2\,dx \rightarrow \frac{x^5}{5} - 20\cdot x^3 + 900\cdot x$$

$I_{4a} := \mathbf{Integrate_4}(5) = 2.625\cdot 10^3 \qquad I_{4b} := \mathbf{Integrate_4}(1) = 880.200$

$I_{4c} := I_{4a} - I_{4b} = 1744.800$

FIGURE 13.9

Indefinite integrals

Definite integrals may be solved both symbolically and numerically. Figure 13.10 shows both methods.

Figure 13.11 shows how to integrate using previously defined functions.

Definite integrals may be solved both symbolically and numerically.

$$\int_1^5 x^2 \, dx \rightarrow \frac{124}{3} \xrightarrow{float} 41.333333333333333333$$

$$\int_1^5 x^2 \, dx = 41.333$$

$$\int_1^5 x^3 + 2 \cdot x^2 + x - 3 \, dx \rightarrow \frac{716}{3} \xrightarrow{float} 238.66666666666666667$$

$$\int_1^5 x^3 + 2 \cdot x^2 + x - 3 \, dx = 238.667$$

$$\int_1^5 \sin(x) \, dx \rightarrow \cos(1) - \cos(5) \xrightarrow{float} 0.25664012040491345293$$

$$\int_1^5 \sin(x) \, dx = 0.257$$

$$\int_1^5 \left(30 - x^2\right)^2 \, dx \rightarrow \frac{8724}{5} \xrightarrow{float,6} 1744.8$$

$$\int_1^5 \left(30 - x^2\right)^2 \, dx = 1744.800$$

FIGURE 13.10

Definite integrals

Define four functions.

$a(x) := x^2 \quad b(x) := x^3 + 2 \cdot x^2 + x - 3 \quad c(x) := \underline{\sin}(x) \quad d(x) := (30 - x^2)^2$

$\underline{\text{Integrate9}}(f) := \int f(x)\,dx$

Create a function for the indefinite integral and assign the Function label.

$\underline{\text{Integrate9}}(a) \rightarrow \frac{x^3}{3}$

$\underline{\text{Integrate9}}(b) \rightarrow \frac{x^4}{4} + \frac{2 \cdot x^3}{3} + \frac{x^2}{2} - 3 \cdot x$

$\underline{\text{Integrate9}}(c) \rightarrow -\underline{\cos}(x)$

$\underline{\text{Integrate9}}(d) \rightarrow \frac{x^5}{5} - 20 \cdot x^3 + 900 \cdot x$

$\underline{\text{Integrate10}}(f, a, b) := \int_a^b f(x)\,dx$

Create a function for the definite integral and assign the Function label.

$\underline{\text{Integrate10}}(a, y, z) \rightarrow \frac{z^3}{3} - \frac{y^3}{3}$

$\underline{\text{Integrate10}}(a, 1, 5) = 41.333$

$\underline{\text{Integrate10}}(b, y, z) \rightarrow 3 \cdot y - \frac{2 \cdot y^3}{3} - \frac{y^2}{2} - \frac{y^4}{4} + \frac{z^4}{4} + \frac{2 \cdot z^3}{3} + \frac{z^2}{2} - 3 \cdot z$

$\underline{\text{Integrate10}}(b, 1, 5) = 238.667$

$\underline{\text{Integrate10}}(c, y, z) \rightarrow \underline{\cos}(y) - \underline{\cos}(z)$

$\underline{\text{Integrate10}}(c, 1, 5) = 0.257$

$\underline{\text{Integrate10}}(d, y, z) \rightarrow 20 \cdot y^3 - \frac{y^5}{5} - 900 \cdot y + \frac{z^5}{5} - 20 \cdot z^3 + 900 \cdot z$

$\underline{\text{Integrate10}}(d, 1, 5) = 1744.800$

FIGURE 13.11

Using previously defined functions

Differential Equations

PTC Mathcad has multiple functions to solve various differential equation systems. These solvers are numeric solutions and approximate the exact solutions. The solutions are a saved series of points and do not provide a smooth function. Units are not allowed in the differential equation solvers.

There are many functions that help solve systems of differential equations. These include: ***Adams***, ***AdamsBDF***, ***BDF***, ***Bulstoer***, ***bvalfit***, ***Jacob***, ***multigrid***, ***numol***, ***odesolve***, ***Radau***, ***relax***, ***Rkadapt***, ***rkfixed***, ***sbval***, ***statespace***, ***Stiffb***, and ***Stiffr***. The discussion of these functions is beyond the scope of this book, but both the PTC Mathcad Help and Tutorials are excellent resources. We will focus our discussion on the ***odesolve*** function.

Ordinary Differential Equations (ODEs)

> The ***Odesolve*** function still works in PTC Mathcad Prime 3.0, but it is being replaced by the lowercase ***odesolve*** function. The two functions are essentially the same, but the names of the arguments have been updated.

The function ***odesolve*** is used in a solve block to solve a single differential equation or a system of differential equations. It returns the solution as a function of the independent variable. It does this by saving solutions at a specific number of points (npoints) equally spaced in the solution interval, and then interpolating between these points using the function ***lspline***. The solve block sets the initial value or boundary constraints. The ODE must be linear in its highest derivative term, and the number of initial and boundary conditions must equal the order(s) of the ODE(s). The function has the form ***odesolve***(*vf, x, b, [intvls]*). The arguments are as follows:

- *vf* is a function or a column vector of functions as they appear within the solve block.
- *x* is the name of the variable of integration.
- *b* is the final point of the solution interval. (The initial point of the solution interval is specified by the initial conditions.)
- *intvls* is optional and is the integer number of equally spaced intervals used to interpolate the solution function. The number of solution points is the number of intervals + 1. The default value of *intvls* is 1000. If *intvls* is increased, then the interpolated solution function is more accurate. The default value is usually adequate, but if you are solving over a large interval, then set *intvls* to a value larger than 1000. Increasing *intvls* increases the calculation time.

The default solver for ***odesolve*** is the ***Adams/BDF*** method. Some functions use non-uniform step sizes internally when they solve differential equations, adding more steps in regions of greater variation of the solution, but they return the solution at the number of equally spaced intervals specified in *intvls*. Figures 13.12, 13.13, and 13.14 provide examples of using the ***odesolve*** function.

A tank is filled with 100 gal of salt solution containing 1 lbf of salt per gallon. Fresh brine containing 2 lbf of salt per gallon runs into the tank at a rate of 5 gal/min, and the mixture, assumed to be kept uniform by stirring, runs out at the same rate. Find the amount of salt in the tank at any time t, and determine the amount of salt after 15 minutes.

Let Q represent the total amount of salt (lbf) in solution in the tank at any point t; therefore, the amount of salt per gallon of solution is Q (lbf)/100 (gal). Salt enters the tank at a rate of (2lbf/gal)*(5gal/min)=10lbf/min. At any interval dt, the gain in salt is (10lbf/min)*(dt min) =10*dt*min.

Salt leaving the tank (at a rate of 5gal/min) in the interval dt is the same as the concentration of salt in the tank (Q (lbf)/100 (gal). Therefore (5gal/min)*(Qlbf/100gal)*dt (min)=(Qlbf/20)*dt.

dQ=10dt - (Q/20)dt=(10-Q/20)dt. It can be rewritten as dQ/dt+Q/20=10.

Set up the differential equation solve block.

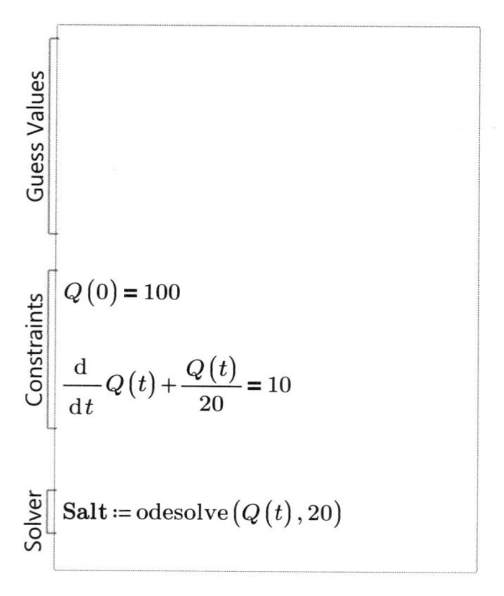

FIGURE 13.12 *(Continued on next page)*

Ordinary differential equation — Salt solution

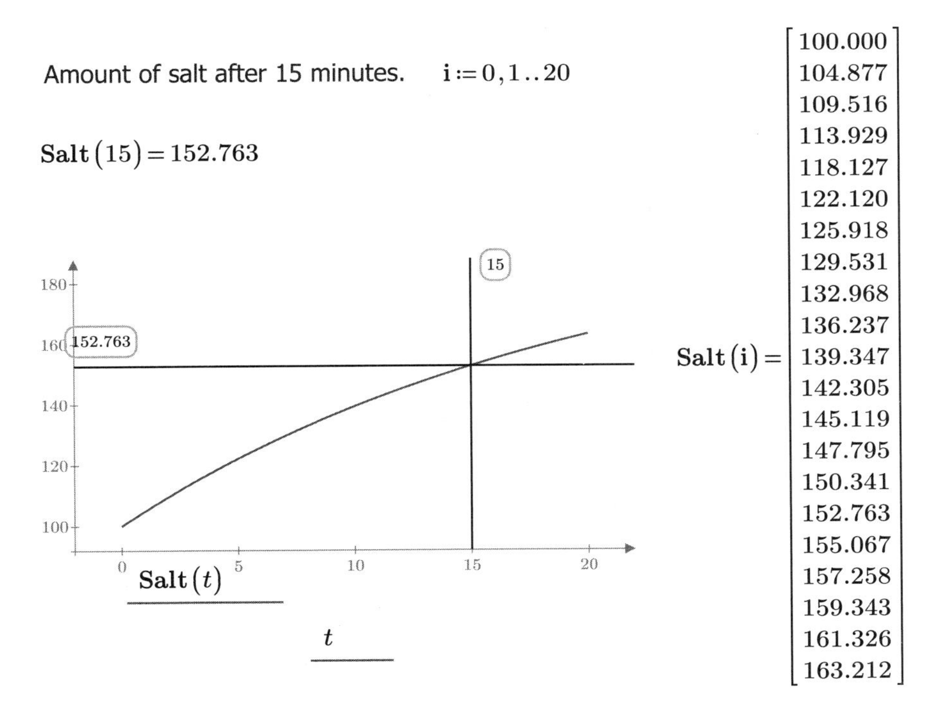

Graph of solution, showing amount of salt as a function of time.

FIGURE 13.12

(continued)

The water flowing out of the bottom of a parabolic shaped tank is modeled by the function: $\pi \cdot h^2 \cdot \frac{dh}{dt} = \frac{-2 \cdot \sqrt{h}}{5}$ where h is the height of water (meters) at time t (hours). The initial depth of water is h(0)=1. Find the depth of water after 2 hours.

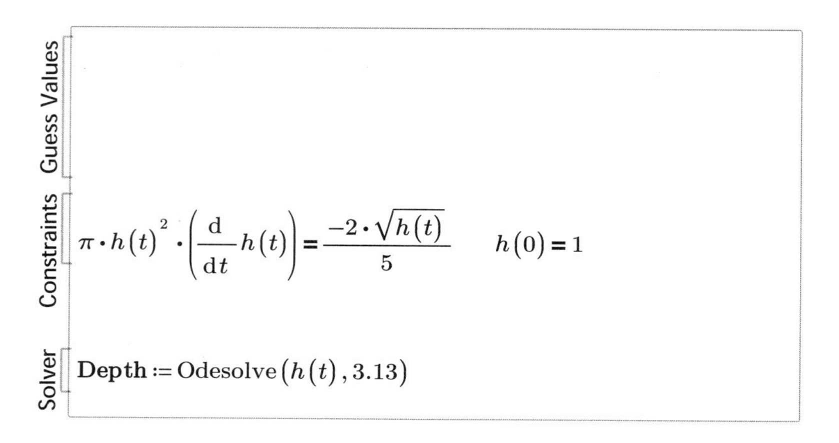

Guess Values
Constraints
Solver
$\pi \cdot h(t)^2 \cdot \left(\frac{d}{dt}h(t)\right) = \frac{-2 \cdot \sqrt{h(t)}}{5}$
$h(0) = 1$
Depth := Odesolve(h(t), 3.13)

Depth(2) = 0.667

i := 0, 0.5 .. 3

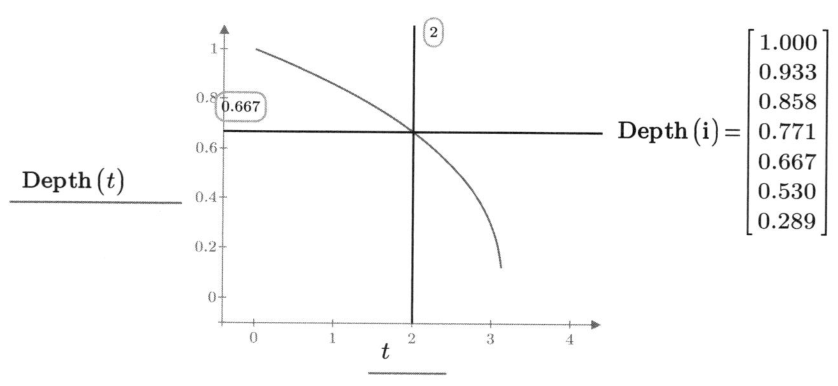

$$\text{Depth}(i) = \begin{bmatrix} 1.000 \\ 0.933 \\ 0.858 \\ 0.771 \\ 0.667 \\ 0.530 \\ 0.289 \end{bmatrix}$$

Graph of solution, showing depth of water as a function of time.

FIGURE 13.13

Ordinary differential equation — Water flow

The half-life of a substance in the process of transformation is the amount of time it takes for one-half of the substance to change. The standard model for this process is: $\frac{dS}{dt} = k \cdot S$ where $S(0) = S_0$. $S(t)$ is the amount of the substance at time t, and $S(0)$ is the amount at time t=0. The model states that the rate of change is dependant upon the amount of of the substance. The smaller the substance, the slower the rate of change.

If the original amount is 10 grams and the half-life is 100 hours, find the amount left after 40 hours.

The challenge with this problem is that we do not know the value of 'k" because it is determined after the solving of the solution equation. In order to get around this, we will create a user-defined function with "k" as the input variable. Once we have solved the solution equation, we can calculate the value of "k," and then we can use the user-defined function with the correct value of "k."

Solver Constraints Guess Values

$$S'(t) - k \cdot S(t) = 0 \qquad S(0) = 10$$

$$\underline{\text{SubstanceEq}}(k) := \text{odesolve}(S(t), 301)$$

The user-defined equation SubstanceEq takes the argument "k" and uses the ***odesolve*** function with the constraints listed in the solve block. This function will be used later in the problem.

Create another function that will set the constraints for a solve block. The constraints are that (Substance at time 0) - 10 = 0, and (Substance at time 100) - 5 = 0.

$$\text{Time} := \begin{bmatrix} 0 \\ 100 \end{bmatrix} \cdot hr \qquad \text{Sub} := \begin{bmatrix} 10 \\ 5 \end{bmatrix} \cdot gm$$

Set values for t=0, 100 and Substance=10, 5.

$$\underline{\text{Constraints}}(k) := \left\| \begin{array}{l} y \leftarrow \underline{\text{SubstanceEq}}(k) \\ \left(y\left(\frac{\text{Time}}{hr}\right) - \frac{\text{Sub}}{gm} \right) \end{array} \right.$$

Use a solve block to solve for the unknown "k."

FIGURE 13.14 *(Continued on next page)*

Ordinary differential equation — Half-life

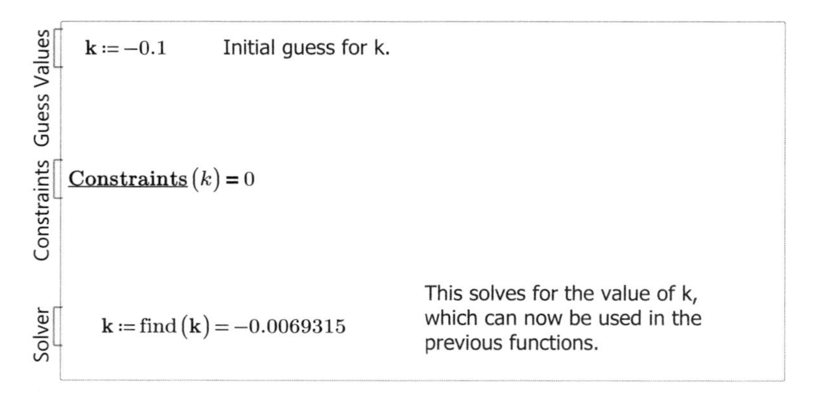

Now that we know "k," we can evaluate the function "Substance" with the value of "k."

$\mathbf{i} := 0, 25 .. 300$

$\mathbf{Substance} := \underline{\mathbf{SubstanceEq}}(\mathbf{k})$

$\mathbf{Substance}(0) = 10.000$

$\mathbf{Substance}(40) = 7.579$ Substance after 40 hours.

$\mathbf{Substance}(100) = 5.000$

$\mathbf{Substance}(200) = 2.500$

$\mathbf{Substance}(300) = 1.250$

$$\mathbf{Substance}(\mathbf{i}) = \begin{bmatrix} 10.000 \\ 8.409 \\ 7.071 \\ 5.946 \\ 5.000 \\ 4.204 \\ 3.536 \\ 2.973 \\ 2.500 \\ 2.102 \\ 1.768 \\ 1.487 \\ 1.250 \end{bmatrix}$$

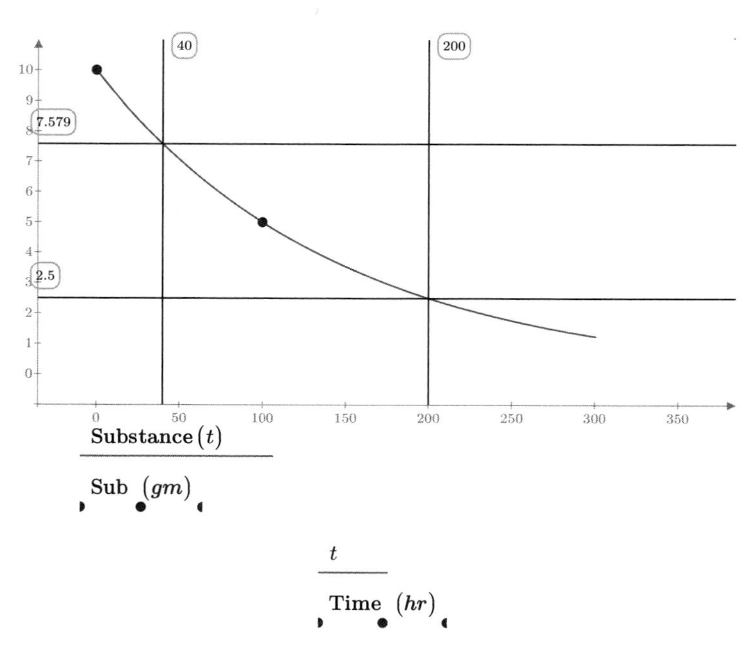

FIGURE 13.14

(*continued*)

Partial Differential Equations (PDEs)

The ***Pdesolve*** function is not available in PTC Mathcad Prime 3.0. Use the ***numol***, ***relax***, or ***multigrid*** functions instead.

The topic of partial differential equations is beyond the scope of this book. PTC Mathcad has three functions available to solve partial differential equations. These are the ***numol***, ***relax***, and ***multigrid*** functions. Information regarding these functions can be found in the Help Center. There are also examples available in the PTC Mathcad Help.

Engineering Examples

Engineering Example 13.1 is an integration problem that calculates the shear, moment, slope, and deflection diagrams for a uniformly loaded simple supported beam.

Engineering Example 13.1

Integration

In Engineering Example 8.1, we plotted the shear, moment, and deflection of a beam with uniform load. In this example, we derive the equations used to plot the shear, moment, slope, and deflection.

Note: PTC Mathcad does not include the integration constant C. Because we are creating functions for multiple solutions, we will need to solve for the integration constant.

In this example, all variables were manually assigned the "Variable" label.

Use the following to check the derived formulas for numeric results.

$\text{Length} := 20 \cdot ft \quad W := 2500 \cdot \frac{lbf}{ft} \quad E_1 := 29000 \cdot ksi \quad I_1 := 428 \cdot in^4$

$$z := 0\,ft, 1\,ft..\text{Length} = \begin{bmatrix} 0.00 \\ 1.00 \\ \vdots \end{bmatrix} ft$$

Load

Since load is downward, use a negative value.

$$\mathbf{Load}(\mathbf{w}) := -\mathbf{w}$$

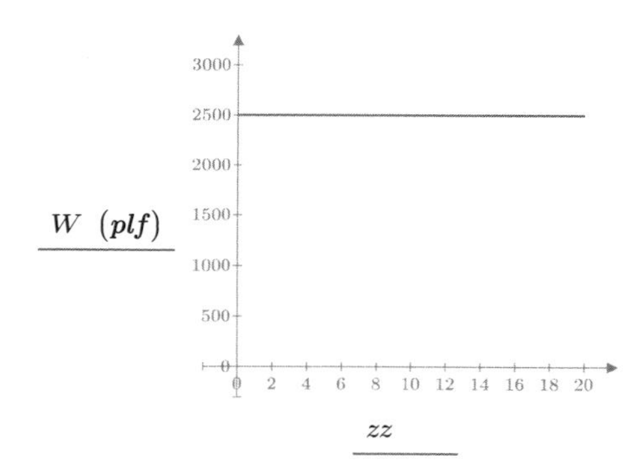

Shear

For this example, the beam is loaded with a uniform load. Calculate the shear by integrating the area under the loading diagram.

Since we do not know C1, include it as a variable to the function.

$$\mathbf{Shear}(\mathbf{x}, \mathbf{w}, \mathbf{L}, \mathbf{C_1}) := \int \mathbf{Load}(\mathbf{w})\, d\mathbf{x} + \mathbf{C_1} \rightarrow \mathbf{C_1} - x \cdot \mathbf{w}$$

Solve for constant of integration C1. The shear at x=0 is equal to w*l/2.

$$\mathbf{C_1} := \mathbf{Shear}(0, \mathbf{w}, \mathbf{L}, \mathbf{C_1}) = \frac{\mathbf{w} \cdot \mathbf{L}}{2} \xrightarrow{solve,\, \mathbf{C_1}} \frac{\mathbf{L} \cdot \mathbf{w}}{2}$$

Now that we know constant C1, eliminate it as a variable in the function.

$$\mathbf{Shear}(\mathbf{x}, \mathbf{w}, \mathbf{L}) := \mathbf{Shear}(\mathbf{x}, \mathbf{w}, \mathbf{L}, \mathbf{C_1}) \rightarrow \frac{\mathbf{L} \cdot \mathbf{w}}{2} - \mathbf{w} \cdot \mathbf{x}$$

$$\mathbf{Shear}(\mathbf{x}, \mathbf{w}, \mathbf{L}) \rightarrow \frac{\mathbf{L} \cdot \mathbf{w}}{2} - \mathbf{w} \cdot \mathbf{x}$$

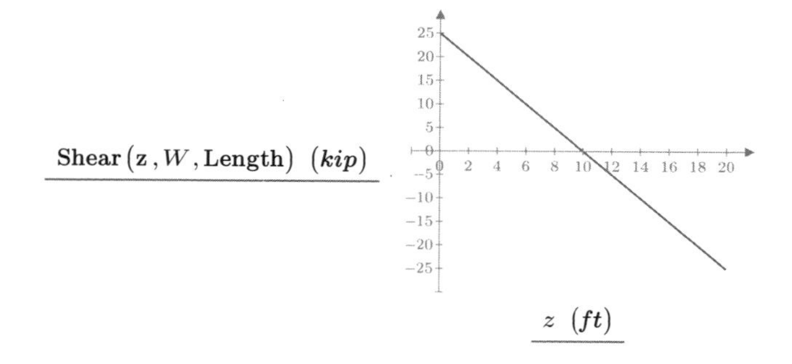

Moment

Calculate the equation for moment by integrating the area under the shear curve.

$$\text{Moment}(x, w, L, C_2) := \int \text{Shear}(x, w, L)\,dx + C_2 \rightarrow C_2 - \frac{\left(w \cdot x - \frac{L \cdot w}{2}\right)^2}{2 \cdot w} \xrightarrow{simplify} \frac{w \cdot L \cdot x}{2} - \frac{w \cdot L^2}{8} - \frac{w \cdot x^2}{2} + C_2$$

Solve for constant of integration C2. The moment at x=0 is equal to 0.

$$C_2 := \text{Moment}(0, w, L, C_2) = 0 \xrightarrow{solve, C_2} \frac{L^2 \cdot w}{8}$$

Now that we know constant C2, eliminate it as a variable in the function.

$$\text{Moment}(x, w, L) := \text{Moment}(x, w, L, C_2) \rightarrow \frac{L \cdot w \cdot x}{2} - \frac{w \cdot x^2}{2} \xrightarrow{simplify} \frac{w \cdot x \cdot (L - x)}{2}$$

$$\text{Moment}(x, w, L) \rightarrow \frac{w \cdot x \cdot (L - x)}{2}$$

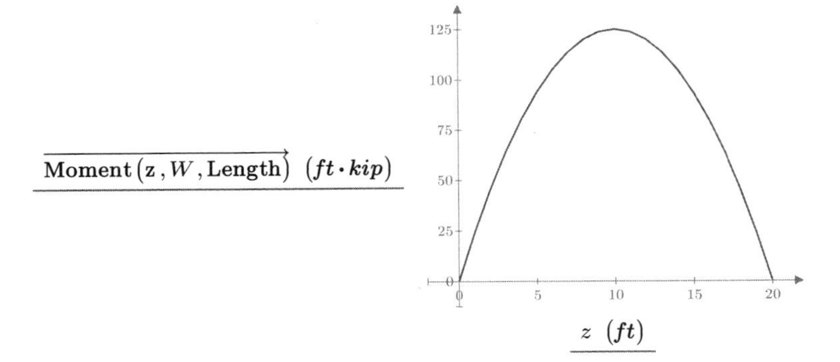

Slope

From mechanics of materials, the relationship for the radius of curvature of a beam p and moment is defined as $\frac{1}{p} = \frac{M}{E \cdot I}$, and the relationship of moment to slope θ is defined as $M = E \cdot I \cdot \frac{d}{dx}\theta$.

Calculate slope θ by integrating M/EI.

$$\text{Slope}(x, w, L, E, I, C_3) := \int \frac{\text{Moment}(x, w, L)}{E \cdot I}\, dx + C_3 \rightarrow C_3 - \frac{w \cdot \left(\frac{x^3}{3} - \frac{L \cdot x^2}{2}\right)}{2 \cdot E \cdot I} \xrightarrow{simplify} C_3 - \frac{2 \cdot w \cdot x^3 - 3 \cdot L \cdot w \cdot x^2}{12 \cdot E \cdot I}$$

Deflection

Calculate deflection by integrating the slope function.

$$\text{Defl}(x, w, L, E, I, C_3, C_4) := \int \text{Slope}(x, w, L, E, I, C_3)\, dx + C_4 \rightarrow C_4 + x \cdot C_3 - \frac{w \cdot x^4}{24 \cdot E \cdot I} + \frac{L \cdot w \cdot x^3}{12 \cdot E \cdot I}$$

Solve for C3 and C4. PTC Mathcad Prime 3.0 does not allow symbolic solutions in a solve block, so use the the keyword "solve" using a column vector to set the constraints.

The deflection at x = 0 is 0, and the deflection at x = L is 0.

$$\begin{bmatrix} C_3 & C_4 \end{bmatrix} := \begin{bmatrix} 0 = \text{Defl}(0, w, L, E, I, C_3, C_4) \\ 0 = \text{Defl}(L, w, L, E, I, C_3, C_4) \end{bmatrix} \xrightarrow{solve, C_3, C_4} \begin{bmatrix} -\frac{L^3 \cdot w}{24 \cdot E \cdot I} & 0 \end{bmatrix}$$

$$C_3 \rightarrow -\frac{L^3 \cdot w}{24 \cdot E \cdot I}$$

C3 and C4 cannot be evaluated numerically because the variables have not yet been defined.

$$C_4 \rightarrow 0$$

Now that we know constants C3 and C4, eliminate them as variables in the formula.

Formulas for slope and deflection at any point "x."

$$\mathrm{Slope}(x, w, L, E, I) := \mathrm{Slope}(x, w, L, E, I, C_3) \xrightarrow{simplify} -\frac{w \cdot (L^3 - 6 \cdot L \cdot x^2 + 4 \cdot x^3)}{24 \cdot E \cdot I}$$

$$\mathbf{Slope}(x, w, L, E, I) \rightarrow -\frac{w \cdot (L^3 - 6 \cdot L \cdot x^2 + 4 \cdot x^3)}{24 \cdot E \cdot I}$$

$$\mathrm{Defl}(x, w, L, E, I) := \mathrm{Defl}(x, w, L, E, I, C_3, C_4) \rightarrow \frac{L \cdot w \cdot x^3}{12 \cdot E \cdot I} - \frac{w \cdot x^4}{24 \cdot E \cdot I} - \frac{L^3 \cdot w \cdot x}{24 \cdot E \cdot I} \xrightarrow{simplify} -\frac{w \cdot x \cdot (L^3 - 2 \cdot L \cdot x^2 + x^3)}{24 \cdot E \cdot I}$$

$$\mathbf{Defl}(x, w, L, E, I) \rightarrow -\frac{w \cdot x \cdot (L^3 - 2 \cdot L \cdot x^2 + x^3)}{24 \cdot E \cdot I}$$

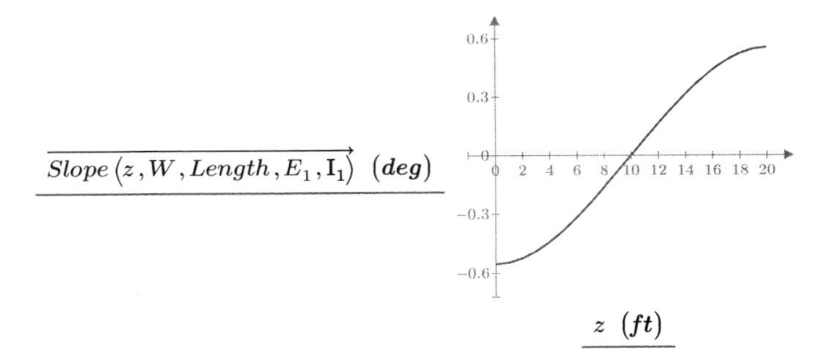

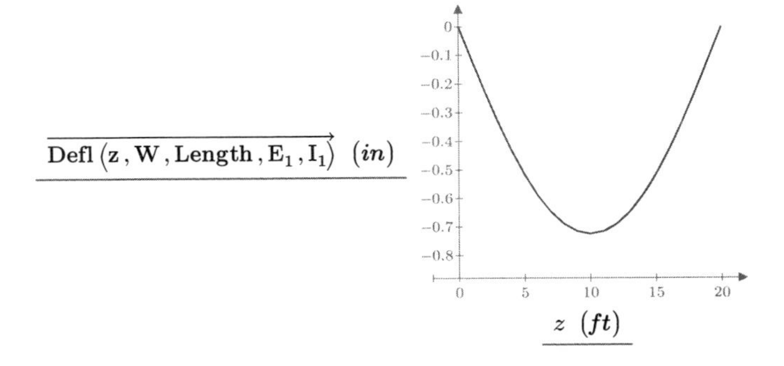

Find the formula for maximum deflection at the point of zero slope.

$$\mathbf{Dist} := \mathbf{Slope}(\mathbf{x}, \mathbf{w}, \mathbf{L}, \mathbf{E}, \mathbf{I}) = 0 \xrightarrow{solve, \mathbf{x}} \begin{bmatrix} -0.36602540378443864676 \cdot \mathbf{L} \\ 1.3660254037844386468 \cdot \mathbf{L} \\ 0.5 \cdot \mathbf{L} \end{bmatrix}$$

$$\mathbf{Dist}_3 \rightarrow 0.5 \cdot \mathbf{L}$$

Use the 3rd solution of 0.5L. Input it as L/2 so that decimal results are not returned.

$$\mathbf{MaxDefl}(\mathbf{w}, \mathbf{L}, \mathbf{E}, \mathbf{I}) := \mathbf{Defl}\left(\frac{\mathbf{L}}{2}, \mathbf{w}, \mathbf{L}, \mathbf{E}, \mathbf{I}\right) \rightarrow -\frac{5 \cdot \mathbf{L}^4 \cdot \mathbf{w}}{384 \cdot \mathbf{E} \cdot \mathbf{I}}$$

$$\mathbf{MaxDefl}(\mathbf{w}, \mathbf{L}, \mathbf{E}, \mathbf{I}) \rightarrow -\frac{5 \cdot \mathbf{L}^4 \cdot \mathbf{w}}{384 \cdot \mathbf{E} \cdot \mathbf{I}}$$

$$\mathbf{Defl}\left(\frac{\mathbf{Length}}{2}, \mathbf{W}, \mathbf{Length}, \mathbf{E}_1, \mathbf{I}_1\right) = -0.73\ in$$

$$\mathbf{MaxDefl}(\mathbf{W}, \mathbf{Length}, \mathbf{E}_1, \mathbf{I}_1) = -0.73\ in$$

Now plot similar values as in Engineering Example 8.1.

$Shear(z, W, Length)\ (kip)$

$\overrightarrow{Moment(z, W, Length)}\ (ft \cdot kip)$

$\overrightarrow{-Defl(z, W, Length, E_1, I_1)}\ \left(\frac{in}{100}\right)$

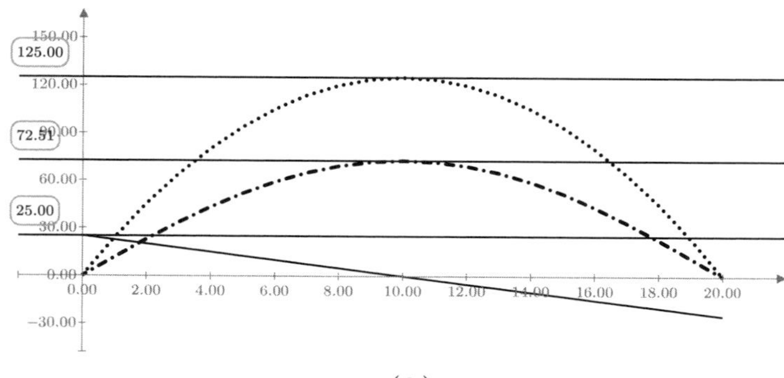

$z\ (ft)$

Engineering Example 13.2 is a differential equation problem that calculates contamination concentration for a two-cell lagoon system.

Engineering Example 13.2

2-cell, well-mixed, lagoon system

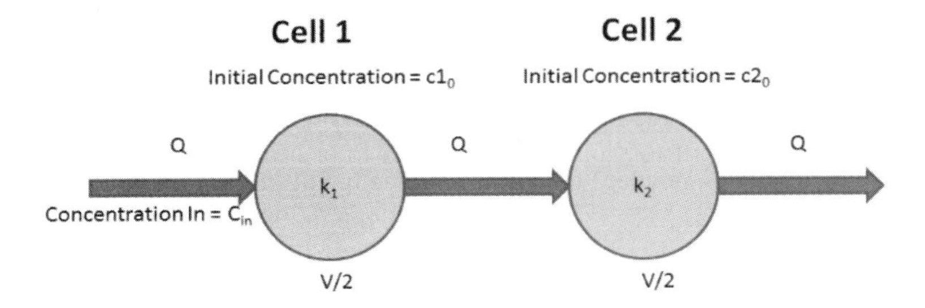

Problem:
A 2-cell, well-mixed sewage lagoon that has a total surface area of 10 hectare and a depth of $1 \cdot m$ is receiving $8640 \cdot \frac{m^3}{day}$ of waste (**Q**) containing $100 \cdot \frac{mg}{liter}$ of biodegradable contaminant. The contaminant is removed by bacteria via a first order reaction term with a reaction rate coefficient of $k_1 = \frac{0.5}{day}$ in Cell 1 and a reaction rate coefficient of $k_2 = \frac{0.1}{day}$ in Cell 2.

1. Write a mass balance equation on the contaminant for this system.
2. Solve the resulting ODE. The contaminant concentration in Cell 1 at time = 0 is $c1_0 = 100 \cdot \frac{mg}{liter}$ and the contaminant concentration in Cell 2 at time = 0 is $c2_0 = 50 \cdot \frac{mg}{liter}$. Plot the contaminant concentration in the lagoon effluent vs time (hours) for 20 days (480 hours).

Mass Balance Equation

A well-mixed sewage lagoon implies that at any point in time the conditions everywhere in the lagoon are the same. The mass balance on biodegradable contaminant for a well-mixed lagoon is:

Contaminant in - Contaminant out - Contaminant used by bacteria = Contaminant accumulation.

The concentration is equal to the Contaminant divided by the volume. The volume in each lagoon is one-half the total volume.

Symbolically the change in Contaminant can be written as:

$$\frac{d}{dt}C(t) = C'(t) = Q \cdot c_{in} - Q \cdot \frac{C(t)}{Vol} - k \cdot C(t)$$ -

All terms in this form of the mass balance have units of mass/time.

We are after the concentration in the lagoon effluent with concentration units of mg/liter. To get concentration, divide the Contaminant by the volume.

$$\frac{d}{dt}\left(\frac{C(t)}{Vol}\right) = \frac{Q \cdot c_{in}}{Vol} - \frac{Q \cdot \frac{C(t)}{Vol}}{Vol} - \frac{k \cdot C(t)}{Vol}$$. Because c=C/Vol, the equation can be rewritten as:

$$\frac{d}{dt}c(t) = \frac{Q}{Vol} \cdot c_{in} - \frac{Q}{Vol} \cdot c(t) - k \cdot c(t).$$

Input Variables

Volume in each lagoon:	$Vol := \frac{10 \cdot hectare \cdot 1 \cdot m}{2}$	$Vol = (5 \cdot 10^7)\ L$
Flow rate:	$Q := 8640 \cdot \frac{m^3}{day}$	$Q = (3.6 \cdot 10^5)\ \frac{liter}{hr}$
Initial concentration in cell 1:	$c1_0 := 100 \cdot \frac{mg}{liter}$	
Initial concentration in cell 2:	$c2_0 := 50 \cdot \frac{mg}{liter}$	
Influent concentration:	$c_{in} := 100 \cdot \frac{mg}{liter}$	
Bacteria reaction rate for cell 1:	$k_1 := 0.50 \cdot \frac{1}{day}$	$k_1 = 0.02083\ \frac{1}{hr}$
Bacteria reaction rate for cell 2:	$k_2 := 0.10 \cdot \frac{1}{day}$	$k_2 = 0.00417\ \frac{1}{hr}$

For this Engineering Example we will use two methods to solve the differential equations. The first method will use the ***odesolve*** function. The second method will use the ***Radau*** function.

For this solution we want to capture the concentration at time of 0, so we need to change the worksheet ORIGIN from 1 to 0.

Method 1: *odesolve*

Cell 1

Set the function and the constraints. In order to calculate Cell 2 at 480 hours, the results of Cell 1 must extend beyond 480 hours.

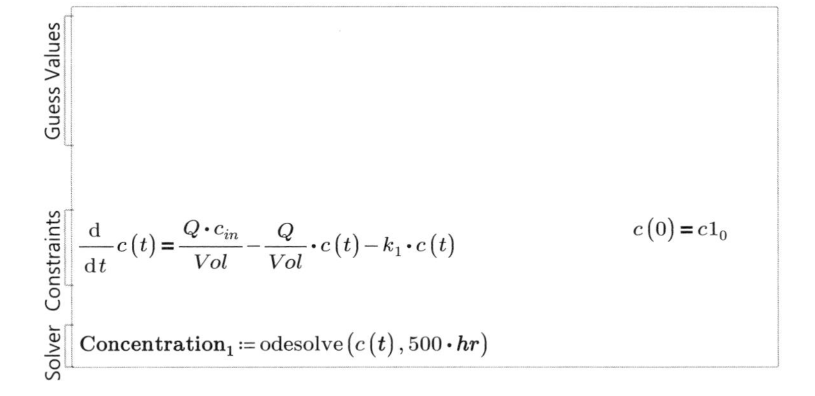

Guess Values
Constraints
Solver
$\frac{d}{dt}c(t) = \frac{Q \cdot c_{in}}{Vol} - \frac{Q}{Vol} \cdot c(t) - k_1 \cdot c(t)$
$c(0) = c1_0$
$Concentration_1 := odesolve(c(t), 500 \cdot hr)$

Display values of Concentration at various times.

$$Concentration_1(0 \cdot hr) = 100 \frac{mg}{liter}$$

$$Concentration_1(24\ hr) = 63.606 \frac{mg}{liter}$$

$$Concentration_1(96 \cdot hr) = 30.722 \frac{mg}{liter}$$

$$Concentration_1(200 \cdot hr) = 25.957 \frac{mg}{liter}$$

$$Concentration_1(480 \cdot hr) = 25.684 \frac{mg}{liter}$$

$$t := 0\ hr, 1\ hr .. 500\ hr$$

Plot the values of Concentration1 over time.

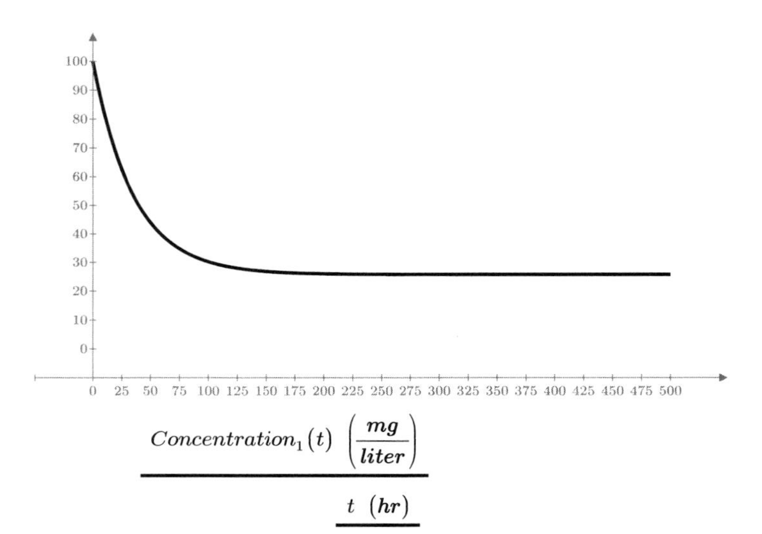

Cell 2

Bacteria reaction rate for Cell 2: $k_2 = 0.00417 \frac{1}{hr}$ $c2_0 = 50 \frac{mg}{liter}$

The input concentration for Cell 2 is equal to the output concentration of Cell 1. Therefore, use the solution to Cell 1 as an input function for Cell 2.

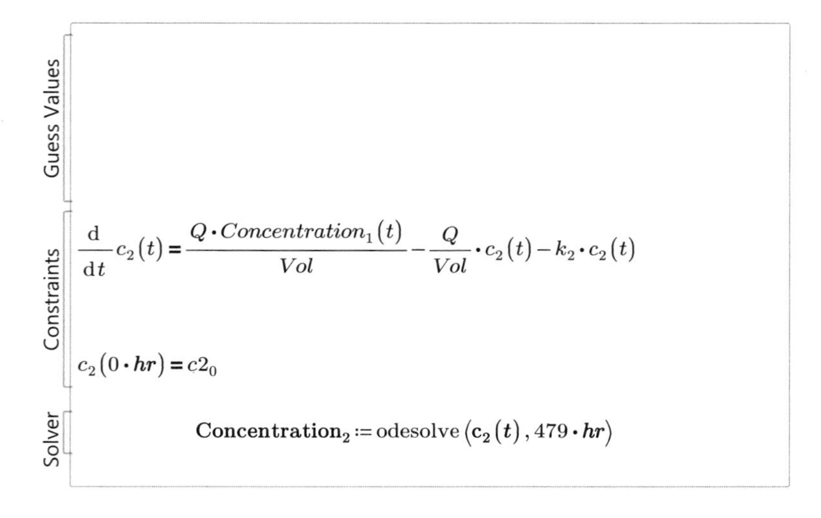

$$\text{Concentration}_2(0 \cdot hr) = 50 \frac{mg}{liter}$$

$$\text{Concentration}_2(24 \cdot hr) = 50.007 \frac{mg}{liter}$$

$$\text{Concentration}_2(96 \cdot hr) = 36.204 \frac{mg}{liter}$$

$$\text{Concentration}_2(200 \cdot hr) = 22.93 \frac{mg}{liter}$$

$$\text{Concentration}_2(479 \cdot hr) = 16.554 \frac{mg}{liter}$$

Plot the values of Concentration1 and Concentration2 over time.

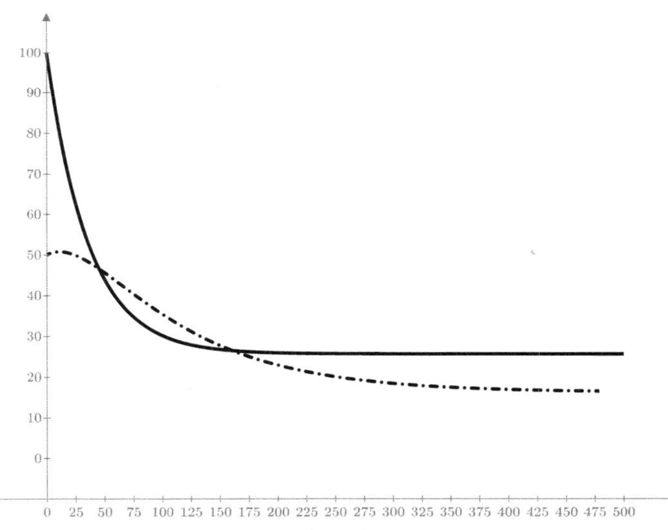

Method 2: *Radau*

Cell 1

We will now use the ***Radau*** function to solve the above problem. The other ODE functions are very similar. PTC Mathcad defines the ***Radau*** function as follows:

Radau (■,■,■,■,■,■,■,■)

Radau(y, x1, x2, npoints, D, [J], [M], [tol])

Returns a matrix of solution values for the stiff differential equation specified by the derivatives in D, and initial conditions y on the interval [x1,x2], using a RADAU5 method. Parameter npoints controls the number of rows in the matrix output.

The arguments J, M, and tol are optional and are not needed for this solution.

Define the function D for use in the ***Radau*** function. Note that in this function the arguments t and c are vector values, and c_0 is an array subscript and not a literal subscript.

$$D1(t,c) := \left[\frac{\frac{Q}{1} \cdot \frac{c_{in}}{1}}{\frac{Vol}{1}} - \frac{\frac{Q}{1}}{Vol} \cdot c_0 - (k_1 \cdot 1) \cdot c_0 \right] \qquad c1_0 = 100 \, \frac{mg}{liter}$$

Assign the result of the ***Radau*** function to the variable Concentration1.

$$Concentration1 := Radau\left(\frac{c1_0}{1}, 0\,hr, 480\,hr, 480, D1\right)$$

Input arguments for the ***Radau*** function

Radau (y, x1, x2, npoints, D, [J], [M], [tol]).

y = Initial condition of 100 mg/liter.

x1 is the first value of the interval (time = 0 hours.)

x2 is the last value of the interval (time = 480 hours.)

$$480 \cdot hr = 20\ day$$

npoints is the number of points to solve for. In this case, we will solve for 480 points — one for each hour.

D is the previously defined function.

The result is a matrix of two columns and 480 rows. The first column is the values of t and the second column is the values of c.

$$\mathbf{Concentration1} = \begin{bmatrix} 0\ s & 0.1\ \frac{kg}{m^3} \\ (3.6 \cdot 10^3)\ s & 0.098\ \frac{kg}{m^3} \\ (7.2 \cdot 10^3)\ s & 0.096\ \frac{kg}{m^3} \\ (1.08 \cdot 10^4)\ s & 0.094\ \frac{kg}{m^3} \\ (1.44 \cdot 10^4)\ s & 0.092\ \frac{kg}{m^3} \\ (1.8 \cdot 10^4)\ s & 0.09\ \frac{kg}{m^3} \\ (2.16 \cdot 10^4)\ s & 0.088\ \frac{kg}{m^3} \\ (2.52 \cdot 10^4)\ s & 0.087\ \frac{kg}{m^3} \\ & \vdots \end{bmatrix}$$

Extract the values of each column.

$\mathbf{Time} := \mathbf{Concentration1}^{\langle 0 \rangle}$

$\mathbf{ConcentrationCell\_1} := \mathbf{Concentration1}^{\langle 1 \rangle}$

$$\mathbf{Time} = \begin{bmatrix} 0 \\ 1 \\ 2 \\ 3 \\ 4 \\ 5 \\ 6 \\ 7 \\ 8 \\ 9 \\ 10 \\ 11 \\ \vdots \end{bmatrix} hr \qquad \mathbf{ConcentrationCell\_1} = \begin{bmatrix} 100 \\ 97.946 \\ 95.948 \\ 94.006 \\ 92.117 \\ 90.28 \\ 88.495 \\ 86.758 \\ 85.07 \\ 83.428 \\ 81.833 \\ 80.283 \\ \vdots \end{bmatrix} \frac{mg}{liter}$$

Plot the values of Cell 1

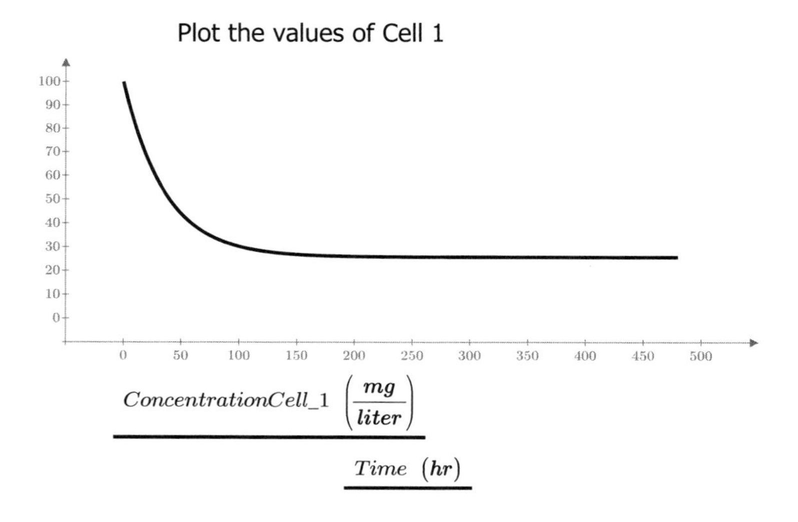

Cell 2

The output from Cell 1 is an array of data for time and concentration. The input for Cell 2 is based on the outflow from Cell 1. It is required to be in the form of a function. An array will not work.

We need to create a function from the Cell 1 data. We will use the PTC Mathcad curve fitting functions ***lspline*** and ***interp*** to create the function. Instructions on how to use these funtions can be found in PTC Mathcad Help.

Create the spline curve vector used by the ***interp*** function.
Here we use the linear spline function ***lspline.*** The other spline functions available are cubic (***cspline***) and parabolic (***pspline***):

$$S := \underline{\text{lspline}}(\text{Time}, \text{ConcentrationCell\_1})$$

Use the spline curve vector and the ***interp*** function to create a function to match the data:

$$\text{ConcentrationFunction}(t) := \underline{\text{interp}}(S, \text{Time}, \text{ConcentrationCell\_1}, t)$$

Define the function D for use in the ***Radau*** function.
Note that in this function the value for cin from Cell 1 is replaced by the ConcentrationFunction calculated above. The arguments t and c are vector values, and c0 is an array subscript and not a literal subscript.

$$D2(t, c) := \left[\frac{Q \cdot \text{ConcentrationFunction}(t)}{\text{Vol}} - \frac{Q}{\text{Vol}} \cdot c_0 - (k_2) \cdot c_0 \right]$$

$$c2_0 = 50 \, \frac{mg}{liter}$$

$$\text{Concentration2} := \underline{\text{Radau}}\left(c2_0, 0\ hr, 480\ hr, 480, D2\right)$$

<u>Input arguments for the ***Radau*** function</u>

Radau (y, x1, x2, npoints, D, [J], [M], [tol])

y = Initial condition of 50 mg/liter.

x1 is the first value of the interval (time = 0 hours.)

x2 is the last value of the interval (time = 480 hours.)

$$480 \cdot hr = 20\ day$$

npoints is the number of points to solve for. In this case, we will solve for 480 points — one for each hour.

D is the previously defined function.

$$\mathbf{Time} := \mathbf{Concentration2}^{\langle 0 \rangle}$$

$$\mathbf{ConcentrationCell\_2} := \mathbf{Concentration2}^{\langle 1 \rangle}$$

$$\mathbf{Time} = \begin{bmatrix} 0 \\ 1 \\ 2 \\ 3 \\ 4 \\ 5 \\ 6 \\ 7 \\ 8 \\ 9 \\ 10 \\ 11 \\ \vdots \end{bmatrix} hr \qquad \mathbf{ConcentrationCell\_2} = \begin{bmatrix} 50 \\ 50.143 \\ 50.271 \\ 50.382 \\ 50.479 \\ 50.562 \\ 50.63 \\ 50.685 \\ 50.727 \\ 50.757 \\ 50.774 \\ 50.779 \\ \vdots \end{bmatrix} \frac{mg}{liter}$$

As you can see from the plot below, the solution using the ***Radau*** function matches the solution using the ***odesolve*** function.

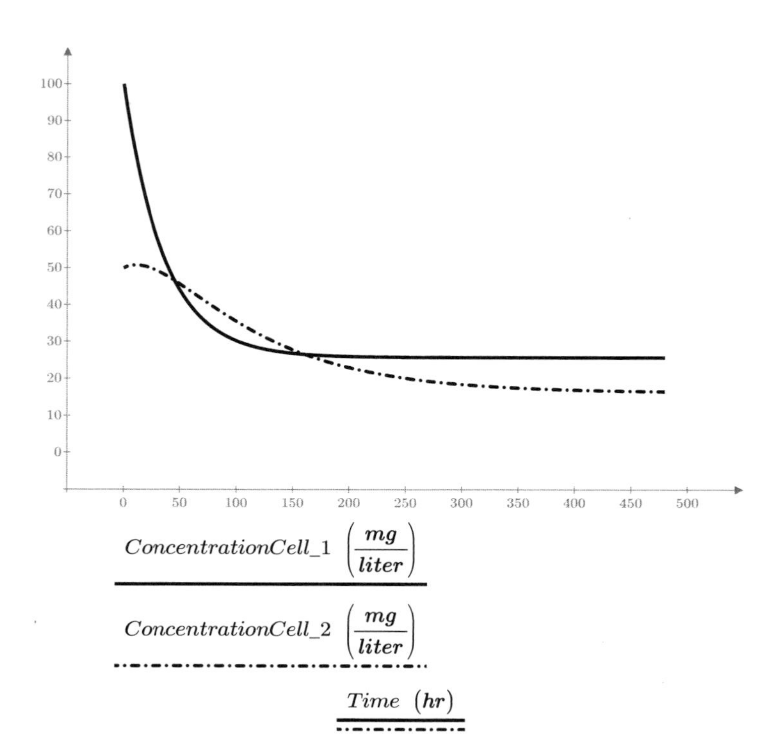

Practice

The PTC Mathcad Prime 3.0 figures and examples used in this book are available for download from the book's website. The reader is encouraged to download the files and use them to practice the concepts learned. Additional examples and problems are also provided. To access this content go to http://store.elsevier.com/9780124104105 *and click on the Resources tab, and then click on the link for the Online Companion Materials.*

1. From your calculus textbook, use the ***symbolic evaluation*** operator to solve 10 differentiation problems.
2. From your calculus textbook, use the ***symbolic evaluation*** operator to solve 10 indefinite integral problems.
3. From your calculus textbook, use the ***symbolic evaluation*** operator to solve 10 definite integrals.
4. Use numeric integration to solve the same problems as above.
5. Use the ***symbolic evaluation*** operator and numeric differentiation (at $x = 5$) to differentiate the following (Note: Assign the numeric values at the end of your worksheet, so that the symbolic evaluations will not be affected by the numeric values.):
   a. acos(x)
   b. sec(x)
   c. x^n (assign $n = 5$ for the numeric differentiation)
   d. $3x^3 + 2x^2 - 4x + 3$
   e. e^x
6. Use the ***symbolic evaluation*** operator to integrate the following:
   a. x^n (assign $n = 5$ for the definite integral in the next problem)
   b. e^x
   c. cos(x)
   d. $x*\sec(x^2)$
   e. tan(a*x) (assign $a = 0.8$ for the definite integral in the next problem)
7. Evaluate the above integrals as definite integrals from the values of $a = 1$ and $b = 5$ (For $x*\sec(x^2)$ use $a = 0.1$ and $b = 0.2$.)
8. From your differential equations text book, solve 5 first order ordinary differential equations using the ***odesolve*** function.

PART

Creating and Organizing Your Engineering Calculations with PTC Mathcad

Part IV discusses the use of PTC Mathcad to create a complete set of engineering calculations. This section uses the phrase "engineering calculations" extensively, but the ideas and concepts covered are just as applicable to scientific or other technical calculations. I hope our friends from the scientific community will bear with the focus on engineering calculations.

You now have a toolbox full of simple tools and power tools. There are many other tools that still can be added, but it is time to discuss how to apply the tools you have. It is time to start building PTC Mathcad calculations.

Part IV will begin with a discussion about customizing PTC Mathcad settings and creating templates in order to get consistent worksheets. It will then discuss ways of assembling calculations from standard calculation worksheets. It finishes with a discussion of how PTC Mathcad interacts with Microsoft Excel, and shows how to use your existing Microsoft Excel spreadsheets in PTC Mathcad.

CHAPTER

Putting It All Together 14

Creating an entire set of engineering project calculations with PTC Mathcad is much more complex than writing a one or two page worksheet. Many issues need to be considered when creating a full set of engineering project calculations. The purpose of this chapter is to briefly review some of the topics covered in earlier chapters and to introduce the topics to be discussed in Part IV.

Chapter 14 will:

- Discuss the concept of the PTC Mathcad toolbox and the tools in it.
- Introduce the concept of using the tools to build a project.
- Review the topics discussed in all the chapters up to this point.
- Paint a picture of what will be accomplished in Part IV, by briefly discussing the topics to be covered in Chapters 15–23.

Introduction

Let's go back to the analogy used at the beginning of this book—teaching you how to build a house. We have helped you build a toolbox—a place where you can store your tools. Your PTC Mathcad toolbox is a thorough understanding of PTC Mathcad basics.

We have also collected many tools and put them into your toolbox. We have taken classes in hammers, screwdrivers, pliers, power saws, power drills, etc. In Part IV, we teach you how to use these tools to build the house.

Guidelines for Naming Variables

As you begin creating and assembling calculations, the variable names you chose to use will begin to be very important.

When you create a simple one or two page worksheet, it usually does not matter what variable names you chose to use. Single letter variables are more than adequate because the entire worksheet can be taken in context and you usually do not run the risk of redefining the variable.

Essential PTC® Mathcad Prime® 3.0.

Variable names are critical when you create a complete set of project calculations that can be over 100 pages long. You may have many conditions where you want to use the same variable name, but you want each variable to be unique, and you do not want to redefine a previously used variable. The use of literal subscripts to make each variable name unique is very useful in this case.

Descriptive variable names are also very useful. They help you remember what the variable refers to, and they also help those reviewing the calculations know what the variable name means. For example, is it easier to understand d:=5 ft or DepthOfWater:= 5 ft? In a single page worksheet either one would be adequate. When your calculations are 50 pages long, if you defined "d" on page 5 and do not use it again until page 50, then it may be more useful to have a descriptive variable name.

Descriptive variable names also help prevent you from redefining a variable name. In the previous example, if you defined the depth of water as d:=5 ft and then later defined the depth of soil as d:=6 ft, you have redefined the variable d. This will cause an error in your calculations if later on you use the variable "d" thinking that it is the depth of water. It is much better to use DepthOfWater:=5 ft and DepthOfSoil:=6 ft.

The following naming guidelines will help you choose variable names to be used in your engineering calculations. There will be exceptions to each of these guidelines, but if they are consistently followed, they will help prevent problems that will be discussed in future chapters.

Naming Guideline 1

Use Descriptive Variable Names

Single letters are appropriate if you are just doing a quick calculation for something that won't be saved. They are also appropriate if the single letter represents a universal constant such as "e." They are also useful if you have an equation that uses single letter variables. But, generally, it is better to use descriptive variable names with more than one letter.

Naming Guideline 2

Use a Combination of Uppercase and Lowercase Letters to Help Make Your Variable Names Easier to Read

Unless there is some specific reason not to, your variable names should begin with an uppercase letter. If your variable name is more than one word, then use a combination of uppercase and lowercase letters to make the variable name easier to read. For example instead of naming a variable "vesselpressure," name it "VesselPressure." See Figure 14.1 for some examples.

Buildingwidth	BuildingWidth
Maximumacceleration	MaximumAcceleration
Momentofinertia	MomentOfInertia
Transversewavevelocity	TransverseWaveVelocity
Areaofcircle	AreaOfCircle

FIGURE 14.1

Using uppercase letters to separate words in variable names

Naming Guideline 3

Use Underscore to Separate Different Names in Your Variable Names

In the previous guideline we discussed using a combination of uppercase and lowercase letters to make variable names easier to read. Another way to do this is to use the underscore to separate names. For example, instead of naming a variable "VesselPressure," you could name it "Vessel_Pressure." See Figure 14.2 for some examples.

BuildingWidth	Building_Width
Maximumacceleration	Maximum_Acceleration
Momentofinertia	Moment_Of_Inertia
Transversewavevelocity	Transverse_Wave_Velocity
Areaofcircle	Area_Of_Circle

FIGURE 14.2

Using underscore to separate words in variable names

Naming Guideline 4

Make Good Use of Subscripts in Your Variable Names

Literal subscripts are created by typing **CTRL + MINUS** within your variable name. Subscripts are useful to distinguish variables which are very similar or related. Remember that the literal subscript is a toggle that you turn on and off to create multiple subscripts in your variable name. Figure 14.3 shows a few examples of using subscripts.

Area1	$Area_1$
Volume1	$Volume_1$
Velocity1	$Velocity_1$
Integral1	$Integral_1$
H2SO4	H_2SO_4

FIGURE 14.3

Using subscripts in variable names

Naming Guideline 5

Use the Single Quote Key ' If You Need to Use a "prime" in Your Variable Name

> The single quote key now works as a prime symbol in variable names. In Mathcad 15, the single quote key added parentheses. ☺

Sometimes your variable name has an apostrophe in it, such as the strength of concrete, f'_c. Use the single quote key for this symbol.

PTC Mathcad Toolbox

Let's quickly review the skills and understanding you should have mastered from Part I.

Variables

By now, you are completely familiar with how to define and use variables, but do you remember how to use the special text mode to add symbols to your variable names? You do this by first creating a string, and then deleting the double quote in the string.

It will be helpful to establish some personal or corporate naming guidelines. There have been many times when I have been very careful to use descriptive variable names, only to be caught later on not remembering what I named a particular variable. Did I name it DepthOfWater or WaterDepth? Was it BuildingLength or LengthOfBuilding or Building_Length? It will be beneficial for you to create some guidelines for your specific needs so that the names used in your calculations have a consistent usage.

The figures for each chapter in this book were created in a single PTC Mathcad file. This required the use of unique variable names for each figure. This book intentionally uses a variety of variable naming methods in the figures to expose you to several different ways to name variables.

Editing

You could not have gotten this far in the book without understanding how to edit an expression. We will assume that you have a good understanding of editing techniques.

User-defined Functions

The discussion of user-defined functions in Chapter 4 was very basic. This was intentional in order to establish a solid understanding of user-defined functions prior to introducing the more powerful and complex features.

Each chapter has been built on this foundation of user-defined functions. In Chapter 5, we introduced units to user-defined functions. In Part II and Part III, we added arrays, programming, and symbolics to user-defined functions.

Units!

Units are powerful! They are essential to engineering calculations! Use them consistently!

You should be very comfortable with using units. If not, go back and review Chapter 5. Units are very easy to use with most engineering equations. For the few equations where units will not work (such as empirical equations), divide the variable by the desired unit to make it a unitless number of the expected value.

Remember to include units in the argument list for user-defined equations. When plotting, remember to add units in the unit placeholders.

PTC Mathcad Settings

Chapter 15 will discuss PTC Mathcad settings and suggest some changes to PTC Mathcad default values.

Customizing PTC Mathcad with Templates

In Chapter 16, we will discuss styles and labels for math variables. We will recommend changing some default PTC Mathcad styles and we will also discuss the use of headers and footers. In Chapter 17, we discuss saving these customizations into a template.

Hand Tools

Arrays are an essential tool for engineering calculations. We have used vectors in numerous examples in order to display results from multiple input variables. Range variables have been used in many examples. If you have forgotten about the ***vectorize*** operator, review Chapter 6 for its use. These are essential tools as you begin to build engineering calculations.

Chapter 7 discussed a few selected functions. These are only the beginning of the functions you can add to your PTC Mathcad toolbox. Review the functions and operators discussed in Chapter 7, and see which functions will be most useful to you.

Plotting is a very useful tool. If you are not comfortable with creating simple XY plots or simple Polar plots, review Chapter 8.

A thorough understanding of simple logic programming is essential. Engineering calculations require the use of logic programs to choose appropriate actions. If you are not yet comfortable with the use of programs in expressions or user-defined programs, review Chapter 9.

Power Tools

Chapters 10 through 13 introduced some topics that may seem confusing at first. That is one reason that their discussion was held off until Part III.

The symbolic calculations, solving tools, and programming tools are extremely powerful. These tools will be very beneficial in your engineering calculations. When it comes to using PTC Mathcad to solve for various equations, there are many ways to arrive at the same answer. We have tried to show you the various ways of solving for solutions. The engineering examples shown in Chapter 11 provide a good summary of how to use the solving functions.

Let's Start Building

Now that you have your toolbox full of tools, it is time to start building. One of the primary purposes of this book is to teach you how to create and organize engineering calculations. The principals can be used for any technical calculations, not just engineering calculations.

PTC Mathcad is perfectly suited to create and organize your project calculations. The reasons for this include:

- It speaks our language—math.
- Calculations and equations are visible, and not hidden in spreadsheet cells.
- The calculations are electronic and can be printed, archived, shared with co-workers around the world, searched, and reused.
- PTC Mathcad calculations can be used over and over again.
- If input variables change, the calculations are automatically updated.

What is Ahead

The chapters in Part IV will help you as you begin using PTC Mathcad to create and organize your engineering calculations. In Chapter 18, we discuss how to save standard calculations and to reuse them in your worksheets. A very powerful feature of PTC Mathcad is its ability to communicate with and transfer information between Microsoft Excel. An entire Chapter (19) is devoted to this topic.

Summary

This chapter is a review of the topics covered in previous chapters. We emphasized the importance of choosing suitable variable names in your calculations. We also encouraged you to go back and review any chapter with which you are not comfortable. We then looked to the future and highlighted the topics we are about to discuss.

Practice

The PTC Mathcad Prime 3.0 figures and examples used in this book are available for download from the book's website. The reader is encouraged to download the files and use them to practice the concepts learned. Additional examples and problems are also provided. To access this content go to http://store.elsevier.com/9780124104105 *and click on the Resources tab, and then click on the link for the Online Companion Materials.*

1. Create five variable names using the special text mode discussed in Chapter 3. Use different names than you used for the practice in Chapter 3.
2. Go back and briefly review the topics covered in each chapter. Look at the practice exercises to see how well you remember the topics.

CHAPTER

PTC Mathcad Settings

15

This chapter will give you detailed information about many of the PTC Mathcad settings. With this information you can make informed decisions as to what setting should be used for your situation.

This chapter will discuss each setting in the order it appears along the ribbon. The changes you make on the ribbon will only be effective for the specific worksheet. Each worksheet may have different ribbon settings. Refer to Chapter 17 to see how to set default settings for all future worksheets. Feel free to skip over any item that is not clear or not useful. After you become more familiar with PTC Mathcad you can refer back to this chapter to review the different settings. Previous chapters have touched on several of the topics covered in this chapter. Much of the information here is taken directly from the PTC Mathcad Help Center with additional comments added.

Chapter 15 will:

- Discuss the PTC Mathcad Options dialog box and show how different settings affect the way PTC Mathcad starts up and how it operates.
- Discuss the various settings on the Ribbon Bar and show how different settings can affect a specific worksheet.
- Recommend specific settings for various features.
- Discuss and show how to control the way results are displayed.

PTC Mathcad Options

The settings in the PTC Mathcad Options dialog box affect all worksheets. This box allows you to control the location of templates, which template your new worksheets are based on, and what items appear when you open PTC Mathcad. The dialog box is opened by clicking the PTC Mathcad Button (M) and then clicking **Options**. See Figure 15.1.

Enable Getting Started Tab

The **Getting Started** tab contains useful information for learning PTC Mathcad. Placing a check in the **Enable Getting Started Tab** checkbox turns on the **Getting Started** tab and makes it visible in the ribbon. Unchecking the box will turn off the **Getting Started** tab. Recommendation: Leave it checked. There are valuable resources available on the tab. See Chapter 1 and Figure 15.2.

Essential PTC® Mathcad Prime® 3.0.

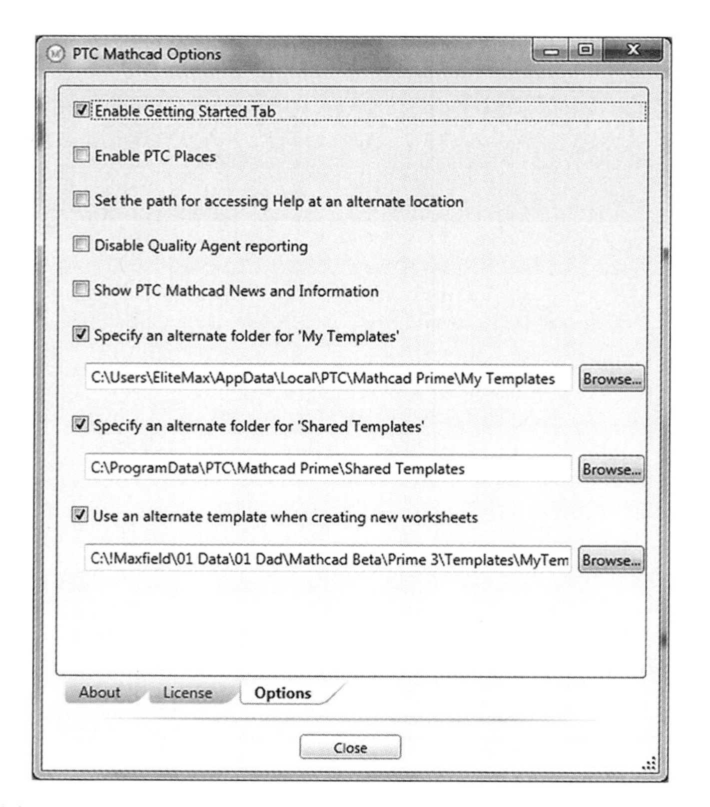

FIGURE 15.1

PTC Mathcad Options dialog box

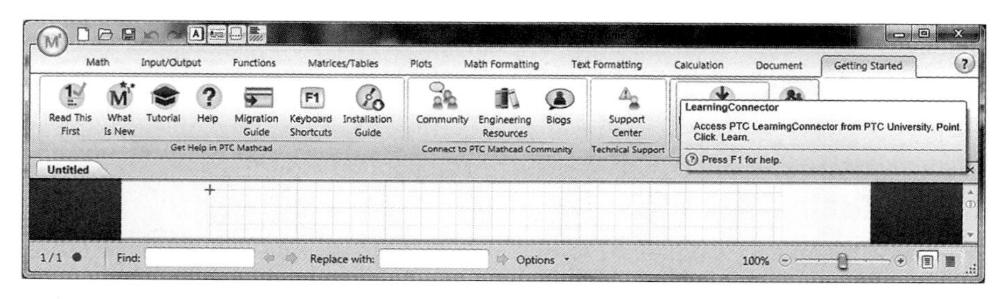

FIGURE 15.2

Getting Started tab

Enable PTC Places

PTC Places is used with PTC Windchill integration. Recommendation: Leave it unchecked if you do not use PTC Windchill.

Set the Path for Accessing Help at an Alternate Location

This is used when help files are located on a network server. Recommendation: Leave it unchecked.

Disable Quality Agent Reporting

The Quality Agent is an application that gathers and sends reports to PTC on the performance and usage of PTC Mathcad. It is always installed and enabled with PTC Mathcad. If you have a full functionality license, you may uncheck this box to stop PTC Mathcad from sending reports.

Show PTC Mathcad News and Information

PTC Mathcad News and Information is an RSS feed that is docked on the right side of the PTC Mathcad main window. It provides helpful information about PTC Mathcad. This check box turns the docked window on or off. Recommendation: It takes up valuable space, so leave it off, but turn it on occasionally to view the current topics. You may learn something new.

Specify an Alternate Folder for My Templates

Clicking this text box allows you to specify a specific folder location for the templates you create. (See Chapter 17 for a discussion of templates.) If this box is unchecked, your personal templates will be stored in the PTC Mathcad default location, which is buried deep in the Windows AppData folder. The path is usually C:\Users\(your user name)\AppData\Local\PTC\Mathcad Prime\My Templates. Recommendation: Check this box and save your templates in a known folder location you understand and will backup.

Specify an Alternate Folder for Shared Templates

Clicking this text box allows you to select a specific network directory and folder where all PTC Mathcad users can access common templates. If this box is unchecked, the default location is usually C:\ProgramData\PTC\MathcadPrime\Shared Templates. Recommendation: Check this box and create or use a common network drive to share PTC Mathcad templates.

Use an Alternate Template When Creating New Worksheets

Chapter 17 recommends creating a custom template to use when creating new worksheets. When this box is unchecked PTC Mathcad creates new worksheets based on the "Default.mctx" template. The default location is usually C:\Program Files\PTC\Mathcad Prime 3.0\Default Templates\en-US\Default.mctx. After checking this box you will be able to create new worksheets based on other templates. Recommendation: Read Chapter 17, create a custom template, save the template to "My Templates" or "Shared Templates," check this box, and then browse to your custom template. Corporate users may want to have all PTC Mathcad users open a corporate template saved in the "Shared Templates" folder.

Ribbon Bar Settings

The following sections discuss the various tabs on the Ribbon Bar and the various associated settings.

Math Tab

> PTC® Mathcad Prime® 3.0 does not allow you to customize units. ☹

The **Math** tab is most often used to insert items into your worksheet such as regions, operators, symbols, constants, symbolic keywords, and units. See Figure 15.3. The **Math** tab is also where you set the default units for your worksheet. These are located in the **Units** group. Your options for unit systems are SI, USCS (United States Customary System), and CGS (Centimeter-Gram-Second). Once you select a system, all results are displayed in the default units of that system. For example, the default unit of volume is L (for liter) in the SI system, and gal (for gallon) in the USCS system. You cannot customize your unit system to automatically display as m^3 or ft^3. In future versions of PTC Mathcad Prime, you will be able to customize your unit system to allow specific types of units to display as you desire.

The **Base Units** control is a toggle. It is either on or off. When it is on, the displayed units are broken down into the seven base unit dimensions of current, length, luminous

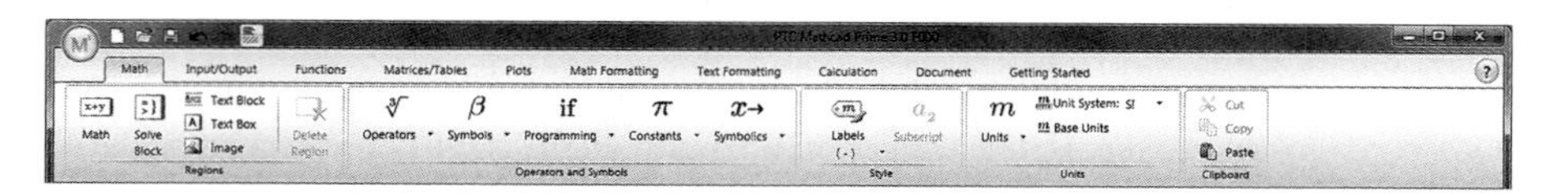

FIGURE 15.3

Math tab

intensity, mass, substance, temperature, and time. For example, a force would be displayed in units of mass*length/s^2 (kg*m/s^2, lbm*ft/s^2, or gm*cm/s^2). Figure 15.4 shows how various results appear with the **Base Units** control turned off. Figure 15.5 shows how the same results appear with the **Base Units** control turned on.

The following results are the default results with the **Base Units** control turned off.

$5\ N = 5\ N$ $\qquad 3\ kg \cdot 10\ \frac{m}{s^2} = 30\ N$

$4\ J = 4\ J$ $\qquad 5\ N \cdot 3\ m = 15\ J$

$100\ gal = 378.5411968\ L$ $\qquad 5\ m \cdot 3\ m \cdot 4\ m = 60000\ L$

$5\ kg \cdot 4\ \frac{m}{s} = 20\ \frac{kg \cdot m}{s}$ $\qquad 3\ N \cdot 4\ s = 12\ \frac{kg \cdot m}{s}$

FIGURE 15.4

Results with Base Units turned off

The following results are the default results with the **Base Units** control turned on.

$5\ N = 5\ \frac{kg \cdot m}{s^2}$ $\qquad 3\ kg \cdot 10\ \frac{m}{s^2} = 30\ \frac{kg \cdot m}{s^2}$

$4\ J = 4\ \frac{kg \cdot m^2}{s^2}$ $\qquad 5\ N \cdot 3\ m = 15\ \frac{kg \cdot m^2}{s^2}$

$100\ gal = 0.379\ m^3$ $\qquad 5\ m \cdot 3\ m \cdot 4\ m = 60\ m^3$

$5\ kg \cdot 4\ \frac{m}{s} = 20\ \frac{kg \cdot m}{s}$ $\qquad 3\ N \cdot 4\ s = 12\ \frac{kg \cdot m}{s}$

FIGURE 15.5

Results with Base Units turned on

Matrices/Tables Tab

The two settings on this tab occur in the **Result Format** group. These settings allow you to show matrix indices and to collapse or expand nested matrices. See Figure 15.6.

$$\text{Matrix} := \begin{bmatrix} 1 & 2 & 3 & 4 & 5 & 6 & 7 & 8 & 9 & 10 & 11 & 12 & 13 \\ 2 & 3 & 4 & 5 & 6 & 7 & 8 & 9 & 10 & 11 & 12 & 13 & 14 \\ 3 & 4 & 5 & 6 & 7 & 8 & 9 & 10 & 11 & 12 & 13 & 14 & 15 \\ 4 & 5 & 6 & 7 & 8 & 9 & 10 & 11 & 12 & 13 & 14 & 15 & 16 \\ 5 & 6 & 7 & 8 & 9 & 10 & 11 & 12 & 13 & 14 & 15 & 16 & 17 \\ 6 & 7 & 8 & 9 & 10 & 11 & 12 & 13 & 14 & 15 & 16 & 17 & 18 \\ 7 & 8 & 9 & 10 & 11 & 12 & 13 & 14 & 15 & 16 & 17 & 18 & 19 \\ 8 & 9 & 10 & 11 & 12 & 13 & 14 & 15 & 16 & 17 & 18 & 19 & 20 \\ 9 & 10 & 11 & 12 & 13 & 14 & 15 & 16 & 17 & 18 & 19 & 20 & 21 \\ 10 & 11 & 12 & 13 & 14 & 15 & 16 & 17 & 18 & 19 & 20 & 21 & 22 \\ 11 & 12 & 13 & 14 & 15 & 16 & 17 & 18 & 19 & 20 & 21 & 22 & 23 \\ 12 & 13 & 14 & 15 & 16 & 17 & 18 & 19 & 20 & 21 & 22 & 23 & 24 \\ 13 & 14 & 15 & 16 & 17 & 18 & 19 & 20 & 21 & 22 & 23 & 24 & 25 \\ 14 & 15 & 16 & 17 & 18 & 19 & 20 & 21 & 22 & 23 & 24 & 25 & 26 \\ 15 & 16 & 17 & 18 & 19 & 20 & 21 & 22 & 23 & 24 & 25 & 23 & 27 \end{bmatrix}$$

$$\text{Matrix} = \begin{bmatrix} 1 & 2 & 3 & 4 & 5 & 6 & 7 & 8 & 9 & \\ 2 & 3 & 4 & 5 & 6 & 7 & 8 & 9 & 10 & \\ 3 & 4 & 5 & 6 & 7 & 8 & 9 & 10 & 11 & \\ 4 & 5 & 6 & 7 & 8 & 9 & 10 & 11 & 12 & \\ 5 & 6 & 7 & 8 & 9 & 10 & 11 & 12 & 13 & \\ 6 & 7 & 8 & 9 & 10 & 11 & 12 & 13 & 14 & \\ & & & & & & & & & \ddots \end{bmatrix}$$

Show Indices turned off

$$\text{Matrix} = \begin{array}{c|cccccccccc|} & 0 & 1 & 2 & 3 & 4 & 5 & 6 & 7 & 8 & \cdots 12 \\ 0 & 1 & 2 & 3 & 4 & 5 & 6 & 7 & 8 & 9 & \\ 1 & 2 & 3 & 4 & 5 & 6 & 7 & 8 & 9 & 10 & \\ 2 & 3 & 4 & 5 & 6 & 7 & 8 & 9 & 10 & 11 & \\ 3 & 4 & 5 & 6 & 7 & 8 & 9 & 10 & 11 & 12 & \\ 4 & 5 & 6 & 7 & 8 & 9 & 10 & 11 & 12 & 13 & \\ 5 & 6 & 7 & 8 & 9 & 10 & 11 & 12 & 13 & 14 & \\ \vdots & & & & & & & & & & \ddots \\ 14 & & & & & & & & & & \end{array}$$

Show Indices turned on

FIGURE 15.6

Show Indices

Plots Tab

Refer to Chapter 8—Plotting for the settings on this tab.

Math Formatting Tab

See Figure 15.7 for the **Math Formatting** tab. The **Math Font** and **Label Styles** groups will be discussed at length in Chapter 16.

The **Results** group allows you to change how numeric results are displayed. See Figure 15.8. There are four controls in the **Results** group. To apply results formatting

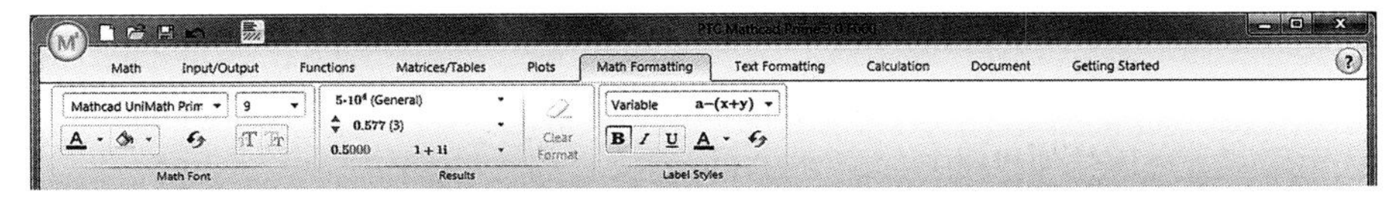

FIGURE 15.7

Math Formatting tab

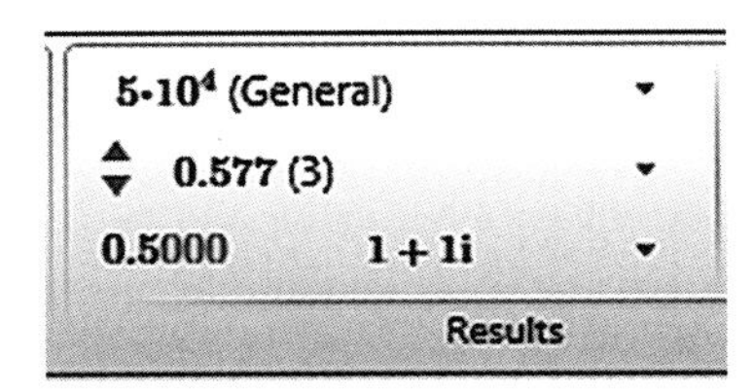

FIGURE 15.8

Results Group

to all math regions in your worksheet, you must click outside a math region. If you have a math region selected, the changes will only affect the selected region. You may select multiple regions to apply the changes to all the selected regions.

Result Format

> PTC Mathcad Prime 3.0 does not allow you to set an exponential threshold. ☹

The upper most control is the **Result Format** control. It provides you with five display options. See Figure 15.9.

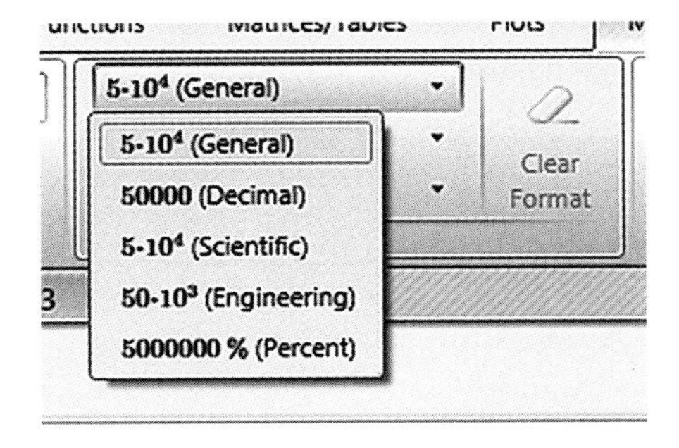

FIGURE 15.9

Result Format options

- **General:** Provides results in exponential notation when the exponential threshold of 3 is exceeded. Hopefully future versions of PTC Mathcad will allow you to customize the exponential threshold.
- **Decimal:** Results are never displayed in exponential notation.
- **Scientific:** Results are always in exponential notation.
- **Engineering:** Results are always in exponential notation, and the exponents are in multiples of three.
- **Percent:** Results are multiplied by 100 and displayed as percent.

Show Trailing Zeros

The bottom left control is the **Show Trailing Zeros** control. It is a toggle control. It controls how many zeros are displayed to the right of the decimal point. When it is turned on, PTC Mathcad will display zeros to the display precision value. When it is turned off, trailing zeros are not displayed.

I like to turn on the **Show Trailing Zeros** control, because if I see a number displayed as 1.6, I don't know how accurate the number is. The number could be 1.649 with the **Display Precision** set to (1), or the number could be 1.600 with the **Show Trailing Zeros** control turned off. If **Show Trailing Zeros** is turned on, then there is not a question. See Figure 15.10.

Example1 := 1.649

Example2 := 1.6

Results with Trailing Zeros turned on.

Example1 = 1.649 Display Precision set to (3) for Example1.

Example2 = 1.600

Results with Trailing Zeros turned off.

Example1 = 1.6 Display Precision set to (1) for Example1.

Example2 = 1.6

The above two results appear to be equal.

FIGURE 15.10

Trailing Zeros

Display Precision

The middle control is the **Display Precision** control. It is used to set how many decimal places you want displayed to the right of the decimal point. This only affects the display, and does not affect the calculation precision. If the **Show Trailing Zeros** control is turned off, then there may be cases when all decimals are not shown. See Figure 15.11. You can set the control by clicking the control button and then selecting one of the 0 to 15 options, or you can use the small triangle buttons to the left of the **Display Precision** control to increase or decrease the number of displayed decimal places.

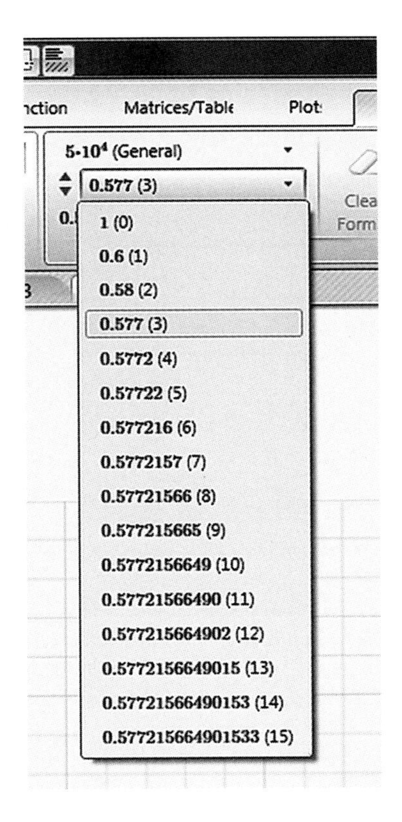

FIGURE 15.11

Display precision and trailing zeros

Numeric Precision

This is a good time to talk about numeric precision. This was discussed briefly in Chapter 10. The PTC Mathcad processor carries 15 significant figures – 14 exact and the 15th an approximate digit caused by rounding. Figure 15.12 illustrates the internal precision of the numeric processor by examining π and the sin(π). When you type `sin(p CTRL + g =` PTC Mathcad displays a result of $1.224*10^{-16}$. What gives? I have been taught that the sin(π) is zero. Why does PTC Mathcad not provide zero as the result? The answer comes with an understanding of the internal precision of the numeric processor. The internal processor views π as a 15 digit approximation of π. It then takes the *sin* of this approximation. The result is $1.225*10^{-16}$, rather than zero. The result is also carried to only 15 significant

Let's examine numeric precision by using the value of pi.
The numeric processor carries 14 exact figures, and the 15th is an approximate digit caused by rounding.

Pi displayed with Display Precision set to 15.

$$\mathbf{pi} := \pi = 3.14159265358979$$

Compare the above to the symbolic result using ***float***, 20.

$$\pi \xrightarrow{float,20} 3.1415926535897932385$$

The last numeric digit of "9" in the numeric processor is a rounding of "...932385..." So the numeric result is not exact, but is accurate to 15 decimal places.

Multiplying pi by a large number does not increase the accuracy. It is still carried to only 15 significant digits.

$$\mathbf{pi} \cdot 10^{10} = 31415926535.897900000000000$$

The above display has Display Precision set to 15, and Show Trailing Zeros turned on. The result is not accurate to 26 significant figures as implied by the trailing zeros.

$$\mathbf{pi} \cdot 10^{10} = 31415926535.8979$$

This result has Display Precision set to 15, but Show Trailing Zeros is turned off.

The above explains why the following occurs. The sin of pi is using an approximate value of pi, but the symbolic uses an exact value of pi.

$$\underline{\sin}(\pi) = 1.22460635382238 \cdot 10^{-16}$$

$$\underline{\sin}(\pi) \rightarrow 0$$

FIGURE 15.12

Internal Precision

figures. No matter how large a number you multiply the result by, there are still only 15 significant figures, with the remaining digits trailing zeros.

If you understand the above concept, you can understand why PTC Mathcad only allows you to set the display precision to (15). If you could set more, it would only display trailing zeros. It also allows you to understand when it may be a good idea to not display trailing zeros. Once PTC Mathcad displays 15 significant digits, if you show the trailing zeros it implies a greater accuracy than is actually occurring.

Complex Values

The **Complex Values** control allows you to set the display form of complex numbers. There are five options available. See Figure 15.13. The first two allow you to set the display of imaginary numbers to Cartesian Form using either i or j as the imaginary unit. The last three controls allow you to display the result in Polar Form. These three options allow you to display the angle as a value of π, in radians, or in degrees. The PTC Mathcad Help has a good example of using complex numbers in polar form. Type `polar form` in the Search box.

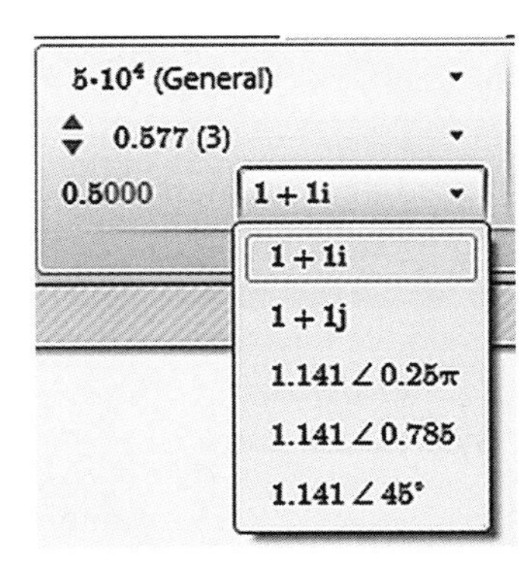

FIGURE 15.13

Complex value options

Calculation Tab

The settings on the **Calculation** tab control when and how PTC Mathcad calculates the worksheet. See Figure 15.14 for an example of the **Calculation** tab.

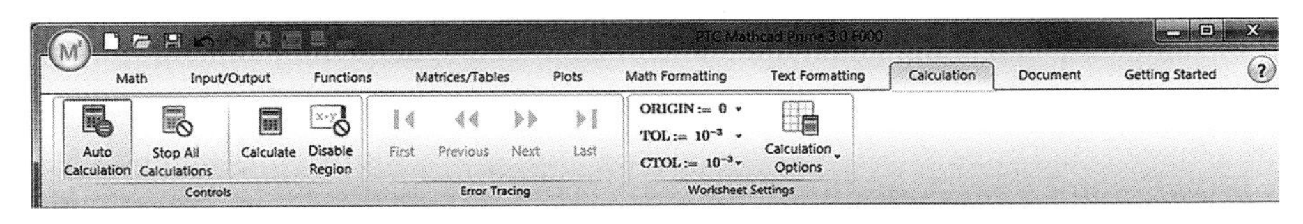

FIGURE 15.14

Calculation tab

Controls

From the **Controls** group you control when PTC Mathcad updates the numerical results. By default PTC Mathcad automatically updates results as you enter or edit regions. You can tell PTC Mathcad to stop the automatic worksheet calculations. You may want to do this if some equations take a long time to calculate, or you are using long algorithms that take a long time to converge. The **Stop All Calculations** control is a toggle with the **Auto Calculation** control. If you click this control, the green dot in the lower left corner of the status bar turns from green to red, warning you that the displayed results are not up-to-date. Recommendation: Stay with the Auto Calculation default, unless the time to calculate is interfering with your work.

The **Calculate** control recalculates the worksheet. If your worksheet contains included worksheets from other files, PTC Mathcad does not automatically include the results of these files when you open the worksheet. You must click the **Calculate** control to bring in the results and update all math regions. When you open a worksheet converted from an older Mathcad version, the worksheet does not automatically recalculate, even if **Auto Calculation** is enabled. You must click **Calculate** in order to calculate the worksheet and update the results.

The **Disable Region** control allows you to stop calculation inside a specific region or regions. You must have one or more math regions selected for this control to work. When a math region is disabled, it is dimmed, and the results are not recalculated. Any regions that are dependent on the definitions or calculation results from the disabled region will report errors.

A warning about **Stop All Calculations**: If you print a worksheet prior to calculating the worksheet when **Stop All Calculations** is enabled, the results displayed on the printout are not necessarily up-to-date. The results that have not been updated are dimmed, but they still display old results. This can be a serious problem if someone is relying on your printed calculations and does not understand what the dimmed results mean. For this reason, it is suggested that you always enable **Auto Calculation**.

Worksheet Settings

See Figure 15.15.

ORIGIN

This control sets the value for the built-in variable ORIGIN to either 0 or 1. ORIGIN represents the starting index of all arrays in your worksheet. The PTC Mathcad default for this variable is 0; earlier in the book we recommended changing the value to 1. You can overwrite this value to another value by defining the variable ORIGIN in your worksheet. This definition will take precedent over the worksheet setting.

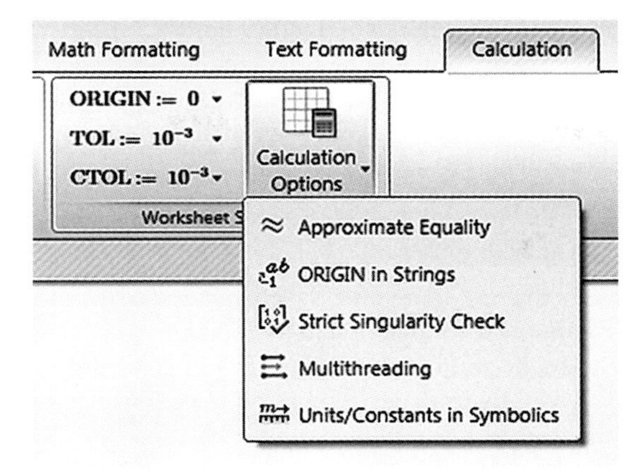

FIGURE 15.15

Worksheet Settings and Calculation Options

Convergence Tolerance (TOL)

This variable controls convergence precision of some functions. For example, it controls the length of the iteration in solve blocks and in the root function. This value was discussed in detail in Chapter 11. It also controls the precision to which integrals and derivatives are evaluated. The PTC Mathcad default is 10^{-3} (0.001). Recommendation: Use the Mathcad default for this variable.

Constraint Tolerance (CTOL)

This variable controls how closely a constraint in a solve block must be met for a solution to be acceptable. Solve blocks were discussed in Chapter 11. The PTC Mathcad default is 10^{-3} (0.001). Recommendation: Use the PTC Mathcad default for this variable.

Calculation Options

Clicking this control displays five additional controls: Approximate Equality, ORIGIN in Strings, Strict Singularity check, Multithreading, and Units/Constants in Symbolics.

Approximate Equality

This controls the standard used in Boolean comparisons and truncation functions. The PTC Mathcad Help states this about Approximate Equality: "When this option is active, the absolute value of the difference between two numbers divided by their average must be $< 10^{-12}$ for them to be considered equal, and only the first 12 decimal places are used in truncation." What this means is that if you have a very small number (i.e. Plank's constant) and another value that differs by a very small amount, they may not be considered equal. However, if you have a very large number (i.e. the speed of light) that differs slightly from another number, these may be considered equal. Another way of saying it would be that absolute difference between numbers matters less as the numbers get larger. The intent of Approximate Equality is to ignore differences between numbers that are most likely due to round off errors in the numeric processor. The PTC Mathcad default is off. Recommendation: Turn this on.

ORIGIN in Strings

PTC Mathcad defaults to considering the first character in a string as 0, similar to the default value for the built-in variable ORIGIN. Clicking this control will consider the first character to be the value of ORIGIN. If you changed the default value of the built-in variable ORIGIN from 0 to 1, you can select this control and PTC Mathcad will consider the first character in the string to be 1 instead of 0. If this control is turned off, the integer associated with the first character in a string will be 0. Recommendation: Change the PTC Mathcad default and turn this control on.

I do not like the first character in a string to be considered 0; I like it to be 1. I like to check this box to match the change I make to ORIGIN.

Use Strict Singularity Checking for Matrices

This control is used when matrix inversion fails. If a matrix has a determinant near zero then the typical inversion algorithms may become unstable and fail to work. This control uses a more rigorous but slower matrix inversion algorithm to help resolve the issue. This control may be set globally or for the selected regions. Recommendation: Stay with the PTC Mathcad default which is off. This is adequate for most applications.

Multithreading

When this control is active, PTC Mathcad can simultaneously perform multiple independent calculations. This may speed up the calculation of some worksheets that contain a number of independent calculations, and better utilizing multiprocessor systems. This is a PTC Mathcad setting, not a worksheet setting. When selected it changes for all open and future worksheets. It would have been better for PTC Mathcad to locate this setting in the PTC Mathcad Options dialog box because it is a PTC Mathcad setting and not a worksheet setting. The PTC Mathcad default is off. Recommendation: Most computers have multiple processors and this setting should be turned on.

Units/Constants in Symbolics

When this control is off PTC Mathcad does not try to identify built-in units or constants when using the ***symbolic evaluation*** operator. When this control is active, PTC Mathcad will identify built-in units or constants in symbolics and add the **Unit** label style or the **Constant** label style. This allows common units to be combined when the ***symbolic evaluation*** operator is used. The PTC Mathcad default is off. Recommendation: Turn this control on. See Figure 10.40 for an example.

Document Tab

The **Document** tab is another location where you can insert text boxes, text blocks, and images (in addition to the **Math** tab). The controls to add and remove space and to separate regions are also located on this tab. The worksheet settings that are located on the **Document** tab are found in the **Page** group and the **View** group. The controls in these groups allow you to determine how the worksheet looks on a screen and on a printed page. See Figure 15.16.

Headers and footers will be discussed in Chapter 16.

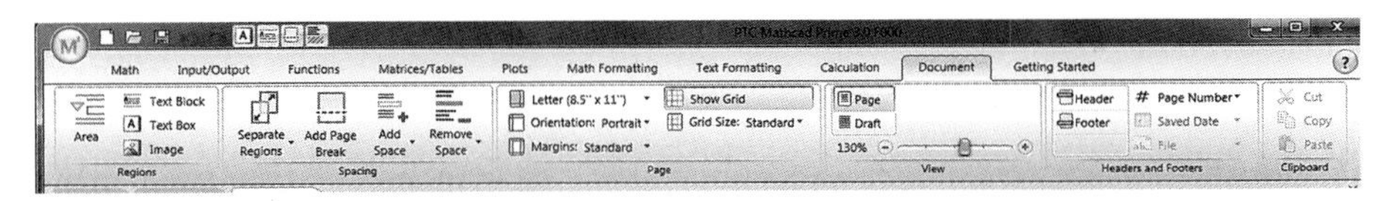

FIGURE 15.16

Document tab

Page

PTC Mathcad by default has a grid showing. Regions will snap to a grid point. You can turn this grid on and off by clicking the **Show Grid** control. You can also change from 5 grids per inch to 10 grids per inch by clicking the **Grid Size** control and selecting **Fine**.

Do not switch from Standard to Fine or from Fine to Standard grid size once you have begun working on a worksheet. PTC Mathcad Prime 3.0 anchors regions based on the number of grids, not on a page location. When you switch from Standard to Fine, all regions get moved much closer together. When you switch from Fine to Standard, all regions get moved much farther apart. Hopefully this will get resolved in future versions of PTC Mathcad.

From the **Page Size** control you can select from six paper sizes: Letter, Legal, Ledger, A3, A4, and A5. See Figure 15.17.

The **Page Orientation** control allows you to change from the default Portrait to Landscape orientation.

You cannot customize margin widths in PTC Mathcad Prime 3.0. You can only select from one of three preset options. The chosen option also sets the height of the header and footer. ☹

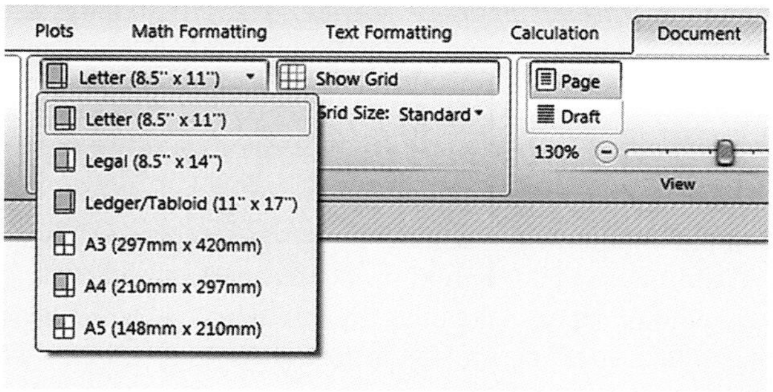

FIGURE 15.17

Page size

The **Margin** control provides only three choices: Standard, Narrow, and Wide. The default is Standard, which provides 1 inch of margin space on the sides, top and bottom. The Narrow option provides 1/2 inch margins at the sides, top and bottom. The Wide option provides 2 inch margins at the sides, top and bottom. Unfortunately, the height allowed for headers and footers is also controlled by the **Margin** control. If you select the Narrow margin, then the header and footer height is reduced to less than 1/2 inch. Recommendation: Stay with the Standard margin.

View

> In PTC Mathcad Prime 3.0 you cannot print anything to the right of the right margin. You do not have the option as you had in Mathcad 15.

The PTC Mathcad default view displays the page as it will be printed—a sort of "print preview" view.

The **Draft** view displays the entire worksheet without headers and footers. The right margin is indicated by a dashed vertical line. The bottom margin is indicated by a dashed horizontal line. The area to the right of the vertical dashed line is a non-printable area. You may add math and text regions in this area, but they will not be printed. When you are in page view, all regions to the right of the right margin will not be visible, but if there are regions in this area, PTC Mathcad displays a small arrow in the right margin indicating that there are regions that are not viewable. If you click the arrow, then the view switches to draft view.

In order to scroll down below the bottom margin, you must first add a page break.

Summary

Chapter 15—PTC Mathcad Settings discussed the settings that affect how PTC Mathcad functions and how PTC Mathcad displays results. The settings in the PTC Mathcad Options affect all current and future worksheets. The settings on the Ribbon Bar affect only the current worksheet. Some of these settings may be applied to the entire worksheet or only applied to selected regions of the worksheet.

In Chapter 15 we:

- Showed how the PTC Mathcad Options dialog box sets global PTC Mathcad settings.
- Discussed the general settings for how PTC Mathcad operates and starts up.
- Set default locations for saving files.
- Showed how the ribbon settings can set worksheet-specific settings.
- Recommended the settings for values of built-in variables.
- Encouraged the use of the automatic calculation setting.
- Told how to set the default unit system.
- Showed how to set the number of displayed decimal places.
- Discussed numeric precision.

Practice

The PTC Mathcad Prime 3.0 figures and examples used in this book are available for download from the book's website. The reader is encouraged to download the files and use them to practice the concepts learned. Additional examples and problems are also provided. To access this content go to http://store.elsevier.com/9780124104105 *and click on the Resources tab, and then click on the link for the Online Companion Materials.*

1. Open the PTC Mathcad Options dialog box. Change the default locations for "My Templates" and "Shared Templates" to a location other than the PTC Mathcad default. Open the Worksheet Options dialog box. Set the Array Origin (ORIGIN) to 1. Next, on the Calculation tab place a check in the "Use ORIGIN for string indexing" box.
2. Type the following on a blank PTC Mathcad worksheet:
   - **a.** 1.6=
   - **b.** 1.06=
   - **c.** 1.006=
   - **d.** 1.0006=
   - **e.** 1.00006=
   - **f.** 1.000006=
   - **g.** 1.0000006=
   - **h.** 1.00000006=
   - **i.** 1=

j. 11=
k. 111=
l. 1111=
m. 11111=
n. 111111=
o. 1111111=
p. 11111111=

3. Now open the **Math Formatting** tab. Choose "General" format, and turn off "Show Trailing Zeros." Change the "Display Precision" to zero and look at the displayed results. Now change the "Display Precision" to 1 and look at the displayed results. Next change the "Display Precision" incrementally from 2 to 9 and see how each number affects the displayed results.
4. Open the **Math Formatting** tab and turn on "Show trailing zeros." Look at the displayed results. Now change the "Display Precision" incrementally from 0 to 9 and see how each number affects the displayed results.
5. On the **Formatting** tab, experiment with the "General," "Decimal," "Scientific," "Engineering," and "Percent" formats. See how the different settings affect the displayed results.

CHAPTER

Customizing PTC Mathcad 16

With some customizing, PTC Mathcad calculations can look as nice as a published textbook. This chapter will teach you how to set up customizations to improve the appearance of your worksheets.

One way to achieve a consistent, professional look to your calculations is with the use of styles. Styles allow variables and results to have different appearances. Text formatting features allow you to change the look of text for titles, headings, explanations, and conclusions.

Headers and footers are also critical to scientific and engineering calculations. They identify you, your company, the project information, and the date the calculations were performed.

Chapter 16 will:

- Discuss PTC Mathcad styles for variables and constants.
- Tell about the advantages of using styles.
- Show how to create and modify math styles.
- Explain how to create headers and footers.
- Describe how to create a standard header that includes a graphic logo.
- Give suggestions about information that should be included in headers and footers.
- Discuss margins, including how to use information located to the right of the right margin.
- Discuss how to customize the icons on the ribbon.

Styles

Whether you know it or not, you always use styles when you use PTC Mathcad. Every time you type a definition or enter text, PTC Mathcad assigns a default style to the typed information. A style is a specific set of formatting characteristics associated with the items displayed on your PTC Mathcad worksheet. The formatting characteristics of a style include such things as: font, font size, font color, bold, underline, and italics.

Essential PTC® Mathcad Prime® 3.0.

Math Formatting and Label Styles

There are two important concepts to grasp in order to understand how math regions appear in your worksheet. The first concept is math formatting. The second is a math style referred to as math label style.

Math formatting is found on the **Math Formatting** tab in the **Math Font** group. See Figure 16.1. These controls allow you to change the font, font size, and highlight color of all math elements in your worksheet. The font color will globally affect the numeric results, not the variable names and units. This is because font color is controlled by the math label style. If you want to increase or decrease the size of all expressions in your worksheet, this is where you can do that. Notice that you cannot change bold, italics, or underline. These are set by the math label styles.

> PTC® Mathcad Prime® 3.0 is able to automatically associate labels with math elements. This is a very nice feature. These labels are similar to the Math Styles used in Mathcad 15. While Mathcad 15 only had two automatically applied styles (Variables and Constants), PTC Mathcad Prime 3.0 has six automatically applied styles, or "labels" as they are referred to. Labels are similar to the Mathcad 15 User styles. For example, in Mathcad 15, you could create user-defined styles and call them "Unit" and "Function." These helped make units and functions stand out more on the worksheet, but in order to use them, you had to manually apply them in every instance. The advantage with PTC Mathcad Prime 3.0 is that it understands the context of the math elements. It automatically associates and applies the labels **Unit** and **Function**. You don't need to do anything. Unfortunately, PTC Mathcad Prime 3.0 does not allow you to add user-defined math labels, but they should show up in future releases.
>
> Mathcad 15 had a math style called "Constants." This set the font formatting for the display of numbers and numeric results (which are "constant"). PTC Mathcad Prime 3.0 uses a label style called **Constant**. The use of this term is different than in Mathcad 15 and may cause some confusion. In PTC Mathcad Prime 3.0, the term "Constant" refers to built-in constants such as e, π, and c (the speed of light). Even though PTC Mathcad Prime 3.0 does not have a "Constants" style, you can change the color of numbers and numeric results by changing the font color in the **Math Fonts** group of the **Math Formatting** tab. In Mathcad 15 you could change the size of only the numbers and numeric results ("Constants"). In PTC Mathcad Prime 3.0 the font color is the only thing you can change to distinguish numeric results from variable names and units. You cannot change the font or font size of only the numeric results because these will change the font and font size of the variable names and units as well.

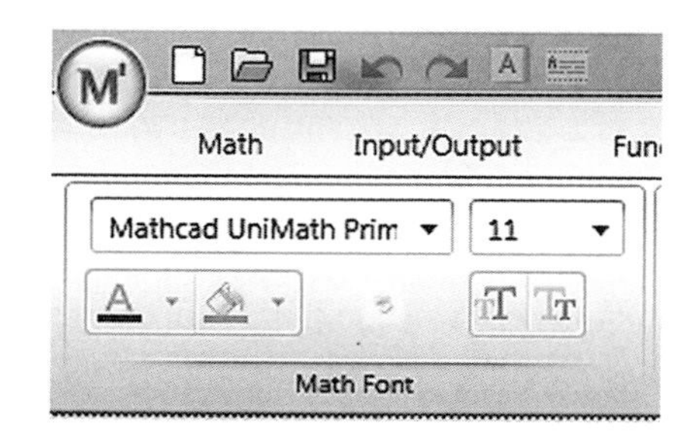

FIGURE 16.1

Math formatting controls

Math label styles are found on the **Math Formatting** tab in the **Label Styles** group. See Figure 16.2. They are applied to specific math elements and variables. They help to identify and classify the math elements. For example, math labels help distinguish unit names from variable names. They also help identify built-in constants and built-in system names. Labels also help to identify functions. PTC Mathcad Prime 3.0 comes with six default math labels. Future versions of PTC Mathcad will allow you to customize your own labels. The six default labels are: **Variable**, **Unit**, **Constant**, **Function**, **System**, and **Keyword**. Each of these labels has a specific math formatting style. They are distinguished by font color, bold, italic, and underline. The font and font size are the same for all labels and are set by the math formatting. The font characteristics of bold, italic, underline and color of each label are set by default, but you may customize the characteristics for each label. This will be discussed later.

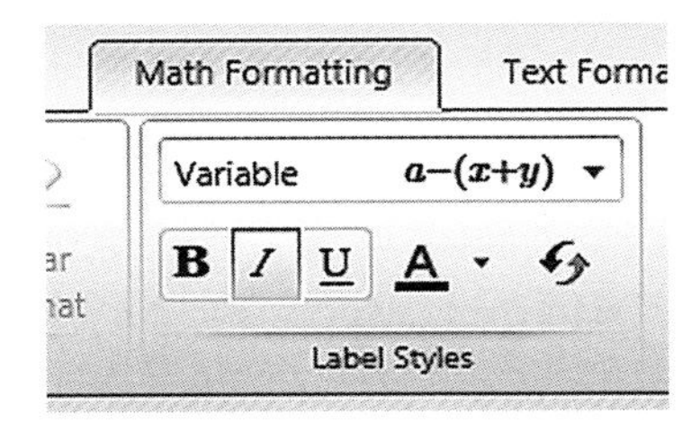

FIGURE 16.2

Label styles

You may already be familiar with most of the default math label styles. The two most common labels are **Variable** and **Unit**. Notice in the following image the difference in font between the units and the variables. The PTC Mathcad default **Variable** label uses italic while the default **Unit** label uses bold, italic, and blue font.

$$a := 4\ \boldsymbol{ft} \quad a = 1.219\ \boldsymbol{m} \quad a = 4\ \boldsymbol{ft}$$

The variable "a" is automatically assigned a **Variable** label, while the "m" and "ft" are automatically assigned **Unit** labels. Built-in constants such as e and π are assigned the label **Constant**. Built-in functions are assigned the label **Function**. You can see which label is assigned to a specific variable type by looking on the **Math** tab in the **Style** group. Under the word **Labels** you will see the name of the label assigned to the variable. Figure 16.3 illustrates each of the six default math labels.

PTC Mathcad automatically assigns one of the label types based on how the context of the variable is used. For example, if you define a new variable name, PTC Mathcad assigns the **Variable** label by default. If you type **m =**, PTC Mathcad assumes that you are referring to the unit "meter," and it displays the default unit of length and assigns a **Unit** label to the result.

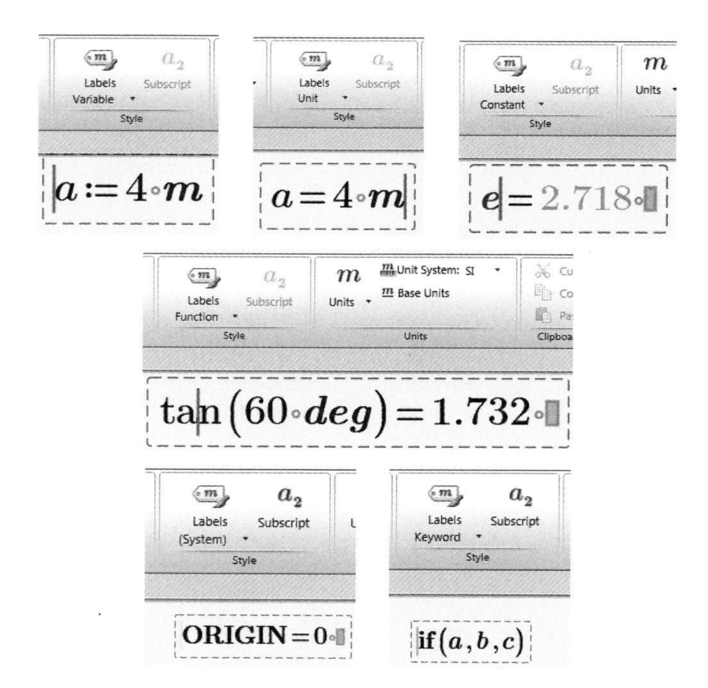

FIGURE 16.3

Default math labels

Differentiating Between Variables with the Same Name

The use of labels makes the worksheet easier to view because it allows you to visually see the difference between the different math elements. Another benefit of labels is that they allow you to use the same variable name in different ways. You are allowed to manually change the default label. For example, the variable "m" is a built-in unit of length and is assigned the label of **Unit**. By using labels, the variable "m" can now have multiple definitions. See Figure 16.4.

Once "m" has been assigned a variable name it no longer has the default label of **Unit**. PTC Mathcad assumes that when you type "m" you mean the definition you

$m = 1\ m$ The variable "m" is assigned the label **Unit** by default.

$\mathrm{m} := 5\ kg$ Assign "m" to be 5 kg. The default label for "m" is now **Variable**.

The variable "m" now has two meanings: A unit of length and 5 kg.

$\mathrm{m} = 5\ kg$ Because "m" has now been defined, when you type "m=" the assumed label is **Variable**.

$m = 1\ m$ To get the other meaning, assign the **Unit** label to "m". To do this, place the curser either before or after "m" and select the label **Unit** from the Labels control, in the Label Styles group. **Math>Style>Labels>Unit**.

The variable "m" can now be used as 1) a unit of length and 2) for the assigned value of 5 kg.

FIGURE 16.4

Using labels

assigned. When you want "m" to represent the unit meter, you will need to assign the **Unit** label to "m." To do this, select "m" and then select the label style **Unit** from the **Label** control on the **Math** tab, **Style** group **Math>Style>Labels> Unit**.

A similar method is used if the name of a built-in function has been used as a variable name (**Math>Style>Labels>Function**). See Figure 16.5.

$\sin\left(\frac{\pi}{4}\right) = 0.707$

$\mathrm{sin} := 50\ N$

$\mathrm{sin} = 50\ N$ Because "sin" has now been defined, when you type "sin=" the assumed label is **Variable**.

$sin\left(\frac{\pi}{2}\right) = ?$ The label **Function** is no longer automatically attached because sin is now user defined. It must be manually assigned.

$\sin\left(\frac{\pi}{2}\right) = 1$ Once the label **Function** is assigned to sin it is recognized as a built-in function.

FIGURE 16.5

Using labels

Applying Labels to User-defined Functions

Figure 16.6 shows how you can manually change the label for a user-defined function from **Variable** to **Function**. Once the label is changed on the user-defined function, PTC Mathcad automatically assigns the **Function** label on future use of the function (as long as that variable name is not reused with a "Variable" label).

PTC Mathcad automatically assigned the label **Variable** to the two user-defined functions. The label for g(x) was manually changed to **Function**. Once this was changed, PTC Mathcad automatically applied the **Function** label when g(x) was evaluated.

$f(x) := x^2$ $g(x) := \sqrt{x}$

$f(2) = 4$ $g(4) = 2$

$f(3\ m) = 9\ m^2$ $g(9\ m^2) = 3\ m$

Let's now also assign g as a variable.

$g := 9\ N$

$g = 9\ N$ When you evaluate "g", PTC Mathcad must decide if you want "g" the **Function**, or "g" the **Variable,** or "g" the constant (acceleration of gravity). It assumes that you want "g" with the label **Variable**.

If you want to use the function g(x), then manually assign the **Function** label.

$g(2) = ?$ PTC Mathcad did not recognize the function g(x), until the **Function** label was applied.

$g(16) = 4$

$g = 9.807\ \frac{m}{s^2}$ When the **Constant** label is assigned to "g", then it is recognized as the constant for the acceleration of gravity.

FIGURE 16.6

Applying labels to user-defined functions

Changing Label Styles

We have just discussed the math labels that come with the standard PTC Mathcad installation. This section focuses on understanding how to change the formatting of math label styles. The changes you make to math labels will only be effective in the current worksheet, not to all worksheets. Chapter 17 will discuss the ways to save your revised math labels to be used over and over again as a template.

Table 16.1 lists the default font attributes of the six PTC Mathcad math labels. The appearance of these are illustrated in Figure 16.3.

Table 16.1 Label Attributes

Label	PTC Mathcad Default Style
Variable	Italic/Black
Unit	Bold/Italic/Blue
Constant (Such as *e*, *c*, and π)	Bold/Italic/Green
Function	Black
System (Such as TOL, CWD, ORIGIN)	Bold/Black
Keyword (Used in solve block, such as find, maximize, minimize)	Bold/Black

It is easy to change the font characteristics of each label from the **Label Styles** group on the **Math Formatting** tab. To do this, simply select the label name and then select or unselect the font attributes you want to change. Let's change the font characteristics of the **Variable** label. Select **Variable** from the **Label Styles** group, then unselect italic and select bold. Figure 16.7 shows the before and after characteristics of the **Variable** label.

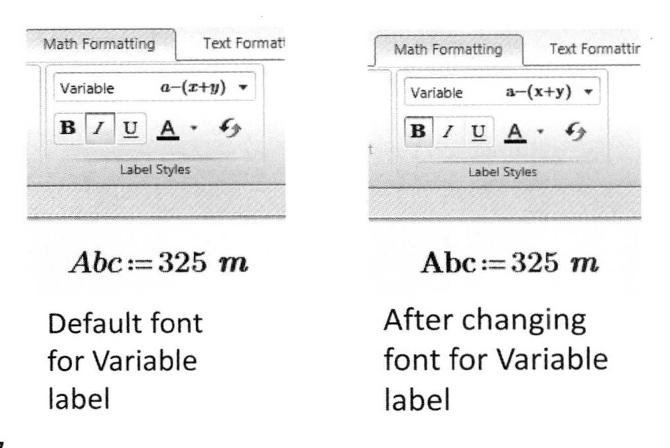

FIGURE 16.7

Changing the **Variable** label

My preference for the **Variable** label is bold with no italics because it more prominently displays the variable names. All examples in this book use bold with no italics for the **Variable** label.

Let's now change the font characteristics of the **Function** label. Select **Function** from the **Label Styles** group, then select bold and underline. Then change the color to maroon.

My preference is to make the **Function** label a different color with underline. This helps distinguish the functions from the variables.

Math Font Changes: Region Specific Verses Global Changes

In Mathcad 15, if you changed the "Variables" font characteristics while within a region, it globally changed all font characteristics in all regions. In PTC Mathcad Prime 3.0, to globally change font characteristics, you must be outside of a region. If you have an active region or regions, only the selected region(s) will be affected.

Up to this point we have been discussing changes that affect all regions of the worksheet. When the curser is outside of a region, any changes you make to math formatting will affect the entire worksheet. If you have one or multiple regions selected, any math font changes you make will only affect the selected regions. This principle applies to both math formatting and text formatting. Note that if you have a specific math region selected, you are able to change the font color of the variable name and unit, as well as the numeric result. This will overwrite the font color associated the math label style, but it will not overwrite the label.

Because you are allowed to change font, font size, and font color of any math region, it is possible to have a variable definition and variable evaluation look different. This does not confuse PTC Mathcad because PTC Mathcad uses the label styles (which have font characteristics associated with them) to differentiate between variable names. It may, however, cause user confusion. See Figure 16.8. The ability to change the font in individual regions without affecting results allows you to reduce the font size of regions that extend beyond the right margin to allow them to fit on one page. See Figure 16.9. You may use the **Remove Format** control in the **Math Font** group to remove customized font characteristics. Select the regions first before using this control.

abc := 3	abc = 3	Variable definition and variable evaluation both match.
cde:=4	cde = 4	PTC Mathcad evaluates the variable name, even though the font, and font size are different. This is because both the definition and the evalutation have the **Variable** label.
efg:=5	efg = 5	PTC Mathcad evaluates the variable name, even though the font, and font size are different. This is because both the definition and the evalutation have the **Function** label.
ghi:=6	ghi = ?	PTC Mathcad does not evaluate the variable name because the definition uses the **Function** label and the evaluation uses the **Variable** label. Mathcad warns to check the label assignements.

FIGURE 16.8

Changing math formatting settings in a single region

The font size in the below math region was changed to 4 point font in order to allow it to fit on the printed page.

$$y := y^3 - 2y^2 + 3y + 2 \xrightarrow{solve,\ y} \begin{bmatrix} \dfrac{\dfrac{2\cdot\left(\dfrac{\sqrt{249}}{9}-\dfrac{46}{27}\right)^{\frac{1}{3}}}{3}+\left(\dfrac{4357}{729}-\dfrac{92\cdot\sqrt{249}}{243}\right)^{\frac{1}{3}}-\dfrac{5}{9}}{\left(\dfrac{\sqrt{249}}{9}-\dfrac{46}{27}\right)^{\frac{1}{3}}} \\ \dfrac{12\cdot\left(\dfrac{\sqrt{249}}{9}-\dfrac{46}{27}\right)^{\frac{1}{3}}-9\cdot\left(\dfrac{4357}{729}-\dfrac{92\cdot\sqrt{249}}{243}\right)^{\frac{1}{3}}+5+5i\cdot\sqrt{3}+9i\cdot\sqrt{3}\cdot\left(\dfrac{4357}{729}-\dfrac{92\cdot\sqrt{249}}{243}\right)^{\frac{1}{3}}}{18\cdot\left(\dfrac{\sqrt{249}}{9}-\dfrac{46}{27}\right)^{\frac{1}{3}}} \\ \dfrac{9\cdot\left(\dfrac{4357}{729}-\dfrac{92\cdot\sqrt{249}}{243}\right)^{\frac{1}{3}}-12\cdot\left(\dfrac{\sqrt{249}}{9}-\dfrac{46}{27}\right)^{\frac{1}{3}}-5+5i\cdot\sqrt{3}+9i\cdot\sqrt{3}\cdot\left(\dfrac{4357}{729}-\dfrac{92\cdot\sqrt{249}}{243}\right)^{\frac{1}{3}}}{18\cdot\left(\dfrac{\sqrt{249}}{9}-\dfrac{46}{27}\right)^{\frac{1}{3}}} \end{bmatrix}$$

FIGURE 16.9

Reducing the font size of results to fit on one page

Text Formatting

PTC Mathcad 3.0 does not have text styles. ☹ You may format text in text regions, but not save it as a style. See Chapter 17 for a proposed workaround.

Text formatting changes are made from the **Text Formatting** tab. Here is a summary of what was discussed in Chapter 3.

Remember: If you want to make changes to text, the **Text Formatting** tab on the ribbon must be selected. The **Math Formatting** tab has similar font controls, and this can be confusing if you are trying to make text changes.

The font characteristics you may change are font, font size, font color, highlighting, bold, italics, and underline. When you make text formatting changes, it is important to understand where the changes apply. It all depends on what is selected.

If your curser is outside of a region, then any changes you make will only apply to all future text regions. All existing text regions will remain as they were. These changes will not become obvious until you begin a new text region.

If you have a single region or multiple regions selected, then all changes will be made only to the selected region(s). Future text regions will continue to use the original characteristics, because the changes only affected the selection region(s).

If you have text within a text region selected, the changes will only affect the selected text within the region. See Figure 16.10.

Text Box 1: This is the default text.

Text Box 2: All text within a single text box may be changed.

Text Box 3: **Individual text** within a text box may be changed.

FIGURE 16.10

Text formatting of text regions

Headers and Footers

We have just discussed math and text formatting. Another essential feature in your engineering calculations is a way to clearly identify the project information at the top and bottom of each page. This can easily be done by using headers and footers.

When using PTC Mathcad to create and organize technical calculations, headers and footers are absolutely critical. This section will discuss how to create headers and footers, and it will discuss what information is important to include in the headers and footers.

Creating Headers and Footers

> Headers and Footers in PTC Mathcad Prime 3.0 are much more basic than in Mathcad 15. You do not have control over side margins and the vertical height of the header, other than the default margins and heights that are set by the "Standard," "Narrow," and "Wide" settings. The only worksheet information that can be added to a header is page numbering, saved date, file path, and file name. ☹

Headers and footers are created from the **Document** tab. The controls are similar to Figure 16.11.

The text in headers and footers is created in text boxes. To add information to a header, click the **Header** control on the **Headers and Footers** group. Insert a text box by using **CTRL + t** or by selecting **Math>Regions>Text Box**. You may insert multiple

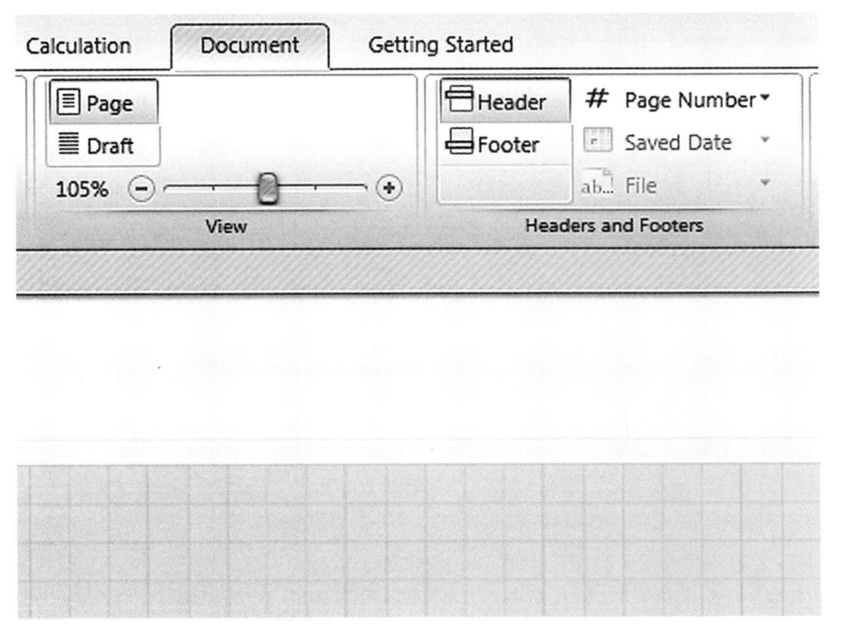

FIGURE 16.11

Header and Footer controls

text boxes and drag and position them anywhere within the header area. You can change the text font characteristics within the text boxes by using the controls in the **Text Font** group of the **Text Formatting** tab.

You may also insert page numbers, saved date, and file. To insert these, use the controls in the **Headers and Footers** group of the **Document** tab. These controls place a box in the header or footer. You may then drag and position these boxes anywhere within the header area. Figure 16.12 show the options that are allowed for the format of the page numbering.

> Images in headers and footers may now be in any image format rather than just bit map (.bmp) images. ☺

The **Image** button on the **Regions** group of the **Document** tab allows you to insert an image in your header or footer. When you click this button, a small button appears on your header or footer that says, "Browse for Image…". When you click this button a dialog box appears that allows you to navigate to the image you want to use. Images are allowed to be in any of the most common image formats.

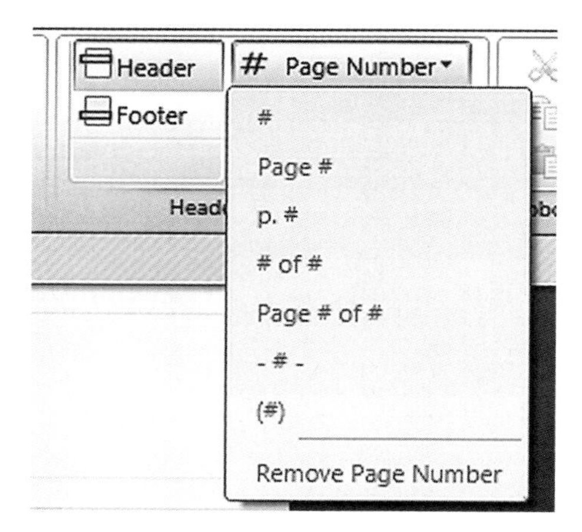

FIGURE 16.12

Page numbering options

Information to Include in Headers and Footers

Headers and footers are essential for printing technical calculations created using PTC Mathcad. Since PTC Mathcad calculations can be changed very easily, it is important to know how current the printed calculations are. Thus, it is important to know when the file was saved. The saved date is important because you will want to be able to make sure that the printed calculations are the same as the saved calculations.

The file name is also useful to have on printed calculations. This will help in locating the file for future reference. It will also help in making sure that the calculations are from the correct file. If you use several network drives, then the path is also a useful item to include on the printed calculations.

Page numbers are essential. It is suggested that you use this format: Page n of nn. This helps ensure that all printed calculations are kept together. If your calculations are contained in several files, it is also helpful to list a subject in front of the page number.

Company logos, photos, or scanned images can be included in a header or footer. The image may be in any common image format. Other information you may want to consider adding to your header or footer includes: Project title, project number, the part of the project the calculations are for, who created the calculations, who checked the calculations, etc.

Examples

Figures 16.13 and 16.14 show a sample header and footer.

Note that the center portion of the header in Figure 16.12 has some font format characteristics that were added. The image of the company logo was inserted from an image file.

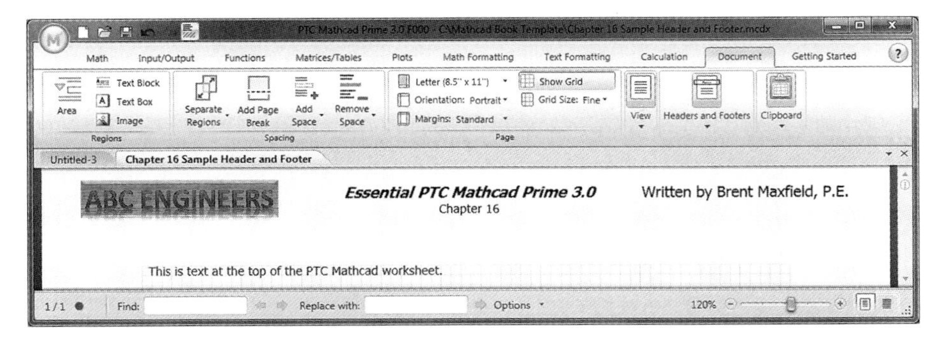

FIGURE 16.13

Sample header

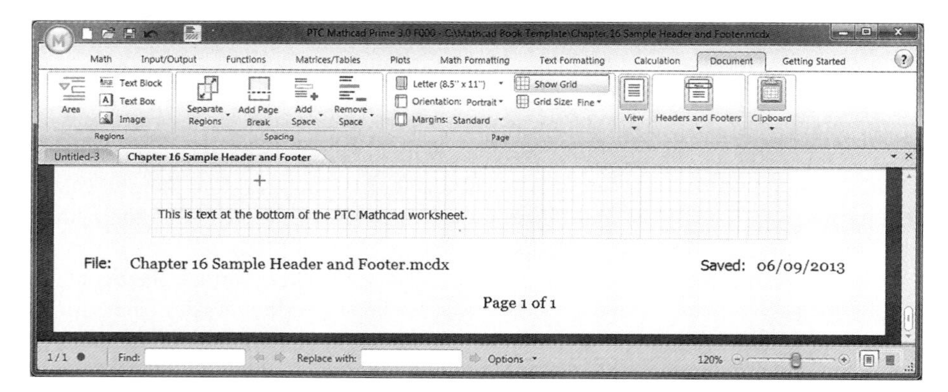

FIGURE 16.14

Sample footer

Margins

> PTC Mathcad Prime 3.0 gives you very little control over margins. ☹

The header and margins in PTC Mathcad Prime 3.0 are very basic. They are preset from the page margins.

The "Standard" margin provides 1 inch side margins and a header and footer that are 1 inch high. The "Narrow" margin provides ½ inch side margins and a header and footer that are ½ inch high. The "Wide" margin provides 2 inch side margins and a header and footer that are 2 inches high.

Quick Access Toolbar Customization

PTC Mathcad allows you to add and remove buttons from the Quick Access Toolbar. This is very handy if you continually use a command. It will save a few extra mouse clicks every time you use the command. To add a command to the Quick Access Toolbar, right click a button on the ribbon. A menu appears. Click on **Add to Quick Access Toolbar**. The control button should now appear on the Quick Access Toolbar. To remove a button from the Quick Access Toolbar, right click on the button you desire to remove, and then select **Remove from Quick Access Toolbar**.

Summary

In Chapter 16 we:

- Learned that PTC Mathcad uses styles for all math regions. These styles are called labels.
- Informed you that text styles do not exist in PTC Mathcad 3.0.
- Showed how you can customize the default PTC Mathcad labels.
- Explained when you may want to use a different math label.
- Explained how to change text font characteristics globally, for selected regions, or for selected text.
- Discussed headers and footers.
- Recommended what information is critical to have in your headers and footers.
- Showed how to customize the Quick Access Toolbar.

Practice

The PTC Mathcad Prime 3.0 figures and examples used in this book are available for download from the book's website. The reader is encouraged to download the files and use them to practice the concepts learned. Additional examples and problems are also provided. To access this content go to http://store.elsevier.com/9780124104105 *and click on the Resources tab, and then click on the link for the Online Companion Materials.*

1. On a blank worksheet type `Sample_1: 25kg`. Then type `Sample_1 = .` Next type `Sample_2: 50kg` and `Sample_2 =`. Notice how all variable names have the "Variable" label. Change the label of Sample_1 to "Unit." Notice the difference between the variable names "Sample_1" and "Sample_2." Now change the label of Sample_1 to "Function." Compare the different look for each different label style.
2. In a text box, type a heading and two paragraphs. Choose any topic to write about. Now experiment with the different text styles. Change the two paragraphs to different text font characteristics. Experiment with different combinations of characteristics.
3. Change all the variable names in exercise 1 back to the "Variable" label. Now experiment with changing the characteristics of the "Variable" label. Change the font color, bold, italic, underline, etc. If the regions overlap, use the **Separate Regions** command from the **Document** tab.
4. Create a header and footer for use with your company or school. Include the items mentioned in this chapter. Also include a graphic with your header or footer. Place some text at the top and bottom of your worksheet. Print your worksheet page. Did the entire header and footer print? Did the text at the top and bottom of your page print?

CHAPTER 17

Templates

In the previous chapters we have covered many different things that will make your worksheets unique, such as how to change many of the default PTC Mathcad features. We have discussed styles and how they give a consistent look to your calculations. We have also discussed how to set specific formats for your numerical results. In most cases, the changes made only affected the specific PTC Mathcad worksheet you were working in. In Chapter 17—Templates, we discuss how to save all of these customizations so that they can be used over and over again. We do this through the use of templates.

Templates are essential in order to have a consistent look for all your engineering calculations, especially if you are working with other engineers. Templates allow PTC Mathcad customizations to be applied consistently to all calculations.

Chapter 17 will:

- Tell what a template is.
- Discuss the type of information that is stored in a template.
- Show the templates that are shipped with PTC Mathcad.
- Explain when to make use of these templates.
- Review the items discussed in Chapters 5, 15, and 16, and show how to include these items in a customized template.
- Suggest items to include in a customized template.
- Create a sample template.
- Show how to modify an existing template.
- Discuss the Default.mctx file, and show how to have PTC Mathcad open a customized template whenever a PTC Mathcad worksheet is created.

Information Saved in a Template

A template is essentially a collection of information that PTC Mathcad uses to set various settings when opening a new document. This information is stored in a template file. Every time PTC Mathcad opens a file based on that template everything is formatted the same way. A template can save the following information:

- Worksheet settings.
- Headers and footers.
- On-screen information.

Essential PTC® Mathcad Prime® 3.0.

- Margins.
- Unit settings.
- Result display settings.
- Styles.
- Fonts.
- Unit system.
- Text regions.
- Math regions with functions or expressions.

PTC Mathcad Templates

PTC Mathcad comes with many templates. Every time you open PTC Mathcad you are actually opening a new worksheet based on the Default.mctx template. You can also open worksheets based on the other templates that are shipped with PTC Mathcad. To do this, click the Mathcad Button and hover your mouse over the **New** button and select **From Default Templates**. This opens the "New from Template" dialog box. See Figure 17.1.

From this dialog box, you can open a new file based on one of several different templates. Take time to open worksheets based on each different template, and explore the differences between the worksheets opened with each template.

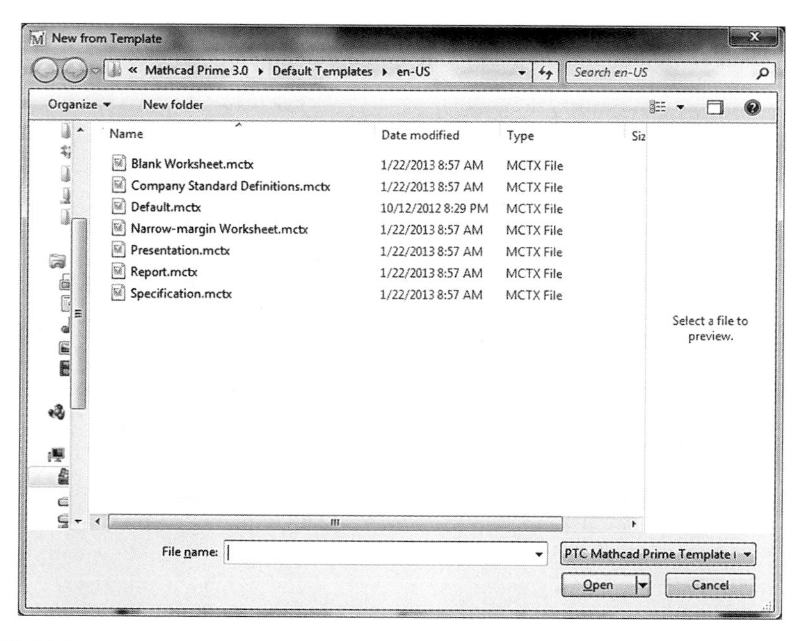

FIGURE 17.1

"New from Template" dialog box for opening a new worksheet based on a template

Compare the math styles, text styles, headers, footers, margins, and tab settings. These templates are more useful as examples than as templates that you would use on a regular basis. The Default.mctx is the default PTC Mathcad template.

Review of Chapters 5, 15, and 16

Let's make a quick review of some of the things discussed in previous chapters. These are things that can be saved in a template.

In Chapter 5 we discussed units. The default unit system can be saved in a template. In Chapter 15 we discussed PTC Mathcad settings. Any of the settings changed on the **Math Formatting**, **Calculation**, or **Document** tabs can be saved in a template. In Chapter 16 we discussed various label styles and text formatting characteristics. We also discussed headers and footers. Each of these items can be saved in a template.

Creating Templates

Saving your own custom template is easy. Simply open a new worksheet based on the Default.mctx or on another existing template. Next, make the changes to the worksheet as discussed above. Once you have the worksheet to a point where you want to use it for a template, simply click the Mathcad Button and hover your mouse over **Save As**. From the flyout select "MCTX." See Figure 17.2. This opens a Save As dialog box. The "Save as type" is "PTC Mathcad Prime Template

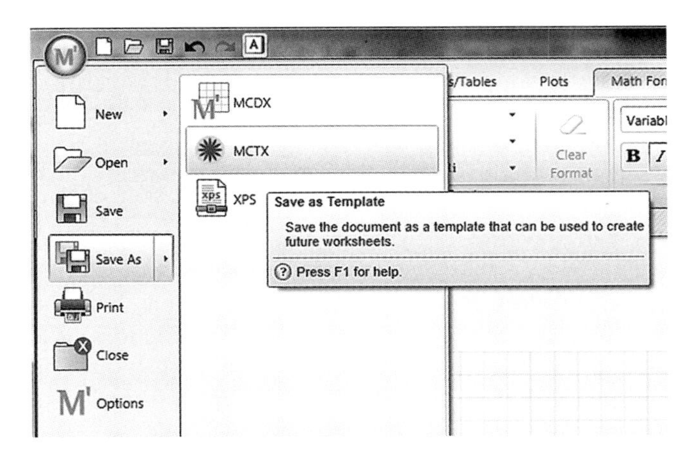

FIGURE 17.2

Saving a template

(*.mctx)." Enter a file name for your template and click **Save**. The default file location for your custom template is in a folder called "My Templates." We will talk more about this location shortly.

To open a new worksheet based on this new template, click the Mathcad Button, hover your mouse over the **New** button, and select **From my Templates.** See Figure 17.3. The name of your new template should appear in the list of templates. Click on the template, and then click on the **OPEN** button. A new worksheet should appear with your customizations included in the worksheet.

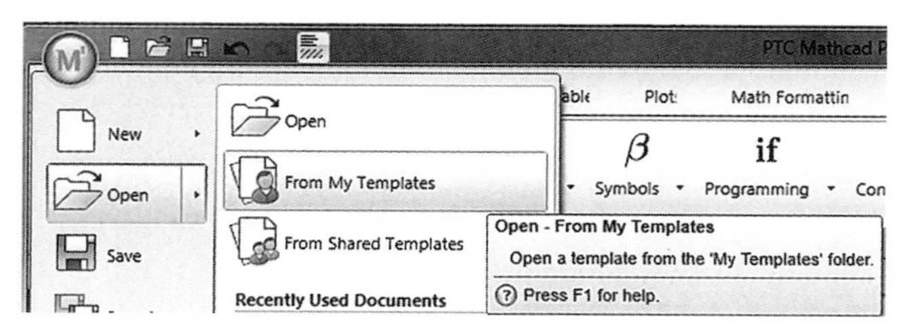

FIGURE 17.3

Opening a worksheet based on a saved template

Where Are Templates Stored?

> PTC® Mathcad Prime® 3.0 now allows you to store template files on a network location so that all users can access a common location! It also allows you to select which template to base new worksheets on rather than always basing them on the default template. ☺

There are three locations where template files (*.mctx) can be stored: The "Default Templates" folder, the "My Templates" folder, and the "Shared Templates" folder. The "Default Templates" folder contains the templates that ship with PTC Mathcad. Any templates that you create can go into the "My Templates" folder or the "Shared Templates" folder. The "My Templates" folder usually resides on your computer, and the "Shared Templates" folder usually resides on a network drive. The location of these two folders can be customized by clicking the Mathcad Button and selecting **Options**. From the PTC Mathcad Options dialog box, select the **Options** tab at the bottom. See Figure 17.4. From this list of options click "Specify an alternate folder for 'My Templates'". This then opens a box where you can type or browse to a folder location. Do the same for the "Specify an alternate folder for 'Shared Templates'".

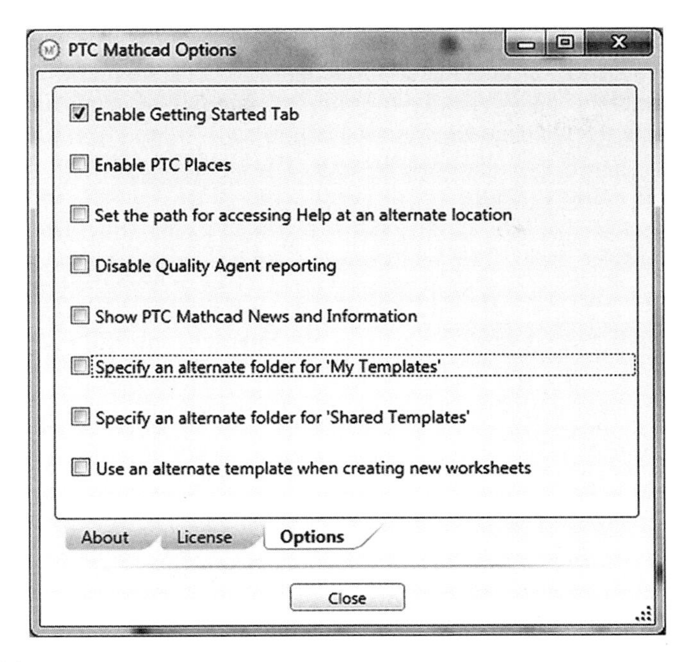

FIGURE 17.4

PTC Mathcad Options dialog box

Creating Your Customized Template

Now that we understand the concepts involved with customizing PTC Mathcad, let's create two customized templates that the figures in this book have been based on.

EM Metric

Let's call the first template EM Metric (for Essential Mathcad). Start by opening a new worksheet based on the "Blank Worksheet" template that ships with PTC Mathcad. This is easy to do; just click on the Mathcad Button, hover over **New**, select **From Templates**, and select **Blank Worksheet.mctx**. We can now begin customizing this worksheet. Once we are done, we will save it as a template.

Document Tab

From this tab we will make changes to how the printed and displayed page will look.

Page Group

From the **Page** group, select **Margins: Standard** by using the drop down arrow. Make sure that **Show Grid** is on by clicking the icon. It is a toggle. Change the grid size to fine by clicking the **Grid Size** icon and selecting **Fine**.

Headers and Footers Group

Let's add a header that will appear on every printed page. Select **Header** from the **Headers and Footers** group. This will open the header and turn the worksheet page gray. Create a text box by typing `CTRL + t`. Type `Essential PTC Mathcad Prime 3.0 EM Template`. Select the **Text Formatting** tab and in the **Paragraph** group, click the **Center Text** control. Select the text and set the font size to 12 pt. Move the text box until it is at the top of the header space and approximately centered between left and right margins.

Create another text box and enter your name. Resize the text box so the right side matches the edge of your name, and move the text box to the upper right corner of the header.

Let's now create a footer. You can double click in the footer space, or select **Footer** from the **Headers and Footers** group. Create a new text box and type `File:`. Adjust the width of the text box to match the text, and then move this text box to the left side of the footer. Next, from the **Headers and Footers** group, select **File>Name**. This inserts a text box with the name of the file. The text in the box will change once the file is saved with a new name. Move this text box adjacent to the File: text box.

Select **Saved Date>dd/mm/yyyy** from the **Headers and Footers** group, and move the box to the far right side of the footer. Create a new text box and type `Saved:`. Resize the text box and move it adjacent to the saved date.

In the center of the footer, select **Page Number>Page # of #**.

The footer should look similar to Figure 17.5.

File: EM Metric.mctx Page 1 of 1 Saved: 06/09/2013

FIGURE 17.5

Footer

Math Tab

Ensure that the unit system is set to SI. From the **Units** group, select the **Unit System** control and be sure that it is set to SI. The **Base Units** control should be unselected.

Math Formatting Tab

This tab changes how variables appear and how the results will be displayed.

Results Group

PTC Mathcad Prime does not allow you to set an exponential threshold. ☹ The default is 3. You can turn off exponential display by selecting the "Decimal" result format. PTC Mathcad Prime does not allow display in mixed fractions. ☹

Let's modify how the results will appear in our worksheet. From the **Results** group, set the following:

Result	Description	PTC Mathcad Default	Revised for Template
Results Format	The default. This displays numbers as exponential starting at 1,000. This is too small an exponential threshold. Change to Decimal.	General	Decimal
Display Precision	Change to only show two decimal places.	(3)	(2)
Show Trailing Zeros	Turn on, so precision is clearly understood.	Off	On
Complex Values	Leave as default.	1 + 1i	1 + 1i

Math Font Group

To allow more information on our worksheet we will change the default math font from 11 point to 9 point. From the **Math Font** group change the text size from 11 to 9 by using the drop down arrow.

Label Styles Group

We want our variable names to be more prominent, so let's change how the various label styles look.

Select each label from the **Label Styles** group and change the following properties:

Label	Description	PTC Mathcad Default	Revised for Template
Variable	Turn on bold; turn off italic; leave color black.	Italic/Black	Bold/Black
Unit	Leave as default.	Bold/Italic/Blue	Bold/Italic/ Blue
Constant (Such as *e* and π)	Leave as default.	Bold/Italic/Green	Bold/Italic/ Green
Function	Change so that it is clearly different than variable. Turn on bold; turn on italic; turn on underline; change color to maroon.	Black	Bold/Italic/ Underline/ Maroon
System (Such as TOL, CWD, ORIGIN)	Change so that it is clearly different than variable. Turn off bold; turn on italic; change color to green.	Bold/Black	Italic/Green
Keyword (Used in solve block, such as find, maximize, minimize)	Turn off bold.	Bold/Black	Black

Text Formatting Tab

Let's change our default text color to blue so that it is clearly different from math text. Also change the font size to 9 to allow more information on the worksheet. From the **Text Font** group, use the drop down arrows to change the font from 11 to 9, and change the color from black to blue.

Text Styles

PTC Mathcad Prime does not have text styles ☹.

Unfortunately PTC Mathcad Prime 3.0 does not allow for text styles. Nevertheless, the use of consistent formatting for various headings and information is critical. As a workaround for this issue, we will create four text boxes each with custom formats. These text boxes can be copied and pasted when the various headings are desired, and the text can then be modified.

The first two headings will be a major heading and a minor heading. Create the first text box by typing `CTRL + t`. Type `Major Heading`. Next select the text and change the following: Bold, Underline, Font size 16, Green color.

Create another text box and type `Minor Heading`. Select the text and change the following: Bold, Font size 14, Green color.

Create a third text box that can be used to make comments about your results. Create a text box and type `Results`. Select the text and change the following: Bold, Font size 12, Red color.

The fourth text box that can be used to make general notes. Create the fourth text box and type `Notes`. Select the text and change the following: Italic, Font size 12, Purple color, font face to Lucida Bright.

Drag select these four text boxes and move them to the upper right corner of the worksheet. When you want to add a major heading, copy the text box, paste it at the desired location, and then modify the text. It is not perfect, but it will allow you to have consistent headings.

Calculation Tab

From the **Worksheet Settings** group change ORIGIN from 0 to 1 by using the drop down arrow.

From the **Worksheet Settings** group and the **Calculation Options** control, turn on **Origin in Strings** by clicking the button. It is a toggle, and should now be highlighted. Also turn on **Approximate Equality** and **Units/Constants in Symbolics**.

Saving the Template

Now that we have made all these customizations, we are ready to save the template. Hover your mouse over **Save As** from the Mathcad Button. A flyout will appear. Select **MCTX**. This will open a Save As dialog box. The default save location is the location of your "My Templates" folder. Type the name `EM Metric` and click **Save**.

Now that we have saved this template, we can create new worksheets based on this template. To do this, hover your mouse over **New** from the **Mathcad** button. Then select **From My Templates**. The EM Metric template should be listed in the dialog box. Click this template and a new worksheet should appear. The new worksheet will be based on the EM Metric template with all of the settings exactly as changed above. Figure 17.6 shows how the new worksheet should look.

Essential PTC Mathcad Prime 3.0 EM Template Brent Maxfield

Major Heading
Minor Heading
Results
Notes

File: Untitled-4 Page 1 of 1 Saved: 00/00/0000

FIGURE 17.6

Worksheet based on EM Metric.mctx

EM US

Now let's create a similar template for U.S. Customary Units. This will be much simpler. We will open the EM Metric template, make one change, and then save it as EM US.

To open the EM Metric template, click on the Mathcad Button, hover over **Open,** and select "From My Templates." This opens an "Open Template" dialog box. Select the EM Metric.mctx file and click **Open**. Once the template is open, we will make the following change:

- Change the default unit system to U.S. Customary Units on the **Math** tab.

That is all we have to do. We want the rest of the template to remain the same. We now need to save the template. Hover your mouse over **Save As** from the Mathcad Button. A flyout will appear. Select **MCTX**. This will open a "Save As" dialog box. Change the file name to `EM US` and click **Save**.

You now have two customized templates that you can use anytime you want.

Alternate Default Template

> You can now have PTC Mathcad create worksheets based on your custom template. ☺

To create a new file based on one of the two customized templates, you must manually select the template as explained earlier. However, there is a way to have the EM Metric template be used every time you start PTC Mathcad.

Let's configure PTC Mathcad so that every default worksheet is based on the EM Metric.mctx template. To do this, open the PTC Mathcad Options dialog box by clicking on the Mathcad button. Place a check mark next to "Use an alternate template when creating new worksheets." Click the Browse button. Hopefully, you have already changed the default location for your "My Templates" folder. Browse to the location of your "My Templates" folder, and select "EM Metric.mctx" from the list. If it does not show up, then you did not save the template file correctly, or you did not browse to the correct location of your "My Templates" folder. Once you have selected "EM Metric.mctx," click Open. That is all you have to do. Now, every time PTC Mathcad opens it will open a new worksheet based on the EM Metric.mctx template.

If you want everyone in your organization to create worksheets based on the same template, then save the template to the "Shared Templates" location. From the PTC Mathcad Options dialog box, place a check mark next to "Use an alternate template when creating new worksheets" and browse to the template in the "Shared Templates" folder.

Summary

Templates are essential to engineering with PTC Mathcad. They will give your calculations a consistent look. They allow a consistent appearance for all corporate calculations. Instructors can create a template and have all students use the same template so that all assignments have a consistent appearance. You may want to create different templates for different clients. There are many uses for templates, but whatever the use, you should create your own template and begin using it.

In Chapter 17 we:

- Discussed what information is stored in a template.
- Told how a template works.
- Reviewed Chapters 5, 15, and 16.
- Showed how to save a template.
- Created customized templates with many unique styles and settings.
- Demonstrated how easy it is to create a new template when based on an existing template.
- Encouraged you to create your own customized template and have PTC Mathcad create all new worksheets based on this template.

Practice

The PTC Mathcad Prime 3.0 figures and examples used in this book are available for download from the book's website. The reader is encouraged to download the files and use them to practice the concepts learned. Additional examples and problems are also provided. To access this content go to http://store.elsevier.com/9780124104105 *and click on the Resources tab, and then click on the link for the Online Companion Materials.*

1. Create the EM Metric template discussed in this chapter.
2. Create the EM US template discussed in this chapter.
3. Create and save your own custom template. Start your template from a worksheet based on the "Blank Worksheet" template. Select a name for your template. Include the following in your template: Header, footer, custom label styles, custom worksheet settings, and custom number format settings. List the characteristics of your new template.
4. Open the template file you created in exercise 3. Refer to the custom units you created in the Chapter 5 practice exercises. Place these custom units on the right side of the right margin, and save your template. Open a new worksheet based on this template. Did the custom units come into the new worksheet?
5. Configure PTC Mathcad to create new worksheets based on your custom template. Now open a new worksheet to see if the new worksheet is based on your custom template.

CHAPTER 18

Assembling Calculations from Standard Calculation Worksheets

Throughout this book, we have seen the benefits of using PTC Mathcad to solve engineering problems. In this chapter, we will focus on the benefits of using PTC Mathcad to organize your engineering calculations.

Once PTC Mathcad worksheets are developed, they can be used over and over again. It is very useful to develop standard PTC Mathcad calculation worksheets. These standard calculation worksheets can be used as independent files, or they can be copied and pasted into new PTC Mathcad calculations. There are times when the PTC Mathcad calculation file becomes so large that it may be necessary to divide the calculation into several different files. These files can be linked together, so they behave as if they were one file.

Chapter 18 will:

- Explore ways of creating worksheets that can be used repeatedly.
- Demonstrate how to copy and paste these standard worksheets into new worksheets.
- Show how to use the reference command to include information from worksheets located in different files.
- Discuss when it is appropriate to break calculations into different files.
- Discuss the potential problems that may occur when assembling calculations from different files.
- Show how variable definitions may change after pasting portions of another PTC Mathcad worksheet.
- Provide examples showing some of the problems that occur after pasting other PTC Mathcad worksheets into calculations.
- Present different solutions to the problems discussed.
- Make recommendations as to how the calculations can be organized so that critical information is not redefined unexpectedly.
- Provide guidelines and examples to help avoid problems associated with variable redefinitions.
- Demonstrate the use of Find and Replace to help solve problems associated with variable names.

Essential PTC® Mathcad Prime® 3.0.

Copying Regions from Other PTC Mathcad Worksheets

The purpose of Part IV of *Essential PTC® Mathcad Prime® 3.0* is to teach you how to use PTC Mathcad to create and organize your engineering calculations. In this chapter, we focus on PTC Mathcad's ability to use previously created PTC Mathcad calculations. You can easily copy and paste regions from previously created worksheets. You can copy a single region or an entire worksheet.

To reuse information from a previous PTC Mathcad worksheet, open the worksheet, select the regions you want to copy, copy the regions (**CTRL + c**), and then paste the regions (**CTRL + v**) into your current worksheet.

If you are pasting between existing regions in your worksheet, you must add enough blank lines to make room for the pasted regions. PTC Mathcad does not automatically add new lines. If you do not have enough space, PTC Mathcad will paste regions on top of one another. If this occurs, undo the paste and add more blank lines.

It is currently not possible to drag and drop regions from one worksheet to another.

Creating Standard Calculation Worksheets

You may have already created calculation worksheets that you use repeatedly. The results from these worksheets are undoubtedly printed and placed in your calculation binder. The power of PTC Mathcad is that you can assemble all of these calculation worksheets into a single calculation file.

PTC Mathcad makes it easy to create and organize project calculations because you can reuse existing calculations and insert them anywhere into your current worksheet. Standard calculation worksheets can make it much easier to create a complete set of engineering project calculations. If you commonly use worksheets stored as standard calculations, you can copy from these worksheets, and paste them in a new worksheet. You can then assemble the calculations for an entire project into a single PTC Mathcad worksheet, just as you can assemble the paper calculations in a binder.

Standard calculation worksheet files can be as small as only a few regions, or they can contain multiple pages. You may only need to use a few regions from a standard calculation worksheet, or you may use the entire worksheet.

A few things should distinguish your standard calculation worksheets from other worksheets:

- They are designated as standard calculation worksheets.
- They have been thoroughly checked.
- They include a worksheet comment, listing who checked the calculation and date that the calculation was checked.

- Additional data can be added such as Author, Department, Dates, etc.
- The file is locked to allow read access only, so that changes are not accidentally made to the checked worksheet file.
- They are stored in a common folder location, accessible to everyone in your organization.
- They are organized with thought about how they will be copied into other calculations.

What type of worksheets should be saved as standard calculation worksheets?

- Repeatedly used equations and functions.
- Repeatedly used calculation worksheets.
- Cover sheets for project calculations.
- Project design criteria.
- Code equations.
- Reference data.
- Almost anything that will be used more than once.

Protecting Information

After working hard to create and verify standard calculations, you need to be able to prevent unwanted changes from being made to your standard worksheets.

The first thing you need to do is write-protect the standard calculation files so that you do not accidentally overwrite them. The easiest way to write-protect a file is to find the file in My Computer, then right-click on the file and click **Properties**. On the General tab of the Properties dialog box, place a check mark in the "Read-only" box under Attributes. This will allow the file to be opened, but not saved.

> PTC Mathcad Prime 3.0 does not allow areas to be locked or worksheets to be protected as you can do in Mathcad 15. ☹

Potential Problems with Inserting Standard Calculation Worksheets and Recommended Solutions

There are many advantages to using standard calculation worksheets. However, there is also one very large disadvantage. When you paste a standard calculation worksheet that contains the same variable name(s) as your current worksheet, you redefine the variable definition(s). This can have disastrous consequences in your calculations.

When creating project calculations, it is wise to define the project criteria only once. This allows you to be able to change the criteria and have the entire project result updated automatically. For example, assume that you need to use the yield strength of steel in your calculations. You define this to be Fy:= 36 ksi on the first page of your calculations. Each time you use Fy in your calculations you do not need to redefine Fy:= 36 ksi. You simply use the variable Fy, and PTC Mathcad knows that Fy = 36 ksi. Now, if you paste a standard calculation worksheet into your project calculations that has Fy:= 36 ksi, you have a new variable definition in your worksheet. If the value is Fy:= 36 ksi, then it appears that there is no problem—all your results are correct. But what happens if at some later time, you change the value of steel from Fy:= 36 ksi to Fy:= 50 ksi? You change this at the beginning of your calculation, but do not change the new definition added when you pasted the standard calculation worksheet. Now, all the results following the new definition are not updated to Fy = 50 ksi. They are still using Fy = 36 ksi.

Another scenario is if the standard calculation worksheet uses the value Fy:= 50 ksi. For our discussion, let's assume that Fy = 50 ksi is appropriate for the specific standard calculation worksheet, so you do not change the value of Fy from 50 ksi to 36 ksi after pasting into your project calculation. If you happen to insert this standard calculation worksheet into the middle of your project calculation, what happens to all the expressions that use Fy for the remainder of the worksheet? They all use the value of Fy = 50 ksi. Now all the project calculation results for the remainder of the worksheet are incorrect.

We have been discussing only a single variable. It may be easy to solve this problem when you are dealing with only a single variable, but if the project calculations and the standard calculation worksheet have many identical variable names, then it becomes very difficult to ensure that you do not have a problem.

This important issue needs to be considered whenever you create project calculations. It needs careful attention. The advantages of using PTC Mathcad outweigh this disadvantage. Let's now look at some ways to avoid the problems discussed.

Guidelines

The following guidelines are suggested as a means to prevent unwanted changes to your project calculations.

> PTC Mathcad Prime 3.0 does not have a redefinition warning to visually alert you about redefined variables. ☹

1. After pasting a standard calculation worksheet into your project calculations, scan the pasted regions and look for redefined variables. If you find a redefined variable, then you need to decide if the variable that is being redefined has the exact same meaning as a previously defined variable.
   a. If the redefined variable has exactly the same definition and value as a previously defined variable, then delete the ***definition*** operator := and

replace it with the ***evaluation*** operator =. This way the value of the earlier defined variable can be seen, but not overwritten.

   b. If the redefined variable has a different value or a different meaning than a previously defined variable, then you need to revise either the project calculations or the standard calculation worksheet so that each variable has a unique name. See the section below on using the ***Find*** and ***Replace*** features.
   c. There may be a situation where redefined variables do not cause a problem. See the discussion in the following section.

2. If possible, place all variable definitions at the top of your standard calculation worksheets. This makes it easier to check for redefinitions after pasting into your project calculations. The most critical variables to place at the top are the ones that may overwrite your project design criteria.
3. Whenever you use previously defined variables in an expression, always display the value of the variables near the expression. For example, if you have an expression Area:= b*d, then always type b= and d= near the expression. (This is assuming that b and d were defined on a previous page.) This makes it easier to check the calculations, and it helps ensure that the proper values are being used in the expression. You may also use the "explicit" keyword with the symbolic equal sign. See Chapter 10 for a discussion of how to use this method of displaying previously defined variables.
4. Always display the resulting value of expressions or functions. For example, if you have an expression Area:= b*d, then always type `Area =` following the definition. This makes it easier to check the calculations, and it helps to ensure that the results are what you expect.
5. Once you define a variable, do not redefine it (usually). Use subscripts or other means to make each variable unique. See the discussion in the next section on reusing variables.
6. Once a variable has been defined, use the variable name. In future expressions, do not use the variable value. This helps ensure that the calculations are correctly updated if the variable changes.
7. Use user-defined functions rather than expressions. See the section below.

How to Use Redefined Variables in Project Calculations

There are times when you need to use redefined variables in your project calculations. This section will give you some ideas on how to do this without causing problems in your project calculations.

Pasting the same standard calculation worksheet into your project calculations multiple times will cause redefinitions to occur. This could cause problems as discussed above. It is not practical to change the variables used in a standard calculation worksheet each time it is pasted into your project calculations. How do you prevent critical project calculation variables from being redefined?

Let's look at an example. Figure 18.1 shows a standard calculation worksheet. Figure 18.2 shows a project calculation with a standard calculation pasted into it two times. Notice that every variable in System 2 is redefining a previously used variable. None of the variables from System 1 can be reused later in the calculations. Figure 18.3 shows the same condition, but demonstrates how you can keep the variables unique in each System.

This standard calculation calculates the total head (Energy/Density of fluid) in a water system using the Bernoulli equation.

Input Variables:

Pressure $P := 80 \cdot psi$

Density $\rho := 62.4 \cdot pcf$

Velocity of fluid $v := 5 \cdot \frac{ft}{sec}$

Height above datum $z := 25 \cdot ft$

Calculate total head in system

$$H := \frac{P}{\rho} + \frac{v^2}{2 \cdot g} + z \qquad H = 210.00\ ft$$

FIGURE 18.1

Example of a standard calculation worksheet

Calculate the head for System 1

Input Variables:

Pressure	$P := 20 \cdot psi$
Density	$\rho := 62.4 \cdot pcf$
Velocity of fluid	$v := 10 \cdot \frac{ft}{sec}$
Height above datum	$z := 40 \cdot ft$

Calculate total head in system

$$H := \frac{P}{\rho} + \frac{v^2}{2 \cdot g} + z \qquad H = 87.71\ ft$$

Calculate the head for System 2

Input Variables:

Pressure	$P := 30 \cdot psi$
Density	$\rho := 62.4 \cdot pcf$
Velocity of fluid	$v := 20 \cdot \frac{ft}{sec}$
Height above datum	$z := 10 \cdot ft$

Calculate total head in system

$$H := \frac{P}{\rho} + \frac{v^2}{2 \cdot g} + z \qquad H = 85.45\ ft$$

In this example the variables for System 2 redefined every variable for System 1. This may not be a problem. It only becomes a problem if one of the variables in System 1 needs to be used later on in the project calculations. Figure 18.3 shows one method of using standard calculation worksheets and still keeping unique variable names.

FIGURE 18.2

Inserting the same standard calculation worksheet multiple times

This example is similar to Figure 18.2 except we attempt to prevent the redefinition of critical variables. Let's assume that the variables P, rho, v, z, and H are not critical variables. They are only useful within the context of the standard calculation. If any of these variable names are critical to your project calculations, then you will need to change the variable names in either the project calculations or in the standard calculation worksheet.

For this example, let's assume that you either calculated or were given the input variables to be used in calculating the total head for System 1 and System 2. Let's also assume that you need to reuse these variables later on in the project so you do not want to redefine them.

Calculate the head for System 1

$P_{System1} := 20 \cdot psi$ $\quad$ $WaterDensity := 62.4 \cdot \frac{lbf}{ft^3}$ $\quad$ $V_{System1} := 10 \cdot \frac{ft}{sec}$

$Height_{System1} := 40 \cdot ft$

We can now rewrite the standard calculation equation and use the above variables, but that would make the equation much more complex-looking. A better way is to assign the above variables to standard calculation variables.

Input Variables:

Pressure	$P := P_{System1}$	$P = 20.00\ psi$
Density	$\rho := WaterDensity$	$\rho = 62.40\ pcf$
Velocity of fluid	$v := V_{System1}$	$v = 10.00\ \frac{ft}{s}$
Height above datum	$z := Height_{System1}$	$z = 40.00\ ft$

Calculate total head in system

$H := \frac{P}{\rho} + \frac{v^2}{2 \cdot g} + z$ $\quad$ $H = 87.71\ ft$ $\quad$ $Head_{System1} := H$ $\quad$ $Head_{System1} = 87.71\ ft$

FIGURE 18.3 *(Continued on next page)*

Inserting the same standard calculation worksheet multiple times and still keeping unique results

Calculate the head for System 2

$P_{System2} := 30 \cdot psi$ WaterDensity $= 62.40\ pcf$ $V_{System2} := 20 \cdot \frac{ft}{sec}$

$Height_{System2} := 10 \cdot ft$

Note: We do not want to redefine the variable WaterDensity so we only display it.

Input Variables:

Pressure	$P := P_{System2}$	$P = 30.00\ psi$
Density	$\rho :=$ WaterDensity	$\rho = 62.40\ pcf$
Velocity of fluid	$v := V_{System2}$	$v = 20.00\ \frac{ft}{s}$
Height above datum	$z := Height_{System2}$	$z = 10.00\ ft$

Calculate total head in system

$H := \frac{P}{\rho} + \frac{v^2}{2 \cdot g} + z$ $H = 85.45\ ft$ $Head_{System2} := H$

$Head_{System2} = 85.45\ ft$

Using this method, we are able use reuse the standard calculations, and still maintain the unique variable names from System 1 and System 2.

FIGURE 18.3

(Continued)

Resetting Variables

There is another problem that may occur in your calculations when you reuse standard calculation worksheets. If for some reason a redefined variable is accidentally deleted, then PTC Mathcad goes up and gets the previously defined value. This may occur without you knowing it. This does not create an error in PTC Mathcad, but your results will be incorrect.

Here is another case that can cause errors. If your displayed result is just slightly higher in the worksheet than the definition for the result, PTC Mathcad will go up and get the result from the previous variable instead of the adjacent variable. See Figure 18.4. There is a way to prevent this from happening. You can clear each variable at the top of your standard calculation worksheet. Then, if a variable is deleted, the values above will not be used. This will cause an error or provide unexpected results, but this way you will be notified that there is a problem, and it prevents PTC Mathcad from using unwanted previously defined variables. See Figure 18.5.

Look for the inaccurate displayed value of H for System 2.

Calculate the head for System 1

Input Variables:

Pressure $P := 20 \cdot psi$

Density $\rho := 62.4 \cdot pcf$

Velocity of fluid $v := 10 \cdot \frac{ft}{sec}$

Height above datum $z := 40 \cdot ft$

Calculate total head in system

$H := \frac{P}{\rho} + \frac{v^2}{2 \cdot g} + z$ $H = 87.71\ ft$

Calculate the head for System 2

Input Variables:

Pressure $P := 30 \cdot psi$

Density $\rho := 62.4 \cdot pcf$

Velocity of fluid $v := 20 \cdot \frac{ft}{sec}$

Height above datum $z := 10 \cdot ft$

Calculate total head in system

If H is just slightly above the redefined definition of H, then PTC Mathcad selects the previous definition of H. This can cause errors to occur in your displayed calculations.

$H = 87.71\ ft$

$H := \frac{P}{\rho} + \frac{v^2}{2 \cdot g} + z$

The display of H above did not use the result from the expression on the left. The correct result should be $H = 85.45\ ft$. Figure 18.5 shows one way to prevent this from occurring.

FIGURE 18.4

Potential problem to avoid when using standard calculations multiple times

In Figure 18.4 we showed how the displayed value of H for System 2 was incorrect. In order to catch this error, you would need to do a very thorough check of the calculations. One way to prevent this error from occurring is to clear the variables at the beginning of each System.

Calculate the head for System 1

$\text{clear}(P)$ $\quad Clear(\rho)$ $\quad Clear(\text{v})$ $\quad \text{clear}(\text{z})$ $\quad \text{clear}(H)$

Input Variables:

Pressure $\quad P := 20 \cdot psi$

Density $\quad \rho := 62.4 \cdot pcf$

Velocity of fluid $\quad \text{v} := 10 \cdot \frac{ft}{sec}$

Height above datum $\quad \text{z} := 40 \cdot ft$

Calculate total head in system

$$H := \frac{P}{\rho} + \frac{\text{v}^2}{2 \cdot g} + \text{z} \qquad H = 87.71\ ft$$

Calculate the head for System 2

$\text{clear}(P)$ $\quad Clear(\rho)$ $\quad Clear(\text{v})$ $\quad \text{clear}(\text{z})$ $\quad \text{clear}(H)$

Input Variables:

Pressure $\quad P := 30 \cdot psi$

Density $\quad \rho := 62.4 \cdot pcf$

Velocity of fluid $\quad \text{v} := 20 \cdot \frac{ft}{sec}$

Height above datum

Calculate total head in system $\quad \text{z} := 10 \cdot ft$

Now, if H is just slightly above the redefined definition of H, then PTC Mathcad gives a text message instead of the previous definition of H. This way PTC Mathcad will not use any previous definitions, and you can detect an error. If any of the variables in System 2 get deleted, you will get the units of Henry (inductance) instead of using the value from System 1.

$$H = 3.28\ \frac{kg \cdot m}{s^2 \cdot A^2} \cdot ft$$

$$H := \frac{P}{\rho} + \frac{\text{v}^2}{2 \cdot g} + \text{z}$$

The correct result should be $H = 85.45\ ft$.

FIGURE 18.5

Resolving a potential problem to avoid when using standard calculations multiple times

Using User-defined Functions in Standard Calculation Worksheets

When you define variables and equations in your standard worksheets, these variables and equations are then pasted into your project calculations. One way to avoid having all these additional variables added to your project calculations is to use user-defined functions.

By doing this, the standard worksheets become more compact, and the project calculations will not have so many additional variables. Let's relook at Figures 18.1 and 18.3 and see how much cleaner it is to use a user-defined function. See Figures 18.6 and 18.7.

This standard calculation calculates the total head (H) (Energy/Density of fluid) in a water system using the Bernoulli equation.

Input Variables: Pressure (P), Density (r), Velocity of fluid (v), and Height above datum (z).

Calculate total head in system.

$$H(P, \rho, v, z) := \frac{P}{\rho} + \frac{v^2}{2 \cdot g} + z$$

The below expression is an example and does not need to be copied to project calculations.

$$H\left(80 \cdot psi, 62.4 \cdot pcf, 5 \cdot \frac{ft}{sec}, 25 \cdot ft\right) = 210.00\ ft$$

FIGURE 18.6

Example of a revised Figure 18.1

Calculate the head for System 1

$P_{System1} = 20.00\ psi$ $\quad$ $\mathbf{WaterDensity} = 62.40\ pcf$ $\quad$ $V_{System1} = 10.00\ \frac{ft}{s}$

$\mathbf{Height}_{System1} = 40.00\ ft$

Input Variables: Pressure (P), Density (r), Velocity of fluid (v), and Height above datum (z).

Calculate total head in system

$$H(P, \rho, v, z) := \frac{P}{\rho} + \frac{v^2}{2 \cdot g} + z$$

This formula is copied from the standard calculation worksheet in Figure 18.5.

$\mathbf{Head}_{system1} := H\left(P_{System1}, \mathbf{WaterDensity}, V_{System1}, \mathbf{Height}_{System1}\right)$

$\mathbf{Head}_{system1} = 87.71\ ft$

Calculate the head for System 2

$P_{System2} = 30.00\ psi$ $\quad$ $\mathbf{WaterDensity} = 62.40\ pcf$ $\quad$ $V_{System2} = 20.00\ \frac{ft}{s}$

$\mathbf{Height}_{System2} = 10.00\ ft$

$\mathbf{Head}_{system2} := H\left(P_{System2}, \mathbf{WaterDensity}, V_{System2}, \mathbf{Height}_{System2}\right)$

$\mathbf{Head}_{system2} = 85.45\ ft$

Notice how much cleaner this worksheet is than Figure 18.3. The standard calculation only contained the user-defined function. There were not extra variables to copy.

Using this method, we are able to reuse the standard calculations, and still maintain the unique variable names from System 1 and System 2.

FIGURE 18.7

Inserting standard calculation containing user-defined equations

Using the *Include* Feature

As you assemble your project calculations, you do not need to have all of the calculations included in a single PTC Mathcad file. The ***Include*** feature allows PTC Mathcad to get variable definitions from another PTC Mathcad worksheet and use them as if they were included in the current worksheet. For example, you could have a project design criteria worksheet where all the key variables for a project are located. Then each project worksheet will reference this one worksheet for all project related criteria. You can also have a corporate worksheet that contains standard definitions and functions. This corporate worksheet can be referenced from your corporate template. Figures 18.8 and 18.9 give an example of using the ***Include*** feature.

The values from this worksheet will be used in Figure 18.9.

Yield Strength of Steel $F_y := 50 \cdot ksi$

Compressive Strength of Concrete $f'_c := 4000 \cdot psi$

FIGURE 18.8

Referencing other worksheets

These variables are not defined in this worksheet.

$F_y = ?\ ksi$ $f'_c = ?\ psi$

Include << C:\Mathcad Book Template\Figures 18 MC3 Assembling Calculations B.mcdx

After adding the reference file, it is possible to access the variables defined in Figure 18.8. Values are available below the *Include* feature.

$F_y = 50.00\ ksi$ $f'_c = 4000.00\ psi$

FIGURE 18.9

Referencing other worksheets

To reference a file, select the **Include Worksheet** control from the **PTC Mathcad Worksheets** Group on the **Input/Output** tab. This inserts an "Include" button on your worksheet. Click the button and browse to the file location, and then select the file to be included. You can then browse to the file location, and select the file to be referenced. Both files must be stored prior to using the ***Include*** feature.

When to Separate Project Calculation Files

When is a project worksheet so large that it needs to be separated into two or more files? Try to stay with one file unless one or more of the following occurs:

- It takes too much time to recalculate the worksheet.
- Your worksheet has many graphics, which makes the files over 10 or 15 MB in size.
- Your worksheet gets to be over 100 pages in length.
- It makes sense to separate the project calculations into separate worksheets with related topics.

If you separate your project calculations, here are a few suggestions:

- Create a master worksheet that contains the critical project design criteria.
- Reference the master worksheet to get the critical project design criteria. Do not define the project design criteria in each worksheet because there is a risk of one worksheet using different criteria than another worksheet.
- If you need to use results from another file, use the ***Include*** feature to add a reference to the file. Do not redefine the data in the new worksheet. If the data in the other file changes, you want to have up-to-date results in the current worksheet.
- After referencing a file, display key variables below the reference line. This will aid in checking calculations.
- Add page numbers to either the header or footer. Add prefixes to the page numbers for each file, so that paper copies of the calculations will have unique page numbers.

Summary

This chapter focused on using PTC Mathcad to assemble project calculations. It showed how to use information from standard calculation worksheets. It also showed the advantages and disadvantages of pasting regions from standard calculation worksheets.

In Chapter 18 we:

- Demonstrated how to copy and paste regions from one worksheet to another.
- Discussed creating standard calculation worksheets.
- Warned of potential problems that may occur when you paste regions from other worksheets.
- Gave suggestions on how to avoid the potential problems.
- Provided guidelines for using standard calculation worksheets.
- Showed how to use the ***Include*** feature.
- Discussed when to split project calculations into separate files.

Practice

The PTC Mathcad Prime 3.0 figures and examples used in this book are available for download from the book's website. The reader is encouraged to download the files and use them to practice the concepts learned. Additional examples and problems are also provided. To access this content go to http://store.elsevier.com/9780124104105 *and click on the Resources tab, and then click on the link for the Online Companion Materials.*

1. Create and save two worksheets that can be used as standard calculations. In the first worksheet use expressions. In the second worksheet use user-defined functions.
2. Create a new worksheet with expressions that uses the same variable names as used in your standard calculation worksheet.
3. Observe what happens to your variable definitions when you copy and paste from the two standard calculation worksheets. See if some of your variable definitions were redefined by copying from your standard calculation worksheets.
4. In your worksheet, use the ***Include*** feature to reference values and user-defined functions from a previously defined worksheet. Use variables and user-defined functions from the included worksheet in your new worksheet to ensure that they are available.

CHAPTER 19

Microsoft® Excel Component

Because so many engineers have spent considerable time and money developing Microsoft® Excel spreadsheets, this entire chapter will focus on integrating Microsoft Excel spreadsheets into PTC Mathcad.

As discussed in Chapter 7, you can read from Excel files using the ***READEXCEL*** and ***READFILE*** functions, and you can write to Excel files using the WRITEEXCEL and WRITEFILE functions. This chapter will show how to take a previously written Excel spreadsheet and have PTC Mathcad provide the input to the spreadsheet, and how to have Excel pass the results back to PTC Mathcad.

Chapter 19 will:

- Explore the use of Microsoft Excel as a PTC Mathcad component.
- Discuss the advantages and disadvantages of Microsoft Excel.
- Provide examples of incorporating input and output from Excel spreadsheets into PTC Mathcad.

Introduction

> The Excel component block in PTC® Mathcad Prime® is a much more straightforward way of both getting data into and from Excel. You no longer need to tell PTC Mathcad how many inputs and outputs you will have as you had to do in Mathcad 15. ☺

Essential PTC® Mathcad Prime® 3.0.

PTC Mathcad communicates with the Excel component through the use of the Excel component block. This block is a PTC Mathcad region that is divided into three areas. See the figure below.

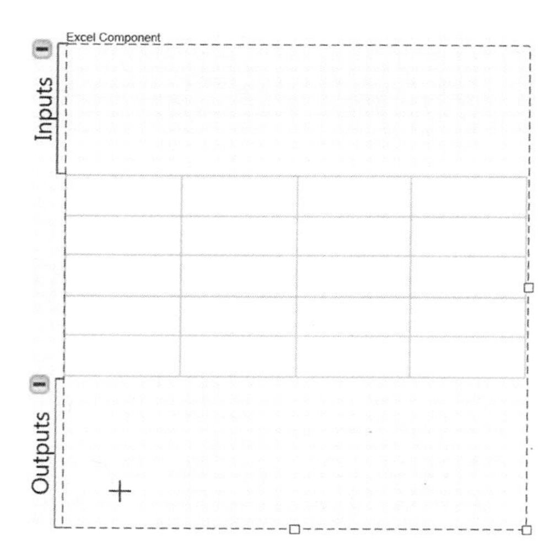

The top area is where input variables are placed. In this area you tell PTC Mathcad what values to place in Excel and which cells to put the values in.

The middle area is referred to as the "component table." This is a small display view of the full Excel worksheet.

The bottom area is where you tell PTC Mathcad which Excel values to bring back into PTC Mathcad. You will assign variable names to the desired Excel values.

You may have multiple inputs into Excel and multiple outputs from Excel. The number is only controlled by your system resources. In order to use the Microsoft Excel component, you must have Microsoft Excel 2003 or later installed on your computer.

Excel Component Block

The control to insert a new Excel component block into your worksheet is located on the **Input/Output** tab in the **Data Import/Export** group (**Input/Output>Data Import/Export>Excel Component>Insert Excel Component**). The keyboard shortcut is **CTRL + SHIFT + E**. You can resize the block by drag selecting the square located in the bottom right of the block. As you make the block larger, more cells are displayed in the component table. You can make both the Inputs and Outputs areas larger by clicking in these areas and pressing **ENTER**. You can make them smaller by pressing **DELETE**. You can move the block by selecting the block and using your mouse, or by using the arrow keys from your keyboard.

Input Context

> The context for getting data into and out of Excel is different in PTC Mathcad Prime. You may now refer to any of the sheets in the excel component. ☺

The context for getting information into Excel and from Excel comprises the variable name "excel" with an array subscript (using the [key) comprised of a string. Remember that strings are text placed between double quotation marks. The information in the string tells PTC Mathcad the Excel sheet name and the cell range. For example, the string would look something like this: "Sheet1! A1:B3." This refers to a spreadsheet called "Sheet1" and cells A1 to B3 in the spreadsheet. You may omit the spreadsheet name if you are referring to the first spreadsheet. When either one of the below figures are placed in the Input area of the Excel component block the value 1234 will be placed in cell A1 of the first spreadsheet.

$\text{excel}_{\text{"Sheet1!A1"}} := 1234$ $\text{excel}_{\text{"A1"}} := 1234$

Excel Component Table

The center portion of the Excel component block is referred to as the Excel "component table." When you double click within the component table, PTC Mathcad opens Microsoft Excel in a new window. Once Excel is open you can add information and formulas. Once you are done working in Excel, you can close Excel by clicking the "x" in the upper right corner of the Excel window. You do not need to save the Excel spreadsheet prior to closing Excel. When you double click to open Excel, the Excel component table turns gray to remind you that you have an open Excel spreadsheet.

Output Context

The bottom portion of the Excel component block is used to get data from Excel back into PTC Mathcad. The context is similar to the input context. First, type the name of the variable and the ***definition*** operator. Next, type **excel["**. This creates the word "excel" with a string in the array subscript. In the string, you must type a range of cells. If you are only referring to a single cell then type the same cell twice as shown in the figure below. If you are referring to the first spreadsheet, you may omit the name of the spreadsheet.

$\text{Result} := excel_{\text{"A2:A2"}}$

Simple Example

Let's create a very simple Excel component to illustrate how PTC Mathcad works with Excel.

In Figure 19.1 we will put the value of 2 into cell A1 of Excel. In cell A2, we will have Excel multiply the value of cell A1 by 5. We will then have PTC Mathcad retrieve the value of cell A2 and assign it to the variable "Result." To do this, type `CTRL + SHIFT + E` to insert an Excel component block. In the Inputs area, type `excel["A1 RIGHT ARROW Spacebar Spacebar:2`. Next, double click the Excel component table to open Excel. In cell A2 type `=A1*5`, and then close Excel. Finally, in the Outputs area, type `Result:excel["A2:A2`. Below the Excel component block type `Result =`.

Figure 19.1 also illustrates what happens when you change the input value from 2 to 7.

We have just demonstrated the key things to understand when using the Excel component. The rest of the chapter will add more detailed information to these key concepts.

Insert a Microsoft Excel spreadsheet by clicking **Excel Component** from the **Data Import/Export** Group on the **Input/Output** tab.

The figure below illustrates what happens when you change the input value from 2 to 7.

Inputs

$\text{excel}_{\text{"A1"}} := 2$

2	
10	

Outputs

$\text{Result} := excel_{\text{"A2:A2"}}$

$\text{Result} = 10.00$

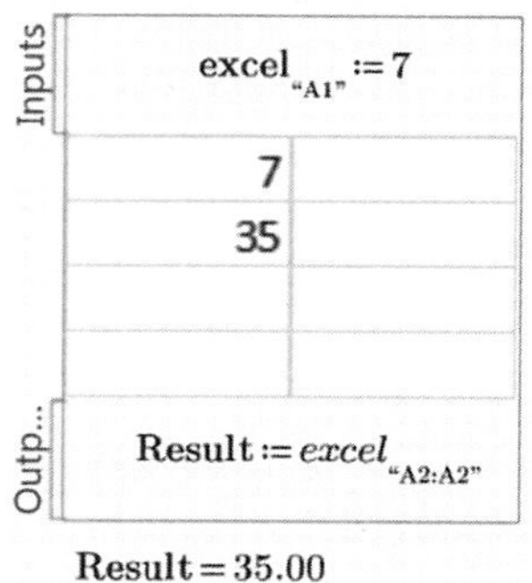

FIGURE 19.1

Simple example

Inputs and Outputs

Inputs

Input values can include previously defined variables. The input values can be scalars, vectors, matrices, or strings. When you insert a vector or matrix you only need to specify the beginning cell. The remaining cells will be filled in to the right and below the beginning cell.

There is a control to insert the input expression. When your cursor is in the Inputs area, click the **Excel Component** control from the **Input/Output** tab and select **Insert Input Expression** (**Input/Output>Data Import/Export>Excel Component>Insert Input Expression**). This will insert the following:

$$excel_{\text{“A1”}} :=$$

The default string is “A1.” You can edit this string to include a different input cell.

The Inputs area can include multiple input expressions. Each expression will populate a different cell in Excel. Use the **ENTER** key to add more space for additional input expressions. You can populate cells in different worksheets by including the worksheet name in the string. See Figure 19.2.

If you open Excel (by double clicking the component table) and select a cell (or cells) prior to using the **Insert Input Expression** control, the string in the input expression will include the sheet name and the cell reference that was selected. You must keep Excel open when you do this. See Figure 19.3.

Outputs

The string for an output must include a range of cells. This was illustrated in Figure 19.1. There is a control to add an output expression. When you cursor is in the Outputs area, click the **Excel Component** control from the **Input/Output** tab and select **Insert Output Expression** (**Input/Output>Data Import/Export>Excel Component>Insert Output Expression**). This will insert the following:

Define variables that will be used in the Excel component block. We will use a scalar, vector, and matrix. The vector includes a text string. The value for "d" is placed on Sheet 2.

$$a := 5 \qquad b := \begin{bmatrix} 2 \\ 21 \\ 298 \\ \text{"This is a text string"} \end{bmatrix} \qquad c := \begin{bmatrix} 5 & 100 & 33 \\ 4 & 76 & 24 \\ 29 & 883 & 1003 \end{bmatrix}$$

$$d := 12345$$

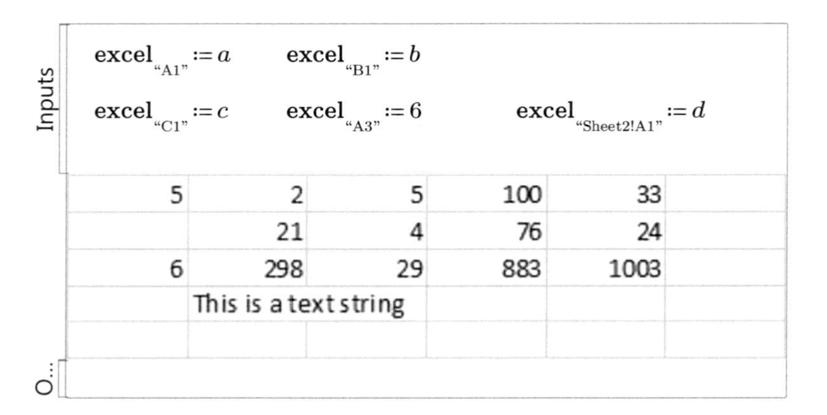

FIGURE 19.2

Input examples

If you double click the Excel compont table to open Excel, select a cell within Excel, and then use the ribbon to insert an input expression, the inserted input expression will have the cell reference in the string.

Close Excel AFTER inserting the input expression.

In the below example, cell B2 was selected prior to using the **Insert Input Expression** control. Note that the component table is gray because Excel is open.

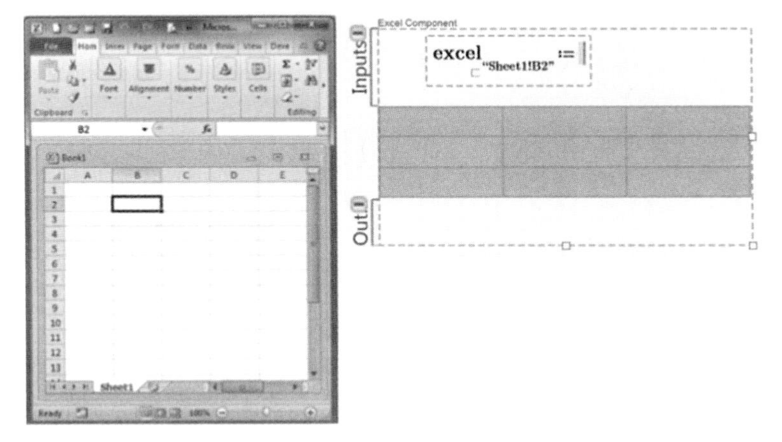

FIGURE 19.3

Selecting cells prior to using the **Insert Input Expression** control

The default string is "A1:A1." You can edit this string to include a different range of cells. The beginning placeholder is for the variable name. If you select a range of cells from the open Excel spreadsheet and then select Insert Output Expression, the string will include the sheet name and the range of cells selected. See Figure 19.4.

In this example, the variable "Multiplier" is placed in cell A1. The values from matrix "F" are placed beginning in cell D2. Excel multiplies each value of the matrix by the value in cell A1.

$$\mathbf{Multiplier} := 2.0 \quad \mathrm{i} := 0..2 \quad \mathrm{j} := 0..2 \qquad \mathbf{F}_{i+ORIGIN,\,j+ORIGIN} := i + j$$

$$\mathbf{F} = \begin{bmatrix} 0.00 & 1.00 & 2.00 \\ 1.00 & 2.00 & 3.00 \\ 2.00 & 3.00 & 4.00 \end{bmatrix}$$

ORIGIN is set to 1 in this worksheet, so we need to add the value of 1 to the array indeces.

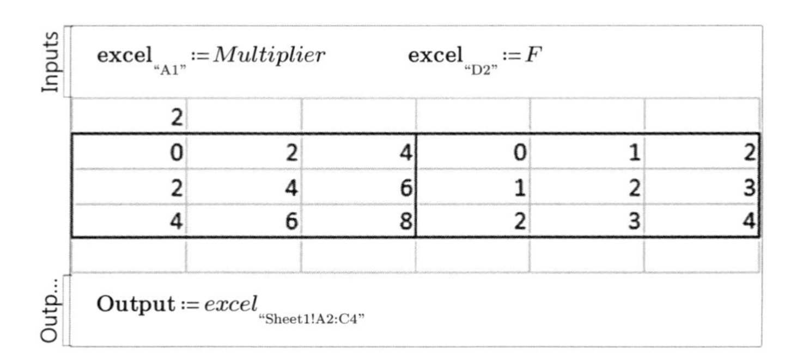

$$\mathbf{Output} = \begin{bmatrix} 0.00 & 2.00 & 4.00 \\ 2.00 & 4.00 & 6.00 \\ 4.00 & 6.00 & 8.00 \end{bmatrix}$$

Select the output cells prior to using the **Insert Output Expression** control.

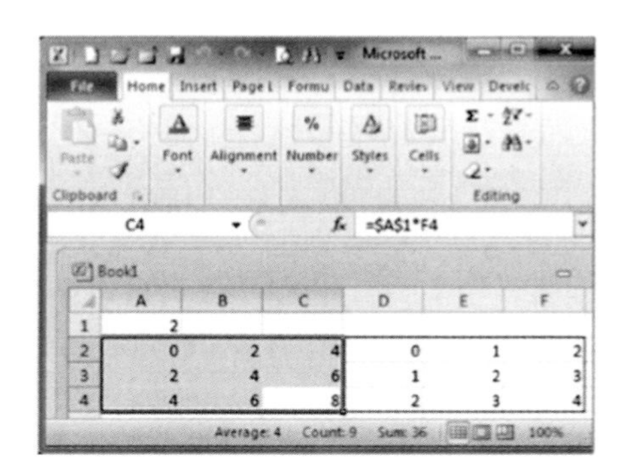

FIGURE 19.4

Selecting cells prior to using the **Insert Output Expression** control

Figure 19.5 illustrates how values change by changing one of the input values. You cannot evaluate the results of the output variable within the Excel component block. You must be outside of the component before you can evaluate the output variable.

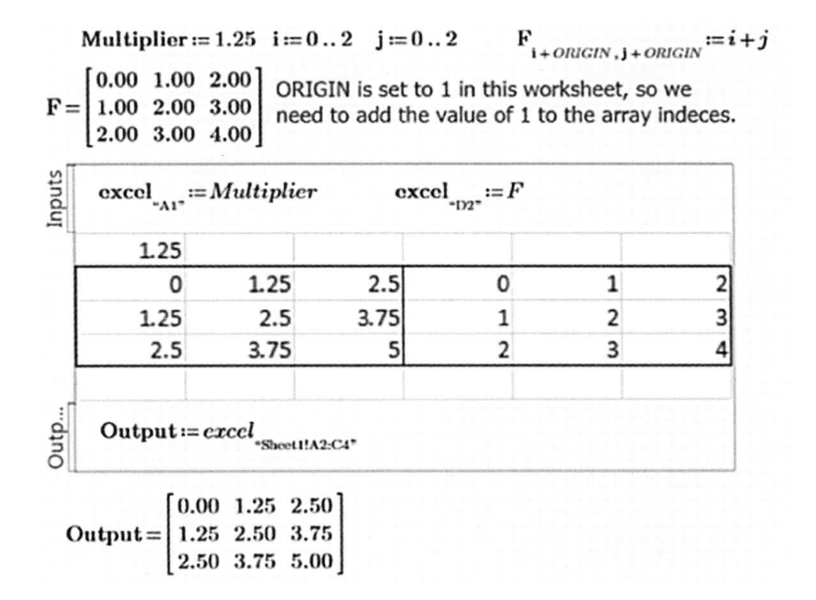

FIGURE 19.5

Effect of changing an input value

Hiding Inputs and Outputs

You can hide input and output expressions by clicking the small ⊟ icon that appears when the Excel component block is active. See Figure 19.6.

The below simple example shows what the Excel component block looks like with the inputs and outputs shown and hidden.

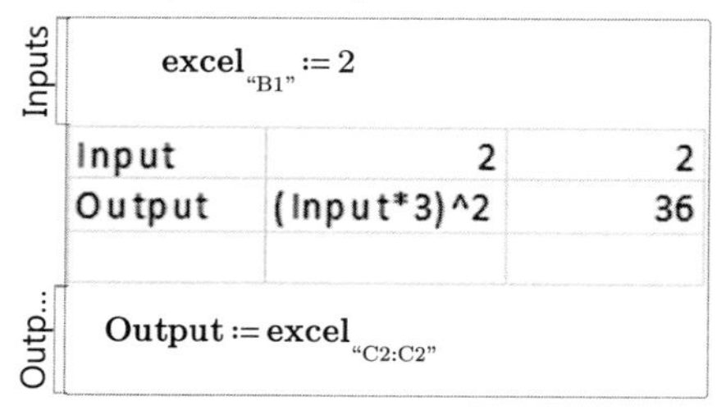

Output = 36.00

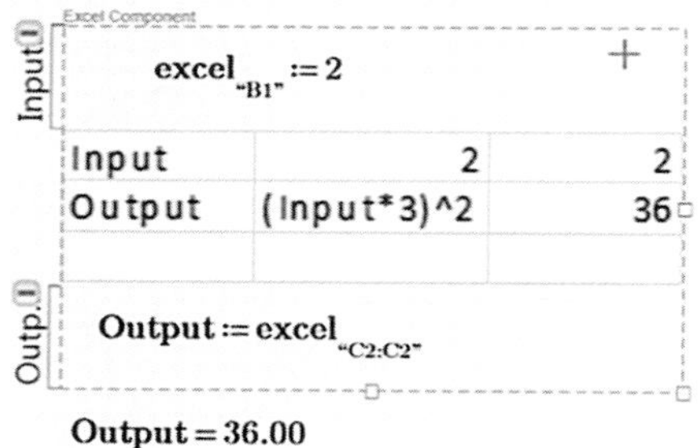

When the Excel component block is active, the hide icons appear.

Output = 36.00

Input	2	2
Output	(Input*3)^2	36

When the hide icons are selected, the input and output areas are hidden.

Excel Component

Input	2	2
Output	(Input*3)^2	36

This is what the Excel component block looks like when it is active.

FIGURE 19.6

Hiding inputs and outputs

Important Concepts

Opening, Closing, and Saving Excel

You open the Excel component by double clicking on the component table. You can also close Excel by double clicking on the component table. This is a toggle for open and close. You can also close Excel by clicking on the "x" in the upper right corner of the Excel window.

You can save a copy of the Excel component as a separate file by using the Save As command in Excel. The new file will not have a link to the Excel component.

Input Values Have Precedence

The input values take precedence over values or formulas entered directly into Excel. This means that if you open Excel by double clicking the Excel component table and then add a value or formula, this value or formula will be overwritten after closing Excel if an input value writes to that cell.

I recommend protecting the Excel worksheet, except for the cells that require input. This will prevent you from accidentally overwriting important cells. You cannot protect the entire worksheet, because that would not allow PTC Mathcad to change the input values.

What Displays in the Component Table

When you are working on an open Excel worksheet the displayed component table may or may not show the updates you are making. Here are some general guidelines about what is updated in the displayed component table.

1. If the area of the Excel spreadsheet you are working on is visible in the component table, these changes will be immediately displayed in the component table.
2. You will not see changes in the component table if the area of the Excel worksheet is not visible in the component table.
3. You can enlarge the component table by drag selecting the boxes on the right side, bottom, or corner of the Excel component block.
4. Formatting changes within Excel such as changing the font sizes, adjusting column widths, adjusting row heights, or scrolling to a different location in the spreadsheet are not immediately displayed in the component table. To update the display of the component table without exiting Excel (which will update the component table), save the Excel spreadsheet by typing **CTRL + S**. Once the spreadsheet has been saved, then the component table will be updated.

5. If you scroll to a new location in Excel, the new location will be shown in the component table after you save the spreadsheet (CTRL+S) or once you close Excel.
6. If you want a different worksheet to display in the component table, then move to the worksheet in Excel and type **CTRL + S**.
7. If you want to show more information in the component table, within Excel reduce the font size, reduce the width of the columns, and reduce the height of the rows. See Figure 19.7.
8. You can hide rows or columns within Excel to allow more information to be shown in the component table.

$i := 0..9 \quad j := 0..9 \qquad G_{i+ORIGIN,\, j+ORIGIN} := i + j$

$$G = \begin{bmatrix} 0.00 & 1.00 & 2.00 & 3.00 & 4.00 & 5.00 & 6.00 & 7.00 & 8.00 & 9.00 \\ 1.00 & 2.00 & 3.00 & 4.00 & 5.00 & 6.00 & 7.00 & 8.00 & 9.00 & 10.00 \\ 2.00 & 3.00 & 4.00 & 5.00 & 6.00 & 7.00 & 8.00 & 9.00 & 10.00 & 11.00 \\ 3.00 & 4.00 & 5.00 & 6.00 & 7.00 & 8.00 & 9.00 & 10.00 & 11.00 & 12.00 \\ 4.00 & 5.00 & 6.00 & 7.00 & 8.00 & 9.00 & 10.00 & 11.00 & 12.00 & 13.00 \\ 5.00 & 6.00 & 7.00 & 8.00 & 9.00 & 10.00 & 11.00 & 12.00 & 13.00 & 14.00 \\ 6.00 & 7.00 & 8.00 & 9.00 & 10.00 & 11.00 & 12.00 & 13.00 & 14.00 & 15.00 \\ 7.00 & 8.00 & 9.00 & 10.00 & 11.00 & 12.00 & 13.00 & 14.00 & 15.00 & 16.00 \\ 8.00 & 9.00 & 10.00 & 11.00 & 12.00 & 13.00 & 14.00 & 15.00 & 16.00 & 17.00 \\ 9.00 & 10.00 & 11.00 & 12.00 & 13.00 & 14.00 & 15.00 & 16.00 & 17.00 & 18.00 \end{bmatrix}$$

Inp...

$excel_{\text{"A1"}} := G$

0	1	2
1	2	3
2	3	4

Outp...

This is the default display.

Inp...

$excel_{\text{"A1"}} := G$

0	1	2	3	4	5	6	7	8	9
1	2	3	4	5	6	7	8	9	10
2	3	4	5	6	7	8	9	10	11
3	4	5	6	7	8	9	10	11	12
4	5	6	7	8	9	10	11	12	13
5	6	7	8	9	10	11	12	13	14
6	7	8	9	10	11	12	13	14	15
7	8	9	10	11	12	13	14	15	16
8	9	10	11	12	13	14	15	16	17
9	10	11	12	13	14	15	16	17	18

Out...

This is the display after reducing the font within Excel and after reducing the width of columns and the height of rows within Excel.

FIGURE 19.7

Showing more information in the component table

Dealing with Empty Cells

To put a blank cell into Excel use the built-in constant NaN (Not a Number). A blank cell returns a value of 0 in PTC Mathcad. See Figure 19.8.

Notice how each of the following are input into Excel:
NaN is input as a blank cell.
0 puts a 0 into Excel.

$A := NaN \qquad B := 55 \qquad C := 0$

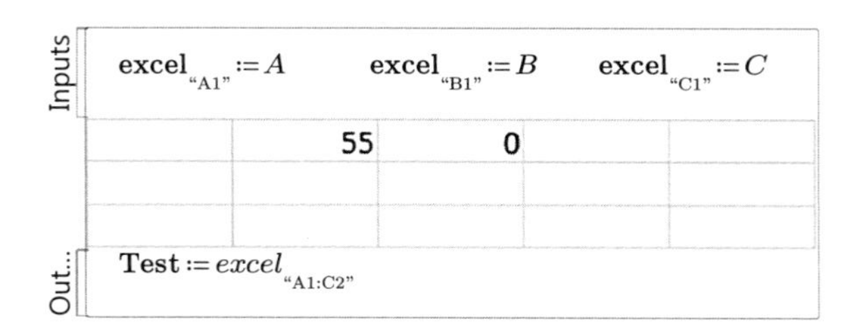

$$\text{Test} = \begin{bmatrix} 0.00 & 55.00 & 0.00 \\ 0.00 & 0.00 & 0.00 \end{bmatrix}$$

For output, blank cells and 0 both return a value of 0.

FIGURE 19.8

Dealing with blank and empty cells

Inputs and Outputs Do Not Track Excel Changes

When you move cells within Excel, the formulas inside of Excel are adjusted to track the changes. The Excel component input and output expressions are not updated, so it is important to understand that if you adjust data inside of Excel, you must also update the input and output expressions. See Figure 19.9.

Figure 19.10 illustrates a similar issue that can occur if you change the output range.

Let's say you create an Excel spreadsheet to calculate the area of a circle.

$\textbf{Radius} := 3$

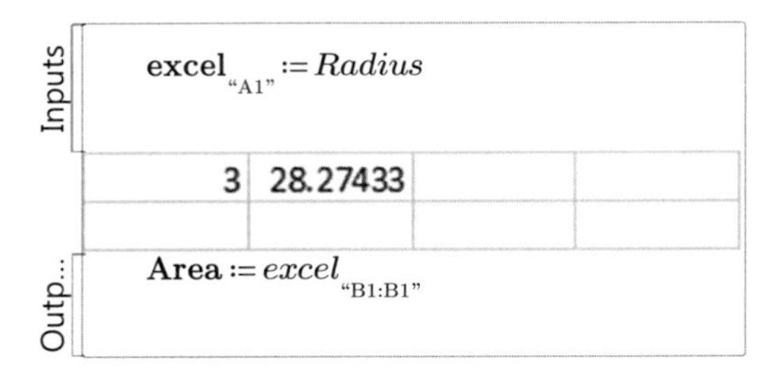

$\textbf{Area} = 28.27$

Now, let's say that you want to add an explanation about what is being calculated, so you insert a row and move the formula one cell to the right. You then add some descriptive text in row one as shown below.

When you close Excel, note what happens.

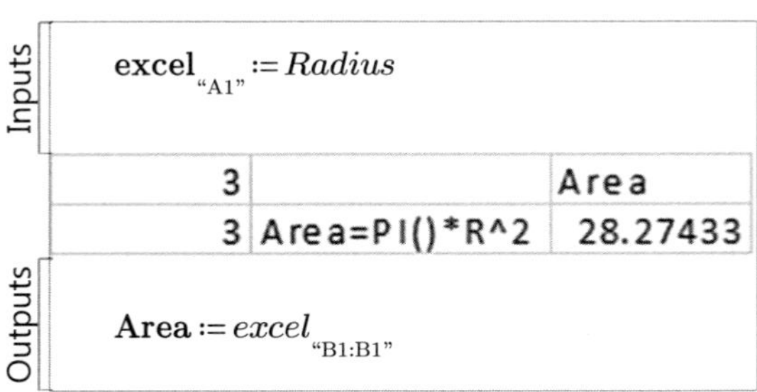

$\textbf{Area} = 0.00$

FIGURE 19.9 *(Continued on next page)*

Adjusting for changes in Excel

The input and output expressions did not automatically update.

The input expression did not update to the new location. It still refers to cell "A1." This input value takes precedence over the change that was made in Excel, so the 3 overwrites "Radius." The 3 in cell A2 is no longer linked to the Excel component. If the value of Radius is changed, it will update cell A1, not A2.

The output expression still refers to "B1:B1." It did not update to the new location, which should be "C2:C2."

The input and output expressions must be manually changed. Note that you must change the input expression to refer to cell A2, prior to adding "Radius" to cell A1, or your change will be overwritten when you close Excel.

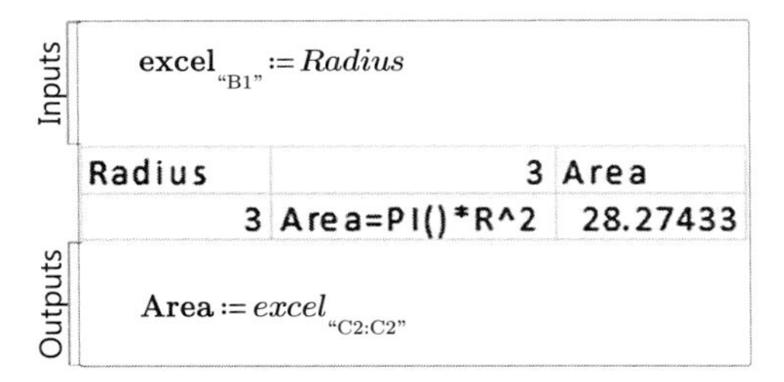

Area $= 28.27$

FIGURE 19.9

(*Continued*)

$\mathbf{Start} := 20$

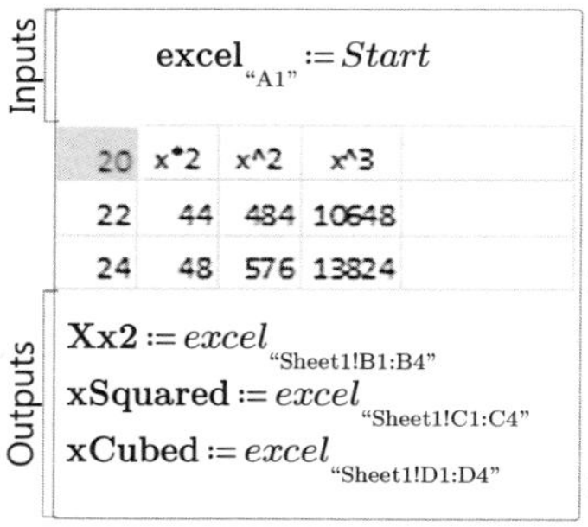

Inputs

$\mathbf{excel}_{\text{"A1"}} := Start$

20	x*2	x^2	x^3
22	44	484	10648
24	48	576	13824

Outputs

$\mathbf{Xx2} := excel_{\text{"Sheet1!B1:B4"}}$

$\mathbf{xSquared} := excel_{\text{"Sheet1!C1:C4"}}$

$\mathbf{xCubed} := excel_{\text{"Sheet1!D1:D4"}}$

$\mathbf{Xx2} = \begin{bmatrix} \text{"x*2"} \\ 44.00 \\ 48.00 \\ 52.00 \end{bmatrix}$

$\mathbf{xSquared} = \begin{bmatrix} \text{"x^2"} \\ 484.00 \\ 576.00 \\ 676.00 \end{bmatrix}$

$\mathbf{xCubed} = \begin{bmatrix} \text{"x^3"} \\ 10648.00 \\ 13824.00 \\ 17576.00 \end{bmatrix}$

Assign the third row of xCubed to the variable "Data."

$\mathbf{Data} := \mathbf{xCubed}_3 = 13824.00$

Now, look what happens if you decide that you do not want the row headers to show in the output, and you change to output range.

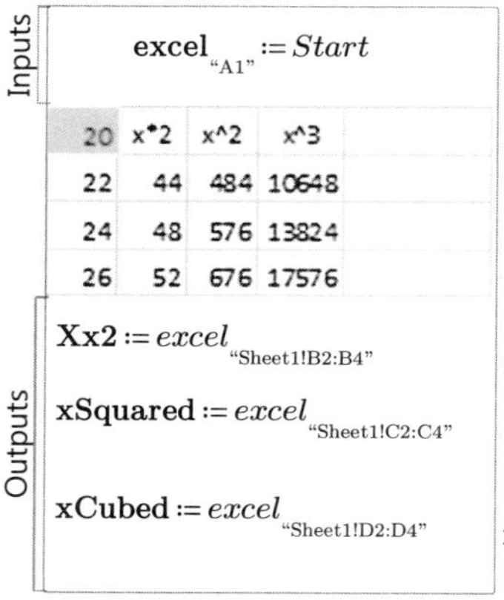

Inputs

$\mathbf{excel}_{\text{"A1"}} := Start$

20	x*2	x^2	x^3
22	44	484	10648
24	48	576	13824
26	52	676	17576

Outputs

$\mathbf{Xx2} := excel_{\text{"Sheet1!B2:B4"}}$

$\mathbf{xSquared} := excel_{\text{"Sheet1!C2:C4"}}$

$\mathbf{xCubed} := excel_{\text{"Sheet1!D2:D4"}}$

$\mathbf{Xx2} = \begin{bmatrix} 44.00 \\ 48.00 \\ 52.00 \end{bmatrix}$

$\mathbf{xSquared} = \begin{bmatrix} 484.00 \\ 576.00 \\ 676.00 \end{bmatrix}$

$\mathbf{xCubed} = \begin{bmatrix} 10648.00 \\ 13824.00 \\ 17576.00 \end{bmatrix}$

"Data" is still assigned to the third row of xCubed, but the third row is now 17,576. The variable "Data" now has an incorrect value.

$\mathbf{Data} := \mathbf{xCubed}_3 = 17576.00$

It is essential to understand the effects of making adjustments within the Excel component box.

FIGURE 19.10

Effect of changing output range

Existing Spreadsheets

You cannot bring an existing spreadsheet into the Excel component as you could in Mathcad 15. ☹

In order to bring an existing spreadsheet or portions of an existing spreadsheet into an Excel component, you must open the worksheet, and then highlight and copy the desired cells. You can then paste these cells into the Excel component. You cannot bring an existing spreadsheet directly into the Excel component.

I recommend protecting the Excel worksheet, except for the cells that require input. Do this prior to copying the spreadsheet into the Excel component. This way you do not accidentally overwrite a critical cell by providing a wrong cell address.

Using Units with Excel

Input

You are probably aware that Excel is unit ignorant. It does not know 12 ft. from 12 m. If you have followed the recommendations in this book, all your PTC Mathcad worksheets use units. The use of units in PTC Mathcad presents a problem when passing values to Excel. Figure 19.11 illustrates this problem. In the case of passing values to Excel, what you see is not what you get. PTC Mathcad must pass a unitless number to Excel. The number that gets passed is the base unit of the SI unit system.

It is important to understand what values you want to input into Excel, and it is equally important to understand how to get the proper values into Excel. The solution is similar to what you need to do for empirical equations. You need to create a unitless number prior to inputting the values. In doing so you can control what number gets passed to Excel. To do this, you divide the input variable by the units you want to be used in Excel. When the output comes back from Excel, you will need to attach the proper units. See Figure 19.12.

Figure 19.13 illustrates the need to understand what values to input into Excel. There are times when it does not matter what units are input as long as consistent units are used. There are many other times when it does matter. The last two examples in Figure 19.13 illustrate this.

Inputs

$excel_{"A1"} := 5\ m$ $excel_{"B1"} := 5\ ft$

$excel_{"A2"} := 5\ kg$ $excel_{"B2"} := 5\ lbm$

5	1.524			
5	2.267962			

Out...

PTC Mathcad must pass a unitless number to Excel. The number that gets passed is the base unit of the SI unit system. Thus 5 meters gets passed as the number 5, but 5 ft gets passed as 1.524, which is the value representing 1.524 meters. 5 kg gets passed as the number 5, but 5 lbm gets passed as 2.267962, which is the value representing 2.267962 kg.

FIGURE 19.11

Using units with the Excel component

Note how each value is passed into Excel based on the unit it is divided by.

Inputs

$excel_{"A1"} := \frac{10\ m}{m}$ $excel_{"B1"} := \frac{5\ ft}{ft}$ $excel_{"C1"} := \frac{5\ ft}{in}$

$excel_{"A2"} := \frac{12\ kg}{kg}$ $excel_{"B2"} := \frac{35\ lbm}{lbm}$ $excel_{"C2"} := \frac{12\ kg}{lbm}$

$excel_{"A3"} := \frac{45\ N}{N}$ $excel_{"B3"} := \frac{75\ lbf}{lbf}$ $excel_{"C3"} := \frac{75\ lbf}{kip}$

10	5	60		
12	35	26.45547		
45	75	0.075		

Out...

FIGURE 19.12

Units continued

For this example, let's use Excel to calculate the velocity of an object given its acceleration (a), distance traveled (Dist) and initial velocity (V0). (Yes, it is much easier to calculate this in PTC Mathcad, but we are using this example to illustrate how important it is to understand units when using an existing Excel spreadsheet.)

Input values are in SI units. V1 converts units to US units and Excel calculates the results in US units of ft/sec. V2 uses SI units to calculate the velocity. Both results give the same final answer if the appropriate units are attached to the Excel output.

$a := g = 9.81 \frac{m}{s^2}$ $\quad$ $\text{Dist} := 100\ m$ $\quad$ $V0 := 50 \frac{m}{s}$

$\text{Velocity} := \sqrt{V0^2 + 2 \cdot a \cdot \text{Dist}} = 66.79 \frac{m}{s}$

$\text{Velocity} = 219.14 \frac{ft}{s}$ $\quad$ $\text{Velocity} = 149.41\ mph$

Inputs

$\text{excel}_{\text{"A2"}} := \frac{a}{\frac{ft}{s^2}}$ $\quad$ $\text{excel}_{\text{"B2"}} := \frac{\text{Dist}}{ft}$ $\quad$ $\text{excel}_{\text{"C2"}} := \frac{V0}{\frac{ft}{s}}$

Accel.	Distance	V0	Calculate Velocity
32.174049	328.08399	164.04199	219.1377568

Outp...

$V1 := excel_{\text{"D2:D2"}}$

$V1 := V1 \cdot \frac{ft}{s} = 219.14 \frac{ft}{s}$ $\quad$ $V1 = 66.79 \frac{m}{s}$ $\quad$ Attach ft/s units

Inputs

$\text{excel}_{\text{"A2"}} := \frac{a}{\frac{m}{s^2}}$ $\quad$ $\text{excel}_{\text{"B2"}} := \frac{\text{Dist}}{m}$ $\quad$ $\text{excel}_{\text{"C2"}} := \frac{V0}{\frac{m}{s}}$

Accel.	Distance	V0	Calculate Velocity
9.80665	100	50	66.79318828

Outp...

$V1 := excel_{\text{"D2:D2"}}$

$V1 := V1 \cdot \frac{m}{s} = 219.14 \frac{ft}{s}$ $\quad$ $V1 = 66.79 \frac{m}{s}$ $\quad$ Attach m/s units

Both worksheets worked because the formula worked for a consistent set of units.

FIGURE 19.13 *(Continued on next page)*

Example of using units in Excel

In the next examples, Excel calculates mph based on USCS units. The first example divides by USCS units; the second divides by SI units. Incorrect results are given if metric values are passed to Excel.

Inputs

$\text{excel}_{\text{"A2"}} := \frac{a}{\frac{ft}{s^2}}$ $\text{excel}_{\text{"B2"}} := \frac{Dist}{ft}$ $\text{excel}_{\text{"C2"}} := \frac{V0}{\frac{ft}{s}}$

Accel.	Distance	V0	Calculate MPH Velocity
32.174049	328.08399	164.04199	149.4121069

Outp...

$V3 := excel_{\text{"D2:D2"}}$

$V3 := V3 \cdot mph = 66.79 \frac{m}{s}$ $V3 = 219.14 \frac{ft}{s}$ $V3 = 149.41\ mph$

Inputs

$\text{excel}_{\text{"A2"}} := \frac{a}{\frac{m}{s^2}}$ $\text{excel}_{\text{"B2"}} := \frac{Dist}{m}$ $\text{excel}_{\text{"C2"}} := \frac{V0}{\frac{m}{s}}$

Accel.	Distance	V0	Calculate MPH Velocity
9.80665	100	50	45.54081019

Out...

$V4 := excel_{\text{"D2:D2"}}$

$V4a := V4 \cdot mph = 20.36 \frac{m}{s}$

Incorrect result. This is a case where all input must represent exactly what Excel expects.

$V4b := V4 \cdot \frac{m}{s} = 45.54 \frac{m}{s}$

FIGURE 19.13

(Continued)

Printing the Excel Component

When you print your PTC Mathcad worksheet, it will only print what is visible within the Excel component table. This has some drawbacks, especially if you want to print the entire spreadsheet. There is a way around this problem. In order to print the entire Excel component table, you must open Excel by double-clicking on the component table, and then use the Excel commands to print.

Summary

The Excel component makes the transition to PTC Mathcad much easier because you do not have to start over. You can easily bring your existing Excel spreadsheets into your project calculations. You can also create new Excel spreadsheets in your project calculations. The two programs can work together by exchanging data back and forth.

In Chapter 19 we:

- Learned how to insert an Excel component block into PTC Mathcad.
- Learned how to reference the Excel cells to transfer information into Excel and extract information from Excel.
- Showed how to get the proper values into Excel when using PTC Mathcad units.
- Discussed ways of attaching units to Excel output.
- Explained how to bring in your existing Excel files into PTC Mathcad including:
  - How to size the region and worksheet
  - How to reference the correct cell addresses
  - How to save and print the component
- Encouraged you to protect your spreadsheet prior to copying it into PTC Mathcad.
- Warned about manually changing variables once you have added PTC Mathcad inputs.
- Warned about mistakes that can occur if you do not divide your input by the proper units.
- Provided examples to illustrate the concepts.

Practice

The PTC Mathcad Prime 3.0 figures and examples used in this book are available for download from the book's website. The reader is encouraged to download the files and use them to practice the concepts learned. Additional examples and problems are also provided. To access this content go to http://store.elsevier.com/9780124104105 *and click on the Resources tab, and then click on the link for the Online Companion Materials.*

1. From your field of study, write five Excel components. Add input from PTC Mathcad, and provide output to PTC Mathcad. Use units.
2. Bring two existing Excel worksheets into PTC Mathcad. Be sure that PTC Mathcad inserts information in the correct cells. List items that need to be considered when using units.

CHAPTER

Conclusion

20

I hope that you have enjoyed learning the essentials of PTC® Mathcad Prime® 3.0. I hope even more that you have followed along with your own version of PTC Mathcad, and have practiced what you have learned. There is no substitute for hands-on practice and application.

Advantages of PTC Mathcad

You may have already known about the mathematical calculation power of PTC Mathcad. After reading this book, you should now see how useful PTC Mathcad is to do everyday scientific and engineering calculations. You can now use PTC Mathcad not only for its powerful mathematical abilities, but also for its ability to help you create and organize your calculations. You can now use PTC Mathcad as the primary tool for your project calculations.

Remember some of the key advantages to using PTC Mathcad:

- Units! PTC Mathcad alerts you when you are using inconsistent units. You can easily get results in SI or U.S. units. You can work in both systems, simultaneously displaying the results in both systems.
- Your formulas and equations are visible and can be easily checked. They are not hidden in cells where only the results are visible.
- You can change an input variable, and your results are immediately updated.
- You can create standard calculations worksheets that can be used over and over again.
- Your calculations can be reused in other projects.
- You can archive the entire project.

Creating Project Calculations

Let's review some of the key topics essential to creating project calculations with PTC Mathcad.

- Use the tools in your PTC Mathcad toolbox. Keep them sharp by reviewing the topics covered in Parts I, II, and III.

- Keep using your existing Excel spreadsheets. Learn how to incorporate them as components in PTC Mathcad so that PTC Mathcad can provide input and receive output from them.
- Use the software programs that make the most sense. Each program has its strengths and weaknesses. PTC Mathcad is not the ideal software program for all applications.
- Use PTC Mathcad to calculate input and to use the output from your other software programs.

Additional Resources

We have just scratched the surface. There is so much more that could have been discussed. Our intent was to point out some useful features, and to get you started using PTC Mathcad for all your project calculations.

The PTC Mathcad Prime 3.0 figures and examples used in this book are available for download from the book's website. We encourage you to download the files and use them to practice the concepts learned. Additional examples and problems are also provided. To access this content go to http://store.elsevier.com/9780124104105 and click on the Resources tab, and then click on the link for the Online Companion Materials.

Here is a list of additional things that you can do to increase your knowledge and PTC Mathcad skills:

- Open the PTC Mathcad Tutorials from the **Getting Started** tab and review each tutorial.
- Review the PTC Mathcad functions from the **Functions** tab. Research additional functions that are not discussed in this book.
- Visit the PTC Mathcad Community and blogs on the internet. There are links on the **Getting Started** tab. Explore the many different topics. Ask a question. Look for answers.
- Look for topics at the PTC Mathcad Community located at: http://communities.ptc.com/community/mathcad

Conclusion

If you are a current Mathcad 15 user, I wish you well as you begin your transition to using PTC Mathcad Prime 3.0. It is not perfect, but it has many advantages over Mathcad 15. If you are new to PTC Mathcad, I wish you well in using PTC Mathcad in your technical calculations. I encourage a continued effort to add more tools to your PTC Mathcad toolbox. There are many tools to add that we have not discussed.

It is also important to keep your tools sharp. Tools that are not used very often get rusty. I encourage you to review the chapters in this book and try to incorporate some of the features in your calculations. Review the chapters from time to time and polish your tools so that they will be sharp and ready to use when the occasion permits.

I wish you the best of success as you enjoy the wonderful world of PTC Mathcad Prime 3.0.

Appendix 1: PTC® Mathcad Prime® 3.0 Keyboard Shortcuts

Regions

Inserting Regions

Operator/Command	Description	Keyboard Shortcut
Area	Inserts a collapsible area you can collapse or expand to toggle the display of your work.	Ctrl+Shift+A
Excel Component	Inserts an Excel component.	Ctrl+Shift+E
Image	Inserts an image.	Ctrl+4
Include Worksheet	Inserts variable and function definitions from another worksheet.	Ctrl+Shift+W
Math	Inserts a math region.	Ctrl+Shift+M
Solve Block	Inserts a solve block.	Ctrl+1
Table	Inserts a table.	Ctrl+6
Text Block	Inserts a text block.	Ctrl+Shift+T
Text Box	Inserts a text box.	Ctrl+T

Working with Regions

Description	Keyboard Shortcut
Moves the cursor to the beginning of the expression.	Home
Moves the cursor to the end of the expression.	End
Moves the cursor to the left of the selected expression or to the beginning of the selected word.	Ctrl+Left Arrow
Moves the cursor to the right of the selected expression or to the end of the selected word.	Ctrl+Right Arrow
Moves the cursor to the next placeholder.	Right Arrow
Moves the cursor to the previous placeholder.	Left Arrow
Recalculates the selected or active region.	F5
Separates the selected regions horizontally.	Ctrl+F8
Separates the selected regions vertically.	Ctrl+F3
Toggles the active region selection.	Ctrl+Enter
Inside a text region, inserts a new line above or below depending on the position of the cursor.	Shift+Enter

Math

Operator/Command	Description	Keyboard Shortcut
Greek Letters	Adds a Greek letter.	corresponding Latin letter, Ctrl+G
i or *j*	Indicates imaginary units.	1i or 1j
Insert String	Inserts a string.	"
Subscripts	Toggles the baseline of a text.	Ctrl+-
Toggle Labels	Toggles the label type.	Ctrl+Q

Constants

Operator/Command	Description	Keyboard Shortcut
γ	Euler's constant/number	g, Ctrl+G
∞	Infinity	Ctrl+Shift+Z
π	Pi	p, Ctrl+G

Plots

Operator/Command	Description	Keyboard Shortcut
XY Plot	Inserts an XY Plot.	Ctrl+2
3D Plot	Inserts a 3D Plot.	Ctrl+3
Contour Plot	Inserts a Contour Plot.	Ctrl+5
Polar Plot	Inserts a Polar Plot.	Ctrl+7
All Plots	Inserts a new axis expression into the active axis expression list.	Shift+Enter

Operators

Algebra Operators

Operator/Command	Description	Keyboard Shortcut
$\lvert x \rvert$	Absolute Value	|
$x + y$	Addition	+
$\lvert x \rvert$	Determinant	|
$\frac{x}{y}$	Division	/
y^x	Exponentiation	^
$n!$	Factorial	!
$x \div y$	In-line Division	Ctrl+/

(*continued*)

—Cont'd

Operator/Command	Description	Keyboard Shortcut
(x)	Matched pair of parentheses	(
$x \cdot y$	Multiplication and Dot Product	*
$x - y$	Negation and Subtraction	–
,	Separate arguments of a function	, Comma
$\sqrt[n]{x}$	Square Root and Nth Root	\
$x\%$	Percent	%

Calculus Operators

Operator/Command	Description	Keyboard Shortcut
$A \circledast B$	Circular Convolution	Ctrl+Shift+V
$\frac{d^2}{dt^2} f(t)$	Derivative	Ctrl+Shift+D
$\int_a^b f(x)dx$	Integral	Ctrl+Shift+I
$A * B$	Linear Convolution	Ctrl+Shift+L
$g := f'$	Prime	Ctrl+′ (Apostrophe)
$\prod_{j=m}^{n} X$	Product	Ctrl+Shift+#
$\sum_{j=m}^{n} X$	Summation	Ctrl+Shift+$

Comparison Operators

Operator/Command	Description	Keyboard Shortcut
$x = y$	Equal To	Ctrl+=
$x > y$	Greater Than	>
$x \geq y$	Greater Than or Equal To	>= or Ctrl+0
$x \neq y$	Inequality	< >
$x \in y$	Is Element Of	Ctrl+F7

(continued)

—Cont'd

Operator/Command	Description	Keyboard Shortcut
$x < y$	Less Than	<
$x \leq y$	Less Than or Equal To	<= Ctrl+9
$x \oplus y$	Logical Exclusive OR	Ctrl+Shift+%
$x \wedge y$	Logical AND	Ctrl+Shift+&
$\neg x$	Logical NOT	Ctrl+Shift+!
$x \vee y$	Logical OR	Ctrl+Shift+@

Definition and Evaluation Operators

Operator/Command	Description	Keyboard Shortcut
$x := a$	Definition	: Colon
$x = 1$	Evaluation	=
$x \equiv b$	Global Definition	Ctrl+Shift+~

Engineering Operators

Operator/Command	Description	Keyboard Shortcut
$\overline{x}$	Complex Conjugate	Ctrl+Shift+_
$x \cdot y$	Scaling	Ctrl+Shift+U
$x \angle y$	Polar	Ctrl+Shift+P

Symbolics

Operator/Command	Description	Keyboard Shortcut
$\rightarrow$	Symbolic Evaluation	Ctrl+. (period)
$\lim_{\blacksquare \rightarrow \blacksquare} \blacksquare$	Limit	Ctrl+L
Cursor at end of a symbolic keyword	Adds a new line with a keyword placeholder	Shift+Enter

Vector and Matrix Operators

Operator/Command	Description	Keyboard Shortcut
$u \times v$	Cross Product	Ctrl+8
$x=1$	Local Definition	=
$[\ \blacksquare\ ]$	Matrix	[Ctrl+M
$M^{\langle j \rangle}$	Matrix Column	Ctrl+Shift+C
$M_{i,j}$	Matrix Index	[
M^{-1}	Matrix Inverse	^−1
$M^{\widehat{\underset{\smile}{i}}}$	Matrix Row	Ctrl+Shift+R
M^{T}	Matrix Transpose	Ctrl+Shift+T
$\|\|M\|\|$	Norm	Ctrl+Shift+|
$x..z$	Range	.. Two periods
$x, y..z$	Step Range	, Comma
$\overrightarrow{M}$	Vectorize	Ctrl+Shift+^

Programming Operators

Operator/Command	Description	Keyboard Shortcut
$\Vert x$	Adds a new line to the program	Enter
also if x $\Vert y$	Also If	Ctrl+Shift+?
break	Break	Ctrl+Shift+{
continue	Continue	Ctrl+Shift+:
else $\Vert y$	Else	Ctrl+Shift+}
else if x $\Vert y$	Else If	Ctrl+;
for $x \in y$ $\Vert z$	For Loop	Ctrl+Shift+"
if x $\Vert y$	If	}
$x \leftarrow y$	Local Assignment	{
return x	Return	Ctrl+\
$\Vert$	Program	]
try $\Vert x$ on error $\Vert y$	Try/On Error	Ctrl+[

(*continued*)

—Cont'd

Operator/Command	Description	Keyboard Shortcut
while x ‖ y	While Loop	Ctrl+]
	Pressing Ctrl+J after a programming word, changes the word into an operator	Ctrl+J

Matrices and Tables

Operator/Command	Description	Keyboard Shortcut
Insert Left/Right	Inserts a column to the left or to the right according to the cursor's location.	Shift+ Space
Insert Above/Below	Inserts a row above or below depending on the location of the cursor.	Shift+Enter

General

Operator/Command	Description	Keyboard Shortcut
Calculate Worksheet	Recalculates the worksheet.	Ctrl+F5
Functions, open dialog box	Opens the **Functions** dialog box.	F2
Add Page Break	Inserts a page break.	Ctrl+Enter
?	Opens the Help Center.	Fl
	Closes the currently open worksheet.	Ctrl+W

Appendix 2: Keyboard Shortcuts for Editing and Worksheet Management

Keyboard Shortcuts for Editing

The following keyboard shortcuts are used for editing within a worksheet.

Enter	Insert blank line. In text, begin a new paragraph.
Delete	Delete blank line. In text or math, remove character to the right of the insertion line.
Shift+Enter	In text, begin a new line within a paragraph.
Ctrl+Enter	Selects the active region.
Ctrl+A	In text, select all the text in the text region.
Ctrl+A	In a blank spot, select all regions in the worksheet.
Ctrl+F	Moves the cursor to the Find box on the Status Bar.
Ctrl+Z	Undo last edit.
Ctrl+Y	Redo. Reverses the action of Undo.
Ctrl+C	Copy selection to clipboard.
Ctrl+V	Paste clipboard contents into worksheet.
Ctrl+X	Cut selection to clipboard.
Ctrl+W	Closes currently open worksheet.

Keyboard Shortcuts for Worksheet Management

The following keys are used for manipulating windows and worksheets as a whole.

Ctrl+F4	Close worksheet.
Ctrl+F6	Make next window active.
Ctrl+N	Create new worksheet.
Ctrl+O	Open worksheet.
Ctrl+P	Print worksheet.
Ctrl+S	Save worksheet.
Alt+F4	Quit.
Ctrl+R	Redraw screen.
F1	Open PTC Mathcad Help.
F5 or F9	Calculates the active math region and all related regions.
Ctrl+F5	Recalculates all regions in the worksheet.
Shift+F1	Enter or exit context sensitive Help.
Esc	Exit context sensitive Help or interrupt a calculation.
Ctrl+F1	Minimizes or maximizes the Ribbon.

Appendix 3: Greek Letters

Greek Toolbar

Enter Roman, then type **CTRL + G** for Greek.

Greek	Uppercase	Lowercase	Roman
Alpha	A	α	A/a
Beta	B	β	B/b
Gamma	Γ	γ	G/g
Delta	Δ	δ	D/d
Epsilon	E	ε	E/e
Zeta	Z	ζ	Z/z
Eta	H	η	H/h
Theta	Θ	θ	Q/q
Theta (alt.)	ϑ		J
Iota	I	ι	I/i
Kappa	K	κ	K/k
Lambda	^	λ	L/l
Mu	M	μ	M/m
Nu	N	ν	N/n
Xi	Ξ	ξ	X/x
Omicron	O	ο	O/o
Pi	Π	π	P/p
Rho	p	ρ	R/r
Sigma	Σ	σ	S/s
Tau	T	τ	T/t

(*continued*)

—Cont'd

Greek	Uppercase	Lowercase	Roman
Upsilon	U	υ	U/u
Phi	Φ	ϕ	F/f
Phi (alt.)		φ	j
Chi	X	χ	C/c
Psi	Ψ	ψ	Y/y
Omega	Ω	ω	W/w

Appendix 4: Built-In Constants and Variables

Math Constants

Constant	Keystroke	Name	Default Value
∞	Ctrl+Shift+Z	Infinity	$1*10^{307}$
i or j	1i or 1j	The imaginary units	Square root of −1
π	p, Ctrl+G	Pi	3.142
NaN	NaN	Not a Number	NaN

Physics Constants

Constant	Description	Value
c	Speed of light in vacuum	$c=(2.998\cdot10^{8})\ \frac{m}{s}$
e_c	Elementary charge	$e_c=(1.602\cdot10^{-19})\ C$
h	Planck's **constant**	$h=(6.626\cdot10^{-34})\ \frac{kg\cdot m^2}{s}$
$\hbar$	Reduced Planck's **constant**	$\hbar=(1.055\cdot10^{-34})\ \frac{kg\cdot m^2}{s}$
k	Boltzmann **constant**	$k=(1.381\cdot10^{-23})\ \frac{kg\cdot m^2}{s^2\cdot K}$
m_u	Atomic mass unit	$m_u=(1.661\cdot10^{-27})\ kg$
N_A	Avogadro's number	$N_A=(6.022\cdot10^{23})\ \frac{1}{mol}$
R	Molar gas **constant**	$R=8.314\ \frac{kg\cdot m^2}{s^2\cdot K\cdot mol}$
R_∞	Rydberg **constant**	$R_\infty=(1.097\cdot10^{7})\ \frac{1}{m}$
α	Fine structure **constant**	$\alpha = 0.007$
γ	Euler-Mascheroni's **constant**	$\gamma = 0.577$
ε_0	Permittivity of free space	$\varepsilon_0=(8.854\cdot10^{-12})\ \frac{s^4\cdot A^2}{kg\cdot m^3}$
μ_0	Magnetic permeability of free space	$\mu_0=(1.257\cdot10^{-6})\ \frac{kg\cdot m}{s^2\cdot A^2}$
σ	Stefan-Boltzmann **constant**	$\sigma=(5.67\cdot10^{-8})\ \frac{kg}{s^3\cdot K^4}$
Φ_0	Magnetic flux quantum	$\Phi_0=(2.068\cdot10^{-15})\ Wb$

System Variables

Name	Default Value	Use
TOL	0.001	Controls the convergence precision of some functions such as integrals, derivatives, odesolve and the root functions.
CTOL	0.001	Controls how closely a constraint in a solve block must be met for a solution to be acceptable when using find, minerr, minimize, or maximize.
ORIGIN	0	Controls array indexing. You can reset ORIGIN to 1 on the Calculation tab, or you can redefine ORIGIN to another value in your worksheet.
PRNPRECISION	4	Controls the number of significant digits to be used when writing to an ASCII data file with the WRITEPRN or APPENDPRN functions.
PRNCOLWIDTH	8	Controls the width of columns in an ASCII data file created by the WRITEPRN or APPENDPRN functions.
CWD	Current working directory in the form of a string variable	Use this system variable as an argument to file handling functions.

Appendix 5: Reference Tables

The PTC Mathcad Reference Tables contain almost 40 tables of scientific and engineering data. You will find information about physics, chemistry, mechanics of materials, mathematics, electronics, and more. They include constants, formulas, and equations. You can access these tables from the PTC Mathcad Help Center at the bottom of the list of topics.

Reference Tables

Basic Science

Fundamental Physical Constants

Universal Constants
Electromagnetic Constants
Atomic Constants
Electron
Muon
Proton
Neutron
Physico-Chemical Constants

Periodic Table of Elements

Physical Formulas — Mechanics

Motion in One Dimension
Motion in Two Dimensions
Newton's Laws of Motion
Work and Energy
Momentum
Rotation
Simple Harmonic Motion

Calculus

Derivative Formulas

Rules
Derivatives of Trigonometric Functions
Derivatives of Logarithms and Exponential Functions
Derivatives of Inverse Trigonometric Functions

Integral Formulas

- Basic Integrals
- Integrals of Trigonometric Functions
- Integration by Parts
- Integrals of Exponential and Logarithmic Functions
- Integrals Containing $a^2 \pm x^2$

Trigonometric Identities

- Basic Identities
- Complex Identities
- Hyperbolic Functions and Exponentials
- Hyperbolic Functions and Trig Relationships

Electromagnetics

Capacitance

- Coaxial Cylinders Capacitor
- Concentric Spheres Capacitor
- Isolated Sphere Capacitor
- Parallel Plates Capacitor

Oscillators

- Clapp Oscillator
- Colpitts Oscillator
- Hartley Oscillator
- Pierce Oscillator
- RC Phase Shift Oacillator
- Tuned Output Oscillator

Geometry

Areas and Perimeters

- Cardoid
- Circle
- Circular Segment
- Circumscribed Circle
- Circumscribed Polygon
- Cycloid
- Ellipse
- Hypocycloid
- Inscribed Circle
- Inscribed Polygon
- Lemniscate
- Parabolic Segment

Index

Note: Page numbers with "*f*" denote figures; "*t*" tables; "b" boxes.

D

L

M

N

O

P